Probability, Random Variables, and Random Signal Principles

McGraw-Hill Series in Electrical and Computer Engineering

Probability, Random Variables, and Random Signal Principles

FOURTH EDITION

Peyton Z. Peebles, Jr., Ph.D.

Professor Emeritus of Electrical and Computer Engineering
University of Florida

McGRAW-HILL, INC.

New York   St. Louis   San Francisco   Auckland   Bogotá
Caracas   Lisbon   London   Madrid   Mexico City   Milan
Montreal   New Delhi   San Juan   Singapore
Sydney   Tokyo   Toronto

McGraw-Hill Higher Education

*A Division of The **McGraw-Hill** Companies*

PROBABILITY, RANDOM VARIABLES, AND RANDOM SIGNAL PRINCIPLES
Published by Irwin/McGraw-Hill, an imprint of The McGraw-Hill Companies, Inc., 1221 Avenue of the Americas, New York, NY, 10020. Copyright © 2001, 1993, 1987, 1980, by The McGraw-Hill Companies, Inc. All rights reserved. No part of this publication may be reproduced or distributed in any form or by any means, or stored in a database or retrieval system, without the prior written consent of The McGraw-Hill Companies, Inc., including, but not limited to, in any network or other electronic storage or transmission, or broadcast for distance learning.
Some ancillaries, including electronic and print components, may not be available to customers outside the United States.

This book is printed on acid-free paper.

1 2 3 4 5 6 7 8 9 0 FGR/FGR 0 9 8 7 6 5 4 3 2 1 0

ISBN 0-07-366007-8

Publisher: *Thomas Casson*
Vice President/Editor-in-Chief: *Kevin T. Kane*
Senior sponsoring editor: *Catherine Fields Shultz*
Development editor: *Michelle Flomenhoft*
Senior marketing manager: *John Wannemacher*
Project manager: *Craig S. Leonard*
Senior production supervisor: *Heather Burbridge*
Senior designer: *Kiera Cunningham*
Senior supplement coordinator: *Cathy L. Tepper*
Interior design: *Z Graphics*
Compositor: *Keyword Publishing Services*
Typeface: *10.5/12 Times Roman*
Printer: *Quebecor Printing Book Company/Fairfield*

Library of Congress Cataloging-in-Publication Data

Peebles, Peyton Z.
    Probability, random variables, and random signal principles / Peyton Z. Peebles, Jr.—4th ed.
        p. cm.—(McGraw-Hill series in electrical and computer engineering)
    ISBN 0-07-366007-8 (alk. paper)
    1. Probabilities. 2. Random variables. 3. Signal theory (Telecommunication) I. Title.
II. Series.

TA340.P43 2001
519.2—dc21   00-034881

www.mhhe.com

ABOUT THE AUTHOR

PEYTON Z. PEEBLES, Jr., is Professor Emeritus in the Department of Electrical and Computer Engineering at the University of Florida, Gainesville, Florida. He has twelve years of industrial experience and over thirty years of teaching experience including time spent at the University of Tennessee and the University of Hawaii, as well as the University of Florida. He earned his Ph.D. degree in 1967 from the University of Pennsylvania where he held a David Sarnoff Fellowship from RCA for two years.

Professor Peebles has authored or coauthored over 50 journal articles and conference papers as well as four other textbooks: *Radar Principles* (John Wiley & Sons, 1998); *Principles of Electrical Engineering* (McGraw-Hill, 1991, with T. A. Giuma); *Digital Communication Systems* (Prentice-Hall, 1987); and *Communication System Principles* (Addison-Wesley, 1976).

Dr. Peebles is a member of Tau Beta Pi, Eta Kappa Nu, Sigma Pi Sigma, and Phi Beta Chi, and is a Life Fellow of the Institute of Electrical and Electronic Engineers (IEEE).

TO MY MOTHER
Maida Erlene Denton Dials

AND STEPFATHER
Ralph Phillip Dials

CONTENTS

PREFACE

In each of the book's earlier editions, all prior prefaces were reproduced to show the developments and purposes or the book over time. In this edition, they have been combined into one preface. It is, therefore, appropriate to first summarize the most important topics of prior editions that have either not changed or have evolved with time. They are:

- The *level* and *amount* of material remains that of a typical undergraduate program for courses of not more than one semester. The book is also applicable to courses at the first-year graduate level if students have had minimal or no prior exposure to the book's topics. Over the past 20 years the book has seen mostly small additions to aid teaching (more problems, more examples, and added topics that seemed to be needed in evolving programs).
- The more advanced material, and problems that require more than typical solution times, remain keyed by a star ($\star$), as before.
- The *need* for the book has not changed. Namely, a well-organized teaching book is needed for lower-level courses (juniors, seniors) as the material has migrated down from the upper graduate level over the years.
- The *background* needed to study the book remains typical of junior or senior undergrate engineering students.

In preparation for this fourth edition, comments and suggestions were sought from many reviewers and instructors who have used the book. The result was the incorporation of a number of additions that range from relatively minor to important. In the former category, additions include: chapter-end summaries, more examples within the text, expanded discussions of probability as a relative frequency, material on permutations and combinations, more detailed material on random variable transformations, a bit more detail on ergodicity, the weak and strong laws of large numbers, sampling and estimation, various important inequalities (Chebychev, Schwarz, Markov, Hölder, Chernoff, etc.), some useful properties of impulse functions (Appendix A), a new appendix on various mathematical topics of interest (Appendix G), a few new problems, and many other small changes. Problems at each chapter end have also been combined into a single list with numbering that corresponds to the chapter's sections to which the problems mainly apply.

However, probably the most important new material relates to discrete-time (DT) random processes and sequences and other topics in the general area of digital signal processing (DSP) such as the DT linear system. There is now coverage of sampling theorems—baseband and bandpass—for random processes to establish a foundation for DT processes and sequences. For clear exposure, this material is placed mainly in Chapter 8, since it requires some understanding of the passing of random signals through networks.

Correlation functions and power spectrums for these processes and sequences are developed and connected through the sampling theorems and the discrete-time Fourier transforms (Chapters 7 and 8). The structure of linear DT digital systems is developed in both the sequence and transform domains. It is hoped this new material will better serve those readers with a strong interest in digital topics.

Other important additions are computer examples and problems scattered throughout the book at key places—mainly where new DT material is located. These examples and problems require no special toolboxes and assume the reader is familiar with the use of version 5.2 of MATLAB software. For the examples, the necessary coding is given, but the reader is expected to provide the coding for the problems. All of these examples and problems are keyed by a "computer" symbol $\boxed{\text{M}}$. With appreciation, I acknowledge the help of Mr. Kenneth Hild, a doctoral student, who worked, coded, and verified all these examples and problems.

Many persons have identified errors in the third edition and have offered suggestions for this new edition. Several of my students are included in this group, and I thank them for their contributions. I also appreciate the detailed comments and suggestions from the following professors who either have used the book, were reviewers, or both: Scott Acton, *Oklahoma State University*; Mahmood R. Azimi, *Colorado State University*; Ross Baldick, *University of Texas, Austin*; Charles Boncelet, *University of Delaware*; Oscar Norberto Bria, *Universidad Nacional de La Plata*; Kevin D. Donohue, *University of Kentucky*; Sammie Giles, Jr., *University of Toledo*; Subhash Kak, *Louisiana State University*; James Kang, *California State Polytechnic University, Pomona*; Venkatarama Krishnan, *University of Massachusetts, Lowell*; Jian Li, *University of Florida*; Rodney Roberts, *Florida A&M University* and *Florida State University*; Sumit Roy, *University of Washington*; Antal Sarkady, *U.S. Naval Academy*; Ness Shroff, *Purdue University*; Emmanouel Varvarigos, *University of California, Santa Barbara*; Donley Winger, *California State Polytechnic University, San Luis Obispo*. My thanks are also extended to Ms. Catherine Fields and Ms. Michelle Flomenhoft, editors at McGraw-Hill who were instrumental in the production of the fourth edition.

Finally, I thank those who have indicated to me that they were pleased that the third edition had relatively few errors. One even said he knew of none. Of course, there were some, but a great deal of effort did go into minimizing the number in print. Those same efforts have also been made in this new edition in hopes that it will prove as free of errors.

Peyton Z. Peebles, Jr.
Gainesville, Florida
March 2000

CHAPTER 1

Probability

1.0
INTRODUCTION TO BOOK AND CHAPTER

The primary goals of this book are to introduce the reader to the principles of random signals and to provide tools whereby one can deal with systems involving such signals. Toward these goals, perhaps the first thing that should be done is define what is meant by random signal. A *random signal* is a time waveform† that can be characterized only in some probabilistic manner. In general, it can be either a desired or undesired waveform.

The reader has no doubt heard background hiss while listening to an ordinary broadcast radio receiver. The waveform causing the hiss, when observed on an oscilloscope, would appear as a randomly fluctuating voltage with time. It is undesirable, since it interferes with our ability to hear the radio program, and is called *noise*.

Undesired random waveforms (noise) also appear in the outputs of other types of systems. In a radio astronomer's receiver, noise interferes with the desired signal from outer space (which itself is a random, but desirable, signal). In a television system, noise shows up in the form of picture interference often called "snow." In a sonar system, randomly generated sea sounds give rise to a noise that interferes with the desired echoes.

The number of desirable random signals is almost limitless. For example, the bits in a computer bit stream appear to fluctuate randomly with time

---

† We shall usually assume random signals to be voltage-time waveforms. However, the theory to be developed throughout the book will apply, in most cases, to random functions other than voltage, of arguments other than time.

1

between the zero and one states, thereby creating a random signal. In another example, the output voltage of a wind-powered generator would be random because wind speed fluctuates randomly. Similarly, the voltage from a solar detector varies randomly due to the randomness of cloud and weather conditions. Still other examples are: the signal from an instrument designed to measure instantaneous ocean wave height; the space-originated signal at the output of the radio astronomer's antenna (the relative intensity of this signal from space allows the astronomer to form radio maps of the heavens); and the voltage from a vibration analyzer attached to an automobile driving over rough terrain.

In Chapters 8 and 9 we shall study methods of characterizing systems having random input signals. However, from the above examples, it is obvious that random signals only represent the behavior of more fundamental underlying random phenomena. Phenomena associated with the desired signals of the last paragraph are: information source for computer bit stream; wind speed; various weather conditions such as cloud density and size, cloud speed, etc.; ocean wave height; sources of outer space signals; and terrain roughness. All these phenomena must be described in some probabilistic way.

Thus, there are actually two things to be considered in characterizing random signals. One is how to describe any one of a variety of random phenomena; another is how to bring time into the problem so as to create the random signal of interest. To accomplish the first item, we shall introduce mathematical concepts in Chapters 2, 3, 4, and 5 (random variables) that are sufficiently general they can apply to any suitably defined random phenomena. To accomplish the second item, we shall introduce another mathematical concept, called a random process, in Chapters 6 and 7. All these concepts are based on probability theory.

The purpose of this chapter is to introduce the elementary aspects of probability theory on which all of our later work is based. Several approaches exist for the definition and discussion of probability. Only two of these are worthy of modern-day consideration, while all others are mainly of historical interest and are not commented on further here. Of the more modern approaches, one uses the relative frequency definition of probability. It gives a degree of physical insight which is popular with engineers, and is often used in texts having principal topics other than probability theory itself (for example, see Peebles, 1976).†

The second approach to probability uses the axiomatic definition. It is the most mathematically sound of all approaches and is most appropriate for a text having its topics based principally on probability theory. The axiomatic approach also serves as the best basis for readers wishing to proceed beyond the scope of this book to more advanced theory. Because of these facts, we mainly adopt the axiomatic approach in this book, but occasionally use the relative frequency method in some practical problems.

---

†References are quoted by name and date of publication. They are listed at the end of the book.

Prior to the introduction of the axioms of probability, it is necessary that we first develop certain elements of set theory.†

1.1
SET DEFINITIONS

A *set* is a collection of objects. The objects are called *elements* of the set and may be anything whatsoever. We may have a set of voltages, a set of airplanes, a set of chairs, or even a set of sets, sometimes called a *class* of sets. A set is usually denoted by a capital letter while an element is represented by a lower-case letter. Thus, if a is an element of set A, then we write

$$a \in A \qquad (1.1\text{-}1)$$

If a is not an element of A, we write

$$a \notin A \qquad (1.1\text{-}2)$$

A set is specified by the content of two braces: $\{\cdot\}$. Two methods exist for specifying content, the tabular method and the rule method. In the tabular method the elements are enumerated explicitly. For example, the set of all integers between 5 and 10 would be $\{6, 7, 8, 9\}$. In the rule method, a set's content is determined by some rule, such as: {integers between 5 and 10}.‡ The rule method is usually more convenient to use when the set is large. For example, {integers from 1 to 1000 inclusive} would be cumbersome to write explicitly using the tabular method.

A set is said to be *countable* if its elements can be put in one-to-one correspondence with the natural numbers, which are the integers 1, 2, 3, etc. If a set is not countable, it is called *uncountable*. A set is said to be *empty* if it has no elements. The empty set is given the symbol $\varnothing$ and is often called the *null set*.

A *finite set* is one that is either empty or has elements that can be counted, with the counting process terminating. In other words, it has a finite number of elements. If a set is not finite, it is called *infinite*. An infinite set having countable elements is called *countably infinite*.

If every element of a set A is also an element in another set B, A is said to be contained in B. A is known as a *subset* of B and we write

$$A \subseteq B \qquad (1.1\text{-}3)$$

If at least one element exists in B which is not in A, then A is a *proper subset* of B, denoted by (Thomas, 1969)

$$A \subset B \qquad (1.1\text{-}4)$$

The null set is clearly a subset of all other sets.

---

†Our treatment is limited to the level required to introduce the desired probability concepts. For additional details the reader is referred to McFadden (1963), or Milton and Tsokos (1976).
‡Sometimes notations such as $\{I | 5 < I < 10, I$ an integer$\}$ or $\{I : 5 < I < 10, I$ an integer$\}$ are seen in the literature.

Two sets, A and B, are called *disjoint* or *mutually exclusive* if they have no common elements.

EXAMPLE 1.1-1. To illustrate the topics discussed above, we identify the sets listed below.

$$A = \{1, 3, 5, 7\} \qquad D = \{0.0\}$$
$$B = \{1, 2, 3, \ldots\} \qquad E = \{2, 4, 6, 8, 10, 12, 14\}$$
$$C = \{0.5 < c \le 8.5\} \qquad F = \{-5.0 < f \le 12.0\}$$

The set A is tabularly specified, countable, and finite. B is also tabularly specified and countable, but is infinite. Set C is rule-specified, uncountable, and infinite, since it contains *all* numbers greater than 0.5 but not exceeding 8.5. Similarly, sets D and E are countably finite, while set F is uncountably infinite. It should be noted that D is *not* the null set; it has one element, the number zero.

Set A is contained in sets B, C, and F. Similarly, $C \subset F$, $D \subset F$, and $E \subset B$. Sets B and F are not subsets of any of the other sets or of each other. Sets A, D, and E are mutually exclusive of each other. The reader may wish to identify which of the remaining sets are also mutually exclusive.

The largest or all-encompassing set of objects under discussion in a given situation is called the *universal set*, denoted S. All sets (of the situation considered) are subsets of the universal set. An example will help clarify the concept of a universal set.

EXAMPLE 1.1-2. Suppose we consider the problem of rolling a die. We are interested in the numbers that show on the upper face. Here the universal set is $S = \{1, 2, 3, 4, 5, 6\}$. In a gambling game, suppose a person wins if the number comes up odd. This person wins for any number in the set $A = \{1, 3, 5\}$. Another person might win if the number shows four or less; that is, for any number in the set $B = \{1, 2, 3, 4\}$.

Observe that both A and B are subsets of S. For any universal set with N elements, there are 2^N possible subsets of S. (The reader should check this for a few values of N.) For the present example, $N = 6$ and $2^N = 64$, so that there are 64 ways one can define "winning" with one die.

It should be noted that winning or losing in the above gambling game is related to a set. The game itself is partially specified by its universal set (other games typically have a different universal set). These facts are not just coincidence, and we shall shortly find that sets form the basis on which our study of probability is constructed.

1.2
SET OPERATIONS

In working with sets, it is helpful to introduce a geometrical representation that enables us to associate a physical picture with sets.

Venn Diagram

Such a representation is the Venn diagram.† Here sets are represented by closed-plane figures. Elements of the sets are represented by the enclosed points (area). The universal set S is represented by a rectangle as illustrated in Figure 1.2-1a. Three sets A, B, and C are shown. Set C is disjoint from both A and B, while set B is a subset of A.

Equality and Difference

Two sets A and B are *equal* if all elements in A are present in B and all elements in B are present in A; that is, if $A \subseteq B$ and $B \subseteq A$. For equal sets we write $A = B$.

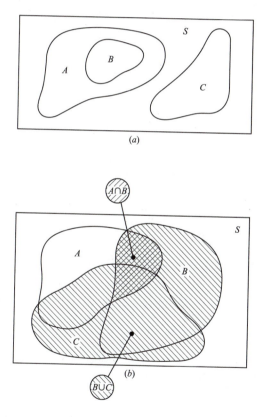

FIGURE 1.2-1
Venn diagrams. (*a*) Illustration of subsets and mutually exclusive sets, and (*b*) illustration of intersection and union of sets. [*Adapted from Peebles (1976), with permission of publishers Addison–Wesley, Advanced Book Program.*]

†After John Venn (1834–1923), an Englishman.

The *difference* of two sets A and B, denoted $A - B$, is the set containing all elements of A that are not present in B. For example, with $A = \{0.6 < a \le 1.6\}$ and $B = \{1.0 \le b \le 2.5\}$, then $A - B = \{0.6 < c < 1.0\}$ or $B - A = \{1.6 < d \le 2.5\}$. Note that $A - B \ne B - A$.

Union and Intersection

The *union* (call it C) of two sets A and B is written

$$C = A \cup B \qquad (1.2\text{-}1)$$

It is the set of all elements of A or B or both. The union is sometimes called the *sum* of two sets.

The *intersection* (call it D) of two sets A and B is written

$$D = A \cap B \qquad (1.2\text{-}2)$$

It is the set of all elements common to both A and B. Intersection is sometimes called the *product* of two sets. For mutually exclusive sets A and B, $A \cap B = \emptyset$. Figure 1.2-1b illustrates the Venn diagram area to be associated with the intersection and union of sets.

By repeated application of (1.2-1) or (1.2-2), the union and intersection of N sets $A_n, n = 1, 2, \ldots, N$, become

$$C = A_1 \cup A_2 \cup \cdots \cup A_N = \bigcup_{n=1}^{N} A_n \qquad (1.2\text{-}3)$$

$$D = A_1 \cap A_2 \cap \cdots \cap A_N = \bigcap_{n=1}^{N} A_n \qquad (1.2\text{-}4)$$

Complement

The *complement* of a set A, denoted by $\bar{A}$, is the set of all elements not in A. Thus,

$$\bar{A} = S - A \qquad (1.2\text{-}5)$$

It is also easy to see that $\bar{\emptyset} = S$, $\bar{S} = \emptyset$, $A \cup \bar{A} = S$, and $A \cap \bar{A} = \emptyset$.

EXAMPLE 1.2-1. We illustrate intersection, union, and complement by taking an example with the four sets

$$S = \{1 \le \text{integers} \le 12\} \qquad B = \{2, 6, 7, 8, 9, 10, 11\}$$
$$A = \{1, 3, 5, 12\} \qquad C = \{1, 3, 4, 6, 7, 8\}$$

Applicable unions and intersections here are:

$$A \cup B = \{1, 2, 3, 5, 6, 7, 8, 9, 10, 11, 12\} \qquad A \cap B = \emptyset$$
$$A \cup C = \{1, 3, 4, 5, 6, 7, 8, 12\} \qquad A \cap C = \{1, 3\}$$
$$B \cup C = \{1, 2, 3, 4, 6, 7, 8, 9, 10, 11\} \qquad B \cap C = \{6, 7, 8\}$$

Complements are:

$$\bar{A} = \{2, 4, 6, 7, 8, 9, 10, 11\}$$
$$\bar{B} = \{1, 3, 4, 5, 12\}$$
$$\bar{C} = \{2, 5, 9, 10, 11, 12\}$$

The various sets are illustrated in Figure 1.2-2.

Algebra of Sets

All subsets of the universal set form an algebraic system for which a number of theorems may be stated (Thomas, 1969). Three of the most important of these relate to laws involving unions and intersections. The *commutative law* states that

$$A \cap B = B \cap A \tag{1.2-6}$$
$$A \cup B = B \cup A \tag{1.2-7}$$

The *distributive law* is written as

$$A \cap (B \cup C) = (A \cap B) \cup (A \cap C) \tag{1.2-8}$$
$$A \cup (B \cap C) = (A \cup B) \cap (A \cup C) \tag{1.2-9}$$

The *associative law* is written as

$$(A \cup B) \cup C = A \cup (B \cup C) = A \cup B \cup C \tag{1.2-10}$$
$$(A \cap B) \cap C = A \cap (B \cap C) = A \cap B \cap C \tag{1.2-11}$$

These are just restatements of (1.2-3) and (1.2-4).

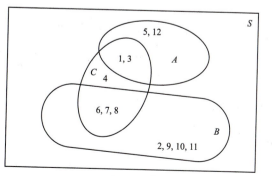

FIGURE 1.2-2
Venn diagram applicable to Example 1.2-1.

De Morgan's Laws

By use of a Venn diagram we may readily prove *De Morgan's laws*†, which state that the complement of a union (intersection) of two sets A and B equals the intersection (union) of the complements $\bar{A}$ and $\bar{B}$. Thus,

$$\overline{(A \cup B)} = \bar{A} \cap \bar{B} \tag{1.2-12}$$

$$\overline{(A \cap B)} = \bar{A} \cup \bar{B} \tag{1.2-13}$$

From the last two expressions one can show that if in an identity we replace unions by intersections, intersections by unions, and sets by their complements, then the identity is preserved (Papoulis, 1965, p. 23).

> **EXAMPLE 1.2-2.** We verify De Morgan's law (1.2-13) by using the example sets $A = \{2 < a \leq 16\}$ and $B = \{5 < b \leq 22\}$ when $S = \{2 < s \leq 24\}$. First, if we define $C = A \cap B$, the reader can readily see from Venn diagrams that $C = A \cap B = \{5 < c \leq 16\}$ so $\bar{C} = \overline{A \cap B} = \{2 < c \leq 5, 16 < c \leq 24\}$. This result is the left side of (1.2-13).
>
> Second, we compute $\bar{A} = S - A = \{16 < a \leq 24\}$ and $\bar{B} = S - B = \{2 < b \leq 5, 22 < b \leq 24\}$. Thus, $C = \bar{A} \cup \bar{B} = \{2 < c \leq 5, 16 < c \leq 24\}$. This result is the right side of (1.2-13) and De Morgan's law is verified.

Duality Principle

This principle (see Papoulis, 1965, for additional reading) states: if in an identity we replace unions by intersections, intersections by unions, S by $\varnothing$, and $\varnothing$ by S, then the identity is preserved. For example, since

$$A \cap (B \cup C) = (A \cap B) \cup (A \cap C) \tag{1.2–14}$$

is a valid identity from (1.2-8), it follows that

$$A \cup (B \cap C) = (A \cup B) \cap (A \cup C) \tag{1.2-15}$$

is also valid, which is just (1.2-9).

> **EXAMPLE 1.2-3.** We demonstrate (1.2-14) for three sets $A = \{1, 2, 4, 6\}$, $B = \{2, 6, 8, 10\}$, and $C = \{3 \leq c \leq 4\}$. Here $B \cup C = \{2, 3 \leq c \leq 4, 6, 8, 10\}$, $A \cap B = \{2, 6\}$, $A \cap C = \{4\}$. For the left side of (1.2-14) $A \cap (B \cup C) = \{2, 4, 6\}$. For the right side of (1.2-14) $A \cap B) \cup (A \cap C) = \{2, 4, 6\}$. This result is the same as for the left side, demonstrating the validity of (1.2-14) for these sets.

---

†After Augustus De Morgan (1806–1871), an English mathematician.

PROBABILITY INTRODUCED THROUGH SETS AND RELATIVE FREQUENCY

In this section we define probability in two ways. The first is based on set theory and fundamental axioms; this approach is the more sound of the two. Also, it is perhaps a bit more difficult to interpret in a practical sense than the second, called *relative frequency*, which is based more on common sense and engineering or scientific observations. We begin with the introduction of probability using set concepts.

Basic to our study of probability is the idea of a physical *experiment*. In this section we develop a mathematical model of an experiment. Of course, we are interested only in experiments that are regulated in some probabilistic way. A single performance of the experiment is called a *trial* for which there is an *outcome*.

Experiments and Sample Spaces

Although there exists a precise mathematical procedure for defining an experiment, we shall rely on reason and examples. This simplified approach will ultimately lead us to a valid mathematical model for any real experiment.†
To illustrate, one experiment might consist of rolling a single die and observing the number that shows up. There are six such numbers and they form all the possible outcomes in the experiment. If the die is "unbiased" our intuition tells us that each outcome is equally likely to occur and the *likelihood* of any one occurring is $\frac{1}{6}$ (later we call this number the *probability* of the outcome). This experiment is seen to be governed, in part, by two *sets*. One is the set of all possible outcomes, and the other is the set of the likelihoods of the outcomes. Each set has six elements. For the present, we consider only the set of outcomes.

The set of all possible outcomes in any given experiment is called the *sample space* and it is given the symbol S. In effect, the sample space is a universal set for the given experiment. S may be different for different experiments, but all experiments are governed by some sample space. The definition of sample space forms the first of three elements in our mathematical model of experiments. The remaining elements are *events* and *probability*, as discussed below.

Discrete and Continuous Sample Spaces

In the earlier die-tossing experiment, S was a finite set of six elements. Such sample spaces are said to be *discrete* and finite. The sample space can also be

---

†Most of our early definitions involving probability are rigorously established only through concepts beyond our scope. Although we adopt a simplified development of the theory, our final results are no less valid or useful than if we had used the advanced concepts.

discrete and *infinite* for some experiments. For example, S in the experiment "choose randomly a positive integer" is the countably infinite set $\{1, 2, 3, \ldots\}$.

Some experiments have an uncountably infinite sample space. An illustration would be the experiment "obtain a number by spinning the pointer on a wheel of chance numbered from 0 to 12." Here any number s from 0 to 12 can result and $S = \{0 < s \leq 12\}$. Such a sample space is called *continuous*.

Events

In most situations, we are interested in some *characteristic* of the outcomes of our experiment as opposed to the outcomes themselves. In the experiment "draw a card from a deck of 52 cards," we might be more interested in whether we draw a spade as opposed to having any interest in individual cards. To handle such situations we define the concept of an event.

An *event* is defined as a subset of the sample space. Because an event is a set, all the earlier definitions and operations applicable to sets will apply to events. For example, if two events have no common outcomes they are *mutually exclusive*.

In the above card experiment, 13 of the 52 possible outcomes are spades. Since any one of the spade outcomes satisfies the event "draw a spade," this event is a set with 13 elements. We have earlier stated that a set with N elements can have as many as 2^N subsets (events defined on a sample space having N possible outcomes). In the present example, $2^N = 2^{52} \approx 4.5(10^{15})$ events.

As with the sample space, events may be either discrete or continuous. The card event "draw a spade" is a discrete, finite event. An example of a discrete, countably infinite event would be "select an odd integer" in the experiment "randomly select a positive integer." The event has a countably infinite number of elements: $\{1, 3, 5, 7, \ldots\}$. However, events defined on a countably infinite sample space do not *have* to be countably infinite. The event $\{1, 3, 5, 7\}$ is clearly not infinite but applies to the integer selection experiment.

Events defined on continuous sample spaces are usually continuous. In the experiment "choose randomly a number a from 6 to 13," the sample space is $S = \{6 \leq s \leq 13\}$. An event of interest might correspond to the chosen number falling between 7.4 and 7.6; that is, the event (call it A) is $A = \{7.4 < a < 7.6\}$.

Discrete events may also be defined on continuous sample spaces. An example of such an event is $A = \{6.13692\}$ for the sample space $S = \{6 \leq s \leq 13\}$ of the previous paragraph. We comment later on this type of event (following Example 1.3-1).

The above definition of an event as a subset of the sample space forms the second of three elements in our mathematical model of experiments. The third element involves defining probability.

Probability Definition and Axioms

To each event defined on a sample space S, we shall assign a nonnegative number called *probability*. Probability is therefore a function; it is a function of the events defined. We adopt the notation $P(A)$† for "the probability of event A." When an event is stated explicitly as a set by using braces, we employ the notation $P\{\cdot\}$ instead of $P(\{\cdot\})$.

The assigned probabilities are chosen so as to satisfy three *axioms*. Let A be any event defined on a sample space S. Then the first two axioms are

axiom 1:
$$P(A) \geq 0 \tag{1.3-1a}$$

axiom 2:
$$P(S) = 1 \tag{1.3-1b}$$

The first only represents our desire to work with nonnegative numbers. The second axiom recognizes that the sample space itself is an event, and, since it is the all-encompassing event, it should have the highest possible probability, which is selected as unity. For this reason, S is known as the *certain event*. Alternatively, the null set $\varnothing$ is an event with no elements; it is known as the *impossible event* and its probability is 0.

The third axiom applies to N events A_n, $n = 1, 2, \ldots, N$, where N may possibly be infinite, defined on a sample space S, and having the property $A_m \cap A_n = \varnothing$ for all $m \neq n$. It is

axiom 3:
$$P\left(\bigcup_{n=1}^{N} A_n\right) = \sum_{n=1}^{N} P(A_n) \qquad \text{if} \qquad A_m \cap A_n = \varnothing \tag{1.3-1c}$$

for all $m \neq n = 1, 2, \ldots, N$, with N possibly infinite. The axiom states that the probability of the event equal to the union of any number of mutually exclusive events is equal to the sum of the individual event probabilities.

An example should help give a physical picture of the meaning of the above axioms.

EXAMPLE 1.3-1. Let an experiment consist of obtaining a number x by spinning the pointer on a "fair" wheel of chance that is labeled from 0 to 100 points. The sample space is $S = \{0 < x \leq 100\}$. We reason that probability of the pointer falling between any two numbers $x_2 \geq x_1$ should be $(x_2 - x_1)/100$ since the wheel is fair. As a check on this assignment, we see that the event $A = \{x_1 < x \leq x_2\}$ satisfies axiom 1 for all x_1 and x_2, and axiom 2 when $x_2 = 100$ and $x_1 = 0$.

Now suppose we break the wheel's periphery into N contiguous segments $A_n = \{x_{n-1} < x \leq x_n\}$, $x_n = (n)100/N$, $n = 1, 2, \ldots, N$, with $x_0 = 0$. Then $P(A_n) = 1/N$, and, for any N,

$$P\left(\bigcup_{n=1}^{N} A_n\right) = \sum_{n=1}^{N} P(A_n) = \sum_{n=1}^{N} \frac{1}{N} = 1 = P(S)$$

from axiom 3.

---

†Occasionally it will be convenient to use brackets, such as $P[A]$ when A is itself an event such as $C - (B \cap D)$.

Example 1.3-1 allows us to return to our earlier discussion of discrete events defined on continuous sample spaces. If the interval $x_n - x_{n-1}$ is allowed to approach zero ($\rightarrow 0$), the probability $P(A_n) \rightarrow P(x_n)$; that is, $P(A_n)$ becomes the probability of the pointer falling exactly on the point x_n. Since $N \rightarrow \infty$ in this situation, $P(A_n) \rightarrow 0$. Thus, the probability of a discrete event defined on a continuous sample space is 0. This fact is true in general.

A consequence of the above statement is that events can occur even if their probability is 0. Intuitively, any number can be obtained from the wheel of chance, but that precise number may never occur again. The infinite sample space has only one outcome satisfying such a discrete event, so its probability is 0. Such events are *not* the same as the impossible event which has *no* elements and *cannot* occur. The converse situation can also happen where events with probability 1 may *not* occur. An example for the wheel of chance experiment would be the event $A = \{$all numbers except the number $x_n\}$. Events with probability 1 (that may not occur) are not the same as the certain event which *must* occur.

Mathematical Model of Experiments

The axioms of probability, introduced above, complete our mathematical model of an experiment. We pause to summarize. Given some real physical experiment having a set of particular outcomes possible, we first defined a *sample space* to mathematically represent the physical outcomes. Second, it was recognized that certain characteristics of the outcomes in the real experiment were of interest, as opposed to the outcomes themselves; *events* were defined to mathematically represent these characteristics. Finally, *probabilities* were assigned to the defined events to mathematically account for the random nature of the experiment.

Thus, a real experiment is defined mathematically by three things: (1) assignment of a sample space; (2) definition of events of interest; and (3) making probability assignments to the events such that the axioms are satisfied. Establishing the correct model for an experiment is probably the single most difficult step in solving probability problems.

EXAMPLE 1.3-2. An experiment consists of observing the sum of the numbers showing up when two dice are thrown. We develop a model for this experiment.

The sample space consists of $6^2 = 36$ points as shown in Figure 1.3-1. Each possible outcome corresponds to a sum having values from 2 to 12.

Suppose we are mainly interested in three events defined by $A = \{$sum $= 7\}$, $B = \{8 <$ sum $\leq 11\}$, and $C = \{10 <$ sum$\}$. In assigning probabilities to these events, it is first convenient to define 36 *elementary events* $A_{ij} = \{$sum for outcome $(i, j) = i + j\}$, where i represents the row and j represents the column locating a particular possible outcome in Figure 1.3-1. An elementary event has only one element.

For probability assignments, intuition indicates that each possible outcome has the same likelihood of occurrence if the dice are fair, so

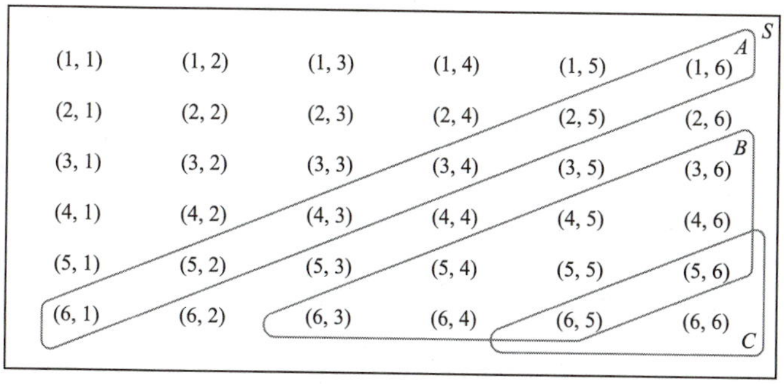

FIGURE 1.3-1
Sample space applicable to Example 1.3-2.

$P(A_{ij}) = \frac{1}{36}$. Now because the events A_{ij}, i and $j = 1, 2, \ldots, N = 6$, are mutually exclusive, they must satisfy axiom 3. But since the events A, B, and C are simply the unions of appropriate elementary events, their probabilities are derived from axiom 3. From Figure 1.3-1 we easily find

$$P(A) = P\left(\bigcup_{i=1}^{6} A_{i,7-i}\right) = \sum_{i=1}^{6} P(A_{i,7-i}) = 6\left(\frac{1}{36}\right) = \frac{1}{6}$$

$$P(B) = 9\left(\frac{1}{36}\right) = \frac{1}{4}$$

$$P(C) = 3\left(\frac{1}{36}\right) = \frac{1}{12}$$

As a matter of interest, we also observe the probabilities of the events $B \cap C$ and $B \cup C$ to be $P(B \cap C) = 2(\frac{1}{36}) = \frac{1}{18}$ and $P(B \cup C) = 10(\frac{1}{36}) = \frac{5}{18}$.

Probability as a Relative Frequency

The use of common sense and engineering and scientific observations leads to a definition of probability as a *relative frequency* of occurrence of some event. For example, everyone can surmise that if a fair coin is flipped several times, the side that shows up will be "heads" about half the time with good "regularity." What reason is saying here is that, if the coin is flipped many times (say n) and heads shows up n_H times out of the n flips, then

$$\lim_{n \to \infty} (n_H/n) = P(H) \tag{1.3-2}$$

where $P(H)$ is interpreted as the probability of the event "heads." The ratio n_H/n is the *relative frequency* (or average number of successes) for this event. The idea of *statistical regularity* is used to account for the fact that relative frequencies approach a fixed value (a probability) as n becomes large (Cooper and McGillem, 1986, p. 9). That such regularity is reasonable is based purely on the fact that many different investigators of numerous physical experiments have observed such regularity (Davenport, 1970, p. 10).

Probability as a relative frequency is intuitively satisfying for many practical problems. In another example, if asked what would be the probability of drawing a six from a thoroughly shuffled regular deck of 52 cards, most would say $4/52 = 1/13$ because four of the 52 cards are sixes. Many trials of this experiment would verify this fact when (1.3-2) is used.

To extend the concept of relative frequency, consider the following example.

EXAMPLE 1.3-3. In a box there are 80 resistors each having the same size and shape. Of the 80 resistors 18 are $10\,\Omega$, 12 are $22\,\Omega$, 33 are $27\,\Omega$, and 17 are $47\,\Omega$. If the experiment is to randomly draw out one resistor from the box with each one being "equally likely" to be drawn, then relative frequency suggests that the following probabilities may reasonably be assumed:

$$P(\text{draw } 10\,\Omega) = 18/80 \qquad P(\text{draw } 22\,\Omega) = 12/80$$
$$P(\text{draw } 27\,\Omega) = 33/80 \qquad P(\text{draw } 47\,\Omega) = 17/80 \qquad (1)$$

Since all resistors are distinct, mutually exclusive, both individually and by type, have only four types, and some resistor *must* be chosen, the sum of the probabilities of all four events must equal 1.

Next, suppose a 22-Ω resistor is drawn from the box and *not replaced*. A second resistor is then drawn from the box. We ask, what are now the probabilities of drawing a resistor of any one of the four values? Since the "population" in the box is now 79 resistors, we have, for the second drawing,

$$P(\text{draw } 10\,\Omega|22\,\Omega) = 18/79 \qquad P(\text{draw } 22\,\Omega|22\,\Omega) = 11/79$$
$$P(\text{draw } 27\,\Omega|22\,\Omega) = 33/79 \qquad P(\text{draw } 47\,\Omega|22\,\Omega) = 17/79 \qquad (2)$$

We use the notation $P(\cdot|22\,\Omega)$ to note that probabilities on the second drawing are now *conditional* on the outcome of the first drawing. This reasoning is readily extended to other drawings in sequence without replacement.

In Example 1.3-3 it was seen that the probability of some event may depend (be conditional) on the occurrence of another event. We consider such probabilities in more detail in the next section.

1.4
JOINT AND CONDITIONAL PROBABILITY

In some experiments, such as in Example 1.3-2 above, it may be that some events are not mutually exclusive because of common elements in the sample space. These elements correspond to the simultaneous or *joint* occurrence of the nonexclusive events. For two events A and B, the common elements form the event $A \cap B$.

Joint Probability

The probability $P(A \cap B)$ is called the *joint probability* for two events A and B which intersect in the sample space. A study of a Venn diagram will readily show that

$$P(A \cap B) = P(A) + P(B) - P(A \cup B) \tag{1.4-1}$$

Equivalently,

$$P(A \cup B) = P(A) + P(B) - P(A \cap B) \leq P(A) + P(B) \tag{1.4-2}$$

In other words, the probability of the union of two events never exceeds the sum of the event probabilities. The equality holds only for mutually exclusive events because $A \cap B = \varnothing$, and therefore, $P(A \cap B) = P(\varnothing) = 0$.

Conditional Probability

Given some event B with nonzero probability

$$P(B) > 0 \tag{1.4-3}$$

we define the *conditional probability* of an event A, given B, by

$$P(A|B) = \frac{P(A \cap B)}{P(B)} \tag{1.4-4}$$

The probability $P(A|B)$ simply reflects the fact that the probability of an event A may depend on a second event B. If A and B are mutually exclusive, $A \cap B = \varnothing$, and $P(A|B) = 0$.

Conditional probability is a defined quantity and cannot be proven. However, as a probability it must satisfy the three axioms given in (1.3-1). $P(A|B)$ obviously satisfies axiom 1 by its definition because $P(A \cap B)$ and $P(B)$ are nonnegative numbers. The second axiom is shown to be satisfied by letting $S = A$:

$$P(S|B) = \frac{P(S \cap B)}{P(B)} = \frac{P(B)}{P(B)} = 1 \tag{1.4-5}$$

The third axiom may be shown to hold by considering the union of A with an event C, where A and C are mutually exclusive. If $P(A \cup C|B) = P(A|B) + P(C|B)$ is true, then axiom 3 holds. Since $A \cap C = \varnothing$ then events $A \cap B$ and $B \cap C$ are mutually exclusive (use a Venn diagram to verify this fact) and

$$P[(A \cup C) \cap B] = P[(A \cap B) \cup (C \cap B)] = P(A \cap B) + P(C \cap B) \tag{1.4-6}$$

Thus, on substitution into (1.4-4)

$$P[(A \cup C)|B] = \frac{P[(A \cup C) \cap B]}{P(B)} = \frac{P(A \cap B)}{P(B)} + \frac{P(C \cap B)}{P(B)}$$
$$= P(A|B) + P(C|B) \tag{1.4-7}$$

and axiom 3 holds.

EXAMPLE 1.4-1. In a box there are 100 resistors having resistance and tolerance as shown in Table 1.4-1. Let a resistor be selected from the box and assume each resistor has the same likelihood of being chosen. Define three events: A as "draw a 47-Ω resistor," B as "draw a resistor with 5% tolerance," and C as "draw a 100-Ω resistor." From the table, the applicable probabilities are†

$$P(A) = P(47\,\Omega) = \frac{44}{100}$$

$$P(B) = P(5\%) = \frac{62}{100}$$

$$P(C) = P(100\,\Omega) = \frac{32}{100}$$

The joint probabilities are

$$P(A \cap B) = P(47\,\Omega \cap 5\%) = \frac{28}{100}$$

$$P(A \cap C) = P(47\,\Omega \cap 100\,\Omega) = 0$$

$$P(B \cap C) = P(5\% \cap 100\,\Omega) = \frac{24}{100}$$

By using (1.4-4) the conditional probabilities become

$$P(A|B) = \frac{P(A \cap B)}{P(B)} = \frac{28}{62}$$

$$P(A|C) = \frac{P(A \cap C)}{P(C)} = 0$$

$$P(B|C) = \frac{P(B \cap C)}{P(C)} = \frac{24}{32}$$

TABLE 1.4-1
Numbers of resistors in a box having given resistance and tolerance

Resistance (Ω)	Tolerance		
	5%	10%	Total
22	10	14	24
47	28	16	44
100	24	8	32
Total	62	38	100

†It is reasonable that probabilities are related to the *number* of resistors in the box that satisfy an event, since each resistor is equally likely to be selected. An alternative approach would be based on elementary events similar to that used in Example 1.3-2. The reader may view the latter approach as more rigorous but less readily applied.

$P(A|B) = P(47\,\Omega|5\%)$ is the probability of drawing a 47-Ω resistor given that the resistor drawn is 5%. $P(A|C) = P(47\,\Omega|100\,\Omega)$ is the probability of drawing a 47-Ω resistor given that the resistor drawn is 100 Ω; this is clearly an impossible event so the probability of it is 0. Finally, $P(B|C) = P(5\%|100\,\Omega)$ is the probability of drawing a resistor of 5% tolerance given that the resistor is 100 Ω.

Total Probability

The probability $P(A)$ of any event A defined on a sample space S can be expressed in terms of conditional probabilities. Suppose we are given N mutually exclusive events B_n, $n = 1, 2, \ldots, N$, whose union equals S as illustrated in Figure 1.4-1. These events satisfy

$$B_m \cap B_n = \varnothing \qquad m \neq n = 1, 2, \ldots, N \tag{1.4-8}$$

$$\bigcup_{n=1}^{N} B_n = S \tag{1.4-9}$$

We shall prove that

$$P(A) = \sum_{n=1}^{N} P(A|B_n)P(B_n) \tag{1.4-10}$$

which is known as the *total probability* of event A.

Since $A \cap S = A$, we may start the proof using (1.4-9) and (1.2-8):

$$A \cap S = A \cap \left(\bigcup_{n=1}^{N} B_n \right) = \bigcup_{n=1}^{N} (A \cap B_n) \tag{1.4-11}$$

Now the events $A \cap B_n$ are mutually exclusive as seen from the Venn diagram (Fig. 1.4-1). By applying axiom 3 to these events, we have

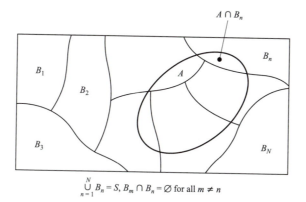

FIGURE 1.4-1
Venn diagram of N mutually exclusive events B_n and another event A.

$$P(A) = P(A \cap S) = P\left[\bigcup_{n=1}^{N}(A \cap B_n)\right] = \sum_{n=1}^{N} P(A \cap B_n) \qquad (1.4\text{-}12)$$

where (1.4-11) has been used. Finally, (1.4-4) is substituted into (1.4-12) to obtain (1.4-10).

Bayes' Theorem†

The definition of conditional probability, as given by (1.4-4), applies to any two events. In particular, let B_n be one of the events defined above in the subsection on total probability. Equation (1.4-4) can be written

$$P(B_n|A) = \frac{P(B_n \cap A)}{P(A)} \qquad (1.4\text{-}13)$$

if $P(A) \neq 0$, or, alternatively,

$$P(A|B_n) = \frac{P(A \cap B_n)}{P(B_n)} \qquad (1.4\text{-}14)$$

if $P(B_n) \neq 0$. One form of Bayes' theorem is obtained by equating these two expressions:

$$P(B_n|A) = \frac{P(A|B_n)P(B_n)}{P(A)} \qquad (1.4\text{-}15)$$

Another form derives from a substitution of $P(A)$ as given by (1.4-10),

$$P(B_n|A) = \frac{P(A|B_n)P(B_n)}{P(A|B_1)P(B_1) + \cdots + P(A|B_N)P(B_N)} \qquad (1.4\text{-}16)$$

for $n = 1, 2, \ldots, N$.

An example will serve to illustrate Bayes' theorem and conditional probability.

> **EXAMPLE 1.4-2.** An elementary binary communication system consists of a transmitter that sends one of two possible symbols (a 1 or a 0) over a channel to a receiver. The channel occasionally causes errors to occur so that a 1 shows up at the receiver as a 0, and vice versa.
>
> The sample space has two elements (0 or 1). We denote by B_i, $i = 1, 2$, the events "the symbol before the channel is 1," and "the symbol before the channel is 0," respectively. Furthermore, define A_i, $i = 1, 2$, as the events "the symbol after the channel is 1," and "the symbol after the channel is 0," respectively. The probabilities that the symbols 1 and 0 are selected for transmission are assumed to be
>
> $$P(B_1) = 0.6 \quad \text{and} \quad P(B_2) = 0.4$$

---

†The theorem is named for Thomas Bayes (1702–1761), an English theologian and mathematician.

Conditional probabilities describe the effect the channel has on the transmitted symbols. The reception probabilities given a 1 was transmitted are assumed to be

$$P(A_1|B_1) = 0.9$$
$$P(A_2|B_1) = 0.1$$

The channel is presumed to affect 0s in the same manner so

$$P(A_1|B_2) = 0.1$$
$$P(A_2|B_2) = 0.9$$

In either case, $P(A_1|B_i) + P(A_2|B_i) = 1$ because A_1 and A_2 are mutually exclusive and are the only "receiver" events (other than the uninteresting events $\varnothing$ and S) possible. The channel is often shown diagrammatically as illustrated in Figure 1.4-2. Because of its form it is usually called a *binary symmetric channel*.

From (1.4-10) we obtain the "received" symbol probabilities

$$P(A_1) = P(A_1|B_1)P(B_1) + P(A_1|B_2)P(B_2)$$
$$= 0.9(0.6) + 0.1(0.4) = 0.58$$
$$P(A_2) = P(A_2|B_1)P(B_1) + P(A_2|B_2)P(B_2)$$
$$= 0.1(0.6) + 0.9(0.4) = 0.42$$

From either (1.4-15) or (1.4-16) we have

$$P(B_1|A_1) = \frac{P(A_1|B_1)P(B_1)}{P(A_1)} = \frac{0.9(0.6)}{0.58} = \frac{0.54}{0.58} \approx 0.931$$
$$P(B_2|A_2) = \frac{P(A_2|B_2)P(B_2)}{P(A_2)} = \frac{0.9(0.4)}{0.42} = \frac{0.36}{0.42} \approx 0.857$$

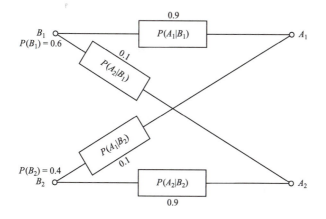

FIGURE 1.4-2
Binary symmetric communication system diagrammatical model applicable to Example 1.4-2.

$$P(B_1|A_2) = \frac{P(A_2|B_1)P(B_1)}{P(A_2)} = \frac{0.1(0.6)}{0.42} = \frac{0.06}{0.42} \approx 0.143$$

$$P(B_2|A_1) = \frac{P(A_1|B_2)P(B_2)}{P(A_1)} = \frac{0.1(0.4)}{0.58} = \frac{0.04}{0.58} \approx 0.069$$

These last two numbers are probabilities of system error while $P(B_1|A_1)$ and $P(B_2|A_2)$ are probabilities of correct system transmission of symbols.

In Bayes' theorem (1.4-16), the probabilities $P(B_n)$ are usually referred to as *a priori probabilities*, since they apply to the events B_n before the performance of the experiment. Similarly, the probabilities $P(A|B_n)$ are numbers typically known prior to conducting the experiment. Example 1.4-2 described such a case. The conditional probabilities are sometimes called *transition probabilities* in a communications context. On the other hand, the probabilities $P(B_n|A)$ are called *a posteriori probabilities*, since they apply after the experiment's performance when some event A is obtained.

EXAMPLE 1.4-3. A student takes a commuter train to get to a school's campus and to class. The probability the student will arrive at class on time is 0.95 provided the train is on time. If the train is known to be on schedule 70% of the time, what is the probability the student will be on time to class? Here we use Bayes' rule and interpret the relative frequency 70% as the probability of the train's being on time. Let C represent the event the student arrives in class and T represent the train arrives on time. The student will arrive on time if the joint event $C \cap T$ is true. The probability of this event is $P(C \cap T) = P(C|T)P(T) = 0.95(0.70) = 0.665$, from (1.4-13).

EXAMPLE 1.4-4. A box contains 6 green balls, 4 black balls, and 10 yellow balls. All balls are equally likely (probable) to be drawn. What is the probability of drawing 2 green balls from the box if the ball on the first draw is not replaced? Let G represent the event "draw a green ball." An application of relative frequency to the first draw suggests that $P(G) = 6/20 = 0.3$. On the second draw we operate with 5 green, 4 black, and 10 yellow balls because the first draw *must* result in a green ball. For the second draw, $P(G|G) = 5/19$ and Bayes' rule gives $P(G \cap G) = P(G|G)P(G) = (5/19)0.3 = 1.5/19 \approx 0.789$.

1.5
INDEPENDENT EVENTS

In this section we introduce the concept of statistically independent events. Although a given problem may involve any number of events in general, it is most instructive to consider first the simplest possible case of two events.

Let two events A and B have nonzero probabilities of occurrence; that is, assume $P(A) \neq 0$ and $P(B) \neq 0$. We call the events *statistically independent* if the probability of occurrence of one event is not affected by the occurrence of the other event. Mathematically, this statement is equivalent to requiring

$$P(A|B) = P(A) \qquad (1.5\text{-}1)$$

for statistically independent events. We also have

$$P(B|A) = P(B) \qquad (1.5\text{-}2)$$

for statistically independent events. By substitution of (1.5-1) into (1.4-4), independence† also means that the probability of the joint occurrence (intersection) of two events must equal the product of the two event probabilities:

$$P(A \cap B) = P(A)P(B) \qquad (1.5\text{-}3)$$

Not only is (1.5-3) [or (1.5-1)] necessary for two events to be independent but it is sufficient. As a consequence, (1.5-3) can, and often does, serve as a test of independence.

Statistical independence is fundamental to much of our later work. When events are independent, it will often be found that probability problems are greatly simplified.

It has already been stated that the joint probability of two mutually exclusive events is 0:

$$P(A \cap B) = 0 \qquad (1.5\text{-}4)$$

If the two events have nonzero probabilities of occurrence, then, by comparison of (1.5-4) with (1.5-3), we easily establish that two events cannot be both mutually exclusive and statistically independent. Hence, in order for two events to be independent they *must* have an intersection $A \cap B \neq \emptyset$.

If a problem involves more than two events, those events satisfying either (1.5-3) or (1.5-1) are said to be *independent by pairs*.

> **EXAMPLE 1.5-1.** In an experiment, one card is selected from an ordinary 52-card deck. Define events A as "select a king," B as "select a jack or queen," and C as "select a heart." From intuition, these events have probabilities $P(A) = \frac{4}{52}$, $P(B) = \frac{8}{52}$, and $P(C) = \frac{13}{52}$.
>
> It is also easy to state joint probabilities. $P(A \cap B) = 0$ (it is not possible to simultaneously select a king and a jack or queen), $P(A \cap C) = \frac{1}{52}$, and $P(B \cap C) = \frac{2}{52}$.
>
> We determine whether A, B, and C are independent by pairs by applying (1.5-3):

---

†We shall often use only the word independence to mean statistical independence.

$$P(A \cap B) = 0 \neq P(A)P(B) = \frac{32}{52^2}$$

$$P(A \cap C) = \frac{1}{52} = P(A)P(C) = \frac{1}{52}$$

$$P(B \cap C) = \frac{2}{52} = P(B)P(C) = \frac{2}{52}$$

Thus, A and C are independent as a pair, as are B and C. However, A and B are not independent, as we might have guessed from the fact that A and B are mutually exclusive.

In many practical problems, statistical independence of events is often *assumed*. The justification hinges on there being no apparent physical connection between the mechanisms leading to the events. In other cases, probabilities assumed for elementary events may lead to independence of other events defined from them (Cooper and McGillem, 1971, p. 24).

Multiple Events

When more than two events are involved, independence by pairs is not sufficient to establish the events as statistically independent, even if *every* pair satisfies (1.5-3).

In the case of three events A_1, A_2, and A_3, they are said to be independent if, and only if, they are independent by all pairs and are also independent as a triple; that is, they must satisfy the *four equations*:

$$P(A_1 \cap A_2) = P(A_1)P(A_2) \qquad (1.5\text{-}5a)$$

$$P(A_1 \cap A_3) = P(A_1)P(A_3) \qquad (1.5\text{-}5b)$$

$$P(A_2 \cap A_3) = P(A_2)P(A_3) \qquad (1.5\text{-}5c)$$

$$P(A_1 \cap A_2 \cap A_3) = P(A_1)P(A_2)P(A_3) \qquad (1.5\text{-}5d)$$

The reader may wonder if satisfaction of (1.5-5d) might be sufficient to guarantee independence by pairs, and therefore, satisfaction of all four conditions? The answer is no, and some further detail on this fact can be found in Davenport (1970, p. 83).

More generally, for N events $A_1, A_2, \ldots, A_N$ to be called statistically independent, we require that all the conditions

$$P(A_i \cap A_j) = P(A_i)P(A_j)$$

$$P(A_i \cap A_j \cap A_k) = P(A_i P(A_j)P(A_k)$$

$$\vdots \qquad (1.5\text{-}6)$$

$$P(A_1 \cap A_2 \cap \cdots \cap A_N) = P(A_1)P(A_2) \cdots P(A_N)$$

be satisfied for all $1 \leq i < j < k < \cdots \leq N$. There are $2^N - N - 1$ of these conditions (Davenport, 1970, p. 83).

EXAMPLE 1.5-2. Consider drawing four cards from an ordinary 52-card deck. Let events A_1, A_2, A_3, A_4 define drawing an ace on the first, second, third, and fourth cards, respectively. Consider two cases. First, draw the cards assuming each is replaced after the draw. Intuition tells us that these events are independent so $P(A_1 \cap A_2 \cap A_3 \cap A_4) = P(A_1)P(A_2)P(A_3)$ $P(A_4) = (4/52)^4 \approx 3.50(10^{-5})$.

On the other hand, suppose we keep each card after it is drawn. We now expect these are not independent events. In the general case we may write

$$P(A_1 \cap A_2 \cap A_3 \cap A_4)$$
$$= P(A_1)P(A_2 \cap A_3 \cap A_4|A_1)$$
$$= P(A_1)P(A_2|A_1)P(A_3 \cap A_4|A_1 \cap A_2)$$
$$= P(A_1)P(A_2|A_1)P(A_3|A_1 \cap A_2)P(A_4|A_1 \cap A_2 \cap A_3)$$
$$= \frac{4}{52} \cdot \frac{3}{51} \cdot \frac{2}{50} \cdot \frac{1}{49} \approx 3.69(10^{-6})$$

Thus, we have approximately 9.5 times better chance of drawing four aces when cards are replaced than when kept. This is an intuitively satisfying result since replacing the ace drawn raises chances for an ace on the succeeding draw.

Properties of Independent Events

Many properties of independent events may be summarized by the statement: If N events $A_1, A_2, \ldots, A_N$ are independent, then any one of them is independent of any event formed by unions, intersections, and complements of the others (Papoulis, 1965, p. 42). Several examples of the application of this statement are worth listing for illustration.

For two independent events A_1 and A_2 it results that A_1 is independent of $\bar{A}_2$, $\bar{A}_1$ is independent of A_2, and $\bar{A}_1$ is independent of $\bar{A}_2$. These statements are proved as a problem at the end of this chapter.

For three independent events A_1, A_2, and A_3 any one is independent of the joint occurrence of the other two. For example

$$P[A_1 \cap (A_2 \cap A_3)] = P(A_1)P(A_2)P(A_3) = P(A_1)P(A_2 \cap A_3) \qquad (1.5\text{-}7)$$

with similar statements possible for the other cases $A_2 \cap (A_1 \cap A_3)$ and $A_3 \cap (A_1 \cap A_2)$. Any one event is also independent of the union of the other two. For example

$$P[A_1 \cap (A_2 \cup A_3)] = P(A_1)P(A_2 \cup A_3) \qquad (1.5\text{-}8)$$

This result and (1.5-7) do not necessarily hold if the events are only independent by pairs.

1.6
COMBINED EXPERIMENTS

All of our work up to this point is related to outcomes from a single experiment. Many practical problems arise where such a constrained approach does not apply. One example would be the simultaneous measurement of wind speed and barometric presure at some location and instant in time. *Two* experiments are actually being conducted; one has the outcome "speed"; the other outcome is "pressure." Still another type of problem involves conducting the *same* experiment several times, such as flipping a coin N times. In this case there are N performances of the same experiment. To handle these situations we introduce the concept of a combined experiment.

A *combined experiment* consists of forming a *single* experiment by suitably combining individual experiments, which we now call *subexperiments*. Recall that an experiment is defined by specifying three quantities. They are: (1) the applicable sample space, (2) the events defined on the sample space, and (3) the probabilities of the events. We specify these three quantities below, beginning with the sample space, for a combined experiment.

Combined Sample Space

Consider only two subexperiments first. Let S_1 and S_2 be the sample spaces of the two subexperiments and let s_1 and s_2 represent the elements of S_1 and S_2, respectively. We form a new space S, called the *combined sample space*,† whose elements are all the ordered pairs (s_1, s_2). Thus, if S_1 has M elements and S_2 has N elements, then S will have MN elements. The combined sample space is denoted

$$S = S_1 \times S_2 \tag{1.6-1}$$

> EXAMPLE 1.6-1. If S_1 corresponds to flipping a coin, then $S_1 = \{H, T\}$, where H is the element "heads" and T represents "tails." Let $S_2 = \{1, 2, 3, 4, 5, 6\}$ corresponding to rolling a single die. The combined sample space $S = S_1 \times S_2$ becomes
>
> $$S = \{(H, 1), (H, 2), (H, 3), (H, 4), (H, 5), (H, 6), (T, 1), (T, 2),$$
> $$(T, 3), (T, 4), (T, 5), (T, 6)\}$$
>
> In the new space, elements are considered to be single objects, each object being a pair of items.

---

†Also called the *cartesian product space* in some texts.

EXAMPLE 1.6-2. We flip a coin twice, each flip being taken as one sub-experiment. The applicable sample spaces are now

$$S_1 = \{H, T\}$$
$$S_2 = \{H, T\}$$
$$S = \{(H, H), (H, T), (T, H), (T, T)\}$$

In this last example, observe that the element (H, T) is considered different from the element (T, H); this fact emphasizes the elements of S are *ordered* pairs of objects.

The more general situation of N subexperiments is a direct extension of the above concepts. For N sample spaces $S_n, n = 1, 2, \ldots, N$, having elements s_n, the combined sample space S is denoted

$$S = S_1 \times S_2 \times \cdots \times S_N \tag{1.6-2}$$

and it is the set of all ordered N-tuples

$$(s_1, s_2, \ldots, s_N) \tag{1.6-3}$$

Events on the Combined Space

Events may be defined on the combined sample space through their relationship with events defined on the subexperiment sample spaces. Consider two subexperiments with sample spaces S_1 and S_2. Let A be any event defined on S_1 and B be any event defined on S_2, then

$$C = A \times B \tag{1.6-4}$$

is an event defined on S consisting of all pairs (s_1, s_2) such that

$$s_1 \in A \quad \text{and} \quad s_2 \in B \tag{1.6-5}$$

Since elements of A correspond to elements of the event $A \times S_2$ defined on S, and elements of B correspond to the event $S_1 \times B$ defined on S, we easily find that

$$A \times B = (A \times S_2) \cap (S_1 \times B) \tag{1.6-6}$$

Thus, the event defined by the subset of S given by $A \times B$ is the intersection of the subsets $A \times S_2$ and $S_1 \times B$. We consider all subsets of S of the form $A \times B$ as events. All intersections and unions of such events are also events (Papoulis, 1965, p. 50).

EXAMPLE 1.6-3. Let $S_1 = \{0 \le x \le 100\}$ and $S_2 = \{0 \le y \le 50\}$. The combined sample space is the set of all pairs of numbers (x, y) with $0 \le x \le 100$ and $0 \le y \le 50$ as illustrated in Figure 1.6-1. For events

$$A = \{x_1 < x < x_2\}$$
$$B = \{y_1 < y < y_2\}$$

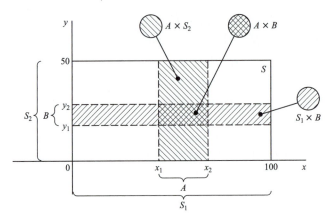

FIGURE 1.6-1
A combined sample space for two subexperiments.

where $0 \leq x_1 < x_2 \leq 100$ and $0 \leq y_1 < y_2 \leq 50$, the events $S_1 \times B$ and $A \times S_2$ are horizontal and vertical strips as shown. The event

$$A \times B = \{x_1 < x < x_2\} \times \{y_1 < y < y_2\}$$

is the rectangle shown. An event $S_1 \times \{y = y_1\}$ would be a horizontal line.

In the more general case of N subexperiments with sample spaces S_n on which events A_n are defined, the events on the combined sample space S will all be sets of the form

$$A_1 \times A_2 \times \cdots \times A_N \tag{1.6-7}$$

and unions and intersections of such sets (Papoulis, 1965, pp. 53–54).

Probabilities

To complete the definition of a combined experiment we must assign probabilities to the events defined on the combined sample space S. Consider only two subexperiments first. Since all events defined on S will be unions and intersections of events of the form $A \times B$, where $A \subset S_1$ and $B \subset S_2$, we only need to determine $P(A \times B)$ for any A and B. We shall only consider the case where

$$P(A \times B) = P(A)P(B) \tag{1.6-8}$$

Subexperiments for which (1.6-8) is valid are called *independent experiments*.

To see what elements of S correspond to elements of A and B, we only need substitute S_2 for B or S_1 for A in (1.6-8):

$$P(A \times S_2) = P(A)P(S_2) = P(A) \tag{1.6-9}$$

$$P(S_1 \times B) = P(S_1)P(B) = P(B) \tag{1.6-10}$$

Thus, elements in the set $A \times S_2$ correspond to elements of A, and those of $S_1 \times B$ correspond to those of B.

For N independent experiments, the generalization of (1.6-8) becomes

$$P(A_1 \times A_2 \times \cdots \times A_N) = P(A_1)P(A_2) \cdots P(A_N) \qquad (1.6\text{-}11)$$

where $A_n \subset S_n$, $n = 1, 2, \ldots, N$.

With independent experiments, the above results show that probabilities for events defined on S are completely determined from probabilities of events defined in the subexperiments.

Permutations

Experiments often involve multiple trials in which outcomes are elements of a finite sample space and they are not replaced after each trial. For example, in drawing four cards from an ordinary 52-card deck, each of the "draws" is not replaced, so the sample spaces for the second, third, and fourth draws have only 51, 50, and 49 elements, respectively. In these and other types of problems, the number of possible sequences of the outcomes is often important.

For n total elements there are n possible outcomes on the first trial, $(n-1)$ on the second, and so forth. For r elements being drawn, the number of possible sequences of r elements from the original n is denoted by P_r^n and is given by

$$\left. \begin{array}{l} \text{Orderings of } r \text{ elements} \\ \text{taken from } n \end{array} \right\} = n(n-1)(n-2)\cdots(n-r+1)$$

$$= \frac{n!}{(n-r)!} = P_r^n, \qquad r = 1, 2, \ldots, n \qquad (1.6\text{-}12)$$

This number is the number of *permutations*, or sequences, of r elements taken from n elements when order of occurrence is important. This last point is clear from the card experiment. Suppose the first two cards are both kings, say, a heart and a spade. Then the king of hearts followed by the king of spades is considered a different sequence from the spade followed by the heart.

EXAMPLE 1.6-4. How many permutations are there for four cards taken from a 52-card deck? From (1.6-12) $P_4^{52} = 52!/(52-4)! = 52(51)50(49) = 6,497,400$.

Combinations

If the order of elements in a sequence is *not* important, we reason that there are now fewer possible sequences of r elements taken from n elements without replacement. In fact, the number of permutations of (1.6-12) is reduced by a factor given by the number of permutations (orderings) of the r things, which is $P_r^r = r!$. The resulting number of sequences where order is not important is called the number of *combinations* of r things taken from n things. The notation $\binom{n}{r}$ is usually used to denote combinations, but other notations are also used in various sources. Thus,

$$\binom{n}{r} = \frac{P_r^n}{P_r^r} = \frac{n!}{(n-r)!r!} \tag{1.6-13}$$

The numbers $\binom{n}{r}$ are called *binomial coefficients* because they are central to the expansion of the binomial $(x + y)^n$ as given by

$$(x + y)^n = \sum_{r=0}^{n} \binom{n}{r} x^r y^{n-r} \tag{1.6-14}$$

In computing factorials in (1.6-12) and (1.6-13) we define $0! = 1$, so $\binom{n}{0} = 1$ and $\binom{n}{n} = 1$.

EXAMPLE 1.6-5. A coach has five athletes from whom a 3-person team is to be selected for a competition. How many such teams could he select? The answer is the number of combinations of (1.6-13) with $r = 3$ and $n = 5$, or $5!/(3!2!) = 10$. Note that the same number occurs for a selection of 2-person teams: $5!/(2!3!) = 10$.

The preceding example points out the symmetry of binomial coefficients:

$$\binom{n}{r} = \binom{n}{n-r} \tag{1.6-15}$$

1.7
BERNOULLI TRIALS

We shall close this chapter on probability by considering a very practical problem. It involves any experiment for which there are only two possible outcomes on any trial. Examples of such an experiment are numerous: flipping a coin, hitting or missing the target in artillery, passing or failing an exam, receiving a 0 or a 1 in a computer bit stream, or winning or losing in a game of chance are just a few.

For this type of experiment, we let A be the elementary event having one of the two possible outcomes as its element. $\bar{A}$ is the only other possible elementary event. Specifically, we shall repeat the basic experiment N times and determine the probability that A is observed exactly k times out of the N trials. Such repeated experiments are called *Bernoulli trials*.† Those readers familiar with combined experiments will recognize this experiment as the combination of N identical subexperiments. For readers who omitted the section on combined experiments, we shall develop the problem so that the omission will not impair their understanding of the material.

---

†After the Swiss mathematician Jacob Bernoulli (1654–1705).

Assume that elementary events are statistically independent for every trial. Let event A occur on any given trial with probability

$$P(A) = p \qquad (1.7\text{-}1)$$

The event $\bar{A}$ then has probability

$$P(\bar{A}) = 1 - p \qquad (1.7\text{-}2)$$

After N trials of the basic experiment, one *particular* sequence of outcomes has A occurring k times, followed by $\bar{A}$ occurring $N - k$ times.† Because of assumed statistical independence of trials, the probability of this one sequence is

$$\underbrace{P(A)P(A)\cdots P(A)}_{k \text{ terms}} \underbrace{P(\bar{A})P(\bar{A})\cdots P(\bar{A})}_{N-k \text{ terms}} = p^k(1-p)^{N-k} \qquad (1.7\text{-}3)$$

Now there are clearly other particular sequences that will yield k events A and $N - k$ events $\bar{A}$.‡ The probability of each of these sequences is given by (1.7-3). Since the sum of all such probabilities will be the desired probability of A occurring exactly k times in N trials, we only need find the number of such sequences. Some thought will reveal that this is the number of ways of taking k objects at a time from N objects. The number is known to be the binomial coefficient of (1.6-13).

$$\binom{N}{k} = \frac{N!}{k!(N-k)!} \qquad (1.7\text{-}4)$$

From the product of (1.7-4) and (1.7-3) we finally obtain

$$P\{A \text{ occurs exactly } k \text{ times}\} = \binom{N}{k} p^k(1-p)^{N-k} \qquad (1.7\text{-}5)$$

EXAMPLE 1.7-1. A submarine attempts to sink an aircraft carrier. It will be successful only if two or more torpedoes hit the carrier. If the sub fires three torpedoes and the probability of a hit is 0.4 for each torpedo, what is the probability that the carrier will be sunk?

Define the event $A = \{$torpedo hits$\}$. Then $P(A) = 0.4$, and $N = 3$. Probabilities are found from (1.7-5):

$$P\{\text{exactly no hits}\} = \binom{3}{0}(0.4)^0(1 - 0.4)^3 = 0.216$$

$$P\{\text{exactly one hit}\} = \binom{3}{1}(0.4)^1(1 - 0.4)^2 = 0.432$$

---

†This particular sequence corresponds to one N-dimensional element in the combined sample space S.

‡ All such sequences define all the elements of S that satisfy the event $\{A$ occurs exactly k times in N trials$\}$ defined on the combined sample space.

$$P\{\text{exactly two hits}\} = \binom{3}{2}(0.4)^2(1 - 0.4)^1 = 0.288$$

$$P\{\text{exactly three hits}\} = \binom{3}{3}(0.4)^3(1 - 0.4)^0 = 0.064$$

The answer we desire is

$$P\{\text{carrier sunk}\} = P\{\text{two or more hits}\}$$
$$= P\{\text{exactly two hits}\} + P\{\text{exactly three hits}\}$$
$$= 0.352$$

EXAMPLE 1.7-2. In a culture used for biological research the growth of unavoidable bacteria occasionally spoils results of an experiment that requires at least three out of four cultures to be unspoiled to obtain a single datum point. Experience has shown that about 6 of every 100 cultures are randomly spoiled by the bacteria. If the experiment requires three simultaneously derived, unspoiled data points for success, we find the probability of success for any given set of 12 cultures (three data points of four cultures each).

We treat individual datum points first as a Bernoulli trial problem with $N = 4$ and $p = P\{\text{good culture}\} = \frac{94}{100} = 0.94$. Here

$$P\{\text{valid datum point}\} = P\{3 \text{ good cultures}\} + P\{4 \text{ good cultures}\}$$

$$= \binom{4}{3}(0.94)^3(1 - 0.94)^1 = \binom{4}{4}(0.94)^4(1 - 0.94)^0 \approx 0.98$$

Finally, we treat the required three data points as a Bernoulli trial problem with $N = 3$ and $p = P\{\text{valid datum point}\} = 0.98$. Now

$$P\{\text{successful experiment}\} = P\{3 \text{ valid data points}\}$$

$$= \binom{3}{3}(0.98)^3(1 - 0.98)^0 \approx 0.941$$

Thus, the given experiment will be successful about 94.1 percent of the time.

When N, k, and $(N - k)$ are large, the factorials in (1.7-5) are difficult to evaluate, so approximations become useful. One approximation, called *Stirling's formula*, is

$$m! \approx (2\pi m)^{1/2} m^m e^{-m} \qquad m \text{ large} \qquad (1.7\text{-}6)$$

It is exact for $m \to \infty$ in the sense that the ratio of $m!$ to the right side of (1.7-6) tends to unity. For other values of m its fractional error is on the order of $1/(12m)$, which is quite good (better than 1 percent) even for m as small as 10.

By applying Stirling's formula to the factorials, in (1.7-5), and then approximating some resulting factors by the first two terms in their series expansions, it can be shown (see Davenport, 1970, pp. 276–278) that

$$P\{A \text{ occurs exactly } k \text{ times}\} = \binom{N}{k} p^k (1-p)^{N-k}$$

$$\approx \frac{1}{\sqrt{2\pi N p(1-p)}} \exp\left[-\frac{(k-Np)^2}{2Np(1-p)}\right] \quad (1.7\text{-}7)$$

This equation, called the *De Moivre–Laplace†* approximation, holds for N, k, and $(N-k)$ large, k near Np such that its deviations from Np (higher or lower) are small in magnitude relative to both Np and $N(1-p)$. We illustrate these restrictions by example.

EXAMPLE 1.7-3. Suppose a certain machine gun fires rounds (cartridges) for 3 seconds at a rate of 2400 per minute, and the probability of any bullet's hitting a large target is 0.4. We find the probability that exactly 50 of the bullets hit the target.

Here $N = 3(2400/60) = 120$, $k = 50$, $p = 0.4$, $Np = 120(0.4) = 48$, and $N(1-p) = 120(0.6) = 72$. Thus, since N, k, and $(N-k) = 70$ are all large, while k is near Np and the deviation of k from Np, which is $50 - 48 = 2$, is much smaller than both $Np = 48$ and $N(1-p) = 72$, we can use (1.7-7):

$$P\{\text{exactly 50 bullets hit the target}\} = \binom{N}{k} p^k (1-p)^{N-k}$$

$$\approx \frac{1}{\sqrt{2\pi(48)0.6}} \exp\left[-\frac{(50-48)^2}{2(48)0.6}\right] = 0.0693$$

The approximation of (1.7-7) fails to be accurate when N becomes very large while p is very small. For these conditions another approximation is helpful. It is called the *Poisson‡* approximation:

$$\binom{N}{k} p^k (1-p)^{N-k} \approx \frac{(Np)^k e^{-Np}}{k!} \qquad N \text{ large and } p \text{ small} \qquad (1.7\text{-}8)$$

1.8
SUMMARY

This chapter has developed the basics of probability, events, and random experiments by successively building on a basic foundation of set theory. It has also defined probability through the concept of a relative frequency of occurrence of events. Specifically, the topics developed were:

---

†Abraham De Moivre (1667–1754) was a French-born scientist who lived most of his life in England and contributed to the mathematics of probability. Marquis Pierre Simon De Laplace (1749–1827) was an outstanding French mathematician.
‡After the French mathematician Siméon Denis Poisson (1781–1840).

- Definitions of sets, characteristics of sets, and how they enter into definitions of probability.
- Introduction of probability defined through sets and through the use of the relative frequency concept.
- Development of special kinds of probability, such as applied to events, multiple events (joint probability), and events dependent on each other (conditional probability).
- Introduction of the statistical independence of events.
- Discussions of how to combine several separate random experiments such that they may be taken as a single (combined) experiment. Through Bernoulli trials these topics were applied to various practical problems involving success and failure outcomes of an experiment.

This chapter's topics form a firm basis to proceed to the important concept of a random variable in the next chapter.

PROBLEMS

1.1-1. Specify the following sets by the rule method.

$$A = \{1, 2, 3\}, B = \{8, 10, 12, 14\}, C = \{1, 3, 5, 7, \ldots\}$$

1.1-2. Use the tabular method to specify a class of sets for the sets of Problem 1.1-1.

1.1-3. State whether the following sets are countable or uncountable, or finite or infinite. $A = \{1\}$, $B = \{x = 1\}$, $C = \{0 < \text{integers}\}$, $D = \{\text{children in public school No. 5}\}$, $E = \{\text{girls in public school No. 5}\}$, $F = \{\text{girls in class in public school No. 5 at 3:00 A.M.}\}$, $G = \{\text{all lengths not exceeding one meter}\}$, $H = \{-25 \le x \le -3\}$, $I = \{-2, -1, 1 \le x \le 2\}$.

1.1-4. For each set of Problem 1.1-3, determine if it is equal to, or a subset of, any of the other sets.

1.1-5. State every possible subset of the set of letters $\{a, b, c, d\}$.

1.1-6. A thermometer measures temperatures from -40 to $130°F$ (-40 to $54.4°C$).
  (a) State a universal set to describe temperature measurements. Specify subsets for:
  (b) Temperature measurements not exceeding water's freezing point, and
  (c) Measurements exceeding the freezing point but not exceeding $100°F$ ($37.8°C$).

*1.1-7.** Prove that a set with N elements has 2^N subsets.

1.1-8. A random noise voltage at a given time may have any value from -10 to $10\,\text{V}$.
  (a) What is the universal set describing noise voltage?
  (b) Find a set to describe the voltages available from a half-wave rectifier for positive voltages that has a linear output-input voltage characteristic.
  (c) Repeat parts (a) and (b) if a dc voltage of $-3\,\text{V}$ is added to the random noise.

1.1-9. Use the tabular method to define a set A that contains all integers with magnitudes not exceeding 7. Define a second set B having odd integers larger than -2 and not larger than 5. Determine if $A \subset B$ and $B \subset A$.

1.1-10. A set A has three elements a_1, a_2, and a_3. Determine all possible subsets of A.

1.1-11. Specify, by both the tabular and rule methods, each of the following sets: (a) all integers between 1 and 9, (b) all integers from 1 to 9, (c) the five values of equivalent resistance for n identical 10-Ω resistors in prallel where $n = 1, 2, \ldots, 5$, and (d) the six values of equivalent resistance for n identical 2.2-Ω resistors in series where $n = 1, 2, \ldots, 6$.

1.1-12. A box contains 100 capacitors (universal set) of which 40 are 0.01 μF with a 100-V voltage rating, 35 are 0.1 μF at a rating of 50 V, and 25 are 1.0 μF and have a 10-V rating. Determine the number of elements in the following sets:
(a) $A = \{\text{capacitors with capacitance} \geq 0.1\,\mu\text{F}\}$
(b) $B = \{\text{capacitors with voltage rating} > 5\,\text{V}\}$
(c) $C = \{\text{capacitors with both capacitance} \geq 0.1\,\mu\text{F} \text{ and voltage rating} \geq 50\,\text{V}\}$.

1.2-1. Show that $C \subset A$ if $C \subset B$ and $B \subset A$.

1.2-2. Two sets are given by $A = \{-6, -4, -0.5, 0, 1.6, 8\}$ and $B = \{-0.5, 0, 1, 2, 4\}$. Find:
(a) $A - B$ (b) $B - A$ (c) $A \cup B$ (d) $A \cap B$

1.2-3. A universal set is given as $S = \{2, 4, 6, 8, 10, 12\}$. Define two subsets as $A = \{2, 4, 10\}$ and $B = \{4, 6, 8, 10\}$. Determine the following:
(a) $\bar{A} = S - A$ (b) $A - B$ and $B - A$ (c) $A \cup B$ (d) $A \cap B$
(e) $\bar{A} \cap B$

1.2-4. Using Venn diagrams for three sets A, B, C, shade the areas corresponding to the sets:
(a) $(A \cup B) - C$ (b) $\bar{B} \cap A$ (c) $A \cap B \cap C$ (d) $(\overline{A \cup B}) \cap C$

1.2-5. Sketch a Venn diagram for three events where $A \cap B \neq \varnothing$, $B \cap C \neq \varnothing$, $C \cap A \neq \varnothing$, but $A \cap B \cap C = \varnothing$.

1.2-6. Use Venn diagrams to show that the following identities are true:
(a) $(\overline{A \cup B}) \cap C = C - [(A \cap C) \cup (B \cap C)]$
(b) $(A \cup B \cup C) - (A \cap B \cap C) = (\bar{A} \cap B) \cup (\bar{B} \cap C) \cup (\bar{C} \cap A)$
(c) $(\overline{A \cap B \cap C}) = \bar{A} \cup \bar{B} \cup \bar{C}$

1.2-7. Use Venn diagrams to prove De Morgan's laws $(\overline{A \cup B}) = \bar{A} \cap \bar{B}$ and $(\overline{A \cap B}) = \bar{A} \cup \bar{B}$.

1.2-8. A universal set is $S = \{-20 < s \leq -4\}$. If $A = \{-10 \leq s \leq -5\}$ and $B = \{-7 < s < -4\}$, find:
(a) $A \cup B$
(b) $A \cap B$
(c) A third set C such that the sets $A \cap C$ and $B \cap C$ are as large as possible while the smallest element in C is -9.

(d) What is the set $A \cap B \cap C$?

1.2-9. Use De Morgan's laws to show that:
(a) $\overline{A \cap (B \cup C)} = (\bar{A} \cup \bar{B}) \cap (\bar{A} \cup \bar{C})$
(b) $\overline{(A \cap B \cap C)} = \bar{A} \cup \bar{B} \cup \bar{C}$
In each case check your results using a Venn diagram.

1.2-10. Shade Venn diagrams to illustrate each of the following sets:
(a) $(A \cup \bar{B}) \cap \bar{C}$,   (b) $\overline{(A \cap B)} \cup \bar{C}$,   (c) $(A \cup \bar{B}) \cup (C \cap D)$,
(d) $(A \cap B \cap \bar{C}) \cup (\bar{B} \cap C \cap D)$.

1.2-11. A universal set S is composed of all points in a rectangular area defined by $0 \le x \le 3$ and $0 \le y \le 4$. Define three sets by $A = \{y \le 3(x-1)/2\}$, $B = \{y \ge 1\}$, and $C = \{y \ge 3 - x\}$. Shade in Venn diagrams corresponding to the sets (a) $A \cap B \cap C$, and (b) $C \cap B \cap \bar{A}$.

1.2-12. The take-off roll distance for aircraft at a certain airport can be any number from 80 m to 1750 m. Propeller aircraft require from 80 m to 1050 m while jets use from 950 m to 1750 m. The overall runway is 2000 m.
(a) Determine sets A, B, and C defined as "propeller aircraft take-off distances," "jet aircraft take-off distances," and "runway length safety margin," respectively.
(b) Determine the set $A \cap B$ and give its physical significance.
(c) What is the meaning of the set $\overline{A \cup B}$?
(d) What are the meanings of the sets $\overline{A \cup B \cup C}$ and $A \cup B$?

1.2-13. Prove that De Morgan's law (1.2-13) can be extended to N events A_i, $i = 1, 2, \ldots, N$ as follows:

$$\overline{(A_1 \cap A_2 \cap \cdots \cap A_N)} = (\bar{A}_1 \cup \bar{A}_2 \cup \cdots \cup \bar{A}_N)$$

1.2-14. Work Problem 1.2-13 for (1.2-12) to prove

$$\overline{(A_1 \cup A_2 \cup \cdots \cup A_N)} = (\bar{A}_1 \cap \bar{A}_2 \cap \cdots \cap \bar{A}_N)$$

1.2-15. Sets $A = \{1 \le s \le 14\}$, $B = \{3, 6, 14\}$, and $C = \{2 < s \le 9\}$ are defined on a sample space S. State if each of the following conditions is true or false.
(a) $C \subset B$,   (b) $C \subset A$,   (c) $B \cap C = \varnothing$,   (d) $C \cup B = S$,
(e) $\bar{S} = \varnothing$,   (f) $A \cap \bar{S} = \varnothing$, and (g) $C \subset A \subset B$.

1.2-16. Draw Venn diagrams and shade the areas corresponding to the sets
(a) $(A \cup B \cup C) \cap (\bar{A} \cup \bar{B} \cup \bar{C})$, and (b) $[(A \cup \bar{B}) \cap C] \cup (\overline{A \cup C})$.

1.2-17. Work Problem 1.2-16 except assume sets (a) $(A \cap B \cap C) \cup (\overline{A \cup B \cup C})$,
(b) $B - (A \cap B)$, and (c) $(A \cap B) \cup (A \cap C) \cup (B \cap C) - (A \cap B \cap C)$.

1.3-1. A die is tossed. Find the probabilities of the events $A = \{$odd number shows up$\}$, $B = \{$number larger than 3 shows up$\}$, $A \cup B$, and $A \cap B$.

1.3-2. In a game of dice, a "shooter" can win outright if the sum of the two numbers showing up is either 7 or 11 when two dice are thrown. What is his probability of winning outright?

1.3-3. A pointer is spun on a fair wheel of chance having its periphery labeled from 0 to 100.

   (*a*) What is the sample space for this experiment?
   (*b*) What is the probability that the pointer will stop between 20 and 35?
   (*c*) What is the probability that the wheel will stop on 58?

1.3-4. An experiment has a sample space with 10 equally likely elements $S = \{a_1, a_2, \ldots, a_{10}\}$. Three events are defined as $A = \{a_1, a_5, a_9\}$, $B = \{a_1, a_2, a_6, a_9\}$, and $C = \{a_6, a_9\}$. Find the probabilities of:
   (*a*) $A \cup C$
   (*b*) $B \cup \bar{C}$
   (*c*) $A \cap (B \cup C)$
   (*d*) $\overline{A \cup B}$
   (*e*) $(A \cup B) \cap C$

1.3-5. Let A be an arbitrary event. Show that $P(\bar{A}) = 1 - P(A)$.

1.3-6. An experiment consists of rolling a single die. Two events are defined as: $A = \{$a 6 shows up$\}$ and $B = \{$a 2 or a 5 shows up$\}$.
   (*a*) Find $P(A)$ and $P(B)$.
   (*b*) Define a third event C so that $P(C) = 1 - P(A) - P(B)$.

1.3-7. In a box there are 500 colored balls: 75 black, 150 green, 175 red, 70 white, and 30 blue. What are the probabilities of selecting a ball of each color?

1.3-8. A single card is drawn from a 52-card deck.
   (*a*) What is the probability that the card is a jack?
   (*b*) What is the probability the card will be a 5 or smaller?
   (*c*) What is the probability that the card is a red 10?

1.3-9. A pair of fair dice are thrown in a gambling problem. Person A wins if the sum of numbers showing up is six or less *and* one of the dice shows four. Person B wins if the sum is five or more *and* one of the dice shows a four. Find: (*a*) The probability that A wins, (*b*) the probability of B winning, and (*c*) the probability that both A and B win.

1.3-10. You (person A) and two others (B and C) each toss a fair coin in a two-step gambling game. In step 1 the person whose toss is not a match to either of the other two is "odd man out." Only the remaining two whose coins match go on to step 2 to resolve the ultimate winner.
   (*a*) What is the probability you will advance to step 2 after the first toss?
   (*b*) What is the probability you will be out after the first toss?
   (*c*) What is the probability that no one will be out after the first toss?

1.3-11. A particular electronic device is known to contain only 10-, 22-, and 48-Ω resistors, but these resistors may have 0.25-, 0.5-, or 1-W ratings, depending on how purchases are made to minimize cost. Historically, it is found that the probabilities of the 10-Ω resistors being 0.25, 0.5, or 1 W are 0.08, 0.10, and 0.01, respectively. For the 22-Ω resistors the similar probabilities are 0.20, 0.26, and 0.05. It is also historically found that the probabilities are 0.40, 0.51, and 0.09 that any resistors are 0.25, 0.50, and 1 W, respectively. What are the probabilities that the 48-Ω resistors are (*a*) 0.25, (*b*) 0.50, and (*c*) 1 W?

1.3-12. For the sample space defined in Example 1.3-2 find the probabilities that: (*a*) one die will show a 2 and the other will show a 3 or larger, and (*b*) the sum of the two numbers showing up will be 4 or less or will be 10 or more.

1.3-13. In a game two dice are thrown. Let one die be "weighted" so that a 4 shows up with probability $\frac{2}{7}$, while its other numbers all have probabilities of $\frac{1}{7}$. The same probabilities apply to the other die except the number 3 is "weighted." Determine the probability the shooter will win outright by having the sum of the numbers showing up be 7. What would be the probability for fair dice?

1.4-1. Two cards are drawn from a 52-card deck (the first is not replaced).
  (*a*) Given the first card is a queen, what is the probability that the second is also a queen?
  (*b*) Repeat part (*a*) for the first card a queen and the second card a 7.
  (*c*) What is the probability that both cards will be a queen?

1.4-2. An ordinary 52-card deck is thoroughly shuffled. You are dealt four cards up. What is the probability that all four cards are sevens?

1.4-3. For the resistor selection experiment of Example 1.4-1, define event D as "draw a 22-Ω resistor," and E as "draw a resistor with 10% tolerance." find $P(D)$, $P(E)$, $P(D \cap E)$, $P(D|E)$, and $P(E|D)$.

1.4-4. For the resistor selection experiment of Example 1.4-1, define two mutually exclusive events B_1 and B_2 such that $B_1 \cup B_2 = S$.
  (*a*) Use the total probability theorem to find the probability of the event "select a 22-Ω resistor," denoted D.
  (*b*) Use Bayes' theorem to find the probability that the resistor selected had 5% tolerance, given it was 22 Ω.

1.4-5. In three boxes there are capacitors as shown in Table P1.4-5. An experiment consists of first randomly selecting a box, assuming each has the same likelihood of selection, and then selecting a capacitor from the chosen box.
  (*a*) What is the probability of selecting a 0.01-μF capacitor, given that box 2 is selected?
  (*b*) If a 0.01-μF capacitor is selected, what is the probability it came from box 3? (*Hint*: Use Bayes' and total probability theorems.)

TABLE P1.4-5
Capacitors

Value (μF)	Number in box			Totals
	1	2	3	
0.01	20	95	25	140
0.1	55	35	75	165
1.0	70	80	145	295
Totals	145	210	245	600

1.4-6. For Problem 1.4-5, list the nine conditional probabilities of capacitor selection, given certain box selections.

1.4-7. Rework Example 1.4-2 if $P(B_1) = 0.6$, $P(B_2) = 0.4$, $P(A_1|B_1) = P(A_2|B_2) = 0.95$, and $P(A_2|B_1) = P(A_1|B_2) = 0.05$.

1.4-8. Rework Example 1.4-2 if $P(B_1) = 0.7$, $P(B_2) = 0.3$, $P(A_1|B_1) = P(A_2|B_2) = 1.0$ and $P(A_2|B_1) = P(A_1|B_2) = 0$. What type of channel does this system have?

1.4-9. A company sells high fidelity amplifiers capable of generating 10, 25, and 50 W of audio power. It has on hand 100 of the 10-W units, of which 15% are defective, 70 of the 25-W units with 10% defective, and 30 of the 50-W units with 10% defective.
(a) What is the probability that an amplifier sold from the 10-W units is defective?
(b) If each wattage amplifier sells with equal likelihood, what is the probability of a randomly selected unit being 50 W and defective?
(c) What is the probability that a unit randomly selected for sale is defective?

1.4-10. A missile can be accidentally launched if two relays A and B both have failed. The probabilities of A and B failing are known to be 0.01 and 0.03, respectively. It is also known that B is more likely to fail (probability 0.06) if A has failed.
(a) What is the probability of an accidental missile launch?
(b) What is the probability that A will fail if B has failed?
(c) Are the events "A fails" and "B fails" statistically independent?

*1.4-11. The communication system of Example 1.4-2 is to be extended to the case of three transmitted symbols 0, 1, and 2. Define appropriate events A_i and B_i, $i = 1, 2, 3$, to represent symbols after and before the channel, respectively. Assume channel transition probabilities are all equal at $P(A_i|B_j) = 0.1$, $i \neq j$, and are $P(A_i|B_j) = 0.8$ for $i = j = 1, 2, 3$, while symbol transmission probabilities are $P(B_1) = 0.5$, $P(B_2) = 0.3$, and $P(B_3) = 0.2$.
(a) Sketch the diagram analogous to Fig. 1.4-2.
(b) Compute received symbol probabilities $P(A_1)$, $P(A_2)$, and $P(A_3)$.
(c) Compute the a posteriori probabilities for this system.
(d) Repeat parts (b) and (c) for all transmission symbol probabilities equal. Note the effect.

1.4-12. A pharmaceutical product consists of 100 pills in a bottle. Two production lines used to produce the product are selected with probabilities 0.45 (line one) and 0.55 (line two). Each line can overfill or underfill bottles by at most 2 pills. Given that line one is observed, the probabilities are 0.02, 0.06, 0.88, 0.03, and 0.01 that the numbers of pills in a bottle will be 102, 101, 100, 99, and 98, respectively. For line two, the similar respective probabilities are 0.03, 0.08, 0.83, 0.04, and 0.02.
(a) Find the probability that a bottle of the product will contain 102 pills. Repeat for 101, 100, 99, and 98 pills.
(b) Given that a bottle contains the correct number of pills, what is the probability it came from line one?
(c) What is the probability that a purchaser of the product will receive less than 100 pills?

1.4-13. A manufacturing plant makes radios that each contain an integrated circuit (IC) supplied by three sources A, B, and C. The probability that the IC in a radio came from one of the sources is $\frac{1}{3}$, the same for all sources. ICs are known to be defective with probabilities 0.001, 0.003, and 0.002 for sources A, B, and C, respectively.
   (a) What is the probability any given radio will contain a defective IC?
   (b) If a radio contains a defective IC, find the probability it came from source A. Repeat for sources B and C.

1.4-14. There are three special decks of cards. The first, deck D_1, has all 52 cards of a regular deck. The second, D_2, has only the 16 face cards of a regular deck (only 4 each of jacks, queens, kings, and aces). The third, D_3, has only the 36 numbered cards of a regular deck (4 twos through 4 tens). A random experiment consists of first randomly choosing one of the three decks, then second, randomly choosing a card from the chosen deck. If $P(D_1) = \frac{1}{2}$, $P(D_2) = \frac{1}{3}$, and $P(D_3) = \frac{1}{6}$, find the probabilities: (a) of drawing an ace, (b) of drawing a three, and (c) of drawing a red card.

1.5-1. Determine whether the three events A, B, and C of Example 1.4-1 are statistically independent.

1.5-2. List the various equations that four events A_1, A_2, A_3, and A_4 must satisfy if they are to be statistically independent.

*1.5-3. Given that two events A_1 and A_2 are statistically independent, show that:
   (a) A_1 is independent of $\bar{A}_2$
   (b) $\bar{A}_1$ is independent of A_2
   (c) $\bar{A}_1$ is independent of $\bar{A}_2$

1.5-4. Show that there are $2^N - N - 1$ equations required in (1.5-6). (*Hint:* Recall that the binomial coefficient is the number of combinations of N things taken n at a time.)

1.5-5. In a communication system the signal sent from point a to point b arrives by two paths in parallel. Over each path the signal passes through two repeaters (in series). Each repeater in one path has a probability of failing (becoming an open circuit) of 0.005. This probability is 0.008 for each repeater on the other path. All repeaters fail independently of each other. Find the probability that the signal will not arrive at point b.

1.5-6. Work Problem 1.5-5, except assume the paths and repeaters of Figure P1.5-6, where the probabilities of the repeaters' failing (independently) are $p_1 = P(R_1) = 0.005$, $p_2 = P(R_2) = P(R_3) = P(R_4) = 0.01$, and $p_3 = P(R_5) = P(R_6) = 0.05$.

1.6-1. An experiment consists of randomly selecting one of five cities on Florida's west coast for a vacation. Another experiment consists of selecting at random one of four acceptable motels in which to stay. Define sample spaces S_1 and S_2 for the two experiments and a combined space $S = S_1 \times S_2$ for the combined experiment having the two subexperiments.

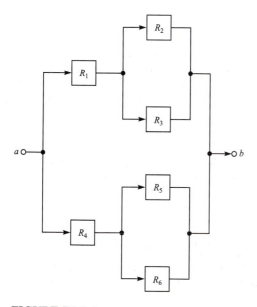

FIGURE P1.5-6

1.6-2. Sketch the area in the combined sample space of Example 1.6-3 corresponding to the event $A \times B$ where:

(a) $A = \{10 < x \le 15\}$ and $B = \{20 < y \le 50\}$
(b) $A = \{x = 40\}$ and $B = \{5 < y \le 40\}$

1.6-3. The six sides of a fair die are numbered from 1 to 6. The die is rolled four times. How many sequences of the four resulting numbers are possible?

1.6-4. In a 5-card poker game, a player is dealt 5 cards. How many poker hands are possible for an ordinary 52-card deck?

1.7-1. A production line manufactures 5-gal (18.93-liter) gasoline cans to a volume tolerance of 5%. The probability of any one can being out of tolerance is 0.03. If four cans are selected at random:

(a) What is the probability they are all out of tolerance?
(b) What is the probability of exactly two being out?
(c) What is the probability that all are in tolerance?

1.7-2. Spacecraft are expected to land in a prescribed recovery zone 80% of the time. Over a period of time, six spacecraft land.

(a) Find the probability that none lands in the prescribed zone.
(b) Find the probability that at least one will land in the prescribed zone.
(c) The landing program is called successful if the probability is 0.9 or more that three or more out of six spacecraft will land in the prescribed zone. Is the program successful?

1.7-3. In the submarine problem of Example 1.7-1, find the probabilities of sinking the carrier when fewer ($N = 2$) or more ($N = 4$) torpedoes are fired.

1.7-4. A student is known to arrive late for a class 40% of the time. If the class meets five times each week find: (a) the probability the student is late for at least three classes in a given week, and (b) the probability the student will not be late at all during a given week.

1.7-5. An airline in a small city has five departures each day. It is known that any given flight has a probability of 0.3 of departing late. For any given day find the probabilities that: (a) no flights depart late, (b) all flights depart late, and (c) three or more depart on time.

1.7-6. The local manager of the airline of Problem 1.7-5 desires to make sure that the probability that all flights leave on time is 0.9. What is the largest probability of being late that the individual flights can have if the goal is to be achieved? Will the operation have to be improved significantly?

1.7-7. A man wins in a gambling game if he gets two heads in five flips of a biased coin. The probability of getting a head with the coin is 0.7.
(a) Find the probability the man will win. Should he play this game?
(b) What is his probability of winning if he wins by getting at least four heads in five flips? Should he play this new game?

*1.7-8. A rifleman can achieve a "marksman" award if he passes a test. He is allowed to fire six shots at a target's bull's eye. If he hits the bull's eye with at least five of his six shots he wins a set. He becomes a marksman only if he can repeat the feat three times straight, that is, if he can win three straight sets. If his probability is 0.8 of hitting a bull's eye on any one shot, find the probabilities of his: (a) winning a set, and (b) becoming a marksman.

1.7-9. A ship can successfully arrive at its destination if its engine and its satellite navigation system do not fail en route. If the engine and navigation system are known to fail independently with respective probabilities of 0.05 and 0.001, what is the probability of a successful arrival?

1.7-10. At a certain military installation six similar radars are placed in operation. It is known that a radar's probability of failing to operate before 500 hours of "on" time have accumulated is 0.06. What are the probabilities that before 500 hours have elapsed, (a) all will operate, (b) all will fail, and (c) only one will fail?

1.7-11. A particular model of automobile is recalled to fix a mechanical problem. The probability that a car will be properly repaired is 0.9. During the week a dealer has eight cars to repair.
(a) What is the probability that two or more of the eight cars will have to be repaired more than once?
(b) What is the probability all eight cars will be properly repaired?

1.7-12. In a large hotel it is known that 99% of all guests return room keys when checking out. If 250 engineers check out after a large conference, what is the probability that not more than three will fail to return their keys? [*Hint*: Use the approximation of (1.7-8).]

CHAPTER 2

The Random Variable

2.0
INTRODUCTION

In the previous chapter we introduced the concept of an event to describe characteristics of outcomes of an experiment. Events allowed us more flexibility in determining properties of an experiment than could be obtained by considering only the outcomes themselves. An event could be almost anything from "descriptive," such as "draw a spade," to numerical, such as "the outcome is 3."

In this chapter, we introduce a new concept that will allow events to be defined in a more consistent manner; they will always be numerical. The new concept is that of a *random variable*, and it will constitute a powerful tool in the solution of practical probabilistic problems.

2.1
THE RANDOM VARIABLE CONCEPT

Definition of a Random Variable

We define a real *random variable*† as a real *function* of the elements of a sample space S. We shall represent a random variable by a capital letter (such as W, X, or Y) and any particular value of the random variable by a lowercase letter (such as w, x, or y). Thus, given an experiment defined by a sample space S with elements s, we assign to every s a real number

---

†Complex random variables are considered in Chapter 5.

$$X(s) \qquad\qquad (2.1\text{-}1)$$

according to some rule and call $X(s)$ a random variable.

A random variable X can be considered to be a function that maps all elements of the sample space into points on the real line or some parts thereof. We illustrate, by two examples, the mapping of a random variable.

EXAMPLE 2.1-1. An experiment consists of rolling a die and flipping a coin. The applicable sample space is illustrated in Figure 2.1-1. Let the random variable be a function X chosen such that (1) a coin head (H) outcome corresponds to positive values of X that are equal to the numbers that show up on the die, and (2) a coin tail (T) outcome corresponds to negative values of X that are equal in magnitude to *twice* the number that shows on the die. Here X maps the sample space of 12 elements into 12 values of X from -12 to 6 as shown in Figure 2.1-1.

EXAMPLE 2.1-2. Figure 2.1-2 illustrates an experiment where the pointer on a wheel of chance is spun. The possible outcomes are the numbers from 0 to 12 marked on the wheel. The sample space consists of the numbers in the set $\{0 < s \leq 12\}$. We define a random variable by the function

$$X = X(s) = s^2$$

Points in S now map onto the real line as the set $\{0 < x \leq 144\}$.

As seen in these two examples, a random variable is a function that maps each point in S into some point on the real line. It is not necessary that the sample-space points map uniquely, however. More than one point in S may map into a single value of X. For example, in the extreme case, we might map all six points in the sample space for the experiment "throw a die and observe the number that shows up" into the one point $X = 2$.

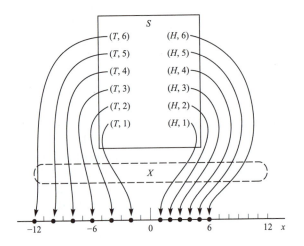

FIGURE 2.1-1
A random variable mapping of a sample space.

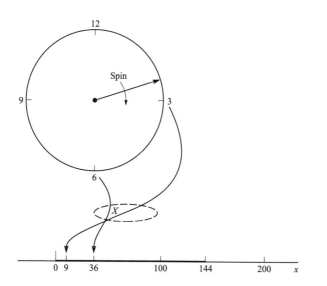

FIGURE 2.1-2
Mapping applicable to Example 2.1-2.

Conditions for a Function to Be a Random Variable

Thus, a random variable may be almost any function we wish. We shall, however, require that it not be multivalued. That is, every point in S must correspond to only one value of the random variable.

  Moreover, we shall require that two additional conditions be satisfied in order that a function X be a random variable (Papoulis, 1965, p. 88). First, the set $\{X \leq x\}$ shall be an event for any real number x. The satisfaction of this condition will be no trouble in practical problems. This set corresponds to those points s in the sample space for which the random variable $X(s)$ does not exceed the number x. The probability of this event, denoted by $P\{X \leq x\}$, is equal to the sum of the probabilities of all the elementary events corresponding to $\{X \leq x\}$.

  The second condition we require is that the probabilities of the events $\{X = \infty\}$ and $\{X = -\infty\}$ be 0:

$$P\{X = -\infty\} = 0 \qquad P\{X = \infty\} = 0 \tag{2.1-2}$$

This condition does not prevent X from being either $-\infty$ or ∞ for some values of s; it only requires that the probability of the set of those s be zero.

Discrete and Continuous Random Variables

A *discrete random variable* is one having only discrete values. Example 2.1-1 illustrated a discrete random variable. The sample space for a discrete random variable can be discrete, continuous, or even a mixture of discrete and continuous points. For example, the "wheel of chance" of Example 2.1-2 has a continuous sample space, but we could define a discrete random variable as having the value 1 for the set of outcomes $\{0 < s \leq 6\}$ and -1 for $\{6 < s \leq 12\}$. The result is a discrete random variable defined on a continuous sample space.

EXAMPLE 2.1-3. A sample space is defined by the set $S = \{1, 2, 3, 4\}$. A random variable X is defined by $X = X(s) = s^3$. Since S is discrete, its points map to discrete points x in the set $\{1, 8, 27, 64\}$. If the probabilities of elements of S are $P(1) = 4/24$, $P(2) = 3/24$, $P(3) = 7/24$, and $P(4) = 10/24$, then the probabilities of the random variable's values become $P\{X = 1\} = 4/24$, $P\{X = 8\} = 3/24$, $P\{X = 27\} = 7/24$, and $P\{X = 64\} = 10/24$ because of one-to-one mapping of the discrete points.

A *continuous random variable* is one having a continuous range of values. It cannot be produced from a discrete sample space because of our requirement that all random variables be single-valued functions of all sample-space points. Similarly, a purely continuous random variable cannot result from a mixed sample space because of the presence of the discrete portion of the sample space. The random variable of Example 2.1-2 is continuous.

EXAMPLE 2.1-4. Suppose the temperature at some geographical point is modeled as a continuous random variable T that is known to always exist from $-60°F$ to $+120°F$. Further, for ease of illustration, let us make the (nonrealistic) assumption that all values $\{-60 \leq t \leq 120\}$ are equally probable. Under these assumptions, we reason, using relative frequency arguments, that values t of T that fall in a small region dt centered anywhere in the range of $-60°F$ to $+120°F$ will have a probability $dt/[120 - (-60)] = dt/180$. This reasoning is extended to find the probability of any *single temperature* within dt by letting $dt \to 0$. It becomes zero.

Example 2.1-4 serves to demonstrate that the probability of occurrence of any discrete value of a continuous random variable is zero.

Mixed Random Variable

A *mixed random variable* is one for which some of its values are discrete and some are continuous. The mixed case is usually the least important type of random variable, but it occurs in some problems of practical significance.

2.2
DISTRIBUTION FUNCTION

The probability $P\{X \leq x\}$ is the probability of the event $\{X \leq x\}$. It is a number that depends on x; that is, it is a function of x. We call this function, denoted $F_X(x)$, the *cumulative probability distribution function* of the random variable X. Thus,

$$F_X(x) = P\{X \leq x\} \tag{2.2-1}$$

We shall often call $F_X(x)$ just the *distribution function* of X. The argument x is
any real number ranging from $-\infty$ to ∞.

The distribution function has some specific properties derived from the
fact that $F_X(x)$ is a probability. These are:†

(1) $F_X(-\infty) = 0$ (2.2-2a)

(2) $F_X(\infty) = 1$ (2.2-2b)

(3) $0 \le F_X(x) \le 1$ (2.2-2c)

(4) $F_X(x_1) \le F_X(x_2)$ if $x_1 < x_2$ (2.2-2d)

(5) $P\{x_1 < X \le x_2\} = F_X(x_2) - F_X(x_1)$ (2.2-2e)

(6) $F_X(x^+) = F_X(x)$ (2.2-2f)

The first three of these properties are easy to justify, and the reader should
justify them as an exercise. The fourth states that $F_X(x)$ is a nondecreasing
function of x. The fifth property states that the probability that X will have
values larger than some number x_1 but not exceeding another number x_2 is
equal to the difference in $F_X(x)$ evaluated at the two points. It is justified from
the fact that the events $\{X \le x_1\}$ and $\{x_1 < X \le x_2\}$ are mutually exclusive, so
the probability of the event $\{X \le x_2\} = \{X \le x_1\} \cup \{x_1 < X \le x_2\}$ is the sum
of the probabilities $P\{X \le x_1\}$ and $P\{x_1 < X \le x_2\}$. The sixth property states
that $F_X(x)$ is a function continuous from the right.

Properties 1, 2, 4, and 6 may be used as tests to determine if some function, say, $G_X(x)$, could be a valid distribution function. If so, all four tests
must be passed. [See Papoulis (1965), p. 99.]

If X is a discrete random variable, consideration of its distribution function defined by (2.2-1) shows that $F_X(x)$ must have a stairstep form, such as
shown in Figure 2.2-1a. The amplitude of a step will equal the probability of
occurrence of the value of X where the step occurs. If the values of X are
denoted x_i, we may write $F_X(x)$ as

$$F_X(x) = \sum_{i=1}^{N} P\{X = x_i\} u(x - x_i) \tag{2.2-3}$$

where $u(\cdot)$ is the unit-step function defined by‡

$$u(x) = \begin{cases} 1 & x \ge 0 \\ 0 & x < 0 \end{cases} \tag{2.2-4}$$

and N may be infinite for some random variables. By introducing the
shortened notation

$$P(x_i) = P\{X = x_i\} \tag{2.2-5}$$

(2.2-3) can be written as

---

†We use the notation x^+ to imply $x + \varepsilon$ where $\varepsilon > 0$ is infinitesimally small; that is, $\varepsilon \to 0$.
‡This definition differs slightly from (A-5) by including the equality so that $u(x)$ satisfies (2.2-2f).

Probability,
Random Variables,
and Random
Signal Principles

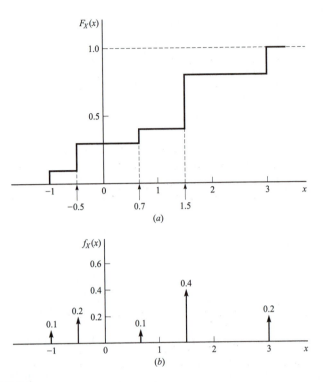

FIGURE 2.2-1
Distribution function (*a*) and density function (*b*) applicable to the discrete random variable of Example 2.2-1. [*Adapted from Peebles (1976) with permission of publishers Addison–Wesley, Advanced Book Program.*]

$$F_X(x) = \sum_{i=1}^{N} P(x_i)u(x - x_i) \tag{2.2-6}$$

We next consider an example that illustrates the distribution function of a discrete random variable.

EXAMPLE 2.2-1. Let X have the discrete values in the set $\{-1, -0.5, 0.7, 1.5, 3\}$. The corresponding probabilities are assumed to be $\{0.1, 0.2, 0.1, 0.4, 0.2\}$. Now $P\{X < -1\} = 0$ because there are no sample space points in the set $\{X < -1\}$. Only when $X = -1$ do we obtain one outcome. Thus, there is an immediate jump in probability of 0.1 in the function $F_X(x)$ at the point $x = -1$. For $-1 < x < -0.5$, there are no additional sample space points so $F_X(x)$ remains constant at the value 0.1. At $x = -0.5$ there is another jump of 0.2 in $F_X(x)$. This process continues until all points are included. $F_X(x)$ then equals 1.0 for all x above the last point. Figure 2.2-1*a* illustrates $F_X(s)$ for this discrete random variable.

A continuous random variable will have a continuous distribution function. We consider an example for which $F_X(x)$ is the continuous function shown in Figure 2.2-2*a*.

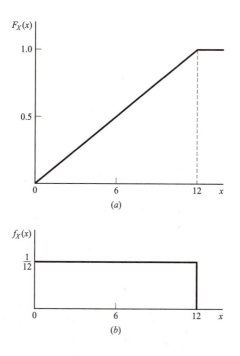

FIGURE 2.2-2
Distribution function (*a*) and density function (*b*) applicable to the continuous random variable of Example 2.2-2. [*Adapted from Peebles (1976) with permission of publishers Addison–Wesley, Advanced Book Program.*]

> **EXAMPLE 2.2-2.** We return to the fair wheel-of-chance experiment. Let the wheel be numbered from 0 to 12 as shown in Figure 2.1-2. Clearly the probability of the event $\{X \le 0\}$ is 0 because there are no sample space points in this set. For $0 < x \le 12$ the probability of $\{0 < X \le x\}$ will increase linearly with x for a fair wheel. Thus, $F_X(x)$ will behave as shown in Figure 2.2-2*a*.

The distribution function of a mixed random variable will be a sum of two parts, one of stairstep form, the other continuous.

2.3
DENSITY FUNCTION

The *probability density function*, denoted by $f_X(x)$, is defined as the derivative of the distribution function:

$$f_X(x) = \frac{dF_X(x)}{dx} \qquad (2.3\text{-}1)$$

We often call $f_X(x)$ just the *density function* of the random variable X.

Existence

If the derivative of $F_X(x)$ exists, then $f_X(x)$ exists and is given by (2.3-1). There may, however, be places where $dF_X(x)/dx$ is not defined. For example, a continuous random variable will have a continuous distribution $F_X(x)$, but $F_X(x)$ may have corners (points of abrupt change in slope). The distribution shown in Figure 2.2-2a is such a function. For such cases, we plot $f_X(x)$ as a function with step-type discontinuities (such as in Figure 2.2-2b).

For discrete random variables having a stairstep form of distribution function, we introduce the concept of the *unit-impulse function* $\delta(x)$ to describe the derivative of $F_X(x)$ at its stairstep points. The unit-impulse function and its properties are reviewed in Appendix A. It is shown there that $\delta(x)$ may be defined by its integral property

$$\phi(x_0) = \int_{-\infty}^{\infty} \phi(x)\delta(x - x_0)\, dx \qquad (2.3\text{-}2)$$

where $\phi(x)$ is any function continuous at the point $x = x_0$; $\delta(x)$ can be interpreted as a "function" with infinite amplitude, area of unity, and zero duration. The unit-impulse and the unit-step functions are related by

$$\delta(x) = \frac{du(x)}{dx} \qquad (2.3\text{-}3)$$

or

$$\int_{-\infty}^{x} \delta(\xi)\, d\xi = u(x) \qquad (2.3\text{-}4)$$

The more general impulse function is shown symbolically as a vertical arrow occurring at the point $x = x_0$ and having an amplitude equal to the amplitude of the step function for which it is the derivative.

We return to the case of a discrete random variable and differentiate $F_X(x)$, as given by (2.2-6), to obtain

$$f_X(x) = \sum_{i=1}^{N} P(x_i)\delta(x - x_i) \qquad (2.3\text{-}5)$$

Thus, the density function for a discrete random variable exists in the sense that we use impulse functions to describe the derivative of $F_X(x)$ at its stairstep points. Figure 2.2-1b is an example of the density function for the random variable having the function of Figure 2.2-1a as its distribution.

A physical interpretation of (2.3-5) is readily achieved. Clearly, the probability of X having one of its particular values, say, x_i, is a number $P(x_i)$. If this probability is assigned to the *point* x_i, then the *density* of probability is infinite because a point has no "width" on the x axis. The infinite "amplitude" of the impulse function describes this infinite density. The "size" of the density of probability at $x = x_i$ is accounted for by the scale factor $P(x_i)$ giving $P(x_i)\delta(x - x_i)$ for the density at the point $x = x_i$.

Several properties that $f_X(x)$ satisfies may be stated:

$$(1) \quad 0 \le f_X(x) \qquad \text{all } x \tag{2.3-6a}$$

$$(2) \quad \int_{-\infty}^{\infty} f_X(x)\, dx = 1 \tag{2.3-6b}$$

$$(3) \quad F_X(x) = \int_{-\infty}^{x} f_X(\xi)\, d\xi \tag{2.3-6c}$$

$$(4) \quad P\{x_1 < X \le x_2\} = \int_{x_1}^{x_2} f_X(x)\, dx \tag{2.3-6d}$$

Proofs of these properties are left to the reader as exercises. Properties 1 and 2 require that the density function be nonnegative and have an area of unity. These two properties may also be used as tests to see if some function, say, $g_X(x)$, can be a valid probability density function (Papoulis, 1965, p. 99). Both tests must be satisfied for validity. Property 3 is just another way of writing (2.3-1) and serves as the link between $F_X(x)$ and $f_X(x)$. Property 4 relates the probability that X will have values from x_1 to, and including, x_2 to the density function.

EXAMPLE 2.3-1. Let us test the function $g_X(s)$ shown in Figure 2.3-1a to see if it can be a valid density function. It obviously satisfies property 1 since it is nonnegative. Its area is $a\alpha$, which must equal unity to satisfy property 2. Therefore, $a = 1/\alpha$ is necessary if $g_X(x)$ is to be a density.

Suppose $a = 1/\alpha$. To find the applicable distribution function we first write

$$g_X(x) = \begin{cases} 0 & x_0 - \alpha > x \ge x_0 + \alpha \\[2mm] \dfrac{1}{\alpha^2}(x - x_0 + \alpha) & x_0 - \alpha \le x < x_0 \\[2mm] \dfrac{1}{\alpha} - \dfrac{1}{\alpha^2}(x - x_0) & x_0 \le x < x_0 + \alpha \end{cases}$$

Next, by using (2.3-6c), we obtain

$$G_X(x) =$$

$$\begin{cases} 0 & x_0 - \alpha > x \\[2mm] \displaystyle\int_{x_0-\alpha}^{x} g_X(\xi)\, d\xi = \dfrac{1}{2\alpha^2}(x - x_0 + \alpha)^2 & x_0 - \alpha \le x < x_0 \\[2mm] \dfrac{1}{2} + \displaystyle\int_{x_0}^{x} g_X(\xi)\, d\xi = \dfrac{1}{2} + \dfrac{1}{\alpha}(x - x_0) - \dfrac{1}{2\alpha^2}(x - x_0)^2 & x_0 \le x < x_0 + \alpha \\[2mm] 1 & x_0 + \alpha \le x \end{cases}$$

This function is plotted in Figure 2.3-1b.

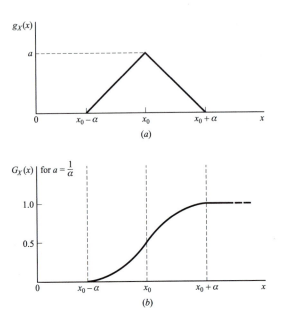

(a)

(b)

FIGURE 2.3-1
A possible probability density function (a) and a distribution function (b) applicable
to Example 2.3-1.

EXAMPLE 2.3-2. Suppose a random variable is known to have the triangular
probability density of the preceding example with $x_0 = 8$, $\alpha = 5$, and
$a = 1/\alpha = \frac{1}{5}$. From the earlier work

$$f_X(x) = \begin{cases} 0 & 3 > x \geq 13 \\ (x-3)/25 & 3 \leq x < 8 \\ 0.2 - (x-8)/25 & 8 \leq x < 13 \end{cases}$$

We shall use this probability density in (2.3-6d) to find the probability
that X has values greater than 4.5 but not greater than 6.7. The probability is

$$P\{4.5 < X \leq 6.7\} = \int_{4.5}^{6.7} [(x-3)/25]\, dx$$

$$= \frac{1}{25}\left[\frac{x^2}{2} - 3x\right]\Bigg|_{4.5}^{6.7} = 0.2288$$

Thus, the event $\{4.5 < X \leq 6.7\}$ has a probability of 0.2288 or 22.88%.

EXAMPLE 2.3-3. A random variable X is known to have a distribution
function

$$F_X(x) = u(x)[1 - e^{-x^2/b}]$$

where $b > 0$ is a constant. Find its density function. By use of (2.3-1)

$$f_X(x) = \frac{dF_X(x)}{dx} = u(x)\frac{d}{dx}[1 - e^{-x^2/b}] + [1 + e^{-x^2/b}]\frac{du(x)}{dx}$$

$$= (1 - e^{-x^2/b})\delta(x) + u(x)\frac{2x}{b}e^{-x^2/b} = u(x)\frac{2x}{b}e^{-x^2/b}$$

The impulse term disappears because its coefficient is zero at $x = 0$ where the impulse "exists." [See (A-29).]

2.4
THE GAUSSIAN RANDOM VARIABLE

A random variable X is called *gaussian*† if its density function has the form

$$f_X(x) = \frac{1}{\sqrt{2\pi\sigma_X^2}}e^{-(x-a_X)^2/2\sigma_X^2} \qquad (2.4\text{-}1)$$

where $\sigma_X > 0$ and $-\infty < a_X < \infty$ are real constants. This function is sketched in Figure 2.4-1a. Its maximum value $(2\pi\sigma_X^2)^{-1/2}$ occurs at $x = a_X$. Its "spread" about the point $x = a_X$ is related to σ_X. The function decreases to 0.607 times its maximum at $x = a_X + \sigma_X$ and $x = a_X - \sigma_X$. It was first derived by De Moivre some 200 years ago and later independently by both Gauss and Laplace (Kennedy and Neville, 1986, p. 175).

The gaussian density is the most important of all densities and it enters into nearly all areas of science and engineering. This importance stems from its accurate description of many practical and significant real-world quantities, especially when such quantities are the result of many small independent random effects acting to create the quantity of interest. For example, the voltage across a resistor at the output of an amplifier can be random (a noise voltage) due to a random current that is the result of many contributions from other random currents at various places within the amplifier. Random thermal agitation of electrons causes the randomness of the various currents. This type of noise is called *gaussian* because the random variable representing the noise voltage has the gaussian density.

The distribution function is found from (2.3-6c) using (2.4-1). The integral is

$$F_X(x) = \frac{1}{\sqrt{2\pi\sigma_X^2}}\int_{-\infty}^{x}e^{-(\xi-a_X)^2/2\sigma_X^2}\,d\xi \qquad (2.4\text{-}2)$$

This integral has no known closed-form solution and must be evaluated by numerical or approximation methods. To make the results generally available, we could develop a set of tables of $F_X(x)$ for various x with a_X and σ_X as

---

†After the German mathematician Johann Friedrich Carl Gauss (1777–1855). The gaussian density is often called the *normal density*.

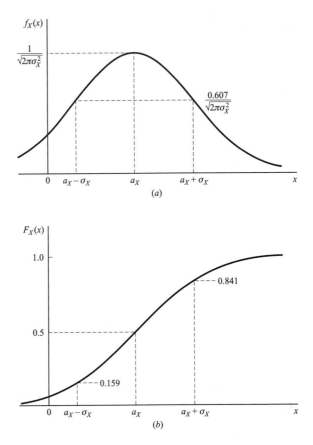

FIGURE 2.4-1
Density (*a*) and distribution (*b*) functions of a gaussian random variable.

parameters. However, this approach has limited value because there is an infinite number of possible combinations of a_X and σ_X, which requires an infinite number of tables. A better approach is possible where only one table of $F_X(x)$ is developed that corresponds to normalized (specific) values of a_X and σ_X. We then show that the one table can be used in the general case where a_X and σ_X can be arbitrary.

We start by first selecting the normalized case where $a_X = 0$ and $\sigma_X = 1$. Denote the corresponding distribution function by $F(x)$. From (2.4-2), $F(x)$ is

$$F(x) = \frac{1}{\sqrt{2\pi}} \int_{-\infty}^{x} e^{-\xi^2/2} \, d\xi \tag{2.4-3}$$

which is a function of x only. This function is tabularized in Appendix B for $x \geq 0$. For a negative value of x we use the relationship

$$F(-x) = 1 - F(x) \tag{2.4-4}$$

To show that the general distribution function $F_X(x)$ of (2.4-2) can be found in terms of $F(x)$ of (2.4-3), we make the variable change

$$u = (\xi - a_X)/\sigma_X \qquad (2.4\text{-}5)$$

in (2.4-2) to obtain

$$F_X(x) = \frac{1}{\sqrt{2\pi}} \int_{-\infty}^{(x-a_X)/\sigma_X} e^{-u^2/2} \, du \qquad (2.4\text{-}6)$$

From (2.4-3), this expression is clearly equivalent to

$$F_X(x) = F\left(\frac{x - a_X}{\sigma_X}\right) \qquad (2.4\text{-}7)$$

Figure 2.4-1b depicts the behavior of $F_X(x)$.

We consider two examples to illustrate the application of (2.4-7).

EXAMPLE 2.4-1. We find the probability of the event $\{X \leq 5.5\}$ for a gaussian random variable having $a_X = 3$ and $\sigma_X = 2$.

Here $(x - a_X)/\sigma_x = (5.5\text{-}3)/2 = 1.25$. From (2.4-7) and the definition of $F_X(x)$

$$P\{X \leq 5.5\} = F_X(5.5) = F(1.25)$$

By using the table in Appendix B

$$P\{X \leq 5.5\} = F(1.25) = 0.8944$$

EXAMPLE 2.4-2. Assume that the height of clouds above the ground at some location is a gaussian random variable X with $a_X = 1830\,\text{m}$ and $\sigma_X = 460\,\text{m}$. We find the probability that clouds will be higher than 2750 m (about 9000 ft). From (2.4-7) and Appendix B:

$$P\{X > 2750\} = 1 - P\{X \leq 2750\} = 1 - F_X(2750)$$

$$= 1 - F\left(\frac{2750 - 1830}{460}\right) = 1 - F(2.0)$$

$$= 1 - 0.9772 = 0.0228$$

The probability that clouds are higher than 2750 m is therefore about 2.28 percent if their behavior is as assumed.

The function $F(x)$ can also be evaluated by approximation. First, we write $F(x)$ of (2.4-3) as

$$F(x) = 1 - Q(x) \qquad (2.4\text{-}8)$$

where

$$Q(x) = \frac{1}{\sqrt{2\pi}} \int_{x}^{\infty} e^{-\xi^2/2} \, d\xi \qquad (2.4\text{-}9)$$

is known as the *Q-function*. As with $F(x)$, $Q(x)$ has no known closed-form solution, but does have an excellent approximation given by

$$Q(x) \approx \left[\frac{1}{(1-a)x + a\sqrt{x^2 + b}} \right] \frac{e^{-x^2/2}}{\sqrt{2\pi}} \qquad x \geq 0 \qquad (2.4\text{-}10)$$

where a and b are constants. This approximation has been found to give minimum absolute relative error, for any $x \geq 0$, when $a = 0.339$ and $b = 5.510$ (see Börjesson and Sundberg, 1979). With these values of a and b, the approximation of (2.4-10) is said to equal the true value of $Q(x)$ within a maximum absolute error of 0.27% of $Q(x)$ for any $x \geq 0$. We consider a simple example.

EXAMPLE 2.4-3. We assume a gaussian random variable for which $a_X = 7$ and $\sigma_X = 0.5$ and find the probability of the event $\{X \leq 7.3\}$. From (2.4-7) and (2.4-8)

$$P\{X \leq 7.3\} = F_X(7.3) = F\left(\frac{7.3 - 7}{0.5}\right) = F(0.6) = 1 - Q(0.6)$$

$$\approx 1 - \left(\frac{1}{0.661(0.6) + 0.339\sqrt{(0.6)^2 + 5.51}}\right) \frac{e^{-(0.6)^2/2}}{\sqrt{2\pi}}$$

$$\approx 0.7264$$

From Table B-1 the answer is $F(0.6) = 0.7257$ so an absolute error of about $|0.7264 - 0.7257|/0.7257 = 0.00096$ (or 0.096%) exists.

2.5
OTHER DISTRIBUTION AND DENSITY EXAMPLES

Many distribution functions are important enough to have been given names. We give five examples. The first two are for discrete random variables; the remaining three are for continuous random variables. Other distributions are listed in Appendix F.

Binomial

Let $0 < p < 1$, and $N = 1, 2, \ldots$, then the function

$$f_X(x) = \sum_{k=0}^{N} \binom{N}{k} p^k (1-p)^{N-k} \delta(x-k) \qquad (2.5\text{-}1)$$

is called the *binomial density function*. The quantity $\binom{N}{k}$ is the binomial coefficient defined in (1.7-4) as

$$\binom{N}{k} = \frac{N!}{k!(N-k)!} \tag{2.5-2}$$

The binomial density can be applied to the Bernoulli trial experiment of Chapter 1. It applies to many games of chance, detection problems in radar and sonar, and many experiments having only two possible outcomes on any given trial.

By integration of (2.5-1), the *binomial distribution function* is found:

$$F_X(x) = \sum_{k=0}^{N} \binom{N}{k} p^k (1-p)^{N-k} u(x-k) \tag{2.5-3}$$

Figure 2.5-1 illustrates the binomial density and distribution functions for $N = 6$ and $p = 0.25$.

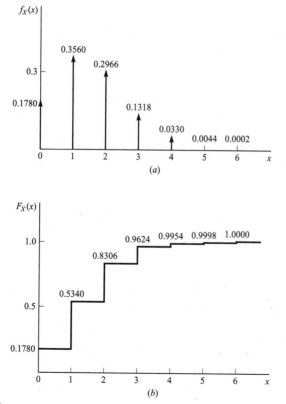

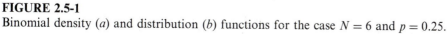

FIGURE 2.5-1
Binomial density (*a*) and distribution (*b*) functions for the case $N = 6$ and $p = 0.25$.

Poisson

The *Poisson* random variable X has a density and distribution given by

$$f_X(x) = e^{-b} \sum_{k=0}^{\infty} \frac{b^k}{k!} \delta(x - k) \tag{2.5-4}$$

$$F_X(x) = e^{-b} \sum_{k=0}^{\infty} \frac{b^k}{k!} u(x - k) \tag{2.5-5}$$

where $b > 0$ is a real constant. When plotted, these functions appear quite similar to those for the binomial random variable (Figure 2.5-1). In fact, if $N \to \infty$ and $p \to 0$ for the binomial case in such a way that $Np = b$, a constant, the Poisson case results.

The Poisson random variable applies to a wide variety of counting-type applications. It describes the number of defective units in a sample taken from a production line, the number of telephone calls made during a period of time, the number of electrons emitted from a small section of a cathode in a given time interval, etc. If the time interval of interest has duration T, and the events being counted are known to occur at an average rate λ and have a Poisson distribution, then b in (2.5-4) is given by

$$b = \lambda T \tag{2.5-6}$$

We illustrate these points by means of an example.

> **EXAMPLE 2.5-1.** Assume automobile arrivals at a gasoline station are Poisson and occur at an average rate of 50/h. The station has only one gasoline pump. If all cars are assumed to require one minute to obtain fuel, what is the probability that a waiting line will occur at the pump?
>
> A waiting line will occur if two or more cars arrive in any one-minute interval. The probability of this event is one minus the probability that either none or one car arrives. From (2.5-6), with $\lambda = \frac{50}{60}$ cars/minute and $T = 1$ minute, we have $b = \frac{5}{6}$. On using (2.5-5)
>
> $$\text{Probability of a waiting line} = 1 - F_X(1) - F_X(0)$$
>
> $$= 1 - e^{-5/6}\left[1 + \frac{5}{6}\right] = 0.2032$$
>
> We therefore expect a line at the pump about 20.32% of the time.

Uniform

The *uniform* probability density and distribution functions are defined by:

$$f_X(x) = \begin{cases} 1/(b - a) & a \leq x \leq b \\ 0 & \text{elsewhere} \end{cases} \tag{2.5-7}$$

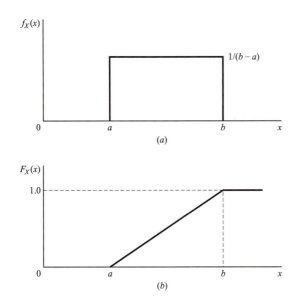

FIGURE 2.5-2
Uniform probability density function (*a*) and its distribution function (*b*).

$$F_X(x) = \begin{cases} 0 & x < a \\ (x-a)/(b-a) & a \le x < b \\ 1 & b \le x \end{cases} \qquad (2.5\text{-}8)$$

for real constants $-\infty < a < \infty$ and $b > a$. Figure 2.5-2 illustrates the behavior of the above two functions.

The uniform density finds a number of practical uses. A particularly important application is in the quantization of signal samples prior to encoding in digital communication systems. Quantization amounts to "rounding off" the actual sample to the nearest of a large number of discrete "quantum levels." The errors introduced in the round-off process are uniformly distributed.

Exponential

The *exponential* density and distribution functions are:

$$f_X(x) = \begin{cases} \dfrac{1}{b}e^{-(x-a)/b} & x > a \\ 0 & x < a \end{cases} \qquad (2.5\text{-}9)$$

$$F_X(x) = \begin{cases} 1 - e^{-(x-a)/b} & x > a \\ 0 & x < a \end{cases} \qquad (2.5\text{-}10)$$

for real numbers $-\infty < a < \infty$ and $b > 0$. These functions are plotted in Figure 2.5-3.

The exponential density is useful in describing raindrop sizes when a large number of rainstorm measurements are made. It is also known to approxi-

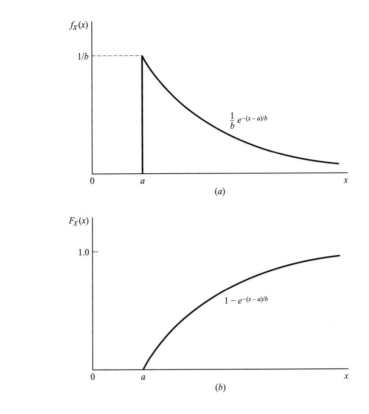

FIGURE 2.5-3
Exponential density (*a*) and distribution (*b*) functions.

mately describe the fluctuations in signal strength received by radar from certain types of aircraft as illustrated by the following example.

EXAMPLE 2.5-2. The power reflected from an aircraft of complicated shape that is received by a radar can be described by an exponential random variable P. The density of P is therefore

$$f_P(p) = \begin{cases} \dfrac{1}{P_0} e^{-p/P_0} & p > 0 \\ 0 & p < 0 \end{cases}$$

where P_0 is the average amount of received power. At some given time P may have a value different from its average value and we ask: what is the probability that the received power is larger than the power received on the average?

We must find $P\{P > P_0\} = 1 - P\{P \le P_0\} = 1 - F_P(P_0)$. From (2.5-10)

$$P\{P > P_0\} = 1 - (1 - e^{-P_0/P_0}) = e^{-1} \approx 0.368$$

In other words, the received power is larger than its average value about 36.8 percent of the time.

Rayleigh

The *Rayleigh*† density and distribution functions are:

$$f_X(x) = \begin{cases} \dfrac{2}{b}(x-a)e^{-(x-a)^2/b} & x \geq a \\ 0 & x < a \end{cases} \qquad (2.5\text{-}11)$$

$$F_X(x) = \begin{cases} 1 - e^{-(x-a)^2/b} & x \geq a \\ 0 & x < a \end{cases} \qquad (2.5\text{-}12)$$

for real constants $-\infty < a < \infty$ and $b > 0$. These functions are plotted in Figure 2.5-4.

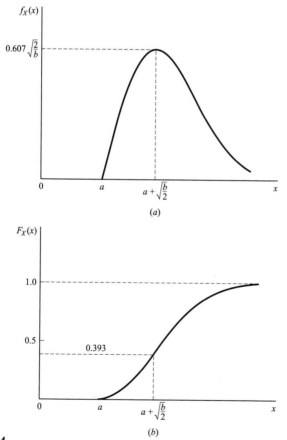

(a)

(b)

FIGURE 2.5-4
Rayleigh density (*a*) and distribution (*b*) functions.

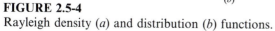

†Named for the English physicist John William Strutt, Lord Rayleigh (1842–1919).

The Rayleigh density describes the envelope of one type of noise when passed through a bandpass filter. It also is important in analysis of errors in various measurement systems.

> **EXAMPLE 2.5-3.** We find the value $x = x_0$ of a Rayleigh random variable for which $P\{X \le x_0\} = P\{x_0 < X\}$. This value of X is called the *median* of the random variable. The probability condition requires $P\{X \le x_0\} = F_X(x_0) = 0.5$. From (2.5-12) $F_X(x_0) = 1 - \exp[-(x_0 - a)^2/b] = 0.5$. The solution for x_0 follows the natural logarithm. We find $x_0 = a + [b\ln(2)]^{1/2}$. The median is similarly defined for random variables other than Rayleigh; it is the value of X for which the probability is 0.5 that values of X do not exceed the median.

2.6
CONDITIONAL DISTRIBUTION AND DENSITY FUNCTIONS

The concept of conditional probability was introduced in Chapter 1. Recall that, for two events A and B where $P(B) \ne 0$, the conditional probability of A given B had occurred was

$$P(A|B) = \frac{P(A \cap B)}{P(B)} \tag{2.6-1}$$

In this section we extend the conditional probability concept to include random variables.

Conditional Distribution

Let A in (2.6-1) be identified as the event $\{X \le x\}$ for the random variable X. The resulting probability $P\{X \le x|B\}$ is defined as the *conditional distribution function* of X, which we denote $F_X(x|B)$. Thus

$$F_X(x|B) = P\{X \le x|B\} = \frac{P\{X \le x \cap B\}}{P(B)} \tag{2.6-2}$$

where we use the notation $\{X \le x \cap B\}$ to imply the joint event $\{X \le x\} \cap B$. This joint event consists of all outcomes s such that

$$X(s) \le x \quad \text{and} \quad s \in B \tag{2.6-3}$$

The conditional distribution (2.6-2) applies to discrete, continuous, or mixed random variables.

Properties of Conditional Distribution

All the properties of ordinary distributions apply to $F_X(x|B)$. In other words, it has the following characteristics:

(1) $F_X(-\infty|B) = 0$ (2.6-4a)

(2) $F_X(\infty|B) = 1$ (2.6-4b)

(3) $0 \le F_X(x|B) \le 1$ (2.6-4c)

(4) $F_X(x_1|B) \le F_X(x_2|B)$ if $x_1 < x_2$ (2.6-4d)

(5) $P\{x_1 < X \le x_2|B\} = F_X(x_2|B) - F_X(x_1|B)$ (2.6-4e)

(6) $F_X\{x^+|B\} = F_X(x|B)$ (2.6-4f)

These characteristics have the same general meanings as described earlier following (2.2-2).

Conditional Density

In a manner similar to the ordinary density function, we define *conditional density function* of the random variable X as the derivative of the conditional distribution function. If we denote this density by $f_X(x|B)$, then

$$f_X(x|B) = \frac{dF_X(x|B)}{dx} \qquad (2.6-5)$$

If $F_X(x|B)$ contains step discontinuities, as when X is a discrete or mixed random variable, we assume that impulse functions are present in $f_X(x|B)$ to account for the derivatives at the discontinuities.

Properties of Conditional Density

Because conditional density is related to conditional distribution through the derivative, it satisfies the same properties as the ordinary density function. They are:

(1) $f_X(x|B) \ge 0$ (2.6-6a)

(2) $\displaystyle\int_{-\infty}^{\infty} f_X(x|B)\,dx = 1$ (2.6-6b)

(3) $\displaystyle F_X(x|B) = \int_{-\infty}^{x} f_X(\xi|B)\,d\xi$ (2.6-6c)

(4) $\displaystyle P\{x_1 < X \le x_2|B\} = \int_{x_1}^{x_2} f_X(x|B)\,dx$ (2.6-6d)

We take an example to illustrate conditional density and distribution.

EXAMPLE 2.6-1. Two boxes have red, green, and blue balls in them; the number of balls of each color is given in Table 2.6-1. Our experiment will be to select a box and then a ball from the selected box. One box (number 2) is slightly larger than the other, causing it to be selected more frequently. Let B_2 be the event "select the larger box" while B_1 is the event "select the smaller box." Assume $P(B_1) = \frac{2}{10}$ and $P(B_2) = \frac{8}{10}$. (B_1 and B_2 are mutually exclusive and $B_1 \cup B_2$ is the certain event, since some box must be selected; therefore, $P(B_1) + P(B_2)$ must equal unity.)

TABLE 2.6-1
Numbers of colored balls in two boxes

		Box		
x_i	Ball color	1	2	Totals
1	Red	5	80	85
2	Green	35	60	95
3	Blue	60	10	70
Totals		100	150	250

Now define a discrete random variable X to have values $x_1 = 1$, $x_2 = 2$, and $x_3 = 3$ when a red, green, or blue ball is selected, and let B be an event equal to either B_1 or B_2. From Table 2.6-1:

$$P(X = 1|B = B_1) = \frac{5}{100} \qquad P(X = 1|B = B_2) = \frac{80}{150}$$

$$P(X = 2|B = B_1) = \frac{35}{100} \qquad P(X = 2|B = B_2) = \frac{60}{150}$$

$$P(X = 3|B = B_1) = \frac{60}{100} \qquad P(X = 3|B = B_2) = \frac{10}{150}$$

The conditional probability density $f_X(x|B_1)$ becomes

$$f_X(x|B_1) = \frac{5}{100}\delta(x - 1) + \frac{35}{100}\delta(x - 2) + \frac{60}{100}\delta(x - 3)$$

By direct integration of $f_X(x|B_1)$:

$$F_X(x|B_1) = \frac{5}{100}u(x - 1) + \frac{35}{100}u(x - 2) + \frac{60}{100}u(x - 3)$$

For comparison, we may find the density and distribution of X by determining the probabilities $P(X = 1)$, $P(X = 2)$, and $P(X = 3)$. These are found from the total probability theorem embodied in (1.4-10):

$$P(X = 1) = P(X = 1|B_1)P(B_1) + P(X = 1|B_2)P(B_2)$$

$$= \frac{5}{100}\left(\frac{2}{10}\right) + \frac{80}{150}\left(\frac{8}{10}\right) = 0.437$$

$$P(X = 2) = \frac{35}{100}\left(\frac{2}{10}\right) + \frac{60}{150}\left(\frac{8}{10}\right) = 0.390$$

$$P(X = 3) = \frac{60}{100}\left(\frac{2}{10}\right) + \frac{10}{150}\left(\frac{8}{10}\right) = 0.173$$

Thus

$$f_X(x) = 0.437\,\delta(x - 1) + 0.390\,\delta(x - 2) + 0.173\,\delta(x - 3)$$

and

$$F_X(x) = 0.437u(x - 1) + 0.390\,u(x - 2) + 0.173u(x - 3)$$

These distributions and densities are plotted in Figure 2.6-1.

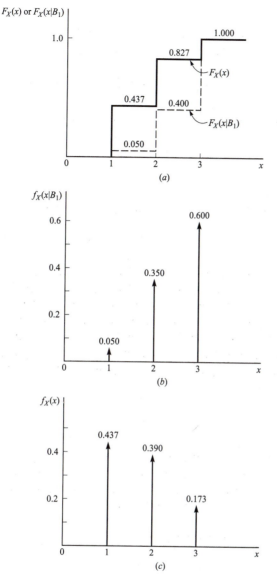

FIGURE 2.6-1
Distributions (*a*) and densities (*b*) and (*c*) applicable to Example 2.6-1.

*Methods of Defining Conditioning Event

The preceding example illustrates how the conditioning event *B* can be defined from some characteristic of the physical experiment. There are several other ways of defining *B* (Cooper and McGillem, 1971, p. 61). We shall consider two of these in detail.

In one method, event *B* is defined in terms of the random variable *X*. We discuss this case further in the next paragraph. In another method, event *B* may depend on some random variable other than X. We discuss this case further in Chapter 4.

One way to define event B in terms of X is to let

$$B = \{X \leq b\} \tag{2.6-7}$$

where b is some real number $-\infty < b < \infty$. After substituting (2.6-7) in (2.6-2), we get†

$$F_X(x|X \leq b) = P\{X \leq x|X \leq b\} = \frac{P\{X \leq x \cap X \leq b\}}{P\{X \leq b\}} \tag{2.6-8}$$

for all events $\{X \leq b\}$ for which $P\{x \leq b\} \neq 0$. Two cases must be considered; one is where $b \leq x$; the second is where $x < b$. If $b \leq x$, the event $\{X \leq b\}$ is a subset of the event $\{X \leq x\}$, so $\{X \leq x\} \cap \{X \leq b\} = \{X \leq b\}$. Equation (2.6-8) becomes

$$F_X(x|X \leq b) = \frac{P\{X \leq x \cap X \leq b\}}{P\{X \leq b\}} = \frac{P\{X \leq b\}}{P\{X \leq b\}} = 1 \qquad b \leq x \tag{2.6-9}$$

When $x < b$ the event $\{X \leq x\}$ is a subset of the event $\{X \leq b\}$, so $\{X \leq x\} \cap \{X \leq b\} = \{X \leq x\}$ and (2.6-8) becomes

$$F_X(x|X \leq b) = \frac{P\{X \leq x \cap X \leq b\}}{P\{X \leq b\}} = \frac{P\{X \leq x\}}{P\{X \leq b\}} = \frac{F_X(x)}{F_X(b)} \qquad x < b \tag{2.6-10}$$

By combining the last two expressions, we obtain

$$F_X(x|X \leq b) = \begin{cases} \dfrac{F_X(x)}{F_X(b)} & x < b \\ 1 & b \leq x \end{cases} \tag{2.6-11}$$

The conditional density function derives from the derivative of (2.6-11):

$$f_X(x|X \leq b) = \begin{cases} \dfrac{f_X(x)}{F_X(b)} = \dfrac{f_X(x)}{\displaystyle\int_{-\infty}^{b} f_X(x)\,dx} & x < b \\ 0 & x \geq b \end{cases} \tag{2.6-12}$$

Figure 2.6-2 sketches possible functions representing (2.6-11) and (2.6-12).

From our assumptions that the conditioning event has nonzero probability, we have $0 < F_X(b) \leq 1$, so the expression of (2.6-11) shows that the conditional distribution function is never smaller than the ordinary distribution function:

$$F_X(x|X \leq b) \geq F_X(x) \tag{2.6-13}$$

A similar statement holds for the conditional density function of (2.6-12) wherever it is nonzero:

---

†Notation used has allowed for deletion of some braces for convenience. Thus, $F_X(x|\{X \leq b\})$ is written $F_X(x|X \leq b)$ and $P(\{X \leq x\} \cap \{X \leq b\})$ becomes $P\{X \leq x \cap X \leq b\}$.

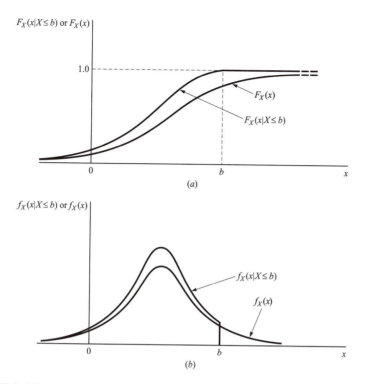

FIGURE 2.6-2
Possible distribution functions (*a*) and density functions (*b*) applicable to a conditioning event $B = \{X \leq b\}$.

$$f_X(x|X \leq b) \geq f_X(x) \qquad x < b \qquad (2.6\text{-}14)$$

The principal results (2.6-11) and (2.6-12) can readily be extended to the more general event $B = \{a < X \leq b\}$ (see Problem 2.6-2).

EXAMPLE 2.6-2. The radial "miss-distance" of landings from parachuting sky divers, as measured from a target's center, is a Rayleigh random variable with $b = 800 \, \text{m}^2$ and $a = 0$. From (2.5-12) we have

$$F_X(x) = [1 - e^{-x^2/800}]u(x)$$

The target is a circle of 50-m radius with a bull's eye of 10-m radius. We find the probability of a parachuter hitting the bull's eye given that the landing is on the target.

The required probability is given by (2.6-11) with $x = 10$ and $b = 50$:

$$P(\text{bull's eye}|\text{landing on target}) = F_X(10)/F_X(50)$$

$$= (1 - e^{-100/800})/(1 - e^{-2500/800}) = 0.1229$$

Parachuter accuracy is such that about 12.29% of landings falling on the target will actually hit the bull's eye.

2.7
SUMMARY

Not all random experiments involve outcomes and events that are numerical. Since engineers and scientists work best with numerical quantities, this chapter is concerned with two things: (1) the definition of a random varaible; it is always a numerical quantity, regardless of the random experiment from which it derives; and (2) the various functions that describe the probabilistic behavior of a random variable. Specific principal topics that were covered are:

- The various types of random variables (discrete, continuous, mixed) were defined.
- The concepts of probability density and cumulative probability distribution functions were introduced to define the probabilistic behavior of a random variable.
- The important gaussian random variable was discussed in detail, while others (binomial, Poisson, etc.) were defined.
- Finally, conditional distribution and density functions were discussed to demonstrate how probabilities that are associated with a random variable can depend on some random event.

The chapter's material is basic to understanding topics of the next chapter, which develops ways of operating on (working with) random variables.

PROBLEMS

2.1-1. The sample space for an experiment is $S = \{0, 1, 2.5, 6\}$. List all possible values of the following random variables:
(a) $X = 2s$
(b) $X = 5s^2 - 1$
(c) $X = \cos(\pi s)$
(d) $X = (1 - 3s)^{-1}$

2.1-2. Work Problem 2.1-1 for $S = \{-2 < s \le 5\}$.

2.1-3. Given that a random variable X has the following possible values, state if X is discrete, continuous, or mixed.
(a) $\{-20 < x < -5\}$
(b) $\{10, 12 < x \le 14, 15, 17\}$
(c) $\{-10 \text{ for } s > 2 \text{ and } 5 \text{ for } s \le 2, \text{ where } 1 < s \le 6\}$
(d) $\{4, 3.1, 1, -2\}$

2.1-4. A random variable X is a function. So is probability P. Recall that the *domain* of a function is the set of values its argument may take on while its *range* is the set of corresponding values of the function. In terms of sets, events, and sample spaces, state the domain and range for X and P.

2.1-5. A man matches coin flips with a friend. He wins $2 if coins match and loses $2 if they do not match. Sketch a sample space showing possible outcomes for this experiment and illustrate how the points map onto the real line x that defines the values of the random variable $X =$ "dollars won on a trial." Show a second mapping for a random variable $Y =$ "dollars won by the friend on a trial."

2.1-6. Temperature in a given city varies randomly during any year from -21 to $49°C$. A house in the city has a thermostat that assumes only three positions: 1 represents "call for heat below $18.3°C$," 2 represents "dead or idle zone," and 3 represents "call for air conditioning above $21.7°C$." Draw a sample space for this problem showing the mapping necessary to define a random variable $X =$ "thermostate setting."

2.1-7. A random voltage can have any value defined by the set $S = \{a \le s \le b\}$. A quantizer divides S into 6 equal-sized contiguous subsets and generates a voltage random variable X having values $\{-4, -2, 0, 2, 4, 6\}$. Each value of X is equal to the midpoint of the subset of S from which it is mapped.
(a) Sketch the sample space and the mapping to the line x that defines the values of X.
(b) Find a and b.

*2.1-8. A random signal can have any voltage value (at a given time) defined by the set $S = \{a_0 < s \le a_N\}$, where a_0 and a_N are real numbers and N is any integer $N \ge 1$. A voltage quantizer divides S into N equal-sized contiguous subsets and converts the signal level into one of a set of discrete levels a_n, $n = 1, 2, \ldots, N$, that correspond to the "input" subsets $\{a_{n-1} < s \le a_n\}$. The set $\{a_1, a_2, \ldots, a_N\}$ can be taken as the discrete values of an "output" random variable X of the quantizer. If the smallest "input" subset is defined by $\Delta = a_1 - a_0$ and other subsets by $a_n - a_{n-1} = 2^{n-1}\Delta$, determine Δ and the quantizer levels a_n in terms of a_0, a_N, and N.

2.1-9. An honest coin is tossed three times.
(a) Sketch the applicable sample space S showing all possible elements. Let X be a random variable that has values representing the number of heads obtained on any triple toss. Sketch the mapping of S onto the real line defining X.
(b) Find the probabilities of the values of X.

2.1-10. Work Problem 2.1-9 for a biased coin for which $P\{\text{head}\} = 0.6$.

2.1-11. Resistor R_2 in Figure P2.1-11 is randomly selected from a box of resistors containing $180\text{-}\Omega$, $470\text{-}\Omega$, $1000\text{-}\Omega$, and $2200\text{-}\Omega$ resistors. All resistor values have the same likelihood of being selected. The voltage E_2 is a discrete random variable. Find the set of values E_2 can have and give their probabilities.

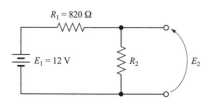

FIGURE P2.1-11

2.1-12. A sample space is defined by $S = \{1, 2 \leq s \leq 3, 4, 5\}$. A random variable is defined by $X = 2$ for $0 \leq s \leq 2.5$, $X = 3$ for $2.5 < s < 3.5$, and $X = 5$ for $3.5 \leq s \leq 6$.

(a) Is X discrete, continuous, or mixed?
(b) Give a set that defines the values X can have.

2.1-13. A gambler flips a fair coin three times.

(a) Draw a sample space S for this experiment. A random variable X representing his winnings is defined as follows: He loses $1 if he gets no heads in three flips; he wins $1, $2, and $3 if he obtains 1, 2, or 3 heads, respectively. Show how elements of S map to values of X.
(b) What are the probabilities of the various values of X?

2.1-14. A random current is described by the sample space $S = \{-4 \leq i \leq 12\}$. A random variable X is defined by

$$X(i) = \begin{cases} -2 & i \leq -2 \\ i & -2 < i \leq 1 \\ 1 & 1 < i \leq 4 \\ 6 & 4 < i \end{cases}$$

(a) Show, by a sketch, the value x into which the values of i are mapped by X.
(b) What type of random variable is X?

2.2-1. Bolts made on a production line are nominally designed to have a 760-mm length. A go-no-go testing device eliminates all bolts less than 650 mm and over 920 mm in length. The surviving bolts are then made available for sale and their lengths are known to be described by a uniform probability density function. A certain buyer orders all bolts that can be produced with a $\pm 5\%$ tolerance about the nominal length. What fraction of the production line's output is he purchasing?

2.2-2. Find and sketch the density and distribution functions for the random variables of parts (a), (b), and (c) in Problem 2.1-1 if the sample space elements have equal likelihoods of occurrence.

2.2-3. If temperature in Problem 2.1-6 is uniformly distributed, sketch the density and distribution functions of the random variable X.

2.2-4. For the uniform random variable defined by (2.5-7) find:

(a) $P\{0.9a + 0.1b < X \leq 0.7a + 0.3b\}$
(b) $P\{(a + b)/2 < X \leq b\}$

2.2-5. Determine which of the following are valid distribution functions:

(a) $G_X(x) = \begin{cases} 1 - e^{-x/2} & x \geq 0 \\ 0 & x < 0 \end{cases}$

(b) $G_X(x) = \begin{cases} 0 & x < 0 \\ 0.5 + 0.5 \sin[\pi(x - 1)/2] & 0 \leq x < 2 \\ 1 & x \geq 2 \end{cases}$

(c) $G_X(x) = \dfrac{x}{a}[u(x - a) - u(x - 2a)]$

2.2-6. A function $G_X(x) = a[1 + (2/\pi)\sin^{-1}(x/c)]\ \mathrm{rect}(x/2c) + (a+b)u(x-c)$ is defined for all $\infty < x < \infty$, where $c > 0$, b, and a are real constants and rect $(\cdot)$ is defined by (E-2). Find any conditions on a, b, and c that will make $G_X(x)$ a valid probability distribution function. Discuss what choices of constants correspond to a continuous, discrete, or mixed random variable.

2.2-7. (*a*) Generalize Problem 2.2-5(*a*) by finding values of real constants a and b such that

$$G_X(x) = [1 - a\exp(-x/b)]u(x)$$

is a valid distribution function.

(*b*) Are there any values of a and b such that $G_X(x)$ corresponds to a mixed random variable X?

2.2-8. (*a*) Find the probabilities associated with all values of the random variable X of Problem 2.1-14.

(*b*) Sketch the probability distribution function of the random variable X.

2.2-9. A random variable X has the distribution function

$$F_X(x) = \sum_{n=1}^{12} \frac{n^2}{650} u(x-n)$$

Find the probabilities: (*a*) $P\{-\infty < X \le 6.5\}$, (*b*) $P\{X > 4\}$, and (*c*) $P\{6 < X \le 9\}$.

2.2-10. If the function

$$G_X(x) = K \sum_{n=1}^{N} n^3 u(x-n)$$

must be a valid probability distribution function, determine K to make it valid. (*Hint*: Use a series from Appendix C.)

2.3-1. Determine the real constant a, for arbitrary real constants m and $0 < b$, such that

$$f_X(x) = ae^{-|x-m|/b}$$

is a valid density function (called the *Laplace density*).

2.3-2. An intercom system master station provides music to six hospital rooms. The probability that any one room will be switched on and draw power at any time is 0.4. When on, a room draws 0.5 W.

(*a*) Find and plot the density and distribution functions for the random variable "power delivered by the master station."

(*b*) If the master-station amplifier is overloaded when more than 2 W is demanded, what is its probability of overload?

*2.3-3. The amplifier in the master station of Problem 2.3-2 is replaced by a 4-W unit that must now supply 12 rooms. Is the probability of overload better than if two independent 2-W units supplied six rooms each?

2.3-4. Jusify that a distribution function $F_X(x)$ satisfies (2.2-2*a*, *b*, *c*).

2.3-5. Use the definition of the impulse function to evaluate the following integrals. (*Hint*: Refer to Appendix A.)

(a) $\int_3^4 (3x^2 + 2x - 4)\delta(x - 3.2)\,dx$

(b) $\int_{-\infty}^{\infty} \cos(6\pi x)\delta(x - 1)\,dx$

(c) $\int_{-\infty}^{\infty} \frac{24(x - 2)\,dx}{x^4 + 3x^2 + 2}$

(d)† $\int_{-\infty}^{\infty} \delta(x - x_0)e^{-j\omega x}\,dx$

(e) $\int_{-3}^3 u(x - 2)\delta(x - 3)\,dx$

2.3-6. Show that the properties of a density function $f_X(x)$, as given by (2.3-6), are valid.

2.3-7. For the random variable defined in Example 2.3-1, find:
(a) $P\{x_0 - 0.6\alpha < X \le x_0 + 0.3\alpha\}$
(b) $P\{X = x_0\}$

2.3-8. Find a constant $b > 0$ so that the function

$$f_X(x) = \begin{cases} e^{3x}/4 & 0 \le x \le b \\ 0 & \text{elsewhere} \end{cases}$$

is a valid probability density.

2.3-9. Given the function

$$g_X(x) = 4\cos(\pi x/2b)\,\text{rect}\,(x/2b)$$

find a value of b so that $g_X(x)$ is a valid probability density.

2.3-10. A random variable X has the density function

$$f_X(x) = (\tfrac{1}{2})u(x)\,\exp(-x/2)$$

Define events $A = \{1 < X \le 3\}$, $B = \{X \le 2.5\}$, and $C = A \cap B$. Find the probabilities of events (a) A, (b) B, and (c) C.

*2.3-11. Let $\phi(x)$ be a continuous, but otherwise arbitrary real function, and let a and b be real constants. Find $G(a, b)$ defined by

$$G(a, b) = \int_{-\infty}^{\infty} \phi(x)\,\delta(ax + b)\,dx$$

(*Hint*: Use the definition of the impulse function.)

---

†The quantity j is the unit-imaginary; that is, $j = \sqrt{-1}$.

2.3-12. For real constants $b > 0$, $c > 0$, and any a, find a condition on constant a and a relationship between c and a (for given b) such that the function

$$f_X(x) = \begin{cases} a[1 - (x/b)] & 0 \le x \le c \\ 0 & \text{elsewhere} \end{cases}$$

is a valid probability density.

2.3-13. Use the properties or definition of the impulse function (Appendix A) to evaluate the following integrals:

(a) $\displaystyle\int_{-\infty}^{\infty} \delta(x+5)\frac{x^2}{1+x^2}\,dx$

(b) $\displaystyle\int_{-\infty}^{\infty} \delta(x-3)\cos(\pi x/6)\,dx$

(c) $\displaystyle\int_{-\infty}^{\infty} e^{-4(x+1)}\delta(x+1)\,dx$

2.3-14. Work Problem 2.3-13 except for the following integrals:

(a) $\displaystyle\int_{-2}^{6} [\delta(x-1) + \delta(x+3) + \delta(x-5)]\,dx$

(b) $\displaystyle\int_{-\infty}^{6} \delta(x-7)u(x+3)\,dx$

(c) $\displaystyle\int_{-3}^{2} [\delta(x-1) - \delta(x+2)]\frac{e^{-2x^2}}{1+x^2+x^4}\,dx$

2.3-15. Find a value for constant A such that

$$f_X(x) = \begin{cases} 0 & x < -1 \\ A(1 - x^2)\cos(\pi x/2) & -1 \le x \le 1 \\ 0 & 1 < x \end{cases}$$

is a valid probability density function.

2.4-1. A random variable X is gaussian with $a_X = 0$ and $\sigma_X = 1$.
(a) What is the probability that $|X| > 2$?
(b) What is the probability that $X > 2$?

2.4-2. Work Problem 2.4-1 if $a_X = 4$ and $\sigma_X = 2$.

2.4-3. For the gaussian density function of (2.4-1), show that

$$\int_{-\infty}^{\infty} x f_X(x)\,dx = a_X$$

2.4-4. For the gaussian density function of (2.4-1), show that

$$\int_{-\infty}^{\infty} (x - a_X)^2 f_X(x)\,dx = \sigma_X^2$$

2.4-5. A production line manufactures 1000-Ω resistors that must satisfy a 10% tolerance.

(a) If resistance is adequately described by a gaussian random variable X for which $a_X = 1000\,\Omega$ and $\sigma_X = 40\,\Omega$, what fraction of the resistors is expected to be rejected?

(b) If a machine is not properly adjusted, the product resistances change to the case where $a_X = 1050\,\Omega$ (5% shift). What fraction is now rejected?

2.4-6. Cannon shell impact position, as measured along the line of fire from the target point, can be described by a gaussian random variable X. It is found that 15.15% of shells fall 11.2 m or farther from the target in a direction toward the cannon, while 5.05% fall farther than 95.6 m beyond the target. what are a_X and σ_X for X?

2.4-7. A gaussian random variable X has $a_X = 2$, and $\sigma_X = 2$.
(a) Find $P\{X > 1.0\}$.
(b) Find $P\{X \le -1.0\}$.

2.4-8. In a certain "junior" olympics, javelin throw distances are well approximated by a gaussian distribution for which $a_X = 30$ m and $\sigma_X = 5$ m. In a qualifying round, contestants must throw farther than 26 m to qualify. In the main event the record throw is 42 m.
(a) What is the probability of being disqualified in the qualifying round?
(b) In the main event what is the probability the record will be broken?

2.4-9. Suppose height to the bottom of clouds is a gaussian random variable X for which $a_X = 4000$ m, and $\sigma_X = 1000$ m. A person bets that cloud height tomorrow will fall in the set $A = \{1000\,\text{m} < X \le 3300\,\text{m}\}$ while a second person bets that height will be satisfied by $B = \{2000\,\text{m} < X \le 4200\,\text{m}\}$. A third person bets they are both correct. Find the probabilities that each person will win the bet.

2.4-10. The output voltage X from the receiver in a particular binary digital communication system, when a binary zero is being received, is gaussian (noise only) as defined by $a_X = 0$ and $\sigma_X = 0.3$. When a binary one is being received it is also gaussian (signal-plus-noise now), but as defined by $a_X = 0.9$ and $\sigma_X = 0.3$. The receiver's decision logic specifies that at the end of a binary (bit) interval, if $X > 0.45$ a binary one is being received. If $X \le 0.45$ a binary zero is decided. If it is given that a binary zero is truly being received, find the probabilities that (a) a binary one (mistake) will be decided, and (b) a binary zero is decided (correct decision).

2.4-11. A gaussian random voltage X for which $a_X = 0$ and $\sigma_X = 4.2$ V appears across a 100-Ω resistor with a power rating of 0.25 W. What is the probability that the voltage will cause an instantaneous power that exceeds the resistor's rating?

2.4-12. Work Problem 2.4-11 except assume a 0.5-W resistor.

2.4-13. For the gaussian random variable, show that the curve's points of inflection (where the first derivative of the probability density function with respect to x has a zero slope) occur at $a_X \pm \sigma_X$.

2.4-14. A random variable X is known to be gaussian with $a_X = 1.6$ and $\sigma_X = 0.4$. Find: (a) $P(1.4 < X \leq 2.0\}$, and (b) $P\{-0.6 < (X - 1.6) \leq 0.6\}$.

2.4-15. The radial distance to the impact points for shells fired over land by a cannon is well-approximated as a gaussian random variable with $a_X = 2000$ m and $\sigma_X = 40$ m when the cannon is aimed at a target located at 1980 m distance.
(a) Find the probability that shells will fall within ± 68 m of the target.
(b) Find the probability that shells will fall at distances of 2050 m or more.

2.4-16. Assume that the time of arrival of birds at a particular place on a migratory route, as measured in days from the first of the year (January 1 is the first day), is approximated as a gaussian random variable X with $a_X = 200$ and $\sigma_X = 20$ days.
(a) What is the probability the birds arrive after 160 days but on or before the 210th day?
(b) What is the probability the birds will arrive after the 231st day?

2.4-17. Assume fluorescent lamps made by a manufacturer have a probability of 0.05 of being inoperable when new. A person purchases eight of the lamps for home use.
(a) Plot the probability distribution function for a random variable "the number of inoperable lamps."
(b) What is the probability that exactly one lamp is inoperable of the eight?
(c) What is the probability that all eight lamps are functional?
(d) Determine the probability that one or more lamps are not operable.

2.5-1. (a) Use the exponential density of (2.5-9) and solve for I_2 defined by

$$I_2 = \int_{-\infty}^{\infty} x^2 f_X(x)\, dx$$

(b) Solve for I_1 defined by

$$I_1 = \int_{-\infty}^{\infty} x f_X(x)\, dx$$

(c) Verify that I_1 and I_2 satisfy the equation $I_2 - I_1^2 = b^2$.

2.5-2. Verify that the maximum value of $f_X(x)$ for the Rayleigh density function of (2.5-11) occurs at $x = a + \sqrt{b/2}$ and is equal to $\sqrt{2/b}\, \exp(-\tfrac{1}{2}) \approx 0.607\sqrt{2/b}$. This value of x is called the *mode* of the random variable. (In general, a random variable may have more than one such value—explain.)

2.5-3. The lifetime of a system expressed in weeks is a Rayleigh random variable X for which

$$f_X(x) = \begin{cases} (x/200)e^{-x^2/400} & 0 \leq x \\ 0 & x < 0 \end{cases}$$

(a) What is the probability that the system will not last a full week?
(b) What is the probability the system lifetime will exceed one year?

2.5-4. The *Cauchy*† random variable has the probability density function

$$f_X(x) = \frac{b/\pi}{b^2 + (x-a)^2}$$

for real numbers $0 < b$ and $-\infty < a < \infty$. Show that the distribution function of X is

$$F_X(x) = \frac{1}{2} + \frac{1}{\pi} \tan^{-1}\left(\frac{x-a}{b}\right)$$

2.5-5. The *log-normal density* function is given by

$$f_X(x) = \begin{cases} \dfrac{\exp\{-[\ln(x-b)-a_X]^2/2\sigma_X^2\}}{\sqrt{2\pi}\sigma_X(x-b)} & x \geq b \\ 0 & x < b \end{cases}$$

for real constants $0 < \sigma_X$, $-\infty < a_X < \infty$, and $-\infty < b < \infty$, where $\ln(x)$ denotes the natural logarithm of x. Show that the corresponding distribution function is

$$F_X(x) = \begin{cases} F\left[\dfrac{\ln(x-b)-a_X}{\sigma_X}\right] & x \geq b \\ 0 & x < b \end{cases}$$

where $F(\cdot)$ is given by (2.4-3).

2.5-6. A random variable X is known to be Poisson with $b = 4$.
(a) Plot the density and distribution functions for this random variable.
(b) What is the probability of the event $\{0 \leq X \leq 5\}$?

2.5-7. The number of cars arriving at a certain bank drive-in window during any 10-min period is a Poisson random variable X with $b = 2$. Find:
(a) The probability that more than 3 cars will arrive during any 10-min period.
(b) The probability that no cars will arrive.

2.5-8. Let X be a Rayleigh random variable with $a = 0$. Find the probability that X will have values larger than its mode (see Problem 2.5-2).

2.5-9. A certain large city averages three murders per week and their occurrences follow a Poisson distribution.
(a) What is the probability that there will be five or more murders in a given week?
(b) On the average, how many weeks a year can this city expect to have no murders?
(c) How many weeks per year (average) can the city expect the number of murders per week to equal or exceed the average number per week?

2.5-10. A certain military radar is set up at a remote site with no repair facilities. If the radar is known to have a *mean-time-between-failures* (MTBF) of 200 h, find

---

†After the French mathematician Augustin Louis Cauchy (1789–1857).

the probability that the radar is still in operation one week later when picked up for maintenance and repairs.

75

CHAPTER 2:
The Random
Variable

2.5-11. If the radar of Problem 2.5-10 is permanently located at the remote site, find the probability that it will be operational as a function of time since its setup.

2.5-12. A computer undergoes downtime if a certain critical component fails. This component is known to fail at an average rate of once per four weeks. No significant downtime occurs if replacement components are on hand because repair can be made rapidly. There are three components on hand, and ordered replacements are not due for six weeks.
(a) What is the probability of significant downtime occurring before the ordered components arrive?
(b) If the shipment is delayed two weeks, what is the probability of significant downtime occurring before the shipment arrives?

2.5-13. The envelope (amplitude) of the output signal of a radar system that is receiving only noise (no signal) is a Rayleigh random voltage X for which $a = 0$ and $b = 2\,\text{V}$. The system gets a false target detection if X exceeds a threshold level V volts. How large must V be to make the probability of false detection 0.001?

2.6-1. Rework Example 2.6-1 to find $f_X(x|B_2)$ and $F_X(x|B_2)$. Sketch the two functions.

*2.6-2. Extend the analysis of the text that leads to (2.6-11) and (2.6-12) to the more general event $B = \{a < X \leq b\}$. Specifically, show that now

$$F_X(x|a < X \leq b) = \begin{cases} 0 & x < a \\ \dfrac{F_X(x) - F_X(a)}{F_X(b) - F_X(a)} & a \leq x < b \\ 1 & b \leq x \end{cases}$$

and

$$f_X(x|a < X \leq b) = \begin{cases} 0 & x < a \\ \dfrac{f_X(x)}{F_X(b) - F_X(a)} = \dfrac{f_X(x)}{\displaystyle\int_a^b f_X(x)\,dx} & a \leq x < b \\ 0 & b \leq x \end{cases}$$

*2.6-3. Consider the system having a lifetime defined by the random variable X in Problem 2.5-3. Given that the system will survive beyond 20 weeks, find the probability that it will survive beyond 26 weeks.

*2.6-4. Assume the lifetime of a laboratory research animal is defined by a Rayleigh density with $a = 0$ and $b = 30$ weeks in (2.5-11) and (2.5-12). If for some clinical reasons it is known that the animal will live at *most* 20 weeks, what is the probability it will live 10 weeks or less?

*2.6-5. Suppose the depth of water, measured in meters, behind a dam is described by an exponential random variable having a density

$$f_X(x) = (1/13.5)u(x)\exp(-x/13.5)$$

There is an emergency overflow at the top of the dam that prevents the depth from exceeding 40.6 m. There is a pipe placed 32.0 m below the overflow (ignore the pipe's finite diameter) that feeds water to a hydroelectric generator.
(a) What is the probability that water is wasted through emergency overflow?
(b) Given that water is not wasted in overflow, what is the probability the generator will have water to drive it?
(c) What is the probability that water will be too low to produce power?

*2.6-6. In Problem 2.6-5 find and sketch the distribution and density functions of water depth given that water will be deep enough to generate power but no water is wasted by emergency overflow. Also sketch for comparisons the distribution and density of water depth without any conditions.

*2.6-7. In Example 2.6-2 a parachuter is an "expert" if he hits the bull's eye. If he falls outside the bull's eye but within a circle of 25-m radius, he is called "qualified" for competition. Given that a parachuter is not an expert but hits the target, what is the probability of being "qualified"?

2.6-8. In a game show contestants choose one of three doors to determine what prize they win. History shows that the three doors, 1, 2, and 3, are chosen with probabilities 0.30, 0.45, and 0.25, respectively. It is also known that given door 1 is chosen, the probabilities of winning prizes of $0, $100, and $1000 are 0.10, 0.20, and 0.70. For door 2 the respective probabilities are 0.50, 0.35, and 0.15, and for door 3 they are 0.80, 0.15, and 0.05. If X is a random variable describing dollars won, and D describes the door selected (values of D are $D_1 = 1, D_2 = 2$, and $D_3 = 3$), find: (a) $F_X(x|D = D_1)$ and $f_X(x|D = D_1)$, (b) $f_X(x|D = D_2)$, (c) $f_X(x|D = D_3)$, and (d) $f_X(x)$.

2.6-9. Again consider the game show of Problem 2.6-8 and find the probabilities of winning (a) $0, (b) $100, and (c) $1000.

*2.6-10. Divers return each day to the site of a sunken treasure ship. Due to random navigational errors, they arrive with a radial positional error (from the true site) described by a random variable X (in kilometers) defined by

$$F_X(x) = [1 - e^{-x^3/2}]u(x)$$

(this is the Weibull† distribution; see Appendix F). If they must arrive with an error of not more than 1.2 km to prevent having to move to a new position, and within 0.6 km for optimum use of air tanks, what is the probability of optimum use of tanks given they arrive on site on the first effort?

*2.6-11. For the navigational errors of Problem 2.6-10 find and plot the conditional density $f_X(x|X \le 1.2\,\text{km})$.

---

†After Ernst Hjalmar Waloddi Weibull (1887–1979), a Swedish applied physicist.

Operations on One Random Variable—Expectation

3.0
INTRODUCTION

The random variable was introduced in Chapter 2 as a means of providing a systematic definition of events defined on a sample space. Specifically, it formed a mathematical model for describing characteristics of some real, physical world random phenomenon. In this chapter, we extend our work to include some important *operations* that may be performed on a random variable. Most of these operations are based on a single concept—expectation.

3.1
EXPECTATION

Expectation is the name given to the process of averaging when a random variable is involved. For a random variable X, we use the notation $E[X]$, which may be read "the mathematical *expectation* of X," "the *expected* value of X," "the *mean* value of X," or "the *statistical average* of X." Occasionally we also use the notation $\bar{X}$, which is read the same way as $E[X]$; that is, $\bar{X} = E[X]$.†

Nearly everyone is familiar with averaging procedures. An example that serves to tie a familiar problem to the new concept of expectation may be the easiest way to proceed.

---

†Up to this point in this book an overbar has represented the complement of a set or event. Henceforth, unless specifically stated otherwise, the overbar will always represent a mean value.

EXAMPLE 3.1-1. Ninety people are randomly selected and the fractional dollar value of coins in their pockets is counted. If the count goes above a dollar, the dollar value is discarded and only the portion from $0\cent$ to $99\cent$ is accepted. It is found that 8, 12, 28, 22, 15, and 5 people had $18\cent$, $45\cent$, $64\cent$, $72\cent$, $77\cent$, and $95\cent$ in their pockets, respectively.

Our everyday experiences indicate that the average of these values is

$$\text{Average } \$ = 0.18\left(\frac{8}{90}\right) + 0.45\left(\frac{12}{90}\right) + 0.64\left(\frac{28}{90}\right) + 0.72\left(\frac{22}{90}\right)$$
$$+ 0.77\left(\frac{15}{90}\right) + 0.95\left(\frac{5}{90}\right)$$
$$\approx \$0.632$$

Expected Value of a Random Variable

The everyday averaging procedure used in the above example carries over directly to random variables. In fact, if X is the discrete random variable "fractional dollar value of pocket coins," it has 100 discrete values x_i that occur with probabilities $P(x_i)$, and its expected value $E[X]$ is found in the same way as in the example:

$$E[X] = \sum_{i=1}^{100} x_i P(x_i) \tag{3.1-1}$$

The values x_i identify with the fractional dollar values in the example, while $P(x_i)$ is identified with the ratio of the number of people for the given dollar value to the total number of people. If a large number of people had been used in the "sample" of the example, all fractional dollar values would have shown up and the ratios would have approached $P(x_i)$. Thus, the average in the example would have become more like (3.1-1) for many more than 90 people.

In general, the expected value of any random variable X is defined by

$$E[X] = \bar{X} = \int_{-\infty}^{\infty} x f_X(x)\, dx \tag{3.1-2}$$

If X happens to be discrete with N possible values x_i having probabilities $P(x_i)$ of occurrence, then

$$f_X(x) = \sum_{i=1}^{N} P(x_i)\delta(x - x_i) \tag{3.1-3}$$

from (2.3-5). Upon substitution of (3.1-3) into (3.1-2), we have

$$E[X] = \sum_{i=1}^{N} x_i P(x_i) \qquad \text{discrete random variable} \tag{3.1-4}$$

Hence, (3.1-1) is a special case of (3.1-4) when $N = 100$. For some discrete random variables, N may be infinite in (3.1-3) and (3.1-4).

EXAMPLE 3.1-2. We determine the mean value of the continuous, exponentially distributed random variable for which (2.5-9) applies:

$$f_X(x) = \begin{cases} \dfrac{1}{b}e^{-x(x-a)/b} & x > a \\ 0 & x < a \end{cases}$$

From (3.1-2) and an integral from Appendix C:

$$E[X] = \int_a^\infty \frac{x}{b}e^{-(x-a)/b}\,dx = \frac{e^{a/b}}{b}\int_a^\infty xe^{-x/b}\,dx = a+b$$

If a random variable's density is symmetrical about a line $x = a$, then $E[X] = a$; that is,

$$E[X] = a \quad \text{if} \quad f_X(x+a) = f_X(-x+a) \tag{3.1-5}$$

Expected Value of a Function of a Random Variable

As will be evident in the next section, many useful parameters relating to a random variable X can be derived by finding the expected value of a real function $g(\cdot)$ of X. It can be shown (see Papoulis, 1965, p. 142) that this expected value is given by

$$E[g(x)] = \int_{-\infty}^\infty g(x)f_X(x)\,dx \tag{3.1-6}$$

If X is a discrete random variable, (3.1-3) applies and (3.1-6) reduces to

$$E[g(X)] = \sum_{i=1}^N g(x_i)P(x_i) \quad \text{discrete random variable} \tag{3.1-7}$$

where N may be infinite for some random variables.

EXAMPLE 3.1-3. It is known that a particular random voltage can be represented as a Rayleigh random variable V having a density function given by (2.5-11) with $a = 0$ and $b = 5$. The voltage is applied to a device that generates a voltage $Y = g(V) = V^2$ that is equal, numerically, to the power in V (in a 1-Ω resistor). We find the average power in V by means of (3.1-6):

$$\text{Power in } V = E[g(V)] = E[V^2] = \int_0^\infty \frac{2v^3}{5}e^{-v^2/5}\,dv$$

By letting $\xi = v^2/5$, $d\xi = 2v\,dv/5$, we obtain

$$\text{Power in } V = 5\int_0^\infty \xi e^{-\xi}\,d\xi = 5\,\text{W}$$

after using (C-46).

Note that if $g(X)$ in (3.1-6) is a sum of N functions $g_n(X)$, $n = 1, 2, \ldots, N$, then the expected value of the sum of N functions of a random variable X is the sum of the N expected values of the individual functions of the random variable.

EXAMPLE 3.1-4. A problem in communication systems is how to define the information of a source. Consider modeling a source capable of issuing any one of L distinct symbols (messages) represented as values x_i, $i = 1, 2, \ldots, L$, of a discrete random variable X ($L = 2$ is the binary case). Let $P(x_i)$ be the probabilities of the symbols $X = x_i$. We ask what is the information contained in this source, on the average. We form three considerations.

First, we reason that information should be largest for source outputs with small probabilities. After all, it conveys little information to predict hot, dry weather for the Sahara desert since these conditions prevail almost all the time. But to predict cool heavy rain carries much "information." Next, information from two independent sources should reasonably add. Finally, information should be a positive quantity (a choice we make) and should be zero for an event that is certain to occur. The only function with these characteristics is the logarithm [Carlson (1975), p. 343]. Now since two quantities represent the smallest measure of choice, the logarithm to the base 2 is chosen for measuring information, and its unit is called the *bit*.

For our source we are led to define the information in symbol x_i as $\log_2[1/P(x_i)] = -\log_2[P(x_i)]$. By use of (3.1-7) we obtain the average information, or *entropy*, of a discrete source as

$$H = -\sum_{i=1}^{L} P(x_i) \log_2[P(x_i)] = \frac{-1}{\ln(2)} \sum_{i=1}^{L} P(x_i) \ln[P(x_i)]$$

where $\ln(\cdot)$ is the natural logarithm (to base e). The unit of H is bits/symbol; it results from averaging the information over all source symbols.

*Conditional Expected Value

If, in (3.1-2), $f_X(x)$ is replaced by the conditional density $f_X(x|B)$, where B is any event defined on the sample space, we have the *conditional expected value* of X, denoted $E[X|B]$:

$$E[X|B] = \int_{-\infty}^{\infty} x f_X(x|B) \, dx \tag{3.1-8}$$

One way to define event B, as shown in Chapter 2, is to let it depend on the random variable X by defining

$$B = \{X \le b\} \qquad -\infty < b < \infty \tag{3.1-9}$$

We showed there that

$$f_X(x|X \le b) = \begin{cases} \dfrac{f_X(x)}{\displaystyle\int_{-\infty}^{b} f_X(x)\,dx} & x < b \\ 0 & x \ge b \end{cases} \qquad (3.1\text{-}10)$$

Thus, by substituting (3.1-10) into (3.1-8):

$$E[X|X \le b] = \frac{\displaystyle\int_{-\infty}^{b} x f_X(x)\,dx}{\displaystyle\int_{-\infty}^{b} f_X(x)\,dx} \qquad (3.1\text{-}11)$$

which is the mean value of X when X is constrained to the set $\{X \le b\}$.

3.2
MOMENTS

An immediate application of the expected value of a function $g(\cdot)$ of a random variable X is in calculating moments. Two types of moments are of interest, those about the origin and those about the mean.

Moments about the Origin

The function

$$g(X) = X^n \qquad n = 0, 1, 2, \ldots \qquad (3.2\text{-}1)$$

when used in (3.1-6) gives the moments about the origin of the random variable X. Denote the nth moment by m_n. Then,

$$m_n = E[X^n] = \int_{-\infty}^{\infty} x^n f_X(x)\,dx \qquad (3.2\text{-}2)$$

Clearly $m_0 = 1$, the area of the function $f_X(x)$, while $m_1 = \bar{X}$, the expected value of X.

Central Moments

Moments about the mean value of X are called *central moments* and are given the symbol μ_n. They are defined as the expected value of the function

$$g(X) = (X - \bar{X})^n \qquad n = 0, 1, 2, \ldots \qquad (3.2\text{-}3)$$

which is

$$\mu_n = E[(X - \bar{X})^n] = \int_{-\infty}^{\infty} (x - \bar{X})^n f_X(x)\,dx \qquad (3.2\text{-}4)$$

The moment $\mu_0 = 1$, the area of $f_X(x)$, while $\mu_1 = 0$. (Why?)

Variance and Skew

The second central moment μ_2 is so important we shall give it the name *variance* and the special notation σ_X^2. Thus, variance is given by†

$$\sigma_X^2 = \mu_2 = E[(X - \bar{X})^2] = \int_{-\infty}^{\infty} (x - \bar{X})^2 f_X(x)\, dx \qquad (3.2\text{-}5)$$

The positive square root σ_X of variance is called the *standard deviation* of X; it is a measure of the spread in the function $f_X(x)$ about the mean.

Variance can be found from a knowledge of first and second moments. By expanding (3.2-5), we have‡

$$\sigma_X^2 = E[X^2 - 2\bar{X}X + \bar{X}^2] = E[X^2] - 2\bar{X}E[X] + \bar{X}^2$$
$$= E[X^2] - \bar{X}^2 = m_2 - m_1^2 \qquad (3.2\text{-}6)$$

> **EXAMPLE 3.2-1.** Let X have the exponential density function given in Example 3.1-2. By substitution into (3.2-5), the variance of X is
>
> $$\sigma_X^2 = \int_a^{\infty} (x - \bar{X})^2 \frac{1}{b} e^{-(x-a)/b}\, dx$$
>
> By making the change of variable $\xi = x - \bar{X}$ we obtain
>
> $$\sigma_X^2 = \frac{e^{-(\bar{X}-a)/b}}{b} \int_{a-\bar{X}}^{\infty} \xi^2 e^{-\xi/b}\, d\xi = (a + b - \bar{X})^2 + b^2$$
>
> after using an integral from Appendix C. However, from Example 3.1-2, $\bar{X} = E[X] = (a + b)$, so
>
> $$\sigma_X^2 = b^2$$
>
> The reader may wish to verify this result by finding the second moment $E[X^2]$ and using (3.2-6).

The third central moment $\mu_3 = E[(X - \bar{X})^3]$ is a measure of the asymmetry of $f_X(x)$ about $x = \bar{X} = m_1$. It will be called the *skew* of the density function. If a density is symmetric about $x = \bar{X}$, it has zero skew. In fact, for this case $\mu_n = 0$ for all odd values of n. (Why?) The normalized third central moment μ_3/σ_X^3 is known as the *skewness* of the density function, or, alternatively, as the *coefficient of skewness*.

> **EXAMPLE 3.2-2.** We continue Example 3.2-1 and compute the skew and coefficient of skewness for the exponential density. From (3.2-4) with $n = 3$ we have

†The subscript indicates that σ_X^2 is the variance of a random variable X. For a random variable Y its variance would be σ_Y^2.

‡We use the fact that the expected value of a sum of functions of X equals the sum of expected values of individual functions, as previously noted.

$$\mu_3 = E[(X - \bar{X})^3] = E[X^3 - 3\bar{X}X^2 + 3\bar{X}^2X - \bar{X}^3]$$
$$= \overline{X^3} - 3\bar{X}\overline{X^2} + 2\bar{X}^3 = \overline{X^3} - 3\bar{X}(\sigma_X^2 + \bar{X}^2) + 2\bar{X}^3$$
$$= \overline{X^3} - 3\bar{X}\sigma_X^2 - \bar{X}^3$$

Next, we have

$$\overline{X^3} = \int_a^\infty \frac{x^3}{b} e^{-(x-a)/b}\, dx = a^3 + 3a^2b + 6ab^2 + 6b^3$$

after using (C-48). On substituting $\bar{X} = a + b$ and $\sigma_X^2 = b^2$ from the earlier example, and reducing the algebra we find

$$\mu_3 = 2b^3$$
$$\frac{\mu_3}{\sigma_X^3} = 2$$

This density has a relatively large coefficient of skewness, as can be seen intuitively from Figure 2.5-3.

Chebychev's Inequality

A useful tool in some probability problems is Chebychev's inequality.† For a random variable X with mean value $\bar{X}$ and variance σ_X^2, it states that

$$P\{|X - \bar{X}| \geq \epsilon\} \leq \sigma_X^2/\epsilon^2 \tag{3.2-7}$$

for any $\epsilon > 0$. This expression can be demonstrated by integration of the probability density, using (2.3-6c):

$$P\{|X - \bar{X}| \geq \epsilon\} = \int_{-\infty}^{\bar{X}-\epsilon} f_X(x)\, dx + \int_{\bar{X}+\epsilon}^\infty f_X(x)\, dx = \int_{|x-\bar{X}|\geq\epsilon} f_X(x)\, dx \tag{3.2-8}$$

But since

$$\sigma_X^2 = \int_{-\infty}^\infty (x - \bar{X})^2 f_X(x)\, dx \geq \int_{|x-\bar{X}|\geq\epsilon} (x - \bar{X})^2 f_X(x)\, dx$$
$$\geq \epsilon^2 \int_{|x-\bar{X}|\geq\epsilon}^\infty f_X(x)\, dx = \epsilon^2 P\{|x - \bar{X}| \geq \epsilon\} \tag{3.2-9}$$

must be true, we solve to show the validity of (3.2-7).

EXAMPLE 3.2-3. We find the largest probability that any random variable's values are smaller than its mean by 3 standard deviations or larger than its mean by the same amount. This probability is $P\{X \geq \bar{X} + 3\sigma_X\} + P\{X \leq \bar{X} - 3\sigma_X\} = P\{|X - \bar{X}| \geq 3\sigma_X\}$. From (3.2-7) with $\epsilon = 3\sigma_X$ we have $P\{|X - \bar{X}| \geq 3\sigma_X\} \leq \sigma_X^2/(3\sigma_X)^2 = 1/9$, or about 11.1%.

---

†After the Russian mathematician Pafnuty Lvovich Chebychev (1821–1894).

By a procedure similar to that above, an alternative form of Chebychev's inequality can be proved. It is

$$P\{|X - \bar{X}| < \epsilon\} \geq 1 - (\sigma_X^2/\epsilon^2) \qquad (3.2\text{-}10)$$

for any $\epsilon > 0$. An interesting result derives immediately from (3.2-10). If $\sigma_X^2 \to 0$ for a random variable, then $P\{|X - \bar{X}| < \epsilon\} \to 1$, for any ϵ. For arbitrarily small ϵ we have $P\{|X - \bar{X}| \to 0\} \to 1$ or $P\{X = \bar{X}\} \to 1$. In other words, if the variance of a random variable X approaches zero, the probability approaches 1 that X will equal its mean value.

Markov's Inequality

Another inequality that is useful in probability problems is *Markov's inequality*, which applies to a nonnegative random variable X; it is

$$P\{X \geq a\} \leq E[X]/a \qquad a > 0 \qquad (3.2\text{-}11)$$

The restriction to nonnegative random variables is relieved by *Chernoff's inequality*, which is developed in Example 3.3-3 below.

*3.3
FUNCTIONS THAT GIVE MOMENTS

Two functions can be defined that allow moments to be calculated for a random variable X. They are the characteristic function and the moment generating function.

*Characteristic Function

The *characteristic function* of a random variable X is defined by

$$\Phi_X(\omega) = E[e^{j\omega X}] \qquad (3.3\text{-}1)$$

where $j = \sqrt{-1}$. It is a function of the real variable $-\infty < \omega < \infty$. If (3.3-1) is written in terms of the density function, $\Phi_X(\omega)$ is seen to be the *Fourier transform*† (with the sign of ω reversed) of $f_X(x)$:

$$\Phi_X(\omega) = \int_{-\infty}^{\infty} f_X(x)e^{j\omega x}\,dx \qquad (3.3\text{-}2)$$

Because of this fact, if $\Phi_X(\omega)$ is known, $f_X(x)$ can be found from the *inverse Fourier transform* (with sign of x reversed)

---

†Readers unfamiliar with Fourier transforms should interpret $\Phi_X(\omega)$ as simply the expected value of the function $g(X) = \exp(j\omega X)$. Appendix D is included as a review for others wishing to refresh their background in Fourier transform theory.

$$f_X(x) = \frac{1}{2\pi} \int_{-\infty}^{\infty} \Phi_X(\omega) e^{-j\omega x} \, d\omega \qquad (3.3\text{-}3)$$

By formal differentiation of (3.3-2) n times with respect to ω and setting $\omega = 0$ in the derivative, we may show that the nth moment of X is given by

$$m_n = (-j)^n \frac{d^n \Phi_X(\omega)}{d\omega^n} \Bigg|_{\omega=0} \qquad (3.3\text{-}4)$$

A major advantage of using $\Phi_X(\omega)$ to find moments is that $\Phi_X(\omega)$ always exists (Davenport, 1970, p. 426), so the moments can always be found if $\Phi_X(\omega)$ is known, provided, of course, both the moments and the derivatives of $\Phi_X(\omega)$ exist.

It can be shown that the maximum magnitude of a characteristic function is unity and occurs at $\omega = 0$; that is,

$$|\Phi_X(\omega)| \leq \Phi_X(0) = 1 \qquad (3.3\text{-}5)$$

(See Problem 3.3-1.)

EXAMPLE 3.3-1. Again we consider the random variable with the exponential density of Example 3.1-2 and find its characteristic function and first moment. By substituting the density function into (3.3-2), we get

$$\Phi_X(\omega) = \int_a^\infty \frac{1}{b} e^{-(x-a)/b} e^{j\omega x} \, dx = \frac{e^{a/b}}{b} \int_a^\infty e^{-(1/b - j\omega)x} \, dx$$

Evaluation of the integral follows the use of an integral from Appendix C:

$$\Phi_X(\omega) = \frac{e^{a/b}}{b} \left[\frac{e^{-(1/b - j\omega)x}}{-(1/b - j\omega)} \Bigg|_a^\infty \right]$$

$$= \frac{e^{j\omega a}}{1 - j\omega b}$$

The derivative of $\Phi_X(\omega)$ is

$$\frac{d\Phi_X(\omega)}{d\omega} = e^{j\omega a} \left[\frac{ja}{1 - j\omega b} + \frac{jb}{(1 - j\omega b)^2} \right]$$

so the first moment becomes

$$m_1 = (-j) \frac{d\Phi_X(\omega)}{d\omega} \Bigg|_{\omega=0} = a + b$$

in agreement with m_1 found in Example 3.1-2.

*Moment Generating Function

Another statistical average closely related to the characteristic function is the *moment generating function*, defined by

$$M_X(v) = E[e^{vX}] \qquad (3.3\text{-}6)$$

where v is a real number $-\infty < v < \infty$. Thus, $M_X(v)$ is given by

$$M_X(v) = \int_{-\infty}^{\infty} f_X(x)e^{vx}\,dx \qquad (3.3\text{-}7)$$

The main advantage of the moment generating function derives from its ability to give the moments. Moments are related to $M_X(v)$ by the expression:

$$m_n = \frac{d^n M_X(v)}{dv^n}\bigg|_{v=0} \qquad (3.3\text{-}8)$$

The main disadvantage of the moment generating function, as opposed to the characteristic function, is that it may not exist for all random variables and all values of v. However, if $M_X(v)$ exists for all values of v in the neighborhood of $v = 0$ the moments are given by (3.3-8) (Wilks, 1962, p. 114).

EXAMPLE 3.3-2. To illustrate the calculation and use of the moment generating function, let us reconsider the exponential density of the earlier examples. On use of (3.3-7) we have

$$M_X(v) = \int_a^{\infty} \frac{1}{b} e^{-(x-a)/b} e^{vx}\,dx$$

$$= \frac{e^{a/b}}{b} \int_a^{\infty} e^{[v-(1/b)]x}\,dx$$

$$= \frac{e^{av}}{1 - bv}$$

In evaluating $M_X(v)$ we have used an integral from Appendix C. By differentiation we have the first moment

$$m_1 = \frac{dM_X(v)}{dv}\bigg|_{v=0}$$

$$= \frac{e^{av}[a(1 - bv) + b]}{(1 - bv)^2}\bigg|_{v=0} = a + b$$

which, of course, is the same as previously found.

*Chernoff's Inequality and Bound

As another example of an application of the moment generating function, we develop Chernoff's inequality through an example.

EXAMPLE 3.3-3. Let X be any random variable, nonnegative or not. For any real $v > 0$ it is clear from some sketches that

$$\exp[v(x - a)] \geq u(x - a) \qquad (1)$$

where $u(\cdot)$ is the unit-step function and a is an arbitrary real constant. Since

$$P\{X \geq a\} = \int_a^{\infty} f_X(x)\,dx = \int_{-\infty}^{\infty} f_X(x)u(x - a)\,dx \qquad (2)$$

we have

$$P\{X \geq a\} \leq \int_{-\infty}^{\infty} f_X(x)e^{v(x-a)}\,dx = e^{-va}M_X(v) \tag{3}$$

from (1), (2), and (3.3-7). Equation (3) is called *Chernoff's inequality*. Because the right side is a function of parameter v, it can be minimized with respect to this parameter. The minimum value is called *Chernoff's bound* [Viniotis (1998), p. 144].

3.4
TRANSFORMATIONS OF A RANDOM VARIABLE

Quite often one may wish to transform (change) one random variable X into a new random variable Y by means of a transformation

$$Y = T(X) \tag{3.4-1}$$

Typically, the density function $f_X(x)$ or distribution function $F_X(x)$ of X is known, and the problem is to determine either the density function $f_Y(y)$ or distribution function $F_Y(y)$ of Y. The problem can be viewed as a "black box" with input X, output Y, and "transfer characteristic" $Y = T(X)$, as illustrated in Figure 3.4-1.

   In general, X can be a discrete, continuous, or mixed random variable. In turn, the transformation T can be linear, nonlinear, segmented, staircase, etc. Clearly, there are many cases to consider in a general study, depending on the form of X and T. In this section we shall consider only three cases: (1) X continuous and T continuous and either monotonically increasing or decreasing with X; (2) X continuous and T continuous but nonmonotonic; (3) X discrete and T continuous. Note that the transformation in all three cases is assumed continuous. The concepts introduced in these three situations are broad enough that the reader should have no difficulty in extending them to other cases (see Problem 3.4-2).

Monotonic Transformations of a Continuous Random Variable

A transformation T is called *monotonically increasing* if $T(x_1) < T(x_2)$ for any $x_1 < x_2$. It is *monotonically decreasing* if $T(x_1) > T(x_2)$ for any $x_1 < x_2$.

   Consider first the increasing transformation. We assume that T is continuous and differentiable at all values of x for which $f_X(x) \neq 0$. Let Y have a

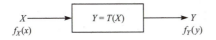

FIGURE 3.4-1
Transformation of a random variable X to a new random variable Y.

particular value y_0 corresponding to the particular value x_0 of X as shown in Figure 3.4-2a. The two numbers are related by

$$y_0 = T(x_0) \qquad \text{or} \qquad x_0 = T^{-1}(y_0) \qquad (3.4-2)$$

where T^{-1} represents the inverse of the transformation T. Now the probability of the event $\{Y \le y_0\}$ must equal the probability of the event $\{X \le x_0\}$ because of the one-to-one correspondence between X and Y. Thus,

$$F_Y(y_0) = P\{Y \le y_0\} = P\{X \le x_0\} = F_X(x_0) \qquad (3.4-3)$$

or

$$\int_{-\infty}^{y_0} f_Y(y)\, dy = \int_{-\infty}^{x_0 = T^{-1}(y_0)} f_X(x)\, dx \qquad (3.4-4)$$

Next, we differentiate both sides of (3.4-4) with respect to y_0 using Leibniz's rule† to get

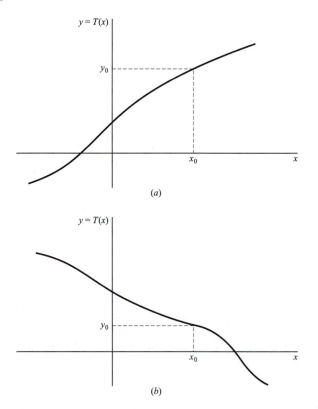

(a)

(b)

FIGURE 3.4-2
Monotonic transformations: (a) increasing, and (b) decreasing. [*Adapted from Peebles (1976) with permission of publishers Addison–Wesley, Advanced Book Program.*]

---

†Leibniz's rule is given in (G-2) of Appendix G.

$$f_Y(y_0) = f_X[T^{-1}(y_0)]\frac{dT^{-1}(y_0)}{dy_0} \qquad (3.4\text{-}5)$$

Since this result applies for any y_0, we may now drop the subscript and write

$$f_Y(y) = f_X[T^{-1}(y)]\frac{dT^{-1}(y)}{dy} \qquad (3.4\text{-}6)$$

or, more compactly,

$$f_Y(y) = f_X(x)\frac{dx}{dy} \qquad (3.4\text{-}7)$$

In (3.4-7) it is understood that x is a function of y through (3.4-2).

A consideration of Figure 3.4-2*b* for the decreasing transformation verifies that

$$F_Y(y_0) = P\{Y \le y_0\} = P\{X \ge x_0\} = 1 - F_X(x_0) \qquad (3.4\text{-}8)$$

A repetition of the steps leading to (3.4-6) will again produce (3.4-6) except that the right side is negative. However, since the slope of $T^{-1}(y)$ is also negative, we conclude that for either type of monotonic transformation

$$f_Y(y) = f_X[T^{-1}(y)]\left|\frac{dT^{-1}(y)}{dy}\right| \qquad (3.4\text{-}9)$$

or simply

$$f_Y(y = f_X(x)\left|\frac{dx}{dy}\right| \qquad (3.4\text{-}10)$$

EXAMPLE 3.4-1. If we take T to be the linear transformation $Y = T(X) = aX+b$
$= aX + b$, where a and b are any real constants, then $X = T^{-1}(Y) = $
$(Y - b)/a$ and $dx/dy = 1/a$. From (3.4-9)

$$f_Y(y) = f_X\left(\frac{y - b}{a}\right)\left|\frac{1}{a}\right|$$

If X is assumed to be gaussian with the density function given by (2.4-1), we get

$$f_Y(y) = \frac{1}{\sqrt{2\pi\sigma_X^2}}e^{-[(y-b)/a-a_X]^2/2\sigma_X^2}\left|\frac{1}{a}\right|$$

$$= \frac{1}{\sqrt{2\pi a^2\sigma_X^2}}e^{-[y-(aa_X+b)]^2/2a^2\sigma_X^2}$$

which is the density function of another gaussian random variable having

$$a_Y = aa_X + b \qquad \text{and} \qquad \sigma_Y^2 = a^2\sigma_X^2$$

Thus, *a linear transformation of a gaussian random variable produces another gaussian random variable.* A linear amplifier having a random voltage X as its input is one example of a linear transformation.

Nonmonotonic Transformations of a Continuous Random Variable

A transformation may not be monotonic in the more general case. Figure 3.4-3 illustrates one such transformation. There may now be more than one interval of values of X that correspond to the event $\{Y \le y_0\}$. For the value of y_0 shown in the figure, the event $\{Y \le y_0\}$ corresponds to the event $\{X \le x_1$ and $x_2 \le X \le x_3\}$. Thus, the probability of the event $\{Y \le y_0\}$ now equals the probability of the event $\{x$ values yielding $Y \le y_0\}$, which we shall write as $\{x|Y \le y_0\}$. In other words

$$F_Y(y_0) = P\{Y \le y_0\} = P\{x|Y \le y_0\} = \int_{\{x|Y \le y_0\}} f_X(x)\,dx \qquad (3.4\text{-}11)$$

Formally, one may differentiate to obtain the density function of Y:

$$f_Y(y_0) = \frac{d}{dy_0} \int_{\{x|Y \le y_0\}} f_X(x)\,dx \qquad (3.4\text{-}12)$$

Although we shall not give a proof, the density function is also given by (Papoulis, 1965, p. 126)

$$f_Y(y) = \sum_n \frac{f_X(x_n)}{\left| \dfrac{dT(x)}{dx} \right|_{x=x_n}} \qquad (3.4\text{-}13)$$

where the sum is taken so as to include all the roots x_n, $n = 1, 2, \ldots$, which are the real solutions of the equation†

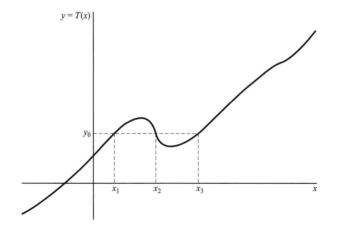

FIGURE 3.4-3
A nonmonotonic transformation. [*Adapted from Peebles (1976) with permission of publishers Addison–Wesley, Advanced Book Program.*]

---

†If $y = T(x)$ has no real roots for a given value of y, then $f_Y(y) = 0$.

$$y = T(x) \tag{3.4-14}$$

We illustrate the above concepts by an example.

EXAMPLE 3.4-2. We find $f_Y(y)$ for the square-law transformation

$$Y = T(X) = cX^2$$

shown in Figure 3.4-4, where c is a real constant $c > 0$. We shall use both the procedure leading to (3.4-12) and that leading to (3.4-13).

In the former case, the event $\{Y \leq y\}$ occurs when $\{-\sqrt{y/c} \leq x \leq \sqrt{y/c}\} = \{x | Y \leq y\}$, so (3.4-12) becomes

$$f_Y(y) = \frac{d}{dy} \int_{-\sqrt{y/c}}^{\sqrt{y/c}} f_X(x)\,dx \qquad y \geq 0$$

Upon use of Leibniz's rule we obtain

$$f_Y(y) = f_X\left(\sqrt{y/c}\right) \frac{d\left(\sqrt{y/c}\right)}{dy} - f_X\left(-\sqrt{y/c}\right) \frac{d\left(-\sqrt{y/c}\right)}{dy}$$

$$= \frac{f_X\left(\sqrt{y/c}\right) + f_X\left(-\sqrt{y/c}\right)}{2\sqrt{cy}} \qquad y \geq 0$$

In the latter case where we use (3.4-13), we have $X = \pm\sqrt{Y/c}$, $Y \geq 0$, so $x_1 = -\sqrt{y/c}$ and $x_2 = \sqrt{y/c}$. Furthermore, $dT(x)/dx = 2cx$ so

$$\left.\frac{dT(x)}{dx}\right|_{x=x_1} = 2cx_1 = -2c\sqrt{\frac{y}{c}} = -2\sqrt{cy}$$

$$\left.\frac{dT(x)}{dx}\right|_{x=x_2} = 2\sqrt{cy}$$

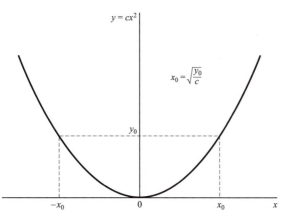

FIGURE 3.4-4
A square-law transformation. [*Adapted from Peebles (1976) with permission of publishers Addison–Wesley, Advanced Book Program.*]

From (3.4-13) we again have

$$f_Y(y) = \frac{f_X(\sqrt{y/c}) + f_X(-\sqrt{y/c})}{2\sqrt{cy}} \qquad y \geq 0$$

Transformation of a Discrete Random Variable

If X is a discrete random variable while $Y = T(X)$ is a continuous transformation, the problem is especially simple. Here

$$f_X(x) = \sum_n P(x_n)\delta(x - x_n) \qquad (3.4\text{-}15)$$

$$F_X(x) = \sum_n P(x_n)u(x - x_n) \qquad (3.4\text{-}16)$$

where the sum is taken to include all the possible values x_n, $n = 1, 2, \ldots$, of X.

If the transformation is monotonic, there is a one-to-one correspondence between X and Y so that a set $\{y_n\}$ corresponds to the set $\{x_n\}$ through the equation $y_n = T(x_n)$. The probability $P(y_n)$ equals $P(x_n)$. Thus,

$$f_Y(y) = \sum_n P(y_n)\delta(y - y_n) \qquad (3.4\text{-}17)$$

$$F_Y(y) = \sum_n P(y_n)u(y - y_n) \qquad (3.4\text{-}18)$$

where

$$y_n = T(x_n) \qquad (3.4\text{-}19)$$

$$P(y_n) = P(x_n) \qquad (3.4\text{-}20)$$

If T is not monotonic, the above procedure remains valid except there now exists the possibility that more than one value x_n corresponds to a value y_n. In such a case $P(y_n)$ will equal the sum of the probabilities of the various x_n for which $y_n = T(x_n)$.

EXAMPLE 3.4-3. Let a discrete random variable X have values $x = -1, 0, 1,$ and 2 with respective probabilities 0.1, 0.3, 0.4, and 0.2, so that $f_X(x) = 0.1\delta(x + 1) + 0.3\delta(x) + 0.4\delta(x - 1) + 0.2\delta(x - 2)$. We assume X is transformed to $Y = 2 - X^2 + (X^3/3)$ and find the density of Y. The values of X map to respective values of Y given by $y = 2/3$, 2, 4/3, and 2/3. The two values $x = -1$ and $x = 2$ map to one value $y = 2/3$. The probability of $\{Y = 2/3\}$ is the sum of probabilities $P\{X = -1\}$ and $P\{X = 2\}$, so

$$f_Y(y) = 0.3\delta[y - (2/3)] + 0.4\delta[y - (4/3)] + 0.3\delta(y - 2)$$

COMPUTER GENERATION OF ONE RANDOM VARIABLE

A digital computer is often used to simulate systems in order to estimate their performance with noise prior to the actual construction of the system. These simulations usually require that random numbers be generated that are values of random variables having prescribed distributions. If software (a computer program, or subroutine, that can be called up on demand) exists for the specified distribution, there is no problem. However, if the computer "library" does not contain the desired program, it is necessary for the simulation to generate its own random numbers. In this section we briefly describe how to generate a random variable with specified probability distribution, given mainly that the computer is able to generate random numbers that are values of a random variable with uniform distribution on (0,1), a commonly satisfied condition in most cases.

The problem, then, is to find the transformation $T(X)$ in Figure 3.4-1 that will create a random variable Y of prescribed distribution function when X has a uniform distribution on (0,1). We assume initially that $T(X)$ is a monotonically nondecreasing function so that (3.4-3) applies. Our work will show that this condition is automatically satisfied. From (3.4-3) we have (for any x and y)

$$F_Y[y = T(x)] = F_X(x) \qquad (3.5\text{-}1)$$

But for uniform X, $F_X(x) = x$ when $0 < x < 1$, from (2.5-8). Thus, we solve for the inverse in (3.5-1) to get

$$y = T(x) = F_Y^{-1}(x) \qquad 0 < x < 1 \qquad (3.5\text{-}2)$$

Since any distribution function $F_Y(y)$ is nondecreasing, its inverse is nondecreasing, and the initial assumption is always satisfied.

Equation (3.5-2) is our principal result. It states that, given a specified distribution $F_Y(y)$ for Y, we find the inverse function by solving $F_Y(y) = x$ for y. The result is $T(x)$. An example will illustrate this simple procedure to create a Rayleigh random variable.

EXAMPLE 3.5-1. We find the transformation required to generate the Rayleigh random variable of (2.5-12) with $a = 0$. On setting

$$F_Y(y) = 1 - e^{-y^2/b} = x \qquad \text{for } 0 < x < 1$$

we solve for y and find

$$y = T(x) = \sqrt{-b \ln(1 - x)} \qquad 0 < x < 1$$

We give another example for the arcsine distribution.

EXAMPLE 3.5-2. A computer software program can generate random numbers on request that are values of a random variable uniformly distributed on (0,1). We find the required transformation to convert these values to those of a random variable with the arcsine distribution of (F-23) in Appendix F. From (F-23) and (3.5-1) we require

$$F_Y(y) = \begin{cases} 0, & y \le -a \\ 0.5 + \dfrac{1}{\pi}\sin^{-1}(y/a), & -a < y < a \\ 1, & a \le y \end{cases} = \begin{cases} 0, & 0 < x < 0 \\ x, & 0 < x < 1 \\ 1, & 1 \le x \end{cases}$$

The solution for y is direct. We obtain

$$y = a\sin[\pi(2x-1)/2] \qquad 0 < x < 1$$

Equation (3.5-2) can be readily applied to any distribution for which its inverse can be analytically determined (see Problems 3.5-1 through 3.5-3 for other examples). For other distributions the required inverse can be stored in the computer for a number of points (y, x), and the simulation can then use interpolation between computed points to obtain values of y for any value of x.

The gaussian random variable is an important example of a distribution for which the inverse cannot be found analytically. Because computer simulations often require gaussian random numbers to be generated, we show in Section 5.6 how this important problem can be solved by extension of the methods of this section.

EXAMPLE 3.5-3. We use MATLAB to generate a sequence of 100 random numbers x_i that correspond to a random variable uniformly distributed on (0,1). The results of Example 3.5-1 are then used to convert the random numbers to a Rayleigh random variable by means of the transformation $y_i = [-\ln(1-x_i)]^{1/2}$. We generate a *histogram* of the uniform values by classifying them into 10 "bins" of width 0.1. A histogram is a plot of the number of values falling in a bin divided by the total number of values generated. Such histograms approximate the probability density of the random variable for which the values apply. For the Rayleigh values a histogram using 22 bins of width 0.1 from 0 to 2.2 is created. Finally, to show that the histograms equal, approximately, the probability densities of the random variables, the true densities are also plotted for comparison.

The applicable MATLAB code is shown in Figure 3.5-1. Results for the uniform random variable are shown in Figure 3.5-2, while the plots for the Rayleigh case are shown in Figure 3.5-3. If more than 100 values of y_i were used, the histograms would tend to approximate the true densities more closely.

```
%%%%%%%%%%%%% Example 3.5-3 %%%%%%%%%%%%%%%%%%%

clear

N = 100; % number of random variables to generate
stp = 0.1; % step size
b = 1; % Rayleigh parameter

x = rand(1,N); % uniformly distributed random numbers
y = sqrt(-b*log(1-x)); % Rayleigh distributed random numbers

f = find(y > 2.2); % find values out of the range of interest
y(f) = []; % remove these values

xcenter = [0.05:stp:1]; % centers of the bins for the histogram
ycenter = [0.05:stp:2.2];

xabscissa = 0:stp:1; % abscissa used for analytic results
yabscissa = 0:stp:2.5;

xhist = hist(x,xcenter); % compute histograms (not normalized)
yhist = hist(y,ycenter);

xtrue = ones(size(xabscissa)); % compute the analytic values
ytrue = 2*yabscissa/b.*exp(-yabscissa.^2/b);

% plot results
clf
bar(xcenter,xhist./(N*stp),1,'w') % plot normalized histogram
hold on
plot(xabscissa,xtrue,'k') % uniform distribution

xlabel('Magnitude Bins')
ylabel('Relative Number of Samples')
title('Histogram of Uniform Distribution')

figure
bar(ycenter,yhist./(N*stp),1,'w')
hold on
plot(yabscissa,ytrue,'k') % Rayleigh distribution

xlabel('Magnitude Bins')
ylabel ('Relative Number of Samples')
title('Histogram of Rayleigh Distribution')
```

FIGURE 3.5-1
MATLAB code used in Example 3.5-3.

Probability,
Random Variables,
and Random
Signal Principles

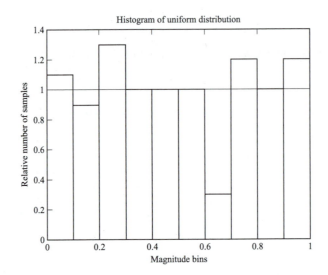

FIGURE 3.5-2
Histogram and true density function for the uniform random variable of Example 3.5-3.

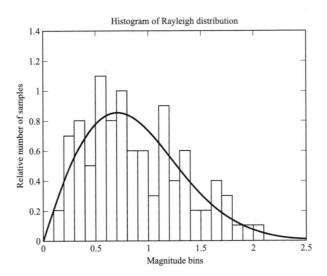

FIGURE 3.5-3
Histogram and true density function for the Rayleigh random variable of Example 3.5-3.

3.6
SUMMARY

Although the preceding chapter discussed the random variable and how to calculate probabilities of its having particular values, these topics are not sufficient for many practical applications. One must also be able to determine useful *characteristics* of random variables, and know how to convert (transform) one random variable into another. These were the subjects of this chapter. All these subjects were developed from a single concept: *expectation*, which is nothing more than an averaging procedure for random quantities. Items covered were:

- Expectation was defined in general for a random variable or some function of a random variable.
- Moments (about the origin, and central) were developed as valuable measures of a random variable's characteristics. Of particular value were mean value, variance, and skew.
- The characteristic function and moment generating function were given as convenient methods for finding moments.
- Methods were given to transform one random variable into another, and to find distribution and density functions of the new random variable.
- The important concepts of how to generate a specified random variable by computer were developed and an example and chapter-end problems were included that are based on use of MATLAB software.

PROBLEMS

3.1-1. A discrete random variable X has possible values $x_i = i^2$, $i = 1, 2, 3, 4, 5$, which occur with probabilities 0.4, 0.25, 0.15, 0.1, and 0.1, respectively. Find the mean value $\bar{X} = E[X]$ of X.

3.1-2. The natural numbers are the possible values of a random variable X: that is, $x_n = n$, $n = 1, 2, \ldots$. These numbers occur with probabilities $P(x_n) = (\frac{1}{2})^n$. Find the expected value of X.

3.1-3. If the probabilities in Problem 3.1-2 are $P(x_n) = p^n$, $0 < p < 1$, show that $p = \frac{1}{2}$ is the only value of p that is allowed for the problem as formulated. (*Hint*: Use the fact that $\int_{-\infty}^{\infty} f_X(x) \, dx = 1$ is necessary.)

3.1-4. (*a*) Find the average amount the gambler in Problem 2.1-13 can expect to win.
(*b*) What is his probability of winning on any given playing of the game?

3.1-5. The *arcsine* probability density is defined by

$$f_X(x) = \frac{\text{rect}(x/2a)}{\pi\sqrt{a^2 - x^2}}$$

for any real constant $a > 0$. Show that $\bar{X} = 0$ and $\overline{X^2} = a^2/2$ for this density.

*3.1-6. For the animal described in Problem 2.6-4 find its expected lifetime given that it will not live beyond 20 weeks.

3.1-7. Find the expected value of the function $g(X) = X^3$ where X is a random variable defined by the density

$$f_X(x) = (\tfrac{1}{2})u(x)\exp(-x/2)$$

3.1-8. A random variable X represents the value of coins (in cents) given in change when purchases are made at a particular store. Suppose the probabilities of $1\cent$, $5\cent$, $10\cent$, $25\cent$, and $50\cent$ being present in change are 0.35, 0.25, 0.20, 0.15, and 0.05, respectively.
(a) Write an expression for the probability density function of X.
(b) Find the mean of X.

3.1-9. In the circuit of Figure P2.1-11 of Chapter 2 let the resistance of R_1 be a random variable uniformly distributed on $(R_0 - \Delta R, R_0 + \Delta R)$ where R_0 and ΔR are constants.
(a) Find an expression for the power dissipated in R_2 for any constant voltage E_1.
(b) Find the mean value of power when R_1 is random.
(c) Evaluate the mean power for $E_1 = 12\,\text{V}$, $R_2 = 1000\,\Omega$, $R_0 = 1500\,\Omega$, and $\Delta R = 100\,\Omega$.

*3.1-10. The power (in milliwatts) returned to a radar from a certain class of aircraft has the probability density function

$$f_P(p) = \frac{1}{10} e^{-p/10} u(p)$$

Suppose a given aircraft belongs to this class but is known to not produce a power larger than 15 mW.
(a) Find the probability density function of P conditional on $P \leq 15\,\text{mW}$.
(b) Find the conditional mean value of P.

3.1-11. A random variable X has a probability density

$$f_X(x) = \begin{cases} (1/2)\cos(x) & -\pi/2 < x < \pi/2 \\ 0 & \text{elsewhere in } x \end{cases}$$

Find the mean value of the function $g(X) = 4X^2$.

3.1-12. Work Problem 3.1-11, except assume a function $g(X) = 4X^4$.

3.1-13. An information source can emit (generate) any one of 128 levels where each is equally probable and independent of all others. What average information does the source represent? (*Hint*: Use the results of Example 3.1-4.)

3.1-14. A random variable X is uniformly distributed on the interval $(-5, 15)$. Another random variable $Y = e^{-X/5}$ is formed. Find $E[Y]$.

3.1-15. A gaussian voltage random variable X [see (2.4-1)] has a mean value $\bar{X} = a_X = 0$ and variance $\sigma_X^2 = 9$. The voltage X is applied to a square-law, full-wave diode detector with a transfer characteristic $Y = 5X^2$. Find the mean value of the output voltage Y.

*3.1-16. For the system having a lifetime specified in Problem 2.5-3 of Chapter 2, determine the expected lifetime of the system given that the system has survived 20 weeks.

3.2-1. Give an example of a random variable where its mean value might not equal any of its possible values.

3.2-2. Find:

(a) the expected value, and

(b) the variance of the random variable with the triangular density of Figure 2.3-1a if $a = 1/\alpha$.

3.2-3. Show that the mean value and variance of the random variable having the uniform density function of (2.5-7) are:

$$\bar{X} = E[X] = (a + b)/2$$

and

$$\sigma_X^2 = (b - a)^2/12$$

3.2-4. A pointer is spun on a fair wheel of chance numbered from 0 to 100 around its circumference.

(a) What is the average value of all possible pointer positions?

(b) What deviation from its average value will the pointer position take on the average; that is, what is the pointer's root-mean-squared deviation from its mean? (*Hint*: Use results of Problem 3.2-3.)

3.2-5. Find:

(a) the mean value, and

(b) the variance of the random variable X defined by Problems 2.1-6 and 2.2-3 of Chapter 2.

*3.2-6. For the *binomial density* of (2.5-1), show that

$$E[X] = \bar{X} = Np$$

and

$$\sigma_X^2 = Np(1 - p)$$

3.2-7. (a) Let resistance be a random variable in Problem 2.1-11 of Chapter 2. Find the mean value of resistance.

(b) What is the output voltage E_2 if an *average* resistor were used in the circuit?

(c) For the resistors specified, what is the mean value of E_2? Does the voltage of part (b) equal this value? Explain your results.

3.2-8. (a) Use the symmetry of the density function given by (2.4-1) to justify that the parameter a_X in the *gaussian density* is the mean value of the random variable: $\bar{X} = a_X$.

(b) Prove that the parameter σ_X^2 is the variance. (*Hint*: Use an equation from Appendix C.)

3.2-9. Show that the mean value $E[X]$ and variance σ_X^2 of the Rayleigh random variable, with density given by (2.5-11), are

$$E[X] = a + \sqrt{\pi b/4}$$

and

$$\sigma_X^2 = b(4 - \pi)/4$$

3.2-10. What is the expected lifetime of the system defined in Problem 2.5-3 of Chapter 2?

3.2-11. Find:
(a) the mean value, and
(b) the variance for a random variable with the *Laplace* density

$$f_X(x) = \frac{1}{2b} e^{-|x-m|/b}$$

where b and m are real constants, $b > 0$, and $-\infty < m < \infty$.

3.2-12. Determine the mean value of the *Cauchy* random variable in Problem 2.5-4 of Chapter 2. What can you say about the variance of this random variable?

*3.2-13. For the *Poisson* random variable defined in (2.5-4) show that:
(a) the mean value is b and
(b) the variance also equals b.

3.2-14. (a) Use (3.2-2) to find the first three moments m_1, m_2, and m_3 for the exponential density of Example 3.1-2.
(b) Find m_1, m_2, and m_3 from the characteristic function found in Example 3.3-1. Verify that they agree with those of part (a).

3.2-15. Find the expressions for all the moments about the origin and central moments for the uniform density of (2.5-7).

3.2-16. Define a function $g(\cdot)$ of a random variable X by

$$g(X) = \begin{cases} 1 & x \geq x_0 \\ 0 & x < x_0 \end{cases}$$

where x_0 is a real number $-\infty < x_0 < \infty$. Show that

$$E[g(X)] = 1 - F_X(x_0)$$

3.2-17. Show that the second moment of any random variable X about an arbitrary point a is minimum when $a = \bar{X}$; that is, show that $E[(X - a)^2]$ is minimum for $a = \bar{X}$.

3.2-18. For any discrete random variable X with values x_i having probabilities of occurrence $P(x_i)$, show that the moments of X are

$$m_n = \sum_{i=1}^{N} x_i^n P(x_i)$$

$$\mu_n = \sum_{i=1}^{N} (x_i - \bar{X})^n P(x_i)$$

where N may be infinite for some X.

3.2-19. Prove that central moments μ_n are related to moments m_k about the origin by

$$\mu_n = \sum_{k=0}^{n} \binom{n}{k} (-\bar{X})^{n-k} m_k$$

3.2-20. A random variable X has a density function $f_X(x)$ and moments m_n. If the density is shifted higher in x by an amount $\alpha > 0$ to a new origin, show that the moments of the shifted density, denoted m_n', are related to the moments m_n by

CHAPTER 3:
Operations on One
Random
Variable—
Expectation

$$m_n' = \sum_{k=0}^{n} \binom{n}{k} \alpha^{n-k} m_k$$

3.2-21. Continue Problem 3.1–14 by finding all moments of Y. (*Hint:* Treat Y^n as a function of Y, not as a transformation.)

3.2-22. Reconsider the production line that manufactures bolts in Problem 2.2-1.
(*a*) What is the average length of bolts that are placed up for sale?
(*b*) What is the standard deviation of length of bolts sold?
(*c*) What percentage of all bolts sold are expected to have a length within one standard deviation of the average length?
(*d*) By what tolerance (as a percentage) does the average length of bolts sold match the nominally desired length of 760 mm?

3.2-23. A random variable X has a probability density

$$f_X(x) = \begin{cases} (\pi/16)\cos(\pi x/8) & -4 \le x \le 4 \\ 0 & \text{elsewhere} \end{cases}$$

Find: (*a*) its mean value $\bar{X}$, (*b*) its second moment $\overline{X^2}$, and (*c*) its variance.

3.2-24. A certain meter is designed to measure small dc voltages but makes errors because of noise. The errors are accurately represented as a gaussian random variable with a mean of zero and a standard deviation of 10^{-3} V. When the dc voltage is disconnected it is found that the probability is 0.5 that the meter reading is positive due to noise. With the dc voltage present, this probability becomes 0.2514. What is the dc voltage?

3.2-25. Find the skew and coefficient of skewness for a Rayleigh random variable for which $a = 0$ in (2.5-11).

3.2-26. A random variable X has the density

$$f_X(x) = \begin{cases} (\frac{3}{32})(-x^2 + 8x - 12) & 2 \le x \le 6 \\ 0 & \text{elsewhere} \end{cases}$$

Find the following moments: (*a*) m_0, (*b*) m_1, (*c*) m_2, and (*d*) μ_2.

3.2-27. The *chi-square density* with N degrees of freedom is defined by

$$f_X(x) = \frac{x^{(N/2)-1}}{2^{N/2}\Gamma(N/2)} u(x) e^{-x/2}$$

where $\Gamma(\cdot)$ is the gamma function

$$\Gamma(z) = \int_0^\infty \xi^{z-1} e^{-\xi} \, d\xi \qquad \text{real part of } z > 0$$

and $N = 1, 2, \ldots$. Show that (*a*) $\bar{X} = N$, (*b*) $\overline{X^2} = N(N+2)$, and (*c*) $\sigma_X^2 = 2N$ for this density.

3.2-28. For the density of Problem 3.2-27 find its arbitrary moment $\overline{X^n}$, $n = 0, 1, 2, \ldots$.

3.2-29. A random variable X is called *Weibull* if its density has the form

$$f_X(x) = abx^{b-1} \exp(-ax^b)u(x)$$

where $a > 0$ and $b > 0$ are real constants. Use the definition of the gamma function of Problem 3.2-27 to find (a) the mean value, (b) the second moment, and (c) the variance of X.

*3.2-30. Show that the characteristic function of a random variable having the binomial density of (2.5-1) is

$$\Phi_X(\omega) = [1 - p + pe^{j\omega}]^N$$

*3.2-31. Show that the characteristic function of a Poisson random variable defined by (2.5-4) is

$$\Phi_X(\omega) = \exp[-b(1 - e^{j\omega})]$$

*3.2-32. The *Erlang†* *random variable* X has a characteristic function

$$\Phi_X(\omega) = \left[\frac{a}{a - j\omega}\right]^N$$

for $a > 0$ and $N = 1, 2, \ldots$. Show that $\bar{X} = N/a$, $\overline{X^2} = N(N + 1)/a^2$, and $\sigma_X^2 = N/a^2$.

3.2-33. A random variable X has $\bar{X} = -3$, $\overline{X^2} = 11$, and $\sigma_X^2 = 2$. For a new random variable $Y = 2X - 3$, find (a) $\bar{Y}$, (b) $\overline{Y^2}$, and (c) σ_Y^2.

3.2-34. A random variable has a probability density

$$f_X(x) = \begin{cases} (5/4)(1 - x^4) & 0 < x \le 1 \\ 0 & \text{elsewhere in } x \end{cases}$$

Find: (a) $E[X]$, (b) $E[4X + 2]$, and (c) $E[X^2]$.

3.2-35. Use the definition of the gamma function as given by (F–1f) in Appendix F to obtain an expression for the moments $E[X^n]$, $n = 0, 1, 2, \ldots$, for the gamma density defined by (F–50). Use the expression to prove that (F-52) and (F-53) are true.

3.2-36. Suppose it is found that the function

$$f_X(x) = \frac{16/\pi}{(4 + x^2)^2}$$

is a good empirical fit to the probability density function of some random experimental data represented by a random variable X. Find the mean, second moment, and variance of X.

---

†A. K. Erlang (1878–1929) was a Danish engineer.

103

CHAPTER 3:
Operations on One
Random
Variable—
Expectation

*3.3-1. Show that any characteristic function $\Phi_X(\omega)$ satisfies

$$|\Phi_X(\omega)| \le \Phi_X(0) = 1$$

*3.3-2. The characteristic function for a gaussian random variable X, having a mean value of 0, is

$$\Phi_X(\omega) = \exp(-\sigma_X^2 \omega^2 / 2)$$

Find all the moments of X using $\Phi_X(\omega)$.

*3.3-3. Work Problem 3.3-2 using the moment generating function

$$M_X(v) = \exp(\sigma_X^2 v^2 / 2)$$

for the zero-mean gaussian random variable.

*3.3-4. A discrete random variable X can have $N + 1$ values $x_k = k\Delta$, $k = 0, 1, \ldots, N$, where $\Delta > 0$ is a real number. Its values occur with equal probability. Show that the characteristic function of X is

$$\Phi_X(\omega) = \frac{1}{N+1} \frac{\sin[(N+1)\omega\Delta/2]}{\sin(\omega\Delta/2)} e^{jN\omega\Delta/2}$$

*3.3-5. The characteristic function of the Laplace density of Problem 3.2-11 is known to be

$$\Phi_X(\omega) = \frac{e^{jm\omega}}{1 + (b\omega)^2}$$

Use this result to find the mean, second moment, and variance of the random variable X.

*3.3-6. The chi-square density of Problem 3.2-27 has a characteristic function

$$\Phi_X(\omega) = \frac{1}{(1 - j2\omega)^{N/2}}$$

Use this function with (3.3-4) to verify the mean and a second moment found in Problem 3.2-27.

*3.3-7. Solve for the Chernoff bound for a gaussian random variable with zero mean and variance one. [*Hint:* First find $M(v)$ by use of (C-51).]

3.4-1. A random variable X is uniformly distributed on the interval $(-\pi/2, \pi/2)$. X is transformed to the new random variable $Y = T(X) = a \tan(X)$, where $a > 0$. Find the probability density function of Y.

3.4-2. Work Problem 3.4-1 if X is uniform on the interval $(-\pi, \pi)$.

3.4-3. A random variable X undergoes the transformation $Y = a/X$, where a is a real number. Find the density function of Y.

3.4-4. A random variable X is uniformly distributed on the interval $(-a, a)$. It is transformed to a new variable Y by the transformation $Y = cX^2$ defined in Example 3.4-2. Find and sketch the density function of Y.

3.4-5. A zero-mean gaussian random variable X is transformed to the random variable Y determined by

$$Y = \begin{cases} cX & X > 0 \\ 0 & X \le 0 \end{cases}$$

where c is a real constant, $c > 0$. Find and sketch the density function of Y.

3.4-6. If the transformation of Problem 3.4-5 is applied to a Rayleigh random variable with $a \ge 0$, what is its effect?

*3.4-7. A random variable Θ is uniformly distributed on the interval (θ_1, θ_2) where θ_1 and θ_2 are real and satisfy

$$0 \le \theta_1 < \theta_2 < \pi$$

Find and sketch the probability density function of the transformed random variable $Y = \cos(\Theta)$.

3.4-8. A random variable X can have values -4, -1, 2, 3, and 4, each with probability $\frac{1}{5}$. Find:
(a) the density function,
(b) the mean, and
(c) the variance of the random $Y = 3X^3$.

3.4-9. A gaussian random variable, for which

$$f_X(x) = (2/\sqrt{\pi})\exp(-4x^2)$$

is applied to a square-law device to produce a new (output) random variable $Y = X^2/2$. (a) Find the density of Y. (b) Find the moments $m_n = E[Y^n]$, $n = 0, 1, \ldots$. (*Hint*: Put your answer in terms of the gamma function defined in Problem 3.2-27.)

3.4-10. A gaussian random variable, for which $\bar{X} = 0.6$ and $\sigma_X = 0.8$, is transformed to a new random variable by the transformation

$$Y = T(X) = \begin{cases} 4 & 1.0 \le X < \infty \\ 2 & 0 \le X < 1.0 \\ -2 & -1.0 \le X < 0 \\ -4 & -\infty < X < -1.0 \end{cases}$$

(a) Find the density function of Y.
(b) Find the mean and variance of Y.

3.4-11. Work Problem 3.4-1 except assume a transformation $Y = T(X) = a\sin(X)$ with $a > 0$.

3.4-12. Let X be a gaussian random variable with density given by (2.4-1). If X is transformed to a new random variable $Y = b + e^X$, where b is a real constant, show that the density of Y is log-normal as defined in Problem 2.5-5. This transformation allows log-normal random numbers to be generated from gaussian random numbers by a digital computer.

3.4-13. A random variable X is uniformly distributed on $(0, 6)$. If X is transformed to a new random variable $Y = 2(X - 3)^2 - 4$, find: (a) the density of Y, (b) $\bar{Y}$, (c) σ_Y^2.

105

CHAPTER 3:
Operations on One
Random
Variable—
Expectation

3.4-14. It is known that the envelope of the bandpass noise that emerges from a communication or radar receiver can be modeled as a Rayleigh random variable X with the probability density of (2.5-11) when $a = 0$, $b = 2\sigma_X^2$, and σ_X^2 is the power in the bandpass noise. If the envelope is transformed to a new variable $Y = cX^2$, where c is a constant, find the density of Y. This transformation is equivalent to a diode envelope detector where the noise level is small and the diode behaves approximately as a square-law device.

3.4-15. A certain "soft" limiter accepts a random input voltage X and limits the amplitudes of an output random variable Y according to

$$Y = \begin{cases} V(1 - e^{-X/a}) & 0 \le X \\ -V(1 - e^{X/a}) & X \le 0 \end{cases}$$

where $V > 0$ and $a > 0$ are constants. Show that the probability density of Y is

$$f_Y(y) = \frac{a}{(V - y)} f_X\left[a \ln\left(\frac{V}{(V - y)}\right)\right] u(y)$$
$$+ \frac{a}{(V + y)} f_X\left[-a \ln\left(\frac{V}{(V + y)}\right)\right] u(-y)$$

where $f_X(x)$ is the probability density of X.

3.4-16. If X in Problem 3.4-15 has the Laplace density of Problem 3.2-11 with $m = 0$, find the density of the output Y. If $a = b$, how is Y distributed?

3.5-1. In a computer simulation it is desired to transform numbers that are values of a random variable uniformly distributed on $(0, 1)$ to numbers that are values of an exponentially distributed random variable, as defined by (2.5-10) with $a = 0$. Find the required transformation.

3.5-2. Work Problem 3.5-1, except to generate a random variable with a *Weibull* distribution as defined by (F-91) in Appendix F.

3.5-3. Work Problem 3.5-1, except to generate a *Cauchy* random variable as defined by (F-31) of Appendix F with $a = 0$.

3.5-4. A random variable Y has the probability density function

$$f_Y(y) = \frac{4a^4 y u(y)}{(a^2 + y^2)^3}$$

where a is a real positive constant and $u(y)$ is the unit-step function of (A-5). Find the mean value, second moment, variance, and cumulative distribution function of Y. Show that the transformation needed to generate Y from a random variable X that is uniformly distributed on $(0, 1)$ is

$$Y = T(X) = a\left\{\left[\frac{1}{1-X}\right]^{1/2} - 1\right\}^{1/2} \qquad 0 < X < 1$$

3.5-5. Extend Example 3.5-3 by repeating the procedures for 1000 random numbers. Compare the accuracy of the results with those of the example.

3.5-6. Work Example 3.5-3 except generate the histogram of a random variable with arcsine distribution (as defined in Example 3.5-2) using 1000 random numbers.

CHAPTER 4

Multiple Random Variables

4.0
INTRODUCTION

In Chapters 2 and 3, various aspects of the theory of a single random variable were studied. The random variable was found to be a powerful concept. It enabled many realistic problems to be described in a probabilistic way such that practical measures could be applied to the problem even though it was random. For example, we have seen that shell impact position along the line of fire from a cannon to a target can be described by a random variable (Problem 2.4-6). From knowledge of the probability distribution or density function of impact position, we can solve for such practical measures as the mean value of impact position, its variance, and skew. These measures are not, however, a complete enough description of the problem in most cases.

Naturally, we may also be interested in how much the impact positions deviate *from* the line of fire in, say, the perpendicular (cross-fire) direction. In other words, we prefer to describe impact position as a point in a plane as opposed to being a point along a line. To handle such situations it is necessary that we extend our theory to include *two* random variables, one for each coordinate axis of the plane in our example. In other problems it may be necessary to extend the theory to include *several* random variables. We accomplish these extensions in this and the next chapter.

Fortunately, many situations of interest in engineering can be handled by the theory of two random variables.† Because of this fact, we emphasize the two-variable case, although the more general theory is also stated in most discussions to follow.

---

†In particular, it will be found in Chapter 6 that such important concepts as autocorrelation, cross-correlation, and covariance functions, which apply to random processes, are based on two random variables.

4.1
VECTOR RANDOM VARIABLES

Suppose two random variables X and Y are defined on a sample space S, where specific values of X and Y are denoted by x and y, respectively. Then any ordered pair of numbers (x, y) may be conveniently considered to be a *random point* in the xy plane. The point may be taken as a specific value of a *vector random variable* or a *random vector*.† Figure 4.1-1 illustrates the mapping involved in going from S to the xy plane.

The plane of all points (x, y) in the ranges of X and Y may be considered a new sample space. It is in reality a vector space where the components of any vector are the values of the random variables X and Y. The new space is sometimes called the *range sample space* (Davenport, 1970) or the *two-dimensional product space*. In this section and all following work we shall just call it a *joint sample space* and give it the symbol S_J.

As in the case of one random variable, let us define an event A by

$$A = \{X \le x\} \tag{4.1-1}$$

A similar event B can be defined for Y:

$$B = \{Y \le y\} \tag{4.1-2}$$

Events A and B refer to the sample space s, while events $\{X \le x\}$ and $\{Y \le y\}$ refer to the joint sample space S_J.‡ Figure 4.1-2 illustrates the correspon-

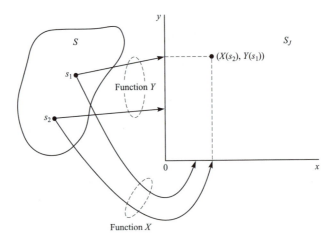

FIGURE 4.1-1
Mapping from the sample space S to the joint sample space S_J (xy plane).

---

†There are some specific conditions that must be satisfied in a complete definition of a random vector (Davenport, 1970, Chapter 5). They are somewhat advanced for our scope and we shall simply assume the validity of our random vectors.
‡Do not forget that elements s of S form the link between the two events since by writing $\{X \le x\}$ we really refer to the set of those s such that $X(s) \le x$ for some real number x. A similar statement holds for the event $\{Y \le y\}$.

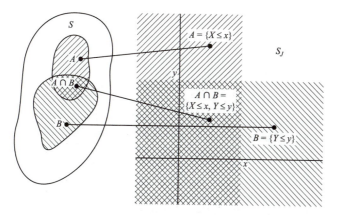

FIGURE 4.1-2
Comparisons of events in s with those in S_J.

dences between events in the two spaces. Event A corresponds to all points in S_J for which the X coordinate values are not greater than x. Similarly, event B corresponds to the Y coordinate values in S_J not exceeding y. Of special interest is to observe that the event $A \cap B$ defined on S corresponds to the *joint event* $\{X \le x$ and $Y \le y\}$ defined on S_J, which we write $\{X \le x, Y \le y\}$. This joint event is shown crosshatched in figure 4.1-2.

In the more general case where N random variables $X_1, X_2, \ldots, X_N$ are defined on a sample space S, we consider them to be components of an *N-dimensional random vector* or *N-dimensional random variable*. The joint sample space S_J is now N-dimensional.

4.2
JOINT DISTRIBUTION AND ITS PROPERTIES

The probabilities of the two events $A = \{X \le x\}$ and $B = \{Y \le y\}$ have already been defined as functions of x and y, respectively, called probability distribution functions:

$$F_X(x) = P\{X \le x\} \tag{4.2-1}$$
$$F_Y(y) = P\{Y \le y\} \tag{4.2-2}$$

We must introduce a new concept to include the probability of the joint event $\{X \le x, Y \le y\}$.

Joint Distribution Function

We define the probability of the joint event $\{X \le x, Y \le y\}$, which is a function of the numbers x and y, by a *joint probability distribution function* and denote it by the symbol $F_{X,Y}(x, y)$. Hence,

$$F_{X,Y}(x, y) = P\{X \le x, Y \le y\} \tag{4.2-3}$$

It should be clear that $P\{X \leq x, Y \leq y\} = P(A \cap B)$, where the joint event $A \cap B$ is defined on S.

To illustrate joint distribution, we take an example where both random variables X and Y are discrete.

EXAMPLE 4.2-1. Assume that the joint sample space S_J has only three possible elements: $(1, 1)$, $(2, 1)$, and $(3, 3)$. The probabilities of these elements are to be $P(1, 1) = 0.2$, $P(2, 1) = 0.3$, and $P(3, 3) = 0.5$. We find $F_{X,Y}(x, y)$.

In constructing the joint distribution function, we observe that the event $\{X \leq x, Y \leq y\}$ has no elements for any $x < 1$ and/or $y < 1$. Only at the point $(1, 1)$ does the function assume a step value. So long as $x \geq 1$ and $y \geq 1$, this probability is maintained so that $F_{X,Y}(x, y)$ has a stair step holding in the region $x \geq 1$ and $y \geq 1$ as shown in Figure 4.2-1a. For larger x and y, the point $(2, 1)$ produces a second stair step of amplitude 0.3 which holds in the region $x \geq 2$ and $y \geq 1$. The second step adds to the first. Finally, a third stair step of amplitude 0.5 is added to the first two when x and y are in the region $x \geq 3$ and $y \geq 3$. The final function is shown in Figure 4.2-1a.

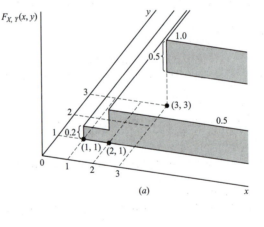

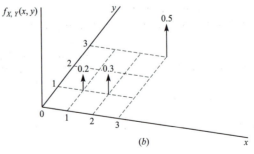

FIGURE 4.2-1
A joint distribution function (a), and its corresponding joint density function (b), that apply to Examples 4.2-1 and 4.2-2.

The preceding example can be used to identify the form of the joint distribution function for two general discrete random variables. Let X have N possible values x_n and Y have M possible values y_m, then

$$F_{X,Y}(x, y) = \sum_{n=1}^{N} \sum_{m=1}^{M} P(x_n, y_m)u(x - x_n)u(y - y_m) \qquad (4.2\text{-}4)$$

where $P(x_n, y_m)$ is the probability of the joint event $\{X = x_n, Y = y_m\}$ and $u(\cdot)$ is the unit-step function. As seen in Example 4.2-1, some couples (x_n, y_m) may have zero probability. In some cases N or M, or both, may be infinite.

If $F_{X,Y}(x, y)$ is plotted for continuous random variables X and Y, the same general behavior as shown in Figure 4.2-1a is obtained except the surface becomes smooth and has no stairstep discontinuities.

For N random variables X_n, $n = 1, 2, \ldots, N$, the generalization of (4.2-3) is direct. The joint distribution function, denoted by $F_{X_1, X_2, \ldots, X_N}(x_1, x_2, \ldots, x_N)$, is defined as the probability of the joint event $\{X_1 \leq x_1, X_2 \leq x_2, \ldots, X_N \leq x_N\}$:

$$F_{X_1, X_2, \ldots, X_N}(x_1, x_2, \ldots, x_N) = P\{X_1 \leq x_1, X_2 \leq x_2, \ldots, X_N \leq x_N\} \qquad (4.2\text{-}5)$$

For a single random variable X, we found in Chapter 2 that $F_X(x)$ could be expressed in general as the sum of a function of stairstep form (due to the discrete portion of a mixed random variable X) and a function that was continuous (due to the continuous portion of X). Such a simple decomposition of the joint distribution when $N > 1$ is not generally true [Cramér, 1946, Section 8.4]. However, it is true that joint density functions in practice often correspond to all random variables being either discrete or continuous. Therefore, we shall limit our consideration in this book almost entirely to these two cases when $N > 1$.

Properties of the Joint Distribution

A joint distribution function for two random variables X and Y has several properties that follow readily from its definition. We list them:

(1) $F_{X,Y}(-\infty, -\infty) = 0 \qquad F_{X,Y}(-\infty, y) = 0 \qquad F_{X,Y}(x, -\infty) = 0$ (4.2-6a)

(2) $F_{X,Y}(\infty, \infty) = 1$ \hfill (4.2-6b)

(3) $0 \leq F_{X,Y}(x, y) \leq 1$ \hfill (4.2-6c)

(4) $F_{X,Y}(x, y)$ is a nondecreasing function of both x and y \hfill (4.2-6d)

(5) $F_{X,Y}(x_2, y_2) + F_{X,Y}(x_1, y_1) - F_{X,Y}(x_1, y_2) - F_{X,Y}(x_2, y_1)$

$\qquad = P\{x_1 < X \leq x_2, y_1 < Y \leq y_2\} \geq 0$ \hfill (4.2-6e)

(6) $F_{X,Y}(x, \infty) = F_X(x) \qquad F_{X,Y}(\infty, y) = F_Y(y)$ \hfill (4.2-6f)

The first five of these properties are just the two-dimensional extensions of the properties of one random variable given in (2.2-2). Properties 1, 2, and 5 may be used as tests to determine whether some function can be a valid distribution function for two random variables X and Y (Papoulis, 1965, p. 169). Property 6 deserves a few special comments.

Marginal Distribution Functions

Property 6 above states that the distribution function of one random variable can be obtained by setting the value of the other variable to infinity in $F_{X,Y}(x, y)$. The functions $F_X(x)$ or $F_Y(y)$ obtained in this manner are called *marginal distribution functions*.

To justify property 6, it is easiest to return to the basic events A and B, defined by $A = \{X \le x\}$ and $B = \{Y \le y\}$, and observe that $F_{X,Y}(x, y) = P\{X \le x, Y \le y\} = P(A \cap B)$. Now if we set y to ∞, this is equivalent to making B the certain event; that is, $B = \{Y \le \infty\} = S$. Furthermore, since $A \cap B = A \cap S = A$, then we have $F_{X,Y}(x, \infty) = P(A \cap S) = P(A) = P\{X \le x\} = F_X(x)$. A similar proof can be stated for obtaining $F_Y(y)$.

EXAMPLE 4.2-2. We find explicit expressions for $F_{X,Y}(x, y)$, and the marginal distributions $F_X(x)$ and $F_Y(y)$ for the joint sample space of Example 4.2-1.

The joint distribution derives from (4.2-4) if we recognize that only three probabilities are nonzero:

$$F_{X,Y}(x, y) = P(1, 1)u(x - 1)u(y - 1)$$
$$+ P(2, 1)u(x - 2)u(y - 1)$$
$$+ P(3, 3)u(x - 3)u(y - 3)$$

where $P(1, 1) = 0.2$, $P(2, 1) = 0.3$, and $P(3, 3) = 0.5$. If we set $y = \infty$:

$$F_X(x) = F_{X,Y}(x, \infty)$$
$$= P(1, 1)u(x - 1) + P(2, 1)u(x - 2) + P(3, 3)u(x - 3)$$
$$= 0.2u(x - 1) + 0.3u(x - 2) + 0.5u(x - 3)$$

If we set $x = \infty$:

$$F_Y(y) = F_{X,Y}(\infty, y)$$
$$= 0.2u(y - 1) + 0.3u(y - 1) + 0.5u(y - 3)$$
$$= 0.5u(y - 1) + 0.5u(y - 3)$$

Plots of these marginal distributions are shown in Figure 4.2-2.

From an N-dimensional joint distribution function we may obtain a k-dimensional *marginal distribution function*, for any selected group of k of the N random variables, by setting the values of the other $N - k$ random variables to infinity. Here k can be any integer $1, 2, 3, \ldots, N - 1$.

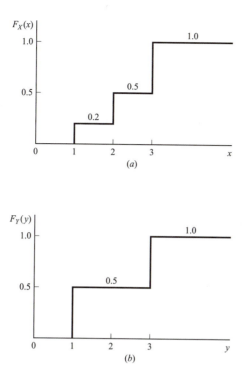

FIGURE 4.2-2
Marginal distributions applicable to Figure 4.2-1 and Example 4.2-2: (*a*) $F_X(x)$ and
(*b*) $F_Y(y)$.

4.3
JOINT DENSITY AND ITS PROPERTIES

In this section the concept of a probability density function is extended to
include multiple random variables.

Joint Density Function

For two random variables X and Y, the *joint probability density function*,
denoted $f_{X,Y}(x, y)$, is defined by the second derivative of the joint distribution
function wherever it exists:

$$f_{X,Y}(x, y) = \frac{\partial^2 F_{X,Y}(x, y)}{\partial x \, \partial y} \tag{4.3-1}$$

We shall refer often to $f_{X,Y}(x, y)$ as the *joint density function*.

   If X and Y are discrete random variables, $F_{X,Y}(x, y)$ will possess step
discontinuities (see Example 4.2-1 and Figure 4.2-1). Derivatives at these discontinuities are normally undefined. However, by admitting impulse functions
(see Appendix A), we are able to define $f_{X,Y}(x, y)$ at these points. Therefore,

the joint density function may be found for any two discrete random variables by substitution of (4.2-4) into (4.3-1):

$$f_{X,Y}(x, y) = \sum_{n=1}^{N} \sum_{m=1}^{M} P(x_n, y_m) \, \delta(x - x_n) \, \delta(y - y_m) \qquad (4.3\text{-}2)$$

An example of the joint density function of two discrete random variables is shown in Figure 4.2-1b.

When N random variables $X_1, X_2, \ldots, X_N$ are involved, the joint density function becomes the N-fold partial derivative of the N-dimensional distribution function:

$$f_{X_1, X_2, \ldots, X_N}(x_2, x_2, \ldots, x_N) = \frac{\partial^N F_{X_1, X_2, \ldots, X_N}(x_1, x_2, \ldots, x_N)}{\partial x_1 \partial x_2 \ldots \partial x_N} \qquad (4.3\text{-}3)$$

By direct integration this result is equivalent to

$$F_{X_1, X_2, \ldots, X_N}(x_1, x_2, \ldots, x_N)$$
$$= \int_{-\infty}^{x_N} \cdots \int_{-\infty}^{x_2} \int_{-\infty}^{x_1} f_{X_1, X_2, \ldots, X_N}(\xi_1, \xi_2, \ldots, \xi_N) d\xi_1 d\xi_2 \ldots d\xi_N \qquad (4.3\text{-}4)$$

Properties of the Joint Density

Several properties of a joint density function may be listed that derive from its definition (4.3-1) and the properties (4.2-6) of the joint distribution function:

(1) $f_{X,Y}(x, y) \geq 0$ $\qquad (4.3\text{-}5a)$

(2) $\int_{-\infty}^{\infty} \int_{-\infty}^{\infty} f_{X,Y}(x, y) \, dx \, dy = 1$ $\qquad (4.3\text{-}5b)$

(3) $F_{X,Y}(x, y) = \int_{-\infty}^{y} \int_{-\infty}^{x} f_{X,Y}(\xi_1, \xi_2) \, d\xi_1 \, d\xi_2$ $\qquad (4.3\text{-}5c)$

(4) $F_X(x) = \int_{-\infty}^{x} \int_{-\infty}^{\infty} f_{X,Y}(\xi_1, \xi_2) \, d\xi_2 \, d\xi_1$ $\qquad (4.3\text{-}5d)$

$F_Y(y) = \int_{-\infty}^{y} \int_{-\infty}^{\infty} f_{X,Y}(\xi_1, \xi_2) \, d\xi_1 \, d\xi_2$ $\qquad (4.3\text{-}5e)$

(5) $P\{x_1 < X \leq x_2, y_1 < Y \leq y_2\} = \int_{y_1}^{y_2} \int_{x_1}^{x_2} f_{X,Y}(x, y) \, dx \, dy$ $\qquad (4.3\text{-}5f)$

(6) $f_X(x) = \int_{-\infty}^{\infty} f_{X,Y}(x, y) \, dy$ $\qquad (4.3\text{-}5g)$

$f_Y(y) = \int_{-\infty}^{\infty} f_{X,Y}(x, y) \, dx$ $\qquad (4.3\text{-}5h)$

Properties 1 and 2 may be used as sufficient tests to determine if some function can be a valid density function. Both tests must be satisfied (Papoulis, 1965, p. 169).

EXAMPLE 4.3-1. Suppose b is a positive constant and we test the function

$$g(x, y) = \begin{cases} b e^{-x} \cos(y) & 0 \leq x \leq 2 \text{ and } 0 \leq y \leq \pi/2 \\ 0 & \text{all other } x \text{ and } y \end{cases}$$

to see if it can be valid probability density function. For the allowed values of x and y the function is not negative and satisfies (4.3-5a). The final test is (4.3-5b)

$$\int_0^{\pi/2} \int_0^2 b e^{-x} \cos(y)\, dx\, dy = b \int_0^2 e^{-x}\, dx \int_0^{\pi/2} \cos(y)\, dy$$

$$= b(1 - e^{-2}) = 1$$

Thus, to be valid $b = 1/[1 - \exp(-2)]$ is necessary.

The first five of the above properties are readily verified from earlier work, and the reader should go through the necessary logic as an exercise. Property 6 introduces a new concept.

Marginal Density Functions

The functions $f_X(x)$ and $f_Y(y)$ of property 6 are called *marginal probability density functions* or just *marginal density functions*. They are the density functions of the single variables X and Y and are defined as the derivatives of the marginal distribution functions:

$$f_X(x) = \frac{dF_X(x)}{dx} \tag{4.3-6}$$

$$f_Y(y) = \frac{dF_Y(y)}{dy} \tag{4.3-7}$$

By substituting (4.3-5d) and (4.3-5e) into (4.3-6) and (4.3-7), respectively, we are able to verify the equations of property 6.

We shall illustrate the calculation of marginal density functions from a given joint density function with an example.

EXAMPLE 4.3-2. We find $f_X(x)$ and $f_Y(y)$ when the joint density function is given by (Clarke and Disney, 1970, p. 108):

$$f_{X,Y}(x, y) = u(x)u(y)xe^{-x(y+1)}$$

From (4.3-5g) and the above equation:

$$f_X(x) = \int_0^\infty u(x)xe^{-x(y+1)}\, dy = u(x)xe^{-x} \int_0^\infty e^{-xy}\, dy$$

$$= u(x)xe^{-x}(1/x) = u(x)e^{-x}$$

after using an integral from Appendix C.

From (4.3-5h):

$$f_Y(y) = \int_0^\infty u(y)xe^{-x(y+1)}\, dx = \frac{u(y)}{(y+1)^2}$$

after using another integral from Appendix C.

For N random variables $X_1, X_2, \ldots, X_N$, the k-*dimensional marginal density function* is defined as the k-fold partial derivative of the k-dimensional marginal distribution function. It can also be found from the joint density function by integrating out all variables except the k variables of interest $X_1, X_2, \ldots, X_k$:

$$f_{X_1,X_2,\ldots,X_k}(x_1, x_2, \ldots, x_k)$$
$$= \int_{-\infty}^\infty \cdots \int_{-\infty}^\infty f_{X_1,X_2,\ldots,X_N}(x_1, x_2, \ldots, x_N)\, dx_{k+1} dx_{k+2} \cdots dx_N \qquad (4.3\text{-}8)$$

4.4
CONDITIONAL DISTRIBUTION AND DENSITY

In Section 2.6, the conditional distribution function of a random variable X, given some event B, was defined as

$$F_x(x|B) = P\{X \le x|B\} = \frac{P\{X \le x \cap B\}}{P(B)} \qquad (4.4\text{-}1)$$

for any event B with nonzero probability. The corresponding conditional density function was defined through the derivative

$$f_X(x|B) = \frac{dF_X(x|B)}{dx} \qquad (4.4\text{-}2)$$

In this section these two functions are extended to include a second random variable through suitable definitions of event B.

Conditional Distribution and Density—Point Conditioning

Often in practical problems we are interested in the distribution function of one random variable X conditioned by the fact that a second random variable Y has some specific value y. This is called *point conditioning*, and we can handle such problems by defining event B by

$$B = \{y - \Delta y < Y \le y + \Delta y\} \qquad (4.4\text{-}3)$$

where Δy is a small quantity that we eventually let approach 0. For this event, (4.4-1) can be written

$$F_X(x|y - \Delta y < Y \le y + \Delta y) = \frac{\int_{y-\Delta y}^{y+\Delta y} \int_{-\infty}^{x} f_{X,Y}(\xi_1, \xi_2) \, d\xi_1 \, d\xi_2}{\int_{y-\Delta y}^{y+\Delta y} f_Y(\xi) \, d\xi} \qquad (4.4\text{-}4)$$

where we have used (4.3-5f) and (2.3-6d).

Consider two cases of (4.4-4). In the first case, assume X and Y are both discrete random variables with values x_i, $i = 1, 2, \ldots, N$, and y_j, $j = 1, 2, \ldots, M$, respectively, while the probabilities of these values are denoted $P(x_i)$ and $P(y_j)$, respectively. The probability of the joint occurrence of x_i and y_j is denoted $P(x_i, y_j)$. Thus,

$$f_Y(y) = \sum_{j=1}^{M} P(y_j) \, \delta(y - y_j) \qquad (4.4\text{-}5)$$

$$f_{X,Y}(x, y) = \sum_{i=1}^{N} \sum_{j=1}^{M} P(x_i, y_j) \, \delta(x - x_i) \, \delta(y - y_j) \qquad (4.4\text{-}6)$$

Now suppose that the specific value of y of interest is y_k. With substitution of (4.4-5) and (4.4-6) into (4.4-4) and allowing $\Delta y \to 0$, we obtain

$$F_X(x|Y = y_k) = \sum_{i=1}^{N} \frac{P(x_i, y_k)}{P(y_k)} u(x - x_i) \qquad (4.4\text{-}7)$$

After differentiation we have

$$f_X(x|Y = y_k) = \sum_{i=1}^{N} \frac{P(x_i, y_k)}{P(y_k)} \delta(x - x_i) \qquad (4.4\text{-}8)$$

EXAMPLE 4.4-1. To illustrate the use of (4.4-8) assume a joint density function as given in Figure 4.4-1a. Here $P(x_1, y_1) = \frac{2}{15}$, $P(x_2, y_1) = \frac{3}{15}$, etc. Since $P(y_3) = (\frac{4}{15}) + (\frac{5}{15}) = \frac{9}{15}$, use of (4.4-8) will give $f_X(x|Y = y_3)$ as shown in Figure 4.4-1b.

The second case of (4.4-4) that is of interest corresponds to X and Y both continuous random variables. As $\Delta y \to 0$ the denominator in (4.4-4) becomes 0. However, we can still show that the conditional density $f_X(x|Y = y)$ may exist. If Δy is very small, (4.4-4) can be written as

$$F_X(x|y - \Delta y < Y \le y + \Delta y) = \frac{\int_{-\infty}^{x} f_{X,Y}(\xi_1, y) \, d\xi_1 \, 2\Delta y}{f_Y(y) 2\Delta y} \qquad (4.4\text{-}9)$$

and, in the limit as $\Delta y \to 0$

$$F_X(x|Y = y) = \frac{\int_{-\infty}^{x} f_{X,Y}(\xi, y) \, d\xi}{f_Y(y)} \qquad (4.4\text{-}10)$$

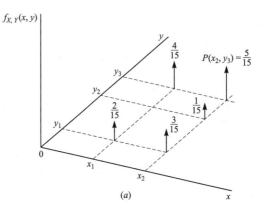

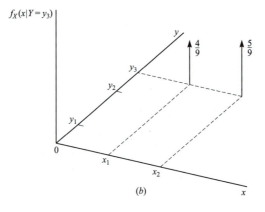

FIGURE 4.4-1
A joint density function (*a*) and a conditional density function (*b*) applicable to
Example 4.4-1.

for every y such that $f_Y(y) \neq 0$. After differentiation of both sides of (4.4-10)
with respect to x:

$$f_X(x|Y = y) = \frac{f_{X,Y}(x, y)}{f_Y(y)} \tag{4.4-11}$$

When there is no confusion as to meaning, we shall often write (4.4-11) as

$$f_X(x|y) = \frac{f_{X,Y}(x, y)}{f_Y(y)} \tag{4.4-12}$$

It can also be shown that

$$f_Y(y|x) = \frac{f_{X,Y}(x, y)}{f_X(x)} \tag{4.4-13}$$

EXAMPLE 4.4-2. We find $f_Y(y|x)$ for the density functions defined in Example 4.3-1. Since

$$f_{X,Y}(x, y) = u(x)u(y)xe^{-x(y+1)}$$

and

$$f_X(x) = u(x)e^{-x}$$

are nonzero only for $0 < y$ and $0 < x$, $f_Y(y|x)$ is nonzero only for $0 < y$ and $0 < x$. It is

$$f_Y(y|x) = u(x)u(y)xe^{-xy}$$

from (4.4-13).

*Conditional Distribution and Density—Interval Conditioning

It is sometimes convenient to define event B in (4.4-1) and (4.4-2) in terms of a random variable Y by

$$B = \{y_a < Y \leq y_b\} \qquad (4.4\text{-}14)$$

where y_a and y_b are real numbers and we assume $P(B) = P\{y_a < Y \leq y_b\} \neq 0$. With this definition it is readily shown that (4.4-1) and (4.4-2) become

$$F_X(x|y_a < Y \leq y_b) = \frac{F_{X,Y}(x, y_b) - F_{X,Y}(x, y_a)}{F_Y(y_b) - F_Y(y_a)}$$

$$= \frac{\int_{y_a}^{y_b} \int_{-\infty}^{x} f_{X,Y}(\xi, y)\, d\xi\, dy}{\int_{y_a}^{y_b} \int_{-\infty}^{\infty} f_{X,Y}(x, y)\, dx\, dy} \qquad (4.4\text{-}15)$$

and

$$f_X(x|y_a < Y \leq y_b) = \frac{\int_{y_a}^{y_b} f_{X,Y}(x, y)\, dy}{\int_{y_a}^{y_b} \int_{-\infty}^{\infty} f_{X,Y}(x, y)\, dx\, dy} \qquad (4.4\text{-}16)$$

These last two expressions hold for X and Y either continuous or discrete random variables. In the discrete case, the joint density is given by (4.3-2). The resulting distribution and density will be defined, however, only for y_a and y_b such that the denominators of (4.4-15) and (4.4-16) are nonzero. This requirement is satisfied so long as the interval $y_a < y \leq y_b$ spans at least one possible value of Y having a nonzero probability of occurrence.

An example will serve to illustrate the application of (4.4-16) when X and Y are continuous random variables.

EXAMPLE 4.4-3. We use (4.4-16) to find $f_X(x|Y \leq y)$ for the joint density function of Example 4.3-1. Since we have here defined $B = \{Y \leq y\}$, then $y_a = -\infty$ and $y_b = y$. Furthermore, since $f_{X,Y}(x, y)$ is nonzero only for $0 < x$ and $0 < y$, we need only consider this region of x and y in finding the conditional density function. The denominator of (4.4-16) can be written as $\int_{-\infty}^{y} f_Y(\xi) \, d\xi$. By using results from Example 4.3-1:

$$\int_{-\infty}^{y} f_Y(\xi) \, d\xi = \int_{-\infty}^{y} \frac{u(\xi) \, d\xi}{(\xi + 1)^2} = \int_{0}^{y} \frac{d\xi}{(\xi + 1)^2} = \frac{y}{y + 1} \qquad y > 0$$

and zero for $y < 0$, after using an integral from Appendix C. The numerator of (4.4-16) becomes

$$\int_{-\infty}^{y} f_{X,Y}(x, \xi) \, d\xi = \int_{0}^{y} u(x) x e^{-x(\xi+1)} \, d\xi$$

$$= u(x) x e^{-x} \int_{0}^{y} e^{-x\xi} \, d\xi$$

$$= u(x) e^{-x} (1 - e^{-xy}) \qquad y > 0$$

and zero for $y < 0$, after using another integral from Appendix C. Thus

$$f_X(x|Y \leq y) = u(x) u(y) \left(\frac{y + 1}{y} \right) e^{-x} (1 - e^{-xy})$$

This function is plotted in Figure 4.4-2 for several values of y.

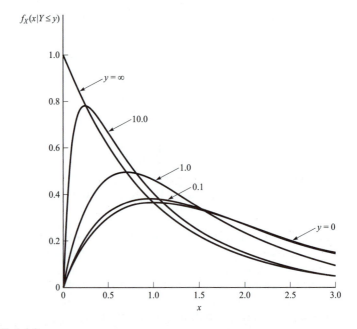

FIGURE 4.4-2
Conditional probability density functions applicable to Example 4.4-3.

4.5
STATISTICAL INDEPENDENCE

It will be recalled from (1.5-3) that two events A and B are statistically independent if (and only if)

$$P(A \cap B) = P(A)P(B) \qquad (4.5\text{-}1)$$

This condition can be used to apply to two random variables X and Y by defining the events $A = \{X \leq x\}$ and $B = \{Y \leq y\}$ for two real numbers x and y. Thus, X and Y are said to be *statistically independent random variables* if (and only if)

$$P\{X \leq x, Y \leq y\} = P\{X \leq x\}P\{Y \leq y\} \qquad (4.5\text{-}2)$$

From this expression and the definitions of distribution functions, it follows that

$$F_{X,Y}(x, y) = F_X(x)F_Y(y) \qquad (4.5\text{-}3)$$

if X and Y are independent. From the definitions of density functions, (4.5-3) gives

$$f_{X,Y}(x, y) = f_X(x)f_Y(y) \qquad (4.5\text{-}4)$$

by differentiation, if X and Y are independent. Either (4.5-3) or (4.5-4) may serve as a sufficient definition of, or test for, independence of two random variables.

The form of the conditional distribution function for independent events is found by use of (4.4-1) with $B = \{Y \leq y\}$:

$$F_X(x|Y \leq y) = \frac{P\{X \leq x, Y \leq y\}}{P\{Y \leq y\}} = \frac{F_{X,Y}(x, y)}{F_Y(y)} \qquad (4.5\text{-}5)$$

By substituting (4.5-3) into (4.5-5), we have

$$F_X(x|Y \leq y) = F_X(x) \qquad (4.5\text{-}6)$$

In other words, the conditional distribution ceases to be conditional and simply equals the marginal distribution for independent random variables. It can also be shown that

$$F_Y(y|X \leq x) = F_Y(y) \qquad (4.5\text{-}7)$$

Conditional density function forms, for independent X and Y, are found by differentiation of (4.5-6) and (4.5-7):

$$f_X(x|Y \leq y) = f_X(x) \qquad (4.5\text{-}8)$$
$$f_Y(y|X \leq x) = f_Y(y) \qquad (4.5\text{-}9)$$

EXAMPLE 4.5-1. For the densities of Example 4.3-1:

$$f_{X,Y}(x, y) = u(x)u(y)xe^{-x(y+1)}$$

$$f_X(x)f_Y(y) = u(x)u(y)\frac{e^{-x}}{(y+1)^2} \neq f_{X,Y}(x, y)$$

Therefore, the random variables X and Y are not independent.

EXAMPLE 4.5-2. The joint density of two random variables X and Y is

$$f_{X,Y}(x, y) = \frac{1}{12} u(x)u(y)e^{-(x/4)-(y/3)}$$

We determine if X and Y are statistically independent. From (4.3-5g) and (4.3-5h)

$$f_X(x) = \int_0^\infty (1/12)u(x)e^{-x/4}e^{-y/3} \, dy = (1/4)u(x)e^{-x/4}$$

$$f_Y(y) = \int_0^\infty (1/12)u(y)e^{-y/3}e^{-x/4} \, dx = (1/3)u(y)e^{-y/3}$$

Since $f_X(x)f_Y(y) = f_{X,Y}(x, y)$, then X and Y are independent.

In the more general study of the statistical independence of N random variables $X_1, X_2, \ldots, X_N$, we define events A_i by

$$A_i = \{X_i \leq x_i\} \qquad i = 1, 2, \ldots, N \tag{4.5-10}$$

where the x_i are real numbers. With these definitions, the random variables X_i are said to be statistically independent if (1.5-6) is satisfied.

It can be shown that if $X_1, X_2, \ldots, X_N$ are statistically independent then any group of these random variables is independent of any other group. Furthermore, a function of any group is independent of any function of any other group of the random variables. For example, with $N = 4$ random variables: X_4 is independent of $X_3 + X_2 + X_1$; X_3 is independent of $X_2 + X_1$, etc. (see Papoulis, 1965, p. 238).

4.6
DISTRIBUTION AND DENSITY OF A SUM OF RANDOM VARIABLES

The problem of finding the distribution and density functions for a sum of *statistically independent* random variables is considered in this section.

Sum of Two Random Variables

Let W be a random variable equal to the sum of two independent random variables X and Y:

$$W = X + Y \tag{4.6-1}$$

This is a very practical problem because X might represent a random signal voltage and Y could represent random noise at some instant in time. The sum W would represent a signal-plus-noise voltage available to some receiver.

The probability distribution function we seek is defined by

$$F_W(w) = P\{W \leq w\} = P\{X + Y \leq w\} \tag{4.6-2}$$

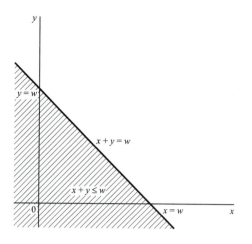

FIGURE 4.6-1
Region in xy plane where $x + y \leq w$.

Figure 4.6-1 illustrates the region in the xy plane where $x + y \leq w$. Now from (4.3-5f), the probability corresponding to an elemental area $dx\,dy$ in the xy plane located at the point (x, y) is $f_{X,Y}(x, y)\,dx\,dy$. If we sum all such probabilities over the region where $x + y \leq w$ we will obtain $F_W(w)$. Thus

$$F_W(w) = \int_{-\infty}^{\infty} \int_{x=-\infty}^{w-y} f_{X,Y}(x, y)\,dx\,dy \qquad (4.6\text{-}3)$$

and, after using (4.5-4):

$$F_W(w) = \int_{-\infty}^{\infty} f_Y(y) \int_{x=-\infty}^{w-y} f_X(x)\,dx\,dy \qquad (4.6\text{-}4)$$

By differentiating (4.6-4), using Leibniz's rule, we get the desired density function

$$f_W(w) = \int_{-\infty}^{\infty} f_Y(y) f_X(w - y)\,dy \qquad (4.6\text{-}5)$$

This expression is recognized as a convolution integral. Consequently, we have shown that *the density function of the sum of two statistically independent random variables is the convolution of their individual density functions.*

EXAMPLE 4.6-1. We use (4.6-5) to find the density of $W = X + Y$ where the densities of X and Y are assumed to be

$$f_X(x) = \frac{1}{a}[u(x) - u(x - a)]$$

$$f_Y(y) = \frac{1}{b}[u(y) - u(y - b)]$$

with $0 < a < b$, as shown in Figure 4.6-2a and b. Now because $0 < X$ and $0 < Y$, we only need examine the case $W = X + Y > 0$. From (4.6-5) we write

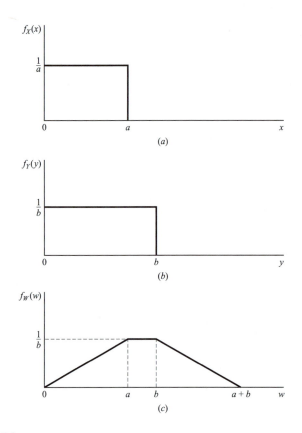

FIGURE 4.6-2
Two density functions (*a*) and (*b*) and their convolution (*c*).

$$f_W(w) = \int_{-\infty}^{\infty} \frac{1}{ab} [u(y) - u(y - b)][u(w - y) - u(w - y - a)] \, dy$$

$$= \frac{1}{ab} \int_0^{\infty} [1 - u(y - b)][u(w - y) - u(w - y - a)] \, dy$$

$$= \frac{1}{ab} \left[\int_0^{\infty} u(w - y) \, dy - \int_0^{\infty} u(w - y - a) \, dy \right.$$

$$\left. - \int_0^{\infty} u(y - b)u(w - y) \, dy + \int_0^{\infty} u(y - b)u(w - y - a) \, dy \right]$$

All these integrands are unity; the values of the integrals are determined by the unit-step functions through their control over limits of integration. After straightforward evaluation we get

$$f_W(w) = \begin{cases} w/ab & 0 \le w < a \\ 1/b & a \le w < b \\ (a + b - w)/ab & b \le w < a + b \\ 0 & w \ge a + b \end{cases}$$

which is sketched in Figure 4.6-2*c*.

When the sum Y of N independent random variables $X_1, X_2, \ldots, X_N$ is to be considered, we may extend the above analysis for two random variables. Let $Y_1 = X_1 + X_2$. Then we know from the preceding work that $f_{Y_1}(y_1) = f_{X_2}(x_2) * f_{X_1}(x_1)$.† Next, we know that X_3 will be independent of $Y_1 = X_1 + X_2$ because X_3 is independent of both X_1 and X_2. Thus, by applying (4.6-5) to the two variables X_3 and Y_1 to find the density function of $Y_2 = X_3 + Y_1$, we get

$$f_{Y_2 = X_1 + X_2 + X_3}(y_2) = f_{X_3}(x_3) * f_{Y_1 = X_1 + X_2}(y_1)$$
$$= f_{X_3}(x_3) * f_{X_2}(x_2) * f_{X_1}(x_1) \qquad (4.6\text{-}6)$$

By continuing the process we find that the density function of $Y = X_1 + X_2 + \cdots + X_N$ is the $(N - 1)$-fold convolution of the N individual density functions:

$$f_Y(y) = f_{X_N}(x_N) * f_{X_{N-1}}(x_{N-1}) * \cdots * f_{X_1}(x_1) \qquad (4.6\text{-}7)$$

The distribution function of Y is found from the integral of $f_Y(y)$ using (2.3-6c).

Another method using characteristic functions can also be employed to find the density function of a sum of random variables. A discussion of the method is given in Section 5.2 for statistically independent random variables.

*4.7
CENTRAL LIMIT THEOREM

Broadly defined, the *central limit theorem* says that the probability distribution function of the sum of a large number of random variables approaches a gaussian distribution. Although the theorem is known to apply to some cases of statistically *dependent* random variables (Cramér, 1946, p. 219), most applications, and the largest body of knowledge, are directed toward statistically independent random variables. Thus, in all succeeding discussions we assume statistically independent random variables.

*Unequal Distributions

Let $\bar{X}_i$ and $\sigma_{X_i}^2$ be the means and variances, respectively, of N random variables X_i, $i = 1, 2, \ldots, N$, which may have arbitrary probability densities. The central limit theorem states that the sum $Y_N = X_1 + X_2 + \cdots + X_N$, which has mean $\bar{Y}_N = \bar{X}_1 + \bar{X}_2 + \cdots \bar{X}_N$ and variance $\sigma_{Y_N}^2 = \sigma_{X_1}^2 + \sigma_{X_2}^2 + \cdots + \sigma_{X_N}^2$, has a probability distribution that asymptotically approaches gaussian as $N \to \infty$. Necessary conditions for the theorem's validity are difficult to state, but sufficient conditions are known to be (Cramér, 1946; Thomas, 1969)

---

†The asterisk denotes convolution.

$$\sigma_{X_i}^2 > B_1 > 0 \qquad i = 1, 2, \ldots, N \qquad (4.7\text{-}1a)$$

$$E[|X_i - \bar{X}_i|^3] < B_2 \qquad i = 1, 2, \ldots, N \qquad (4.7\text{-}1b)$$

where B_1 and B_2 are positive numbers. These conditions guarantee that no one random variable in the sum dominates.

The reader should observe that the central limit theorem guarantees only that the *distribution* of the sum of random variables becomes gaussian. It does not follow that the probability *density* is always gaussian. For continuous random variables there is usually no problem, but certain conditions imposed on the individual random variables (Cramér, 1946; Papoulis, 1965 and 1984) will guarantee that the density is gaussian.

For discrete random variables X_i the sum Y_N will also be discrete so its density will contain impulses and is, therefore, not gaussian, even though the distribution approaches gaussian. When the possible discrete values of each random variable are kb, $k = 0, \pm 1, \pm 2, \ldots$, with b a constant,† the envelope of the impulses in the density of the sum will be gaussian (with mean $\bar{Y}_N$ and variance $\sigma_{Y_N}^2$). This case is discussed in some detail by Papoulis (1965).

The practical usefulness of the central limit theorem does not reside so much in the exactness of the gaussian distribution for $N \to \infty$ because the variance of Y_N becomes infinite from (4.7-1a). Usefulness derives more from the fact that Y_N for *finite* N may have a distribution that is closely approximated as gaussian. The approximation can be quite accurate, even for relatively small values of N, in the central region of the gaussian curve near the mean. However, the approximation can be very inaccurate in the tail regions away from the mean, even for large values of N (Davenport, 1970; Melsa and Sage, 1973). Of course, the approximation is made more accurate by increasing N.

*Equal Distributions

If all of the statistically independent random variables being summed are continuous and have the same distribution function, and therefore the same density, the proof of the central limit theorem is relatively straightforward and is next developed.

Because the sum $Y_N = X_1 + X_2 + \cdots + X_N$ has an infinite variance as $N \to \infty$, we shall work with the zero-mean, unit-variance random variable

$$W_N = (Y_N - \bar{Y}_N)/\sigma_{Y_N} = \sum_{i=1}^{N}(X_i - \bar{X}_i) \bigg/ \left[\sum_{i=1}^{N} \sigma_{X_i}^2 \right]^{1/2}$$

$$= \frac{1}{\sqrt{N}\sigma_X} \sum_{i=1}^{N}(X_i - \bar{X}) \qquad (4.7\text{-}2)$$

---

†These are called *lattice-type* discrete random variables (Papoulis, 1965).

instead. Here we define $\bar{X}$ and $\sigma_{\bar{X}}^2$ by

$$\bar{X}_i = \bar{X} \qquad \text{all } i \qquad\qquad (4.7\text{-}3)$$

$$\sigma_{\bar{X}_i}^2 = \sigma_{\bar{X}}^2 \qquad \text{all } i \qquad\qquad (4.7\text{-}4)$$

since all the X_i have the same distribution.

The theorem's proof consists of showing that the characteristic function of W_N is that of a zero-mean, unit-variance gaussian random variable, which is

$$\Phi_{W_N}(\omega) = \exp(-\omega^2/2) \qquad\qquad (4.7\text{-}5)$$

from Problem 3.3-2. If this is proved the density of W_N must be gaussian from (3.3-3) and the fact that Fourier transforms are unique. The characteristic function of W_N is

$$\Phi_{W_N}(\omega) = E[e^{j\omega W_N}] = E\left[\exp\left\{\frac{j\omega}{\sqrt{N}\sigma_X}\sum_{i=1}^{N}(X_i - \bar{X})\right\}\right]$$

$$= \left\langle E\left\{\exp\left[\frac{j\omega}{\sqrt{N}\sigma_X}(X_i - \bar{X})\right]\right\}\right\rangle^N \qquad\qquad (4.7\text{-}6)$$

The last step in (4.7-6) follows from the independence and equal distribution of the X_i. Next, the exponential in (4.7-6) is expanded in a Taylor polynomial with a remainder term R_N/N:

$$E\left\{\exp\left[\frac{j\omega}{\sqrt{N}\sigma_X}(X_i - \bar{X})\right]\right\}$$

$$= E\left\{1 + \left(\frac{j\omega}{\sqrt{N}\sigma_X}\right)(X_i - \bar{X}) + \left(\frac{j\omega}{\sqrt{N}\sigma_X}\right)^2\frac{(X_i - \bar{X})^2}{2} + \frac{R_N}{N}\right\}$$

$$= 1 - (\omega^2/2N) + E[R_N]/N \qquad\qquad (4.7\text{-}7)$$

where $E[R_N]$ approaches zero as $N \to \infty$ (Davenport, 1970, p. 442). On substitution of (4.7-7) into (4.7-6) and forming the natural logarithm, we have

$$\ln[\Phi_{W_N}(\omega)] = N\ln\{1 - (\omega^2/2N) + E[R_N]/N\} \qquad\qquad (4.7\text{-}8)$$

Since

$$\ln(1 - z) = -\left[z + \frac{z^2}{2} + \frac{z^3}{3} + \cdots\right] \qquad |z| < 1 \qquad\qquad (4.7\text{-}9)$$

we identify z with $(\omega^2/2N) - E[R_N]/N$ and write (4.7-8) as

$$\ln[\Phi_{W_N}(\omega)] = -(\omega^2/2) + E[R_N] - \frac{N}{2}\left[\frac{\omega^2}{2N} - \frac{E[R_N]}{N}\right]^2 + \cdots \qquad\qquad (4.7\text{-}10)$$

so

$$\lim_{N\to\infty}\{\ln[\Phi_{W_N}(\omega)]\} = \ln\left\{\lim_{N\to\infty}\Phi_{W_N}(\omega)\right\} = -\omega^2/2 \qquad\qquad (4.7\text{-}11)$$

Finally, we have

$$\lim_{N \to \infty} \Phi_{W_N}(\omega) = e^{-\omega^2/2} \tag{4.7-12}$$

which was to be shown.

We illustrate the use of the central limit theorem through an example.

EXAMPLE 4.7-1. Consider the sum of just two independent uniformly distributed random variables X_1 and X_2 having the same density

$$f_X(x) = \frac{1}{a}[u(x) - u(x - a)]$$

where $a > 0$ is a constant. The means and variances of X_1 and X_2 are $\bar{X} = a/2$ and $\sigma_X^2 = a^2/12$, respectively. The density of the sum $W = X_1 + X_2$ is available from Example 4.6-1 (with $b = a$):

$$f_X(w) = \frac{1}{a}\text{tri}\left(\frac{w}{a}\right)$$

where the function tri($\cdot$) is defined in (E-4). The gaussian approximation to W has variance $\sigma_W^2 = 2\sigma_X^2 = a^2/6$ and mean $\bar{W} = 2(a/2) = a$:

$$\text{Approximation to } f_W(w) = \frac{e^{-(w-a)^2/(a^2/3)}}{\sqrt{\pi(a^2/3)}}$$

Figure 4.7-1 illustrates $f_W(w)$ and its gaussian approximation. Even for the case of only two random variables being summed the gaussian approximation is a fairly good one. For other densities the approximation may be very poor (see Problem 4.7-1).

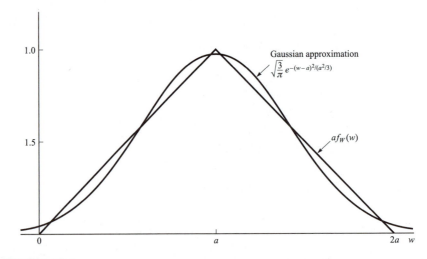

FIGURE 4.7-1
The triangular density function of Example 4.7-1 and its gaussian approximation.

4.8
SUMMARY

One random variable is inadequate to represent many practical problems. The theory of multiple random variables is needed and was developed in this chapter. The main points covered were:

- Multiple (vector) random variables were defined and related through examples to real problems.
- The earlier concepts of joint density and distribution functions, and their properties, were extended to include several random variables.
- Conditional density and distribution functions were developed for several random variables.
- Multiple-variable statistical independence was developed.
- Methods were given to find the distribution and density functions of two or more statistically independent random variables. For some cases the problem can be approximated by the central limit theorem, which is developed in some detail.

This chapter consisted mainly of the extension of the one-variable theory of Chapter 2 to the multiple-random-variable case. The next logical step is to extend the one-variable operations of Chapter 3 to cover several random variables. This extension follows in the next chapter.

PROBLEMS

4.1-1. Two events A and B defined on a sample space S are related to a joint sample space through random variables X and Y and are defined by $A = \{X \leq x\}$ and $B = \{y_1 < Y \leq y_2\}$. Make a sketch of the two sample spaces showing areas corresponding to both events and the event $A \cap B = \{X \leq x, y_1 < Y \leq y_2\}$.

4.1-2. Work Problem 4.1-1 for the two events $A = \{x_1 < X \leq x_2\}$ and $B = \{y_1 < Y \leq y_2\}$.

4.1-3. Work Problem 4.1-1 for the two events $A = \{x_1 < X \leq x_2$ or $x_3 < X \leq x_4\}$ and $B = \{y_1 < Y \leq y_2\}$.

4.1-4. Three events A, B, and C satisfy $C \subset B \subset A$ and are defined by $A = \{X \leq x_a, Y \leq y_a\}$, $B = \{X \leq x_b, Y \leq y_b\}$, and $C = \{X \leq x_c, Y \leq y_c\}$ for two random variables X and Y.
    (a) Sketch the two sample spaces S and S_J and show the regions corresponding to the three events.
    (b) What region corresponds to the event $A \cap B \cap C$?

4.1-5. In a gambling game two fair dice are tossed and the sum of the numbers that show up determines who wins among two players. Random variables X and Y represent the winnings of the first and second numbered players, respectively. The first wins \$3 if the sum is 4, 5, or 6, and loses \$2 if the sum is 11 or 12; he neither wins nor loses for all other sums. The second player wins \$2 for a sum of 8 or more, loses \$3 for a sum of 5 or less, and neither wins nor loses for other sums.

(a) Draw sample spaces S and S_J and show how elements of S map to elements of S_J.

(b) Find the probabilities of all joint outcomes possible in S_J.

4.1-6. Sketch the joint sample space for two random variables X and Y and define the regions that correspond to the events $A = \{Y \leq 2X\}$, $B = \{X \leq 4\}$, and $C = \{Y > -2\}$. Indicate the region defined by $A \cap B \cap C$.

4.2-1. A joint sample space for two random variables X and Y has four elements $(1, 1), (2, 2), (3, 3)$, and $(4, 4)$. Probabilities of these elements are 0.1, 0.35, 0.05, and 0.5, respectively.

(a) Determine through logic and sketch the distribution function $F_{X,Y}(x, y)$.

(b) Find the probability of the event $\{X \leq 2.5, Y \leq 6\}$.

(c) Find the probability of the event $\{X \leq 3\}$.

4.2-2. Write a mathematical equation for $F_{X,Y}(x, y)$ of Problem 4.2-1.

4.2-3. The joint distribution function for two random variables X and Y is

$$F_{X,Y}(x, y) = u(x)u(y)[1 - e^{-ax} - e^{-ay} + e^{-a(x+y)}]$$

where $u(\cdot)$ is the unit-step function and $a > 0$. Sketch $F_{X,Y}(x, y)$.

4.2-4. By use of the joint distribution function in Problem 4.2-3, and assuming $a = 0.5$ in each case, find the probabilities:

(a) $P\{X \leq 1, Y \leq 2\}$

(b) $P\{0.5 < X < 1.5\}$

(c) $P\{-1.5 < X \leq 2, 1 < Y \leq 3\}$.

4.2-5. Find and sketch the marginal distribution functions for the joint distribution function of Problem 4.2-1.

4.2-6. Find and sketch the marginal distribution functions for the joint distribution function of Problem 4.2-3.

4.2-7. Given the function

$$G_{X,Y}(x, y) = u(x)u(y)[1 - e^{-(x+y)}]$$

Show that this function satisfies the first four properties of (4.2-6) but fails the fifth one. The function is therefore not a valid joint probability distribution function.

4.2-8. Random variables X and Y are components of a two-dimensional random vector and have a joint distribution

$$F_{X,Y}(x, y) = \begin{cases} 0 & x < 0 \quad \text{or} \quad y < 0 \\ xy & 0 \leq x < 1 \quad \text{and} \quad 0 \leq y < 1 \\ x & 0 \leq x < 1 \quad \text{and} \quad 1 \leq y \\ y & 1 \leq x \quad \text{and} \quad 0 \leq y < 1 \\ 1 & 1 \leq x \quad \text{and} \quad 1 \leq y \end{cases}$$

(a) Sketch $F_{X,Y}(x, y)$.

(b) Find and sketch the marginal distribution functions $F_X(x)$ and $F_Y(y)$.

4.2-9. Show the function

$$G_{X,Y}(x, y) = \begin{cases} 0 & x < y \\ 1 & x \geq y \end{cases}$$

cannot be a valid joint distribution function. [*Hint:* Use (4.26e).]

4.2-10. Discrete random variables X and Y have a joint distribution function

$$F_{X,Y}(x, y) = 0.10u(x + 4)u(y - 1) + 0.15u(x + 3)u(y + 5)$$
$$+ 0.17u(x + 1)y(y - 3) + 0.05u(x)u(y - 1)$$
$$+ 0.18u(x - 2)u(y + 2) + 0.23u(x - 3)u(y - 4)$$
$$+ 0.12u(x - 4)u(y + 3)$$

Find: (a) the marginal distributions $F_X(x)$ and $F_Y(y)$ and sketch the two functions, (b) $\bar{X}$ and $\bar{Y}$, and (c) the probability $P\{-1 < X \leq 4, -3 < Y \leq 3\}$.

4.2-11. Random variables X and Y have the joint distribution

$$F_{X,Y}(x, y) = \begin{cases} \dfrac{5}{4}\left(\dfrac{x + e^{-(x+1)y^2}}{x + 1} - e^{-y^2}\right)u(y) & 0 \leq x < 4 \\ 0 & x < 0 \text{ or } y < 0 \\ 1 + \dfrac{1}{4}e^{-5y^2} - \dfrac{5}{4}e^{-y^2} & 4 \leq x \text{ and any } y \geq 0 \end{cases}$$

Find: (a) The marginal distribution functions of X and Y, and (b) the probability $P\{3 < X \leq 5, 1 < Y \leq 2\}$.

4.2-12. Find the joint distribution function of the random variables having the joint density of Problem 4.3-16.

4.2-13. The function

$$F_{X,Y}(x, y) = a\left[\frac{\pi}{2} + \tan^{-1}\left(\frac{x}{2}\right)\right]\left[\frac{\pi}{2} + \tan^{-1}\left(\frac{y}{3}\right)\right]$$

is a valid joint distribution function for random variables X and Y if the constant a is chosen properly. What should be the value of a?

4.2-14. Work Problem 4.2-13, except assume the function

$$F_{X,Y}(x, y) = a\left[\frac{\pi}{2} + \frac{\sqrt{3}x}{3 + x^2} + \tan^{-1}\left(\frac{x}{\sqrt{3}}\right)\right]$$
$$\cdot \left[\frac{\pi}{2} + \frac{\sqrt{5}y}{5 + y^2} + \tan^{-1}\left(\frac{y}{\sqrt{5}}\right)\right]$$

4.2-15. Suppose that a pair of random numbers generated by a computer are represented as values of random variables X and Y having the joint distribution function

$$F_{X,Y}(x, y) = \begin{cases} 0 & x < 0 \text{ or } y < 0 \\ \dfrac{27}{26}x\left(1 - \dfrac{x^2}{27}\right) & 0 \le x < 1 \text{ and } 1 \le y \\ \dfrac{27}{26}y\left(1 - \dfrac{y^2}{27}\right) & 1 \le x \text{ and } 0 \le y < 1 \\ \dfrac{27}{26}xy\left(1 - \dfrac{x^2 y^2}{27}\right) & 0 \le x < 1 \text{ and } 0 \le y < 1 \\ 1 & 1 \le x \text{ and } 1 \le y \end{cases}$$

    (a) Determine the marginal distribution functions of X and Y.
    (b) Find the probability of the event $\{0 < X \le 0.5, 0 < Y \le 0.25\}$.

4.3-1. A fair coin is tossed twice. Define random variables by: $X = $ "number of heads on the first toss" and $Y = $ "number of heads on the second toss" (note that X and Y can have only the values 0 or 1).
    (a) Find and sketch the joint density function of X and Y.
    (b) Find and sketch the joint distribution function.

4.3-2. A joint probability density function is

$$f_{X,Y}(x, y) = \begin{cases} 1/ab & 0 < x < a \quad \text{and} \quad 0 < y < b \\ 0 & \text{elsewhere} \end{cases}$$

Find and sketch $F_{X,Y}(x, y)$.

4.3-3. If $a < b$ in Problem 4.3-2, find:
    (a) $P\{X + Y \le 3a/4\}$ (b) $P\{Y \le 2bX/a\}$

4.3-4. Find the joint distribution function applicable to Example 4.3-2.

4.3-5. Sketch the joint density function $f_{X,Y}(x, y)$ applicable to Prolem 4.2-1. Write an equation for $f_{X,Y}(x, y)$.

4.3-6. Determine the joint density and both marginal density functions for Problem 4.2-3.

4.3-7. Find and sketch the joint density function for the distribution function in Problem 4.2-8.

4.3-8. (a) Find a constant b (in terms of a) so that the function

$$f_{X,Y}(x, y) = \begin{cases} be^{-(x+y)} & 0 < x < a \quad \text{and} \quad 0 < y < \infty \\ 0 & \text{elsewhere} \end{cases}$$

    is a valid joint density function.
    (b) Find an expression for the joint distribution function.

4.3-9. (a) By use of the joint density function of Problem 4.3-8, find the marginal density functions.

(b) What is $P\{0.5a < X \le 0.75a\}$ in terms of a and b?

4.3-10. Determine a constant b such that each of the following are valid joint density functions:

(a) $f_{X,Y}(x, y) = \begin{cases} 3xy & 0 < x < 1 \quad \text{and} \quad 0 < y < b \\ 0 & \text{elsewhere} \end{cases}$

(b) $f_{X,Y}(x, y) = \begin{cases} bx(1-y) & 0 < x < 0.5 \quad \text{and} \quad 0 < y < 1 \\ 0 & \text{elsewhere} \end{cases}$

(c) $f_{X,Y}(x, y) = \begin{cases} b(x^2 + 4y^2) & 0 \le |x| < 1 \quad \text{and} \quad 0 \le y < 2 \\ 0 & \text{elsewhere} \end{cases}$

*4.3-11. Given the function

$$f_{X,Y}(x, y) = \begin{cases} (x^2 + y^2)/8\pi & x^2 + y^2 < b \\ 0 & \text{elsewhere} \end{cases}$$

(a) Find a constant b so that this is a valid joint density function.
(b) Find $P\{0.5b < X^2 + Y^2 \le 0.8b\}$. (*Hint:* Use polar coordinates in both parts.)

*4.3-12. On a firing range the coordinates of bullet strikes relative to the target bull's-eye are random variables X and Y having a joint density given by

$$f_{X,Y}(x, y) = \frac{e^{-(x^2+y^2)/2\sigma^2}}{2\pi\sigma^2}$$

Here σ^2 is a constant related to the accuracy of manufacturing a gun's barrel. What value of σ^2 will allow 80% of all bullets to fall inside a circle of diameter 6 cm? (*Hint:* Use polar coordinates.)

4.3-13. Given the function

$$f_{X,Y}(x, y) = \begin{cases} b(x + y)^2 & -2 < x < 2 \quad \text{and} \quad -3 < y < 3 \\ 0 & \text{elsewhere} \end{cases}$$

(a) Find the constant b such that this is a valid joint density function.
(b) Determine the marginal density functions $f_X(x)$ and $f_Y(y)$.

4.3-14. Find a value of the constant b so that the function

$$f_{X,Y}(x, y) = bxy^2 \exp(-2xy)u(x - 2)u(y - 1)$$

is a valid joint probability density.

4.3-15. The locations of hits of darts thrown at a round dartboard of radius r are determined by a vector random variable with components X and Y. The joint density of X and Y is uniform, that is,

$$f_{X,Y}(x, y) = \begin{cases} 1/\pi r^2 & x^2 + y^2 < r^2 \\ 0 & \text{elsewhere} \end{cases}$$

Find the densities of X and Y.

4.3-16. Two random variables X and Y have a joint density

$$f_{X,Y}(x, y) = \tfrac{10}{4}[u(x) - u(x - 4)]u(y) y^3 \exp[-(x + 1) y^2]$$

Find the marginal densities and distributions of X and Y.

4.3-17. Find the marginal densities of X and Y using the joint density

$$f_{X,Y}(x, y) = 2u(x)u(y) \exp\left[-\left(4y + \frac{x}{2}\right)\right]$$

4.3-18. Random variables X and Y have the joint density of Problem 4.3-17. Find the probability that the values of Y are not greater than twice the values of X for $x \le 3$.

4.3-19. The joint density of two random variables X and Y is

$$f_{X,Y}(x, y) = 0.1\delta(x)\delta(y) + 0.12\delta(x - 4)\delta(y)$$
$$+ 0.05\delta(x)\delta(y - 1) + 0.25\delta(x - 2)\delta(y - 1)$$
$$+ 0.3\delta(x - 2)\delta(y - 3) + 0.18\delta(x - 4)\delta(y - 3)$$

Find and plot the marginal distributions of X and Y.

4.3-20. Assume a has the proper value in Problem 4.2-13 and determine the joint density of X and Y. Find the marginal densities of X and Y.

4.3-21. Work Problem 4.3-20 but assume the distribution of Problem 4.2-14.

4.3-22. (a) Find the joint probability density function for the computer-generated numbers of Problem 4.2-15.
(b) Find the marginal densities of X and Y.
(c) Find the probability of the event $\{Y < 1 - X\}$.

4.3-23. The joint density function of random variables X and Y is

$$f_{X,Y}(x, y) = \begin{cases} \dfrac{25}{23ab} \left(\dfrac{y}{a}\right)\left[1 - \left(\dfrac{x}{b}\right)^4\left(\dfrac{y}{a}\right)^3\right] & -b < x < b \text{ and } 0 < y < a \\ 0 & \text{elsewhere} \end{cases}$$

where $a > 0$ and $b > 0$ are constants. Find the marginal densities of X and Y.

4.4-1. Find the conditional density functions $f_X(x|y_1)$, $f_X(x|y_2)$, $f_Y(y|x_1)$, and $f_Y(y|x_2)$ for the joint density defined in Example 4.4-1.

4.4-2. Find the conditional density function $f_X(x|y)$ applicable to Example 4.4-2.

4.4-3. By using the results of Example 4.4-2, calculate the probability of the event $\{Y \le 2|X = 1\}$.

4.4-4. Find the conditional densities $f_X(x|Y = y)$ and $f_Y(y|X = x)$ applicable to the joint density of Problem 4.3-15.

4.4-5. For the joint density of Problem 4.3-16 determine the conditional densities
$f_X(x|Y = y)$ and $f_Y(y|X = x)$.

*4.4-6. The time it takes a person to drive to work is a random variable Y. Because of
traffic, driving time depends on the (random) time of departure, denoted X,
which occurs in an interval of duration T_0 that begins at 7:30 A.M. each day.
There is a minimum driving time T_1 required, regardless of the time of departure. The joint density of X and Y is known to be

$$f_{X,Y}(x, y) = c(y - T_1)^3 u(y - T_1)[u(x) - u(x - T_0)] \exp[-(y - T_1)(x + 1)]$$

where

$$c = (1 + T_0)^3 / 2[(1 + T_0)^3 - 1]$$

(a) Find the average driving time that results when it is given that departure
occurs at 7:30 A.M. Evaluate your results for $T_0 = 1$ h.
(b) Repeat part (a) given that departure time is 7:30 A.M. plus T_0.
(c) What is the average time of departure if $T_0 = 1$ h? (*Hint*: Note that point
conditioning applies.)

*4.4-7. Start with the expressions

$$F_Y(y|B) = P\{Y \le y|B\} = \frac{P\{Y \le y \cap B\}}{P(B)}$$

$$f_Y(y|B) = \frac{dF_Y(y|B)}{dy}$$

which are analogous to (4.4-1) and (4.4-2), and derive $F_Y(y|x_a < X \le x_b)$ and
$f_Y(y|x_a < X \le x_b)$ which are analogous to (4.4-15) and (4.4-16).

*4.4-8. Extend the procedures of the text that lead to (4.4-16) to show that the joint
distribution and density of random variables X and Y, conditional on the
event $B = \{y_a < Y \le y_b\}$, are

$$F_{X,Y}(x, y|y_a < Y \le y_b) = \begin{cases} 0 & y \le y_a \\ \dfrac{F_{X,Y}(x, y) - F_{X,Y}(x, y_a)}{F_Y(y_b) - F_Y(y_a)} & y_a < y \le y_b \\ \dfrac{F_{X,Y}(x, y_b) - F_{X,Y}(x, y_a)}{F_Y(y_b) - F_Y(y_a)} & y_b < y \end{cases}$$

and

$$f_{X,Y}(x, y|y_a < Y \le y_b) = \begin{cases} 0 & y \le y_a \quad \text{and} \quad y > y_b \\ \dfrac{f_{X,Y}(x, y)}{F_Y(y_b) - F_Y(y_a)} & y_a < y \le y_b \end{cases}$$

*4.4-9. Assume that transoceanic aircraft arrive at a random point x (value of random
variable X) within a strip of coastal region of width 10 km centered on a small
city. Aircraft altitude at the time of arrival is not more than 25 km and is a
random variable Y. If X and Y have the joint density of Problem 4.3-23, find
the probability density of arrival altitude, given that aircraft arrive on one side
of the city. Repeat for arrivals on the other side of the city.

*4.4-10. Work Problem 4.4-9 except find the probability density of arrival point X, given that arrival altitude is above 10 km.

4.5-1. Random variables X and Y are *joint gaussian and normalized* if

$$f_{X,Y}(x, y) = \frac{1}{2\pi\sqrt{1 - \rho^2}} \exp\left[-\frac{x^2 - 2\rho xy + y^2}{2(1 - \rho^2)}\right] \qquad \text{where} \quad -1 \le \rho \le 1$$

(*a*) Show that the marginal density functions are

$$f_X(x) = \frac{1}{\sqrt{2\pi}} \exp(-x^2/2) \qquad f_Y(y) = \frac{1}{\sqrt{2\pi}} \exp(-y^2/2)$$

(*Hint*: Complete the square and use the fact that the area under a gaussian density is unity.)

(*b*) Are X and Y statistically independent?

4.5-2. By use of the joint density of Problem 4.5-1, show that

$$f_X(x|Y = y) = \frac{1}{\sqrt{2\pi(1 - \rho^2)}} \exp\left[-\frac{(x - \rho y)^2}{2(1 - \rho^2)}\right]$$

4.5-3. Given the joint distribution function

$$F_{X,Y}(x, y) = u(x)u(y)[1 - e^{-ax} - e^{-ay} + e^{-a(x+y)}]$$

find:
(*a*) The conditional density functions $f_X(x|Y = y)$ and $f_Y(y|X = x)$.
(*b*) Are the random variables X and Y statistically independent?

4.5-4. For two independent random variables X and Y show that

$$P\{Y \le X\} = \int_{-\infty}^{\infty} F_Y(x)f_X(x)\,dx$$

or

$$P\{Y \le X\} = 1 - \int_{-\infty}^{\infty} F_X(y)f_Y(y)\,dy$$

4.5-5. Two random variables X and Y have a joint probability density function

$$f_{X,Y}(x, y) = \begin{cases} \dfrac{5}{16}x^2 y & 0 < y < x < 2 \\ 0 & \text{elsewhere} \end{cases}$$

(*a*) Find the marginal density functions of X and Y.
(*b*) Are X and Y statistically independent?

4.5-6. Determine if random variables X and Y of Problem 4.4-6 are statistically independent.

4.5-7. Determine if X and Y of Problem 4.3-17 are statistically independent.

4.5-8. The joint density of four random variables X_i, $i = 1, 2, 3$, and 4, is

$$f_{X_1,X_2,X_3,X_4}(x_1, x_2, x_3, x_4) = \prod_{i=1}^{4} \exp(-2|x_i|)$$

Find densities
(a) $f_{X_1,X_2,X_3}(x_1, x_2, x_3|x_4)$
(b) $f_{X_1,X_2}(x_1, x_2|x_3, x_4)$, and
(c) $f_{X_1}(x_1|x_2, x_3, x_4)$

4.5-9. Assume that random variables X and Y have the joint density

$$f_{X,Y}(x, y) = \begin{cases} k\cos^2\left(\frac{\pi}{2}xy\right) & -1 < x < 1 \text{ and } -1 < y < 1 \\ 0 & \text{elsewhere} \end{cases}$$

where

$$k = \frac{\pi/2}{\pi + \text{Si}(\pi)} \approx 0.315$$

and the *sine integral* is defined by

$$\text{Si}(x) = \int_0^x \frac{\sin(\xi)}{\xi} d\xi$$

(see Abramowitz and Stegun, 1964). By use of (4.5-4), determine whether X and Y are statistically independent.

4.5-10. Random variables X and Y have the joint density

$$f_{X,Y}(x, y) = \frac{1}{12}u(x)u(y)e^{-(x/4)-(y/3)}$$

Find:
(a) $P\{2 < X \le 4, -1 < Y \le 5\}$ and
(b) $P\{0 < X < \infty, -\infty < Y \le -2\}$

4.6-1. Show, by use of (4.4-13), that the area under $f_Y(y|x)$ is unity.

*4.6-2. Two random variables R and Θ have the joint density function

$$f_{R,\Theta}(r, \theta) = \frac{u(r)[u(\theta) - u(\theta - 2\pi)]r}{2\pi}e^{-r^2/2}$$

(a) Find $P\{0 < R \le 1, 0 < \Theta \le \pi/2\}$.
(b) Find $f_R(r|\Theta = \pi)$.
(c) Find $f_R(r|\Theta \le \pi)$ and compare to the result found in part (b), and explain the comparison.

4.6-3. Random variables X and Y have respective density functions

$$f_X(x) = \frac{1}{a}[u(x) - u(x - a)]$$

$$f_Y(y) = bu(y)e^{-by}$$

where $a > 0$ and $b > 0$. Find and sketch the density function of $W = X + Y$ if X and Y are statistically independent.

4.6-4. Random variables X and Y have respective density functions

$$f_X(x) = 0.1\delta(x-1) + 0.2\delta(x-2) + 0.4\delta(x-3) + 0.3\delta(x-4)$$
$$f_Y(y) = 0.4\delta(y-5) + 0.5\delta(y-6) + 0.1\delta(y-7)$$

Find and sketch the density function of $W = X + Y$ if X and Y are independent.

4.6-5. Find and sketch the density function of $W = X + Y$, where the random variable X is that of Problem 4.6-3 with $a = 5$ and Y is that of Problem 4.6-4. Assume X and Y are independent.

4.6-6. Find the density function of $W = X + Y$, where the random variable X is that of Problem 4.6-4 and Y is that of Problem 4.6-3. Assume A and Y are independent. Sketch the density function for $b = 1$ and $b = 4$.

*4.6-7. Three statistically independent random variables X_1, X_2, and X_3 all have the same density function

$$f_{X_i}(x_i) = \frac{1}{a}[u(x_i) - u(x_i - a)] \qquad i = 1, 2, 3$$

Find and sketch the density function of $Y = X_1 + X_2 + X_3$ if $a > 0$ is constant.

4.6-8. If the difference $W = X - Y$ is formed instead of the sum in (4.6-1), develop the probability density of W. Compare the result with (4.6-5). Is the density still a convolution of the densities of X and Y? Discuss.

4.6-9. Statistically independent random variables X and Y have respective densities

$$f_X(x) = [u(x+12) - u(x-12)][1 - |x/12|]/12$$
$$f_Y(y) = (1/4)u(y)\exp(-y/4)$$

Find the probabilities of the events:
(a) $\{Y \le 8 - (2|X|/3)\}$, and (b) $\{Y \le 8 + (2|X|/3)\}$.
Compare the two results.

4.6-10. Statistically independent random variables X and Y have respective densities

$$f_X(x) = 5u(x)\exp(-5x)$$
$$f_Y(y) = 2u(y)\exp(-2y)$$

Find the density of the sum $W = X + Y$.

*4.6-11. N statistically independent random variables X_i, $i = 1, 2, \ldots, N$, all have the same density

$$f_{X_i}(x_i) = au(x_i)\exp(-ax_i)$$

where $a > 0$ is a constant. Find an expression for the density of the sum $W = X_1 + X_2 + \cdots + X_N$ for any N.

*4.6-12. Statistically independent random variables X and Y have probability densities

$$f_X(x) = \frac{3}{2a^3}[u(x+a) - u(x-a)]x^2 \qquad a > \pi/2$$

$$f_Y(y) = \frac{1}{2} \text{ rect } \left(\frac{y}{\pi}\right) \cos(y)$$

Find the exact probability density of the sum $W = X + Y$.

4.6-13. The probability density functions of two statistically independent random variables X and Y are

$$f_X(x) = \tfrac{1}{2}u(x-1)e^{-(x-1)/2}$$

$$f_Y(y) = \tfrac{1}{4}u(y-3)e^{-(y-3)/4}$$

Find the probability density of the sum $W = X + Y$.

4.6-14. Statistically independent random variables X and Y have probability densities

$$f_X(x) = \begin{cases} \dfrac{3}{32}(4 - x^2) & -2 \leq x \leq 2 \\ 0 & \text{elsewhere in } x \end{cases}$$

$$f_Y(y) = \tfrac{1}{2}[u(y+1) - u(y-1)]$$

Find the exact probability density of the sum $W = X + Y$.

*4.7-1. Find the exact probability density for the sum of two statistically independent random variables each having the density

$$f_X(x) = 3[u(x+a) - u(x-a)]x^2/2a^3$$

where $a > 0$ is a constant. Plot the density along with the gaussian approximation (to the density of the sum) that has variance $2\sigma_X^2$ and mean $2\bar{X}$. Is the approximation a good one?

*4.7-2. Work Problem 4.7-1 except assume

$$f_X(x) = (1/2)\cos(x) \text{ rect } (x/\pi)$$

*4.7-3. Three statistically independent random variables X_1, X_2, and X_3 are defined by

$$\bar{X}_1 = -1 \qquad \sigma_{X_1}^2 = 2.0$$
$$\bar{X}_2 = 0.6 \qquad \sigma_{X_2}^2 = 1.5$$
$$\bar{X}_3 = 1.8 \qquad \sigma_{X_3}^2 = 0.8$$

Write the equation describing the gaussian approximation for the density function of the sum $X = X_1 + X_2 + X_3$. (*Hint:* Refer to the text on the central limit theorem.)

*4.7-4. Two statistically independent random variables X_1 and X_2 have the same probability density given by

$$f_{X_i}(x_i) = \begin{cases} 2x_i/a^2 & 0 \leq x_i < a \\ 0 & \text{elsewhere in } x_i \end{cases}$$

for $i = 1$ and 2, where $a > 0$ is a constant.

(a) Find the exact density of the sum $W = X_1 + X_2$.

(b) Compute the mean and variance of W and find a gaussian approximation for the density of W having the computed mean and variance.

(c) Plot the density of W and the gaussian approximation to see the accuracy of the approximation.

*4.7-5. The probability density functions of statistically independent random variables X and Y are

$$f_X(x) = \begin{cases} \dfrac{2x}{a^2} & 0 \le x < a \\ 0 & \text{elsewhere in } x \end{cases}$$

$$f_Y(y) = bu(y)e^{-by}$$

where $a > 0$ and $b > 0$ are constants.

(a) Find the probability density function of the sum $W = X + Y$.

(b) Find a gaussian approximation for W that has the same mean and variance as W.

(c) Plot the approximation and the density of W for products $ab = 1/2$, 1, and 2.

CHAPTER 5

Operations on Multiple Random Variables

5.0
INTRODUCTION

After establishing some of the basic theory of several random variables in the previous chapter, it is appropriate to now extend the operations described in Chapter 3 to include multiple random variables. This chapter is dedicated to these extensions. Mainly, the concept of expectation is enlarged to include two or more random variables. Other operations involving moments, characteristic functions, and transformations are all special applications of expectation.

5.1
EXPECTED VALUE OF A FUNCTION OF RANDOM VARIABLES

When more than a single random variable is involved, expectation must be taken with respect to all the variables involved. For example, if $g(X, Y)$ is some function of two random variables X and Y the expected value of $g(\cdot, \cdot)$ is given by

$$\bar{g} = E[g(X, Y)] = \int_{-\infty}^{\infty} \int_{-\infty}^{\infty} g(x, y) f_{X, Y}(x, y) \, dx \, dy \qquad (5.1\text{-}1)$$

This expression is the two-variable extension of (3.1-6).

For N random variables $X_1, X_2, \ldots, X_N$ and some function of these variables, denoted $g(X_1, \ldots, X_N)$, the expected value of the function becomes

$$\bar{g} = E[g(X_1, \ldots, X_N)]$$

$$= \int_{-\infty}^{\infty} \cdots \int_{-\infty}^{\infty} g(x_1, \ldots, x_N) f_{X_1,\ldots,X_N}(x_1, \ldots, x_N) \, dx_1 \ldots dx_N \qquad (5.1\text{-}2)$$

Thus, expectation in general involves an N-fold integration when N random variables are involved. It should be clear to the reader from (5.1-2) that the expected value of a *sum of functions* is equal to the sum of the expected values of the individual functions.

We illustrate the application of (5.1-2) with an example that will develop an important point.

EXAMPLE 5.1-1. We shall find the mean (expected) value of a sum of N weighted random variables. If we let

$$g(X_1, \ldots, X_N) = \sum_{i=1}^{N} \alpha_i X_i$$

where the "weights" are the constants α_i, the mean value of the weighted sum becomes

$$E[g(X_1, \ldots, X_N) = E\left[\sum_{i=1}^{N} \alpha_i X_i\right]$$

$$= \sum_{i=1}^{N} \int_{-\infty}^{\infty} \cdots \int_{-\infty}^{\infty} \alpha_i x_i f_{X_1,\ldots,X_N}(x_1, \ldots, x_N) \, dx_1 \ldots dx_N$$

from (5.1-2). After using (4.3-8), the terms in the sum all reduce to the form

$$\int_{-\infty}^{\infty} \alpha_i x_i f_{X_i}(x_i) \, dx_i = E[\alpha_i X_i] = \alpha_i E[X_i]$$

so

$$E\left[\sum_{i=1}^{N} \alpha_i X_i\right] = \sum_{i=1}^{N} \alpha_i E[X_i]$$

which says that *the mean value of a weighted sum of random variables equals the weighted sum of mean values.*

The above extensions (5.1-1) and (5.1-2) of expectation do not invalidate any of our single random variable results. For example, let

$$g(X_1, \ldots, X_N) = g(X_1) \qquad (5.1\text{-}3)$$

and substitute into (5.1-2). After integrating with respect to all random variables except X_1, (5.1-2) becomes

$$\bar{g} = E[g(X_1)] = \int_{-\infty}^{\infty} g(x_1) f_{X_1}(x_1) \, dx_1 \qquad (5.1\text{-}4)$$

which is the same as previously given in (3.1-6) for one random variable. Some reflection on the reader's part will verify that (5.1-4) also validates such earlier topics as moments, central moments, characteristic function, etc., for a single random variable.

Joint Moments about the Origin

One important application of (5.1-1) is in defining *joint moments* about the origin. They are denoted by m_{nk} are are defined by

$$m_{nk} = E[X^n Y^k] = \int_{-\infty}^{\infty} \int_{-\infty}^{\infty} x^n y^k f_{X,Y}(x, y) \, dx \, dy \qquad (5.1\text{-}5)$$

for the case of two random variables X and Y. Clearly $m_{n0} = E[X^n]$ are the moments m_n of X, while $m_{0k} = E[Y^k]$ are the moments of Y. The sum $n + k$ is called the *order* of the moments. Thus m_{02}, m_{20}, and m_{11} are all second-order moments of X and Y. The first-order moments $m_{01} = E[Y] = \bar{Y}$ and $m_{10} = E[X] = \bar{X}$ are the expected values of Y and X, respectively, and are the co-ordinates of the "center of gravity" of the function $f_{X,Y}(x, y)$.

The second-order moment $m_{11} = E[XY]$ is called the *correlation* of X and Y. It is so important to later work that we give it the symbol R_{XY}. Hence,

$$R_{XY} = m_{11} = E[XY] = \int_{-\infty}^{\infty} \int_{-\infty}^{\infty} xy f_{X,Y}(x, y) \, dx \, dy \qquad (5.1\text{-}6)$$

If correlation can be written in the form

$$R_{XY} = E[X]E[Y] \qquad (5.1\text{-}7)$$

then X and Y are said to be *uncorrelated*. Statistical independence of X and Y is sufficient to guarantee they are uncorrelated, as is readily proven by (5.1-6) using (4.5-4). The converse of this statement, that is, that X and Y are independent if X and Y are uncorrelated, is *not* necessarily true in general.†

If

$$R_{XY} = 0 \qquad (5.1\text{-}8)$$

for two random variables X and Y, they are called *orthogonal*.

A simple example is next developed that illustrates the important new topic of correlation.

> **EXAMPLE 5.1-2.** Let X be a random variable that has a mean value $\bar{X} = E[X] = 3$ and variance $\sigma_X^2 = 2$. From (3.2-6) we easily determine the second moment of X about the origin: $E[X^2] = m_{20} = \sigma_X^2 + \bar{X}^2 = 11$.
>
> Next, let another random variable Y be defined by
>
> $$Y = -6X + 22$$

---

†Uncorrelated *gaussian* random variables are, however, known to also be independent (see Section 5.3).

The mean value of Y is $\bar{Y} = E[Y] = E[-6X + 22] = -6\bar{X} + 22 = 4$. The correlation of X and Y is found from (5.1-6)

$$R_{XY} = m_{11} = E[XY] = E[-6X^2 + 22X] = -6E[X^2] + 22\bar{X}$$
$$= -6(11) + 22(3) = 0$$

Since $R_{XY} = 0$, X and Y are orthogonal from (5.1-8). On the other hand, $R_{XY} \neq E[X]E[Y] = 12$, so X and Y are *not* uncorrelated [see (5.1-7)].

We note that two random variables can be orthogonal even though correlated when one, Y, is related to the other, X, by the linear function $Y = aX + b$. It can be shown that X and Y are always correlated if $|a| \neq 0$, regardless of the value of b (see Problem 5.1-9). They are uncorrelated if $a = 0$, but this is not a case of much practical interest. Orthogonality can likewise be shown to occur when a and b are related by $b = -aE[X^2]/E[X]$ whenever $E[X] \neq 0$. If $E[X] = 0$, X and Y cannot be orthogonal for any value of a except $a = 0$, a noninteresting problem. The reader may wish to verify these statements as an exercise.

For N random variables $X_1, X_2, \ldots, X_N$, the $(n_1 + n_2 + \cdots + n_N)$-order joint moments are defined by

$$m_{n_1 n_2 \ldots n_N} = E[X_1^{n_1} X_2^{n_2} \ldots X_N^{n_N}]$$
$$= \int_{-\infty}^{\infty} \cdots \int_{-\infty}^{\infty} X_1^{n_1} \ldots X_N^{n_N} f_{X_1,\ldots,X_N}(x_1, \ldots, x_N) \, dx_1 \ldots dx_N \tag{5.1-9}$$

where $n_1, n_2, \ldots, n_N$ are all integers $= 0, 1, 2, \ldots$.

Joint Central Moments

Another important application of (5.1-1) is in defining *joint central moments*. For two random variables X and Y, these moments, denoted by μ_{nk}, are given by

$$\mu_{nk} = E[(X - \bar{X})^n (Y - \bar{Y})^k]$$
$$= \int_{-\infty}^{\infty} \int_{-\infty}^{\infty} (x - \bar{X})^n (y - \bar{Y})^k f_{X,Y}(x, y) \, dx \, dy \tag{5.1-10}$$

The second-order central moments

$$\mu_{20} = E[(X - \bar{X})^2] = \sigma_X^2 \tag{5.1-11}$$
$$\mu_{02} = E[(Y - \bar{Y})^2] = \sigma_Y^2 \tag{5.1-12}$$

are just the variances of X and Y.

The second-order joint moment μ_{11} is very important. It is called the *covariance* of X and Y and is given the symbol C_{XY}. Hence

$$C_{XY} = \mu_{11} = E[(X - \bar{X}))Y - \bar{Y})]$$
$$= \int_{-\infty}^{\infty} \int_{-\infty}^{\infty} (x - \bar{X})(y - \bar{Y}) f_{X,Y}(x, y) \, dx \, dy \tag{5.1-13}$$

By direct expansion of the product $(x - \bar{X})(y - \bar{Y})$, this integral reduces to the form

$$C_{XY} = R_{XY} - \bar{X}\bar{Y} = R_{XY} - E[X]E[Y] \tag{5.1-14}$$

when (5.1-6) is used. If X and Y are either independent or uncorrelated, then (5.1-7) applies and (5.1-14) shows their covariance is zero:

$$C_{XY} = 0 \qquad X \text{ and } Y \text{ independent or uncorrelated} \tag{5.1-15}$$

If X and Y are orthogonal random variables, then

$$C_{XY} = -E[X]E[Y] \qquad X \text{ and } Y \text{ orthogonal} \tag{5.1-16}$$

from use of (5.1-8) with (5.1-14). Clearly, $C_{XY} = 0$ if either X or Y also has zero mean value.

The normalized second-order moment

$$\rho = \mu_{11}/\sqrt{\mu_{20}\mu_{02}} = C_{XY}/\sigma_X\sigma_Y \tag{5.1-17a}$$

given by

$$\rho = E\left[\frac{(X - \bar{X})}{\sigma_X} \frac{(Y - \bar{Y})}{\sigma_Y}\right] \tag{5.1-17b}$$

is known as the *correlation coefficient* of X and Y. It can be shown (see Problem 5.1-10) that

$$-1 \leq \rho \leq 1 \tag{5.1-18}$$

For N random variables $X_1, X_2, \ldots, X_N$ the $(n_1 + n_2 + \cdots + n_N)$-order joint central moment is defined by

$$\mu_{n_1 n_2 \ldots n_N} = E[(X_1 - \bar{X}_1)^{n_1}(X_2 - \bar{X}_2)^{n_2} \ldots (X_N - \bar{X}_N)^{n_N}]$$

$$= \int_{-\infty}^{\infty} \cdots \int_{-\infty}^{\infty} (x_1 - \bar{X}_1)^{n_1} \ldots$$

$$(x_N - \bar{X}_N)^{n_N} f_{X_1,\ldots,X_N}(x_1, \ldots, x_N) \, dx_1 \ldots x_N \tag{5.1-19}$$

An example is next developed that involves the use of covariances.

EXAMPLE 5.1-3. Again let X be a weighted sum of N random variables X_i; that is, let

$$X = \sum_{i=1}^{N} \alpha_i X_i$$

where the α_i are real weighting constants. The variance of X will be found. From Example 5.1-1,

$$E[X] = \sum_{i=1}^{N} \alpha_i E[X_i] = \sum_{i=1}^{N} \alpha_i \bar{X}_i = \bar{X}$$

so we have

$$X - \bar{X} = \sum_{i=1}^{N} \alpha_i (X_i - \bar{X}_i)$$

and

$$\sigma_X^2 = E[(X - \bar{X})^2] = E\left[\sum_{i=1}^{N} \alpha_i (X_i - \bar{X}_i) \sum_{j=1}^{N} \alpha_j (X_j - \bar{X}_j)\right]$$

$$= \sum_{i=1}^{N} \sum_{j=1}^{N} \alpha_i \alpha_j E[(X_i - \bar{X}_i)(X_j - \bar{X}_j)] = \sum_{i=1}^{N} \sum_{j=1}^{N} \alpha_i \alpha_j C_{X_i X_j}$$

Thus, the variance of a weighted sum of N random variables X_i (weights α_i) equals the weighted sum of all their covariances $C_{X_i X_j}$ (weights $\alpha_i \alpha_j$). For the special case of uncorrelated random variables, where

$$C_{X_i X_j} = \begin{cases} 0 & i \neq j \\ \sigma_{X_i}^2 & i = j \end{cases}$$

is true, we get

$$\sigma_X^2 = \sum_{i=1}^{N} \alpha_i^2 \sigma_{X_i}^2$$

In words: *the variance of a weighted sum of uncorrelated random variables (weights α_i) equals the weighted sum of the variances of the random variables (weights α_i^2).*

*5.2
JOINT CHARACTERISTIC FUNCTIONS

The *joint characteristic function* of two random variables X and Y is defined by

$$\Phi_{X,Y}(\omega_1, \omega_2) = E[e^{j\omega_1 X + j\omega_2 Y}] \qquad (5.2\text{-}1)$$

where ω_1 and ω_2 are real numbers. An equivalent form is

$$\Phi_{X,Y}(\omega_1, \omega_2) = \int_{-\infty}^{\infty} \int_{-\infty}^{\infty} f_{X,Y}(x, y) e^{j\omega_1 x + j\omega_2 y} \, dx \, dy \qquad (5.2\text{-}2)$$

This expression is recognized as the two-dimensional Fourier transform (with signs of ω_1 and ω_2 reversed) of the joint density function. From the inverse Fourier transform we also have

$$f_{X,Y}(x, y) = \frac{1}{(2\pi)^2} \int_{-\infty}^{\infty} \int_{-\infty}^{\infty} \Phi_{X,Y}(\omega_1, \omega_2) e^{-j\omega_1 x - j\omega_2 y} \, d\omega_1 \, d\omega_2 \qquad (5.2\text{-}3)$$

By setting either $\omega_2 = 0$ or $\omega_1 = 0$ in (5.2-2), the characteristic functions of
X or Y are obtained. They are called *marginal characteristic functions*:

$$\Phi_X(\omega_1) = \Phi_{X,Y}(\omega_1, 0) \tag{5.2-4}$$

$$\Phi_Y(\omega_2) = \Phi_{X,Y}(0, \omega_2) \tag{5.2-5}$$

Joint moments m_{nk} can be found from the joint characteristic function as follows:

$$m_{nk} = (-j)^{n+k} \frac{\partial^{n+k} \Phi_{X,Y}(\omega_1, \omega_2)}{\partial \omega_1^n \partial \omega_2^k} \bigg|_{\omega_1=0, \omega_2=0} \tag{5.2-6}$$

This expression is the two-dimensional extension of (3.3-4).

EXAMPLE 5.2-1. Two random variables X and Y have the joint characteristic function

$$\Phi_{X,Y}(\omega_1, \omega_2) = \exp(-2\omega_1^2 - 8\omega_2^2)$$

We show that X and Y are both zero-mean random variables and that they are uncorrelated.

The means derive from (5.2-6):

$$\bar{X} = E[X] = m_{10} = -j \frac{\partial \Phi_{X,Y}(\omega_1, \omega_2)}{\partial \omega_1} \bigg|_{\omega_1=0, \omega_2=0}$$

$$= -j(-4\omega_1) \exp(-2\omega_1^2 - 8\omega_2^2) \bigg|_{\omega_1=0, \omega_2=0} = 0$$

$$\bar{Y} = E[Y] = m_{01} = -j(-16\omega_2) \exp(-2\omega_1^2 - 8\omega_2^2) \bigg|_{\omega_1=0, \omega_2=0} = 0$$

Also from (5.2-6);

$$R_{XY} = E[XY] = m_{11} = (-j)^2 \frac{\partial^2}{\partial \omega_1 \partial \omega_2} [\exp(-2\omega_1^2 - 8\omega_2^2)] \bigg|_{\omega_1=0, \omega_2=0}$$

$$= -(-4\omega_1)(-16\omega_2) \exp(-2\omega_1^2 - 8\omega_2^2) \bigg|_{\omega_1=0, \omega_2=0} = 0$$

Since means are zero, $C_{XY} = R_{XY}$ from (5.1-14). Therefore, $C_{XY} = 0$ and X and Y are uncorrelated.

The joint characteristic function for N random variables $X_1, X_2, \ldots, X_N$ is defined by

$$\Phi_{X_1, \ldots, X_N}(\omega_1, \ldots, \omega_N) = E[e^{j\omega_1 X_1 + \cdots + j\omega_N X_N}] \tag{5.2-7}$$

Joint moments are obtained from

$$m_{n_1 n_2 \ldots n_N} = (-j)^R \frac{\partial^R \Phi_{X_1, \ldots, X_N}(\omega_1, \ldots, \omega_N)}{\partial \omega_1^{n_1} \partial \omega_2^{n_2} \ldots \partial \omega_N^{n_N}} \bigg|_{\text{all } \omega_i=0} \tag{5.2-8}$$

where

$$R = n_1 + n_2 + \cdots + n_N \qquad (5.2\text{-}9)$$

The joint characteristic function is especially useful in certain practical problems where the probability density function is needed for the sum of N statistically independent random variables. We use an example to show how the desired probability density is found.

EXAMPLE 5.2-2. Let $Y = X_1 + X_2 + \cdots + X_N$ be the sum of N statistically independent random variables X_i, $i = 1, 2, \ldots, N$. Denote their probability densities and characteristic functions, respectively, by $f_{X_i}(x_i)$ and $\Phi_{X_i}(\omega_i)$. Because of independence the joint probability density is the product of all the individual densities and (5.2-7) can be written as

$$\Phi_{X_1,\ldots,X_N}(\omega_1,\ldots,\omega_N)$$

$$= \int_{-\infty}^{\infty} \cdots \int_{-\infty}^{\infty} \left[\prod_{i=1}^{N} f_{X_i}(x_i) \right] \exp\left[j \sum_{i=1}^{N} \omega_i x_i \right] dx_1 \ldots dx_N$$

$$= \prod_{i=1}^{N} \int_{-\infty}^{\infty} f_{X_i}(x_i) e^{j\omega_i x_i}\, dx_i = \prod_{i=1}^{N} \Phi_{X_i}(\omega_i)$$

Next, we write the characteristic function of Y using (3.3-1) and note that it is the same as (5.2-7) with $\omega_i = \omega$, all i. Hence,

$$\Phi_Y(\omega) = E[e^{j\omega Y}] = E\left[\exp\left(j \sum_{i=1}^{N} \omega X_i \right) \right]$$

$$= \Phi_{X_1,\ldots,X_N}(\omega,\ldots,\omega) = \prod_{i=1}^{N} \Phi_{X_i}(\omega)$$

Finally, we use (3.3-3) to obtain the desired density of Y:

$$f_Y(y) = \frac{1}{2\pi} \int_{-\infty}^{\infty} \left[\prod_{i=1}^{N} \Phi_{X_i}(\omega) \right] e^{-j\omega y}\, d\omega$$

In the special case where the X_i are identically distributed such that $\Phi_{X_i}(\omega) = \Phi_X(\omega)$, all i, our result reduces to

$$f_Y(y) = \frac{1}{2\pi} \int_{-\infty}^{\infty} [\Phi_X(\omega)]^N e^{-j\omega y}\, d\omega$$

5.3
JOINTLY GAUSSIAN RANDOM VARIABLES

Gaussian random variables are very important because they show up in nearly every area of science and engineering. In this section, the case of two gaussian random variables is first examined. The more advanced case of N random variables is then introduced.

Two Random Variables

149

CHAPTER 5:
Operations on
Multiple Random
Variables

Two random variables X and Y are said to be *jointly gaussian* if their joint density function is of the form

$$f_{X,Y}(x, y) = \frac{1}{2\pi\sigma_X\sigma_Y\sqrt{1 - \rho^2}}$$

$$\cdot \exp\left\{\frac{-1}{2(1 - \rho^2)}\left[\frac{(x - \bar{X})^2}{\sigma_X^2} - \frac{2\rho(x - \bar{X})(y - \bar{Y})}{\sigma_X\sigma_Y} + \frac{(y - \bar{Y})^2}{\sigma_Y^2}\right]\right\}$$

$$(5.3-1)$$

which is sometimes called the *bivariate gaussian density*. Here

$$\bar{X} = E[X] \tag{5.3-2}$$

$$\bar{Y} = E[Y] \tag{5.3-3}$$

$$\sigma_X^2 = E[(X - \bar{X})^2] \tag{5.3-4}$$

$$\sigma_Y^2 = E[(Y - \bar{Y})^2] \tag{5.3-5}$$

$$\rho = E[(X - \bar{X})(Y - \bar{Y})]/\sigma_X\sigma_Y \tag{5.3-6}$$

Figure 5.3-1*a* illustrates the appearance of the joint gaussian density function (5.3-1). Its maximum is located at the point $(\bar{X}, \bar{Y})$. The maximum value is obtained from

$$f_{X,Y}(x, y) \leq f_{X,Y}(\bar{X}, \bar{Y}) = \frac{1}{2\pi\sigma_X\sigma_Y\sqrt{1 - \rho^2}} \tag{5.3-7}$$

The locus of constant values of $f_{X,Y}(x, y)$ will be an ellipse† as shown in Figure 5.3-1*b*. This is equivalent to saying that the line of intersection formed by slicing the function $f_{X,Y}(x, y)$ with a plane parallel to the xy plane is an ellipse.

Observe that if $\rho = 0$, corresponding to uncorrelated X and Y, (5.3-1) can be written as

$$f_{X,Y}(x, y) = f_X(x)f_Y(y) \tag{5.3-8}$$

where $f_X(x)$ and $f_Y(y)$ are the marginal density functions of X and Y given by

$$f_X(x) = \frac{1}{\sqrt{2\pi\sigma_X^2}} \exp\left[-\frac{(x - \bar{X})^2}{2\sigma_X^2}\right] \tag{5.3-9}$$

$$f_Y(y) = \frac{1}{\sqrt{2\pi\sigma_Y^2}} \exp\left[-\frac{(y - \bar{Y})^2}{2\sigma_Y^2}\right] \tag{5.3-10}$$

Now the form of (5.3-8) is sufficient to guarantee that X and Y are statistically independent. Therefore, we conclude that *any uncorrelated gaussian random*

---

†When $\sigma_X = \sigma_Y$ and $\rho = 0$ the ellipse degenerates into a circle; when $\rho = +1$ or -1 the ellipses degenerate into axes rotated by angles $\pi/4$ and $-\pi/4$ respectively that pass through the point $(\bar{X}, \bar{Y})$.

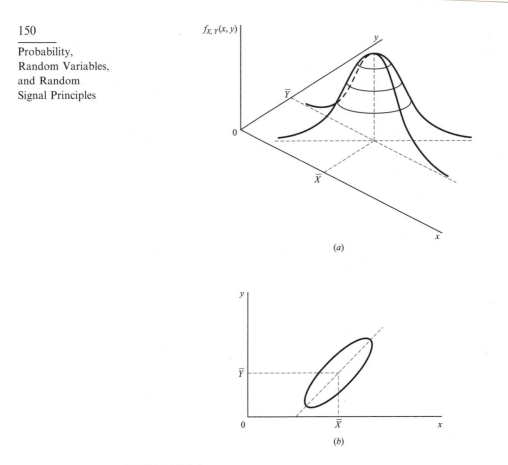

$f_{X,Y}(x,y)$

(a)

(b)

FIGURE 5.3-1
Sketch of the joint density function of two gaussian random variables.

variables are also statistically independent. It results that a coordinate rotation (linear transformation of X and Y) through an angle

$$\theta = \frac{1}{2} \tan^{-1}\left[\frac{2\rho\sigma_X\sigma_Y}{\sigma_X^2 - \sigma_Y^2}\right] \tag{5.3-11}$$

is sufficient to convert correlated random variables X and Y, having variances σ_X^2 and σ_Y^2, respectively, correlation coefficient ρ, and the joint density of (5.3-1), into two statistically independent gaussian random variables.[†]

By direct application of (4.4-12) and (4.4-13), the conditional density functions $f_X(x|Y = y)$ and $f_Y(y|X = x)$ can be found from the above expressions (see Problem 5.3-2).

---

†Wozencraft and Jacobs (1965), p. 155.

EXAMPLE 5.3-1. We show by example that (5.3-11) applies to arbitrary as well as gaussian random variables. Consider random variables Y_1 and Y_2 related to arbitrary random variables X and Y by the coordinate rotation

$$Y_1 = X\cos(\theta) + Y\sin(\theta)$$
$$Y_2 = -X\sin(\theta) + Y\cos(\theta)$$

If $\bar{X}$ and $\bar{Y}$ are the means of X and Y, respectively, the means of Y_1 and Y_2 are clearly $\bar{Y}_1 = \bar{X}\cos(\theta) + \bar{Y}\sin(\theta)$ and $\bar{Y}_2 = -\bar{X}\sin(\theta) + \bar{Y}\cos(\theta)$, respectively. The covariance of Y_1 and Y_2 is

$$\begin{aligned}
C_{Y_1 Y_2} &= E[(Y_1 - \bar{Y}_1)(Y_2 - \bar{Y}_2)] \\
&= E[\{(X - \bar{X})\cos(\theta) + (Y - \bar{Y})\sin(\theta)\} \\
&\quad \cdot \{-(X - \bar{X})\sin(\theta) + (Y - \bar{Y})\cos(\theta)\}] \\
&= (\sigma_Y^2 - \sigma_X^2)\sin(\theta)\cos(\theta) + C_{XY}[\cos^2(\theta) - \sin^2(\theta)] \\
&= (\sigma_Y^2 - \sigma_X^2)(\tfrac{1}{2})\sin(2\theta) + C_{XY}\cos(2\theta)
\end{aligned}$$

Here $C_{XY} = E[(X - \bar{X})(Y - \bar{Y})] = \rho\sigma_X\sigma_Y$. If we require Y_1 and Y_2 to be uncorrelated, we must have $C_{Y_1 Y_2} = 0$. By equating the above equation to zero we obtain (5.3-11). Thus, (5.3-11) applies to arbitrary as well as gaussian random variables.

*N Random Variables

N random variables $X_1, X_2, \ldots, X_N$ are called *jointly gaussian* if their joint density function can be written as†

$$f_{X_1,\ldots,X_N}(x_1, \ldots, x_N) = \frac{|[C_X]^{-1}|^{1/2}}{(2\pi)^{N/2}} \exp\left\{-\frac{[x - \bar{X}]^t [C_X]^{-1} [x - \bar{X}]}{2}\right\} \quad (5.3\text{-}12)$$

where we define matrices

$$[x - \bar{X}] = \begin{bmatrix} x_1 - \bar{X}_1 \\ x_2 - \bar{X}_2 \\ \vdots \\ x_N - \bar{X}_N \end{bmatrix} \quad (5.3\text{-}13)$$

and

$$[C_X] = \begin{bmatrix} C_{11} & C_{12} & \cdots & C_{1N} \\ C_{21} & C_{22} & \cdots & C_{2N} \\ \vdots & \vdots & & \vdots \\ C_{N1} & C_{N2} & \cdots & C_{NN} \end{bmatrix} \quad (5.3\text{-}14)$$

---

†We denote a matrix symbolically by use of heavy brackets $[\cdot]$.

We use the notation $[\cdot]^t$ for the matrix transpose, $[\cdot]^{-1}$ for the matrix inverse, and $|[\cdot]|$ for the determinant. Elements of $[C_X]$, called the *covariance matrix* of the N random variables, are given by

$$C_{ij} = E[(X_i - \bar{X}_i)(X_j - \bar{X}_j)] = \begin{cases} \sigma_{X_i}^2 & i = j \\ C_{X_i X_j} & i \neq j \end{cases} \quad (5.3\text{-}15)$$

The density (5.3-12) is often called the *N-variate gaussian density* function. For the special case where $N = 2$, the covariance matrix becomes

$$[C_X] = \begin{bmatrix} \sigma_{X_1}^2 & \rho\sigma_{X_1}\sigma_{X_2} \\ \rho\sigma_{X_1}\sigma_{X_2} & \sigma_{X_2}^2 \end{bmatrix} \quad (5.3\text{-}16)$$

so

$$[C_X]^{-1} = \frac{1}{(1-\rho^2)} \begin{bmatrix} 1/\sigma_{X_1}^2 & -\rho/\sigma_{X_1}\sigma_{X_2} \\ -\rho/\sigma_{X_1}\sigma_{X_2} & 1/\sigma_{X_2}^2 \end{bmatrix} \quad (5.3\text{-}17)$$

$$|[C_X]^{-1}| = 1/\sigma_{X_1}^2\sigma_{X_2}^2(1-\rho^2) \quad (5.3\text{-}18)$$

On substitution of (5.3-17) and (5.3-18) into (5.3-12), and letting $X_1 = X$ and $X_2 = Y$, it is easy to verify that the bivariate density of (5.3-1) results.

*Some Properties of Gaussian Random Variables

We state without proof some of the properties exhibited by N jointly gaussian random variables $X_1, \ldots, X_N$.

1. Gaussian random variables are completely defined through only their first- and second-order moments; that is, by their means, variances, and covariances. This fact is readily apparent since only these quantities are needed to completely determine (5.3-12).
2. If the random variables are uncorrelated, they are also statistically independent. This property was given earlier for two variables.
3. Random variables produced by a linear transformation of $X_1, \ldots, X_N$ will also be gaussian, as will be proven in Section 5.5.
4. Any k-dimensional (k-variate) marginal density function obtained from the N-dimensional density function (5.3-12) by integrating out $N - k$ random variables will be gaussian. If the variables are ordered so that $X_1, \ldots, X_k$ occur in the marginal density and $X_{k+1} \ldots, X_N$ are integrated out, then the covariance matrix of $X_1, \ldots, X_k$ is equal to the leading $k \times k$ submatrix of the covariance matrix of $X_1, \ldots, X_N$ (Wilks, 1962, p. 168).
5. The conditional density $f_{X_1,\ldots,X_k}(x_1, \ldots, x_k | X_{k+1} = x_{k+1}, \ldots, X_N = x_N)$ is gaussian (Papoulis, 1965, p. 257). This holds for any $k < N$.

*5.4
TRANSFORMATIONS OF MULTIPLE RANDOM VARIABLES

153

CHAPTER 5:
Operations on
Multiple Random
Variables

The function g in either (5.1-1) or (5.1-2) can be considered a transformation involving more than one random variable. By defining a new variable $Y = g(X_1, X_2, \ldots, X_N)$, we see that (5.1-2) is the expected value of Y. In calculating expected values it was not necessary to determine the density function of the new random variable Y. It may be, however, that the density function Y is required in some practical problems, and its determination is briefly considered in this section. First we consider a single functional transformation of more than one random variable. Then we develop the case of several functions of several random variables.

*One Function

Here $Y = g(X_1, X_2, \ldots, X_N)$. We seek to first define the probability distribution of Y and then the probability density. The distribution is $F_Y(y) = P\{Y \le y\} = P\{g(X_1, X_2, \ldots, X_N) \le y\}$. This probability is associated with all points in the $(x_1, x_2, \ldots, x_N)$ hyperspace that map such that $g(x_1, x_2, \ldots, x_N) \le y$ for any y. Formally, we integrate over all such points according to

$$F_Y(y) = P\{g(X_1, X_2, \ldots, X_N) \le y\}$$

$$= \int \cdots \int f_{X_1, X_2, \ldots, X_N}(x_1, x_2, \ldots, x_N) \, dx_1 dx_2 \ldots dx_N \qquad (5.4\text{-}1)$$

$$\{g(x_1, x_2, \ldots, x_N) \le y\}$$

The density follows differentiation

$$f_Y(y) = \frac{dF_Y(y)}{dy}$$

$$= \frac{d}{dy} \int \cdots \int f_{X_1, X_2, \ldots, X_N}(x_1, x_2, \ldots, x_N) \, dx_1 dx_2 \ldots dx_N \qquad (5.4\text{-}2)$$

$$\{g(x_1, x_2, \ldots, x_N) \le y\}$$

Perhaps the use of (5.4-1) and (5.4-2) is best demonstrated by example. We take two cases.

EXAMPLE 5.4-1. We find the density function for the ratio $Y = g(X_1, X_2) = X_1/X_2$ of two positive random variables X_1 and X_2. The event $\{Y = X_1/X_2 \le y\}$ corresponds to points $0 < x_1/x_2 \le y$ in the $x_1 x_2$ plane as shown shaded in Figure 5.4-1. Now the distribution of Y is $F_Y(y) = P\{Y = X_1/X_2 \le y\}$; this probability equals the integral of the joint density of X_1 and X_2 over the shaded areas. On integrating, when using the horizontal strip as shown, we have (for $y > 0$)

$$F_Y(y) = P\{X_1/X_2 \le y\} = \int_0^\infty \cdots \int_0^{yx_2} f_{X_1, X_2}(x_1, x_2) \, dx_1 dx_2$$

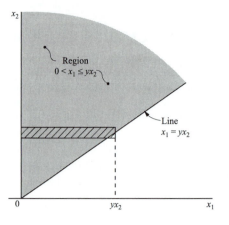

FIGURE 5.4-1
Regions in $x_1 x_2$ plane applicable to Example 5.4-1.

The density results from differentiation according to (2.3-1) and Leibniz's rule

$$f_Y(y) = \frac{dF_Y(y)}{dy} = \int_0^\infty x_2 f_{X_1, X_2}(yx_2, x_2)\, dx_2$$

To progress further requires a specific density be specified.

EXAMPLE 5.4-2. As a second example consider the function $Y = (X_1^2 + X_2^2)^{1/2}$, which is instructive because it involves using Leibniz's rule with a double integral. Here $F_Y(y) = P\{Y = (X_1^2 + X_2^2)^{1/2} \leq y\}$ is the probability of all points in the $x_1 x_2$ plane that fall on, and inside, a circle of radius y.

It is

$$F_Y(y) = \int_{x_2=-y}^{y} \int_{x_1=-(y^2-x_2^2)^{1/2}}^{(y^2-x_2^2)^{1/2}} f_{X_1, X_2}(x_1, x_2)\, dx_1 dx_2$$

$$= \int_{-y}^{y} I(y, x_2)\, dx_2 \qquad (1)$$

where we define

$$I(y, x_2) = \int_{-(y^2-x_2^2)^{1/2}}^{(y^2-x_2^2)^{1/2}} f_{X_1, X_2}(x_1, x_2)\, dx_1 \qquad (2)$$

From Leibniz's rule applied to the last form of (1):

$$f_Y(y) = \frac{dF_Y(y)}{dy} = I(y, y) + I(y, -y) + \int_{-y}^{y} \frac{\partial I(y, x_2)}{\partial y}\, dx_2 \qquad (3)$$

Direct use of (2) proves the first two right-side terms in (3) are zero. On applying Leibniz's rule to the last term in (3), we have

$$f_Y(y) = \int_{-y}^{y} \left\{ f_{X_1,X_2}[(y^2 - x_2^2)^{1/2}, x_2] \frac{y}{(y^2 - x_2^2)^{1/2}} \right.$$

$$+ f_{X_1,X_2}[-(y^2 - x_2^2)^{1/2}, x_2] \frac{y}{(y^2 - x_2^2)^{1/2}}$$

$$\left. + \int_{-(y^2-x_2^2)^{1/2}}^{(y^2-x_2^2)^{1/2}} \frac{\partial f_{X_1,X_2}(x_1, x_2)}{\partial y} \, dx_1 \right\} dx_2 \tag{4}$$

The last term in (4) is zero since the joint density is not dependent on y. Finally, the result is

$$f_Y(y) = \int_{-y}^{y} \left\{ f_{X_1,X_2}[(y^2 - x_2^2)^{1/2}, x_2] \right.$$

$$\left. + f_{X_1,X_2}[-(y^2 - x_2^2)^{1/2}, x_2] \right\} \frac{y}{(y^2 - x_x^2)^{1/2}} \, dx_2 \tag{5}$$

This result is evaluated in Problem 5.4-3 for jointly gaussian X_1 and X_2.

*Multiple Functions

More generally, we are interested in finding the joint density function of a set of functions that defines a set of random variables

$$Y_i = T_i(X_1, X_2, \ldots, X_N) \qquad i = 1, 2, \ldots, N \tag{5.4-3}$$

defined by functional transformations T_i. Now all the possible cases described in Chapter 3 for one random variable carry over to the N-dimensional problem. That is, the X_i can be continuous, discrete, or mixed, while the functions T_i can be linear, nonlinear, continuous, segmented, etc. Because so many cases are possible, many of them being beyond our scope, we shall discuss only one representative problem.

We shall assume that the new random variables Y_i, given by (5.4-3), are produced by single-valued continuous functions T_i having continuous partial derivatives everywhere. It is further assumed that a set of inverse continuous functions T_j^{-1} exists such that the old variables may be expressed as single-valued continuous functions of the new variables:

$$X_j = T_j^{-1}(Y_1, Y_2, \ldots, Y_N) \qquad j = 1, 2, \ldots, N \tag{5.4-4}$$

These assumptions mean that a point in the joint sample space of the X_i maps into only one point in the space of the new variables Y_j, and vice versa.

Let R_X be a closed region of points in the space of the X_i and R_Y be the corresponding region of mapped points in the space of the Y_j, then the probability that a point falls in R_X will equal the probability that its mapped point falls in R_Y. These probabilities, in terms of joint densities, are given by

$$\int_{R_X} \cdots \int f_{X_1, \ldots, X_N}(x_1, \ldots, x_N),\, dx_1 \ldots dx_N$$

$$= \int_{R_Y} \cdots \int f_{Y_1, \ldots, Y_N}(y_1, \ldots, y_N)\, dy_1 \ldots dy_N \tag{5.4-5}$$

This equation may be solved for $f_{Y_1, \ldots, Y_N}(y_1, \ldots, y_N)$ by treating it as simply a multiple integral involving a change of variables.

By working on the left side of (5.4-5) we change the variables x_i to new variables y_j by means of the variable changes (5.4-4). The integrand is changed by direct functional substitution. The limits change from the region R_X to the region R_Y. Finally, the differential hypervolume $dx_1 \ldots dx_N$ will change to the value $|J|dy_1 \ldots dy_N$ (Speigel, 1963, p. 182), where $|J|$ is the magnitude of the jacobian† J of the transformations. The jacobian is the determinant of a matrix of derivatives defined by

$$J = \begin{vmatrix} \dfrac{\partial T_1^{-1}}{\partial Y_1} & \cdots & \dfrac{\partial T_1^{-1}}{\partial Y_N} \\ \vdots & & \vdots \\ \dfrac{\partial T_N^{-1}}{\partial Y_1} & \cdots & \dfrac{\partial T_N^{-1}}{\partial Y_N} \end{vmatrix} \tag{5.4-6}$$

Thus, the left side of (5.4-5) becomes

$$\int_{R_X} \cdots \int f_{X_1, \ldots, X_N}(x_1, \ldots, x_N)\, dx_1 \ldots dx_N$$

$$= \int_{R_Y} \cdots \int f_{X_1, \ldots, X_N}(x_1 = T_1^{-1}, \ldots, x_N = T_N^{-1})|J|\, dy_1 \ldots dy_N \tag{5.4-7}$$

Since this result must equal the right side of (5.4-5), we conclude that

$$f_{Y_1, \ldots, Y_N}(y_1, \ldots, y_N) = f_{X_1, \ldots, X_N}(x_1 = T_1^{-1}, \ldots, x_N = T_N^{-1})|J| \tag{5.4-8}$$

When $N = 1$, (5.4-8) reduces to (3.4-9) previously derived for a single random variable.

The solution (5.4-8) for the joint density of the new variables Y_j is illustrated here with an example.

> **EXAMPLE 5.4-3.** Let the transformations be linear and given by
>
> $$Y_1 = T_1(X_1, X_2) = aX_1 + bX_2$$
> $$Y_2 = T_2(X_1, X_2) = cX_1 + dX_2$$
>
> where a, b, c, and d are real constants. The inverse functions are easy to obtain by solving these two equations for the two variables X_1 and X_2:

---

†After the German mathematician Karl Gustav Jakob Jacobi (1804–1851).

$$X_1 = T_1^{-1}(Y_1, Y_2) = (dY_1 - bY_2)/(ad - bc)$$

$$X_2 = T_2^{-1}(Y_1, Y_2) = (-cY_1 + aY_2)/(ad - bc)$$

where we shall assume $(ad - bc) \neq 0$. From (5.4-6):

$$J = \begin{vmatrix} d/(ad - bc) & -b/(ad - bc) \\ -c/(ad - bc) & a/(ad - bc) \end{vmatrix} = \frac{1}{(ad - bc)}$$

Finally, from (5.4-8),

$$f_{Y_1, Y_2}(y_1, y_2) = \frac{f_{X_1, X_2}\left(\dfrac{dy_1 - by_2}{ad - bc}, \dfrac{-cy_1 + ay_2}{ad - bc}\right)}{|ad - bc|}$$

*5.5
LINEAR TRANSFORMATION OF GAUSSIAN RANDOM VARIABLES

Equation (5.4-8) can be readily applied to the problem of linearly transforming a set of gaussian random variables $X_1, X_2, \ldots, X_N$ for which the joint density of (5.3-12) applies. The new variables $Y_1, Y_2, \ldots, Y_N$ are

$$\begin{aligned} Y_1 &= a_{11}X_1 + a_{12}X_2 + \cdots + a_{1N}X_N \\ Y_2 &= a_{21}X_1 + a_{22}X_2 + \cdots + a_{2N}X_N \\ &\vdots \\ Y_N &= a_{N1}X_1 + a_{N2}X_2 + \cdots + a_{NN}X_N \end{aligned} \tag{5.5-1}$$

where the coefficients a_{ij}, i and $j = 1, 2, \ldots, N$, are real numbers. Now if we define the following matrices:

$$[T] = \begin{bmatrix} a_{11} & a_{12} & \cdots & a_{1N} \\ a_{21} & a_{22} & \cdots & a_{2N} \\ \vdots & \vdots & & \vdots \\ a_{N1} & a_{N2} & \cdots & a_{NN} \end{bmatrix} \tag{5.5-2}$$

$$[Y] = \begin{bmatrix} Y_1 \\ \vdots \\ Y_N \end{bmatrix} \quad [\bar{Y}] = \begin{bmatrix} \bar{Y}_1 \\ \vdots \\ \bar{Y}_N \end{bmatrix} \quad [X] = \begin{bmatrix} X_1 \\ \vdots \\ X_N \end{bmatrix} \quad [\bar{X}] = \begin{bmatrix} \bar{X}_1 \\ \vdots \\ \bar{X}_N \end{bmatrix} \tag{5.5-3}$$

then it is clear from (5.5-1) that

$$[Y] = [T][X] \qquad [Y - \bar{Y}] = [T][X - \bar{X}] \tag{5.5-4}$$

$$[X] = [T]^{-1}[Y] \qquad [X - \bar{X}] = [T]^{-1}[Y - \bar{Y}] \tag{5.5-5}$$

so long as T is nonsingular. Thus,

$$X_i = T_i^{-1}(Y_1, \ldots, Y_N) = a^{i1}Y_1 + a^{i2}Y_2 + \cdots + a^{iN}Y_N \tag{5.5-6}$$

$$\frac{\partial X_i}{\partial Y_j} = \frac{\partial T_i^{-1}}{\partial Y_j} = a^{ij} \tag{5.5-7}$$

$$X_i - \bar{X}_i = a^{i1}(Y_1 - \bar{Y}_1) + \cdots + a^{iN}(Y_N - \bar{Y}_N) \tag{5.5-8}$$

from (5.5-5). Here a^{ij} represents the ijth element of $[T]^{-1}$.

The density function of the new variables $Y_1, \ldots, Y_N$ is found by solving the right side of (5.4-8) in two steps. The first step is to determine $|J|$. By using (5.5-7) with (5.4-6) we find that J equals the determinant of the matrix $[T]^{-1}$. Hence,†

$$|J| = \left| |[T]^{-1}| \right| = \frac{1}{||[T]||} \tag{5.5-9}$$

The second step in solving (5.4-8) proceeds by using (5.5-8) to obtain

$$C_{X_i X_j} = E[(X_i - \bar{X}_i)(X_j - \bar{X}_j)] = \sum_{k=1}^{N} a^{ik} \sum_{m=1}^{N} a^{jm} E[(Y_k - \bar{Y}_k)(Y_m - \bar{Y}_m)]$$

$$= \sum_{k=1}^{N} a^{ik} \sum_{m=1}^{N} a^{jm} C_{Y_k Y_m} \tag{5.5-10}$$

Since $C_{X_i X_j}$ is the ijth element in the covariance matrix $[C_X]$ of (5.3-12) and $C_{Y_k Y_m}$ is the kmth element of the covariance matrix of the new variables Y_i, which we denote $[C_Y]$, (5.5-10) can be written in the form

$$[C_X] = [T]^{-1}[C_Y]([T]^t)^{-1} \tag{5.5-11}$$

Here $[T]^t$ represents the transpose of $[T]$. The inverse of (5.5-11) is

$$[C_X]^{-1} = [T]^t[C_Y]^{-1}[T] \tag{5.5-12}$$

which has a determinant

$$\left| [C_X]^{-1} \right| = \left| [C_Y]^{-1} \right| ||[T]||^2 \tag{5.5-13}$$

On substitution of (5.5-13) and (5.5-12) into (5.3-12):

$$f_{X_1, \ldots, X_N}(x_1 = T_1^{-1}, \ldots, x_N = T_N^{-1})$$

$$= \frac{\left| ||[T]|| \, |[C_Y]^{-1}| \right|^{1/2}}{(2\pi)^{N/2}} \exp \left\{ -\frac{[x - \bar{X}]^t [T]^t [C_Y]^{-1} [T][x - \bar{X}]}{2} \right\} \tag{5.5-14}$$

Finally, (5.5-14) and (5.5-9) are substituted into (5.4-8), and (5.5-4) is used to obtain

$$f_{Y_1, \ldots, Y_N}(y_1, \ldots, y_N) = \frac{\left| [C_Y]^{-1} \right|^{1/2}}{(2\pi)^{N/2}} \exp \left\{ -\frac{[y - \bar{Y}]^t [C_Y]^{-1} [y - \bar{Y}]}{2} \right\} \tag{5.5-15}$$

---

†We represent the magnitude of the determinant of a matrix by $||[\cdot]||$.

This result shows that the new random variables $Y_1, Y_2, \ldots, Y_N$ are jointly gaussian because (5.5-15) is of the required form.

In summary, (5.5-15) shows that a linear transformation of gaussian random variables produces gaussian random variables. The new variables have mean values

$$\bar{Y}_j = \sum_{k=1}^{N} a_{jk}\bar{X}_k \qquad (5.5\text{-}16)$$

from (5.5-1) and covariances given by the elements of the covariance matrix

$$[C_Y] = [T][C_X][T]^t \qquad (5.5\text{-}17)$$

as found from (5.5-11).

EXAMPLE 5.5-1. Two gaussian random variables X_1 and X_2 have zero means and variances $\sigma_{X_1}^2 = 4$ and $\sigma_{X_2}^2 = 9$. Their covariance $C_{X_1 X_2}$ equals 3. If X_1 and X_2 are linearly transformed to new variables Y_1 and Y_2 according to

$$Y_1 = X_1 - 2X_2$$
$$Y_2 = 3X_1 + 4X_2$$

we use the above results to find the means, variances, and covariance of Y_1 and Y_2.

Here

$$[T] = \begin{bmatrix} 1 & -2 \\ 3 & 4 \end{bmatrix} \quad \text{and} \quad [C_X] = \begin{bmatrix} 4 & 3 \\ 3 & 9 \end{bmatrix}$$

Since X_1 and X_2 are zero-mean and gaussian, Y_1 and Y_2 will also be zero-mean and gaussian, thus $\bar{Y}_1 = 0$ and $\bar{Y}_2 = 0$. From (5.5-17):

$$[C_Y] = [T][C_X][T]^t = \begin{bmatrix} 1 & -2 \\ 3 & 4 \end{bmatrix}\begin{bmatrix} 4 & 3 \\ 3 & 9 \end{bmatrix}\begin{bmatrix} 1 & 3 \\ -2 & 4 \end{bmatrix} = \begin{bmatrix} 28 & -66 \\ -66 & 252 \end{bmatrix}$$

Thus $\sigma_{Y_1}^2 = 28$, $\sigma_{Y_2}^2 = 252$, and $C_{Y_1 Y_2} = -66$.

*5.6
COMPUTER GENERATION OF MULTIPLE RANDOM VARIABLES

In Section 3.5 we discussed the generation of a single random variable of prescribed probability density by transformation of a random variable that was uniformly distributed on (0,1). Here, we shall utilize results of the preceding two sections to show how some usefully distributed random variables can be generated by computer when the generation initially requires either two uniformly distributed random variables or two gaussian variables. We describe several examples, the first based on transformation of two statistically independent random variables X_1 and X_2, both uniformly distributed on (0,1).

One common problem in the simulation of systems by a digital computer is the generation of gaussian random variables. As a first example, we note that two statistically independent gaussian random variables Y_1 and Y_2, each with zero mean and unit variance, can be generated by the transformations (see Dillard, 1967)

$$Y_1 = T_1(X_1, X_2) = \sqrt{-2 \ln(X_1)} \, \cos(2\pi X_2) \qquad (5.6\text{-}1a)$$

$$Y_2 = T_2(X_1, X_2) = \sqrt{-2 \ln(X_1)} \, \sin(2\pi X_2) \qquad (5.6\text{-}1b)$$

It can be shown (Problem 5.6-1) that the joint density of Y_1 and Y_2 is

$$f_{Y_1, Y_2}(y_1, y_2) = \frac{e^{-y_1^2/2}}{\sqrt{2\pi}} \frac{e^{-y_2^2/2}}{\sqrt{2\pi}} \qquad (5.6\text{-}2)$$

as it should be for statistically independent Y_1 and Y_2. Our example can be generalized to include arbitrary means and variances (Problem 5.6-2).

As another example, assume we start with two zero-mean, unit-variance, statistically independent gaussian random variables Y_1 and Y_2 (perhaps generated as in our first example above), and seek to transform them to two zero-mean gaussian variates W_1 and W_2 that have arbitrary variances, $\sigma_{W_1}^2$ and $\sigma_{W_2}^2$, and arbitrary correlation coefficient ρ_W. From (5.3-16) applied to W_1 and W_2, and from (5.5-17) for a linear transformation, we have

$$[C_W] = \begin{bmatrix} \sigma_{W_1}^2 & \rho_W \sigma_{W_1} \sigma_{W_2} \\ \rho_W \sigma_{W_1} \sigma_{W_2} & \sigma_{W_2}^2 \end{bmatrix} = [T][T]^t \qquad (5.6\text{-}3)$$

The covariance matrix of Y_1 and Y_2 does not explicitly appear in (5.5-17) because it is a unit matrix due to the unit-variance assumption about Y_1 and Y_2. Our goal is obtained if we solve for $[T]$ that makes (5.6-3) true for arbitrarily specified $\sigma_{W_1}^2$, $\sigma_{W_2}^2$ and ρ_W. As long as $[C_W]$ is nonsingular (the usual case), $[T]$ can be expressed as a lower triangular matrix of the form

$$[T] = \begin{bmatrix} T_{11} & 0 \\ T_{21} & T_{22} \end{bmatrix} \qquad (5.6\text{-}4)$$

On using (5.6-4) in (5.6-3), and solving for the elements, we have

$$T_{11} = \sigma_{W_1} \qquad (5.6\text{-}5a)$$

$$T_{21} = \rho_W \sigma_{W_2} \qquad (5.6\text{-}5b)$$

$$T_{22} = \sigma_{W_2} \sqrt{1 - \rho_W^2} \qquad (5.6\text{-}5c)$$

The final transformations yielding W_1 and W_2 become

$$W_1 = T_{11} Y_1 = \sigma_{W_1} Y_1 \qquad (5.6\text{-}6a)$$

$$W_2 = T_{21} Y_1 + T_{22} Y_2 = \rho_W \sigma_{W_2} Y_1 + \sigma_{W_2} \sqrt{1 - \rho_W^2} \, Y_2 \qquad (5.6\text{-}6b)$$

from the *form* of (5.5-4). Thus, if zero-mean, unit-variance, statistically independent gaussian random variables Y_1 and Y_2 are transformed according to (5.6-6), then W_1 and W_2 are correlated gaussian random variables having zero means, respective variances of $\sigma_{W_1}^2$ and $\sigma_{W_2}^2$, and correlation coefficient ρ_W.

EXAMPLE 5.6-1. We use MATLAB to generate $N = 100$ values x_{1n}, $n = 1, 2, \ldots, N$, of a random variable X_1 uniform on $(0, 1)$. We then repeat the process for a second random variable X_2 with values x_{2n}. Next, we successively use (5.6-1) and (5.6-6) to create two sets of values w_{1n} and w_{2n}, $n = 1, 2, \ldots, N$, of two zero-mean gaussian random variables W_1 and W_2 having respective variances $\sigma_{W_1}^2 = 4$ and $\sigma_{W_2}^2 = 9$, and normalized correlation coefficient $\rho_W = -0.4$. To determine the quality of our random variable's values, we find their means according to

$$\hat{\bar{W}}_i = \frac{1}{N} \sum_{n=1}^{N} w_{in}, \qquad i = 1 \text{ and } 2 \tag{1}$$

their variances according to

$$\widehat{\sigma_{W_i}^2} = \frac{1}{N} \sum_{n=1}^{N} (w_{in} - \hat{\bar{W}}_i)^2, \qquad i = 1 \text{ and } 2 \tag{2}$$

and their normalized correlation coefficient according to

$$\hat{\rho}_W = \frac{(\widehat{\sigma_{W_1}^2} \widehat{\sigma_{W_2}^2})^{-1/2}}{N} \sum_{n=1}^{N} (w_{1n} - \hat{\bar{W}}_1)(w_{2n} - \hat{\bar{W}}_2) \tag{3}$$

The applicable MATLAB code is shown in Figure 5.6-1. Our results are tabulated in Table 5.6-1, where the standard deviations and normalized correlation coefficient are found to be in error by -7.5%, -2.3%, and 45.2%, respectively. For $N = 1000$ values, these errors improve (see Problem 5.6-5).

If arbitrary means $\bar{W}_1$ and $\bar{W}_2$ are desired for W_1 and W_2 in the preceding example, we only need to add these to right sides of (5.6-6):

$$W_1 = \bar{W}_1 = \sigma_{W_1} Y_1 \tag{5.6-7a}$$

$$W_2 = \bar{W}_2 + \rho_W \sigma_{W_2} Y_1 + \sigma_{W_2} \sqrt{1 - \rho_W^2}\, Y_2 \tag{5.6-7b}$$

TABLE 5.6-1
Results applicable to Example 5.6-1

	Mean		Standard deviation		Correlation coefficient
	m_1	m_2	σ_1	σ_2	ρ
True values	0	0	2	3	−0.4
Estimated ($N = 100$)	−0.02	−0.17	1.85	2.93	−0.57
Percent error	—	—	−7.5%	−2.3%	42.5%

```
%%%%%%%%%%%%%% Example 5.6-1 %%%%%%%%%%%%%%%%%%%

clear

N = 100; % number of random variables to generate

sigw1 = sqrt(4); % standard deviation
sigw2 = sqrt(9);
rho = -0.4; % normalized correlation coefficient

x1 = rand(1,N); % uniformly distributed random numbers
x2 = rand(1,N);

y1 = sqrt(-2*log(x1)).*cos(2*pi*x2); % independent Gaussian
% random variables
y2 = sqrt(-2*log(x1)).*sin(2*pi*x2);

T11 = sigw1; % constants
T21 = rho*sigw2;
T22 = sigw2*sqrt(1-rho^2);

w1 = T11*y1; % correlated Gaussian random variables
w2 = T21*y1 + T22*y2;

wmean = [mean(w1) mean(w2)]; % sample mean
wcov = [cov(w1,1) cov(w2,1)]; % (biased) sample covariance

tmp = corrcoef(w1,w2);
rho_hat = tmp(2,1); % estimate of normalized correlation
% coefficient

cov_err = 100*(sqrt(wcov) - ([sigw1 sigw2].^2))./ ...
([sigw1 sigw2].^2) % percent error
rho_err = 100*(rho_hat - rho)./rho
```

FIGURE 5.6-1
MATLAB code used in Example 5.6-1.

The foregoing transformations can be extended to generate any number of zero-mean correlated gaussian random variables by transforming the same number of zero-mean, unit-variance, independent gaussian random variables. For N random variables, $[C_W]$ becomes an $N \times N$ specified (arbitrary) symmetric matrix and the form of (5.6-3) again applies. The elements of $[T]$ can be found from the Cholesky method of factoring matrices, as described in Ralston and Wilf (1967).

As a final example, suppose two statistically independent gaussian random variables W_1 and W_2, with respective means $\bar{W}_1$ and $\bar{W}_2$ and variances both equal to σ^2, are subjected to the transformations

$$R = T_1(W_1, W_2) = \sqrt{W_1^2 + W_2^2} \qquad (5.6-8)$$

$$\Theta = T_2(W_1, W_2) = \tan^{-1}(W_2/W_1) \qquad (5.6-9)$$

From the inverse transformations

$$W_1 = T_1^{-1}(R, \Theta) = R\cos(\Theta) \tag{5.6-10}$$

$$W_2 = T_2^{-1}(R, \Theta) = R\sin(\Theta) \tag{5.6-11}$$

and the use of (5.5-4), we find the Jacobian equals R. Since

$$f_{W_1, W_2}(w_1, w_2) = \frac{1}{2\pi\sigma^2} e^{-[(w_1 - \bar{W}_1)^2 + (w_2 - \bar{W}_2)^2]/(2\sigma^2)} \tag{5.6-12}$$

(5.4-8) yields

$$f_{R,\Theta}(r, \theta) = \frac{ru(r)}{2\pi\sigma^2} \exp\left\{-[[r\cos(\theta) - \bar{W}_1]^2 + [r\sin(\theta) - \bar{W}_2]^2]/(2\sigma^2)\right\}$$

$$= \frac{ru(r)}{2\pi\sigma^2} \exp\left\{-\frac{1}{2\sigma^2}[r^2 + (\bar{W}_1^2 + \bar{W}_2^2) - 2r\bar{W}_1\cos(\theta) - 2r\bar{W}_2\sin(\theta)]\right\}$$

$$\tag{5.6-13}$$

where $u(r)$ is the unit-step function. If we now define

$$A_0 = \sqrt{\bar{W}_1^2 + \bar{W}_2^2} \tag{5.6-14}$$

$$\theta_0 = \tan^{-1}(\bar{W}_2/\bar{W}_1) \tag{5.6-15}$$

(5.6-13) can be written as

$$f_{R,\Theta}(r, \theta) = \frac{ru(r)}{2\pi\sigma^2} \exp\left\{-\frac{1}{2\sigma^2}[r^2 + A_0^2 - 2rA_0\cos(\theta - \theta_0)]\right\} \tag{5.6-16}$$

Equation (5.6-16) is our principal result. It is important in system simulations because it is the joint density of the envelope (R) and phase (Θ) of the sum of a sinusoidal signal (with peak amplitude A_0 and phase θ_0) and a zero-mean gaussian *bandpass* noise of power σ^2. This density is developed further in Section 10.6.

5.7
SAMPLING AND SOME LIMIT THEOREMS

In this section we briefly introduce some basic concepts of sampling. The topic will be expanded further in Chapter 8. Although we shall develop the topics around an example practical problem, the results will apply to much more general situations.

Sampling and Estimation

Engineers and scientists are frequently confronted with the problem of measuring some quantity. For example, if we need to measure a dc voltage, we use a dc voltmeter, which provides a scale indication of the voltage. Now regardless of the mechanism used by the meter to provide its indication, one typically "reads" this scale to obtain a "value" we say is the measurement of the voltage. In other words, we *sample* the indication to get our measurement.

The measurement can only be considered as an *estimate* of voltage, however, because of meter drifts, accuracy tolerances, etc. In fact, any measurement can only be considered as an estimate of the quantity of interest. In our example, our estimate uses only one sample. More generally, we may estimate (measure) a quantity by using more than one sample (observation).

To quantify these practical thoughts further, consider the problem of measuring the average (dc) value of some random noise voltage. If we had a large number of identical such sources, we could imagine sampling the voltage of each (at a given time) and form an estimate of the dc voltage by averaging the samples. For N sources, each sample could be considered a value of one of N random variables that form the N dimensions of a combined experiment, as in Section 1.6. Here samples from each of N subexperiments are combined.

In our practical world we usually must take another approach because we never have multiple identical sources with which to work. We seek another model. Suppose we now take a sequence of N samples over time under the assumption (model) that the voltage's statistical properties stay unchanged with time.† Again, each sample is taken as the value of one of N statistically independent random variables, all having the same probability distribution. We again have a combined experiment, but it is now N repetitions of one basic experiment.

Estimation of Mean, Power, and Variance

For either of the above approaches the N samples x_n represent values of identically distributed, random variables X_n, $n = 1, 2, \ldots, N$. Assume the X_n are independent at least by pairs; they have the same mean value $\bar{X}$ and variance σ_X^2 because of identical distributions. Since we wish to estimate (measure) the mean noise voltage, intuition indicates we should form the average of the sample values as follows:‡

$$\hat{\bar{x}}_N = \text{estimate of average of } N \text{ samples} = \frac{1}{N} \sum_{n=1}^{N} x_n \qquad (5.7\text{-}1)$$

Equation (5.7-1) is a function of the set of specific samples $\{x_n\}$; it gives a number which we call an *estimate* or measurement of the mean of the random variables. Another set of specific samples would produce a different number $\hat{\bar{x}}_N$. When all possible sample sets are considered, we form the function

$$\hat{\bar{X}}_N = \frac{1}{N} \sum_{n=1}^{N} X_n \qquad (5.7\text{-}2)$$

---

†More is said about this model in the next chapter (Section 6.2).
‡The circumflex is notation to imply an estimate, or estimator; in this case an estimate of the time average of N samples denoted by $\bar{x}_N$.

to represent the effect of averaging over the random variables. Equation (5.7-2) is called an *estimator*; it produces a specific estimate of $\bar{X}$ for a specific set of samples. We refer to (5.7-2) as the *sample mean*.

Of great interest is: How does our estimator of the sample mean perform? To seek an answer, we find the mean and variance of our estimator.

CHAPTER 5:
Operations on
Multiple Random
Variables

$$E[\hat{\bar{X}}_N] = E\left[\frac{1}{N}\sum_{n=1}^{N} X_n\right] = \frac{1}{N}\sum_{n=1}^{N} E[X_n] = \bar{X}, \qquad \text{any } N \qquad (5.7\text{-}3)$$

Any estimator (measurement function) for which the mean of the estimator of some quantity equals the quantity being estimated is called *unbiased*. For the variance:

$$E[(\hat{\bar{X}}_N - \bar{X})^2] = \sigma_{\hat{X}_N}^2 = E[\hat{\bar{X}}_N^2 - 2\bar{X}\hat{\bar{X}}_N + \bar{X}^2]$$

$$= E[\hat{\bar{X}}_N^2] - \bar{X}^2 = -\bar{X}^2 + E\left[\frac{1}{N}\sum_{n=1}^{N} X_n \frac{1}{N}\sum_{m=1}^{N} X_m\right]$$

$$= -\bar{X}^2 + \frac{1}{N^2}\sum_{n=1}^{N}\sum_{m=1}^{N} E[X_n X_m] \qquad (5.7\text{-}4)$$

But $E[X_n X_m] = E[X^2]$ for $n = m$ and equals $\bar{X}^2$ for $n \neq m$ because of assumed independence by pairs. Thus,

$$\sigma_{\hat{X}_N}^2 = -\bar{X}^2 + \frac{1}{N^2}[NE(X^2) + (N^2 - N)\bar{X}^2]$$

$$= \frac{1}{N}[E(X^2) - \bar{X}^2] = \sigma_X^2/N \qquad (5.7\text{-}5)$$

From (5.7-5) the variance of our sample mean estimator goes to zero as $N \to \infty$ for finite source variance σ_X^2. This fact implies that for large N our estimator will give an estimate nearly equal to the quantity being estimated with high probability. To prove the implication, we use Chebychev's inequality of (3.2-10). For our notation it says

$$P\{|\hat{\bar{X}}_N - \bar{X}| < \epsilon\} \geq 1 - (\sigma_{\hat{X}_N}^2/\epsilon^2) = 1 - \frac{\sigma_X^2}{N\epsilon^2} \qquad (5.7\text{-}6)$$

Which tends to 1 as $N \to \infty$ for any finite $\epsilon > 0$ and finite σ_X^2. This result indicates that $\hat{\bar{X}}_N$ converges to $\bar{X}$ with probability 1 as $N \to \infty$. Such estimators are called *consistent*.

EXAMPLE 5.7-1. Suppose the mean of our example noise voltage is to be estimated to within 5% of its true value with a probability of 0.95 when $N = 50$ samples are used. We find what mean and variance are allowed. From (5.7-6) with $\epsilon = 0.05\bar{X}$ we require

$$1 - \frac{\sigma_X^2}{50(0.05\bar{X})^2} \leq 0.95$$

which means $\bar{X} \leq (160)^{1/2}\sigma_X$ for the accuracies desired.

Thus far, our discussion has centered on estimating the mean of some random quantity. Estimates of functions of random quantities are also possible. For example, an estimator for the power in a random voltage can be defined as

$$\widehat{X_N^2} = \frac{1}{N}\sum_{n=1}^{N} X_n^2 \qquad (5.7-7)$$

while the estimator for the variance of the voltage can be defined as

$$\widehat{\sigma_X^2} = \frac{1}{N-1}\sum_{n=1}^{N}(X_n - \hat{\hat{X}}_N)^2 \qquad (5.7-8)$$

Here $\hat{\hat{X}}_N$ is defined in (5.7-2). The estimator of (5.7-7) is found in Problem 5.7-1 to be unbiased. Its variance is found in Problem 5.7-2. Similarly, the mean of the variance estimator of (5.7-8) is unbiased, but becomes biased if the factor $1/(N-1)$ is changed to $1/N$ as in the mean and power estimators (Problem 5.7-3).

> **EXAMPLE 5.7-2.** A random noise voltage behaves approximately as an exponential random variable with a mean value of 4 and a variance of 16. Eleven samples are taken having values 0.1 V, 0.4, 0.9, 1.4, 2.0, 2.8, 3.7, 4.8, 6.4, 9.2, and 12.0 V. We use (5.7-2) and (5.7-8) to find the respective mean and variance of these samples. From (5.7-2)
>
> $$\hat{\hat{X}}_N = \frac{1}{11}(0.1 + 0.4 + 0.9 + \cdots + 9.2 + 12.0) = 3.973 \text{ V}$$
>
> From (5.7-8)
>
> $$\widehat{\sigma_X^2} = \frac{1}{10}[(0.1 - 3.973)^2 + (0.4 - 3.973)^2$$
> $$+ \cdots + (12.0 - 3.973)^2]$$
> $$= 14.75 \text{ V}^2$$
>
> Here the sample mean is in error by less than 1% for the given set of sample values, but the estimate of variance is in error by about 7.8%. The reader should be aware that any other set of 11 samples may give different values and different percentage errors.

> **EXAMPLE 5.7-3.** The random variable Y of Problem 3.5-4 can be generated by the transformation $Y = a[(1 - X)^{-1/2} - 1]^{1/2}$, $0 < X < 1$. We use MATLAB to generate 250 values of Y from which the sample mean of (5.7-2), the second moment of (5.7-7), and the variance of (5.7-8) are then calculated. These values are compared to the true mean, second moment, and variance of Y, which are known to be $\pi a/4$, a^2, and $a^2(16 - \pi^2)/16$, respectively. For the calculations we assume $a = 2$.
>
> The MATLAB code for this example is given in Figure 5.7-1. Calculated data, shown in Table 5.7-1, reveals errors in estimating mean, second moment, and variance to be -5.7%, -10.3%, and -8.5%, respectively. Problem 5.7-4 reconsiders this example, but for $N = 1000$ values of Y.

TABLE 5.7-1
Data applicable to Example 5.7-3

	Mean	Second moment	Variance
True values	1.57	4.00	1.53
Estimated ($N = 250$)	1.48	3.59	1.40
Percent error	-5.7%	-10.3%	-8.5%

```
%%%%%%%%%%%%%% Example 5.7-3 %%%%%%%%%%%%%%%%%%

clear

N = 250; % number of random variables to generate
a = 2; % constant

x = rand(1,N);
y = a*sqrt(sqrt(1./(1-x))-1); % random variable

ymean = mean(y) % sample mean
y2moment = mean(y.^2) % second moment
yvar = cov(y) % variance

yestimated = [ymean y2moment yvar];
ytrue = [pi/2 4.0 (16-pi^2)/4];

per_error = 100*(yestimated - ytrue)./ytrue
```

FIGURE 5.7-1
MATLAB code used in Example 5.7-3.

Weak Law of Large Numbers

The preceding developments have shown that the sample mean estimator of
(5.7-2), where the random variables X_n are identically distributed (same mean
and same finite variance) and are at least pairwise statistically independent,
satisfies

$$\lim_{N \to \infty} [P\{\hat{\bar{X}}_N - \bar{X}| < \epsilon\}] = 1 \qquad \text{any } \epsilon > 0 \tag{5.7-9}$$

Expression (5.7-9) is known as the *weak law of large numbers*.

Strong Law of Large Numbers

Another important relationship is the *strong law of large numbers*. For N
random variables, X_n defined as for the weak law, it states that, as $N \to \infty$,

$$P\left\{ \lim_{N \to \infty} (\hat{\bar{X}}_N) = \bar{X} \right\} = 1 \tag{5.7-10}$$

*5.8
COMPLEX RANDOM VARIABLES

A *complex random variable* Z can be defined in terms of real random variables X and Y by

$$Z = X + jY \qquad (5.8-1)$$

where $j = \sqrt{-1}$. In considering expected values involving Z, the joint density of X and Y must be used. For instance, if $g(\cdot)$ is some function (real or complex) of Z, the expected value of $g(Z)$ is obtained from

$$E[g(Z)] = \int_{-\infty}^{\infty} \int_{-\infty}^{\infty} g(z) f_{X,Y}(x, y)\, dx\, dy \qquad (5.8-2)$$

Various important quantities such as the mean value and variance are obtained through application of (5.8-2). The mean value of Z is

$$\bar{Z} = E[Z] = E[X] + jE[Y] = \bar{X} + j\bar{Y} \qquad (5.8-3)$$

The variance σ_Z^2 of Z is defined as the mean value of the function $g(Z) = |Z - E[Z]|^2$; that is,

$$\sigma_Z^2 = E[|Z - E[Z]|^2] \qquad (5.8-4)$$

Equation (5.8-2) can be extended to include functions of two random variables

$$Z_m = X_m + jY_m \qquad (5.8-5)$$

and

$$Z_n = X_n + jY_n \qquad (5.8-6)$$

$n \neq m$, if expectation is taken with respect to four random variables X_m, Y_m, X_n, Y_n through their joint density function $f_{X_m, Y_m, X_n, Y_n}(x_m, y_m, x_n, y_n)$. If this density satisfies

$$f_{X_m, Y_m, X_n, Y_n}(x_m, y_m, x_n, y_n) = f_{X_m, Y_m}(x_m, y_m) f_{X_n, Y_n}(x_n, y_n) \qquad (5.8-7)$$

then Z_m and Z_n are called *statistically independent*. The extension to N random variables is straightforward.

The *correlation* and *covariance* of Z_m and Z_n are defined by

$$R_{Z_m Z_n} = E[Z_m^* Z_n] \qquad n \neq m \qquad (5.8-8)$$

and

$$C_{Z_m Z_n} = E[\{Z_m - E[Z_m]\}^* \{Z_n - E[Z_n]\}] \qquad n \neq m \qquad (5.8-9)$$

respectively, where the superscripted asterisk* represents the complex conjugate. If the covariance is 0, Z_m and Z_n are said to be *uncorrelated random variables*. By setting (5.8-9) to 0, we find that

$$R_{Z_m Z_n} = E[Z_m^*] E[Z_n] \qquad m \neq n \qquad (5.8-10)$$

for uncorrelated random variables. Statistical independence is sufficient to guarantee that Z_m and Z_n are uncorrelated.

Finally, we note that two complex random variables are called *orthogonal* if their correlation, given by (5.8-8), equals 0.

5.9
SUMMARY

This chapter extended the operations performed on a single random variable in Chapter 3 to include operations on multiple random variables. Topics extended were:

- Expected values were developed of functions of random variables, which included both joint moments about the origin and central moments, as well as joint characteristic functions that are useful in finding moments. New moments of special interest were correlation and covariance.
- Multiple gaussian random variables were defined.
- Single and multiple functional transformations of several random variables were developed.
- Transformation results were used to show how linear transformation of jointly gaussian random variables is especially important, as it produces random variables that are also joint gaussian.
- The important technique of how to generate multiple random variables by computer was next introduced. The material was illustrated by a computer example using MATLAB software.
- Some new material on the basics of sampling and estimation of mean, power, and variance was given. It was supported by both regular and a computer example and problem (MATLAB).
- Finally, some more advanced material was given that defines complex random variables and their characteristics.

PROBLEMS

5.1-1. Random variables X and Y have the joint density

$$f_{X,Y}(x, y) = \begin{cases} \dfrac{1}{24} & 0 < x < 6 \quad \text{and} \quad 0 < y < 4 \\ 0 & \text{elsewhere} \end{cases}$$

What is the expected value of the function $g(X, Y) = (XY)^2$?

5.1-2. Extend Problem 5.1-1 by finding the expected value of $g(X_1, X_2, X_3, X_4) = X_1^{n_1} X_2^{n_2} X_3^{n_3} X_4^{n_4}$, where n_1, n_2, n_3, and n_4 are integers ≥ 0 and

$$f_{X_1,X_2,X_3,X_4}(x_1, x_2, x_3, x_4) = \begin{cases} \dfrac{1}{abcd} & 0 < x_1 < a \text{ and } 0 < x_2 < b \\ & \text{and } 0 < x_3 < c \text{ and } 0 < x_4 < d \\ 0 & \text{elsewhere} \end{cases}$$

5.1-3. The density function of two random variables X and Y is

$$f_{X,Y}(x, y) = u(x)u(y)16e^{-4(x+y)}$$

Find the mean value of the function

$$g(X, Y) = \begin{cases} 5 & 0 < X \le \dfrac{1}{2} \quad \text{and} \quad 0 < Y \le \dfrac{1}{2} \\ -1 & \dfrac{1}{2} < X \quad \text{and/or} \quad \dfrac{1}{2} < Y \\ 0 & \text{all other } X \text{ and } Y \end{cases}$$

5.1-4. For the random variables in Problem 5.1-3, find the mean value of the function

$$g(X, Y) = e^{-2(X^2+Y^2)}$$

5.1-5. Three statistically independent random variables X_1, X_2, and X_3 have mean values $\bar{X}_1 = 3$, $\bar{X}_2 = 6$, and $\bar{X}_3 = -2$. Find the mean values of the following functions:
(a) $g(X_1, X_2, X_3) = X_1 + 3X_2 + 4X_3$
(b) $g(X_1, X_2, X_3) = X_1 X_2 X_3$
(c) $g(X_1, X_2, X_3) = -2X_1 X_2 - 3X_1 X_3 + 4X_2 X_3$
(d) $g(X_1, X_2, X_3) = X_1 + X_2 + X_3$

5.1-6. Find the mean value of the function

$$g(X, Y) = X^2 + Y^2$$

where X and Y are random variables defined by the density function

$$f_{X,Y}(x, y) = \frac{e^{-(x^2+y^2)/2\sigma^2}}{2\pi\sigma^2}$$

with σ^2 a constant.

5.1-7. Two statistically independent random variables X and Y have mean values $\bar{X} = E[X] = 2$ and $\bar{Y} = E[Y] = 4$. They have second moments $\overline{X^2} = E[X^2] = 8$ and $\overline{Y^2} = E[Y^2] = 25$. Find:
(a) the mean value   (b) the second moment and
(c) the variance of the random variable $W = 3X - Y$.

5.1-8. Two random variables X and Y have means $\bar{X} = 1$ and $\bar{Y} = 2$, variances $\sigma_X^2 = 4$ and $\sigma_Y^2 = 1$, and a correlation coefficient $\rho_{XY} = 0.4$. New random variables W and V are defined by

$$V = -X + 2Y \qquad W = X + 3Y$$

Find:
(a) the means   (b) the variances   (c) the correlations and
(d) the correlation coefficient ρ_{VW} of V and W.

5.1-9. Two random variables X and Y are related by the expression

$$Y = aX + b$$

where a and b are any real numbers.

(a) Show that their correlation coefficient is

$$\rho = \begin{cases} 1 & \text{if } a > 0 \text{ for any } b \\ -1 & \text{if } a < 0 \text{ for any } b \end{cases}$$

(b) Show that their covariance is

$$C_{XY} = a\sigma_X^2$$

where σ_X^2 is the variance of X.

*5.1-10. Show that the correlation coefficient satisfies the expression

$$|\rho| = \frac{|\mu_{11}|}{\sqrt{\mu_{02}\mu_{20}}} \leq 1$$

5.1-11. Find all the second-order moments and central moments for the density function given in Problem 5.1-3.

5.1-12. Random variables X and Y have the joint density function

$$f_{X,Y}(x, y) = \begin{cases} (x+y)^2/40 & -1 < x < 1 \quad \text{and} \quad -3 < y < 3 \\ 0 & \text{elsewhere} \end{cases}$$

(a) Find all the second-order moments of X and Y.
(b) What are the variances of X and Y?
(c) What is the correlation coefficient?

5.1-13. Find all the third-order moments by using (5.1-5) for X and Y defined in Problem 5.1-12.

5.1-14. For discrete random variables X and Y, show that:
(a) Joint moments are

$$m_{nk} = \sum_{i=1}^{N} \sum_{j=1}^{M} P(x_i, y_j) x_i^n y_j^k$$

(b) Joint central moments are

$$\mu_{nk} = \sum_{i=1}^{N} \sum_{j=1}^{M} P(x_i, y_j)(x_i - \bar{X})^n (y_j - \bar{Y})^k$$

where $P(x_i, y_j) = P\{X = x_i, Y = y_j\}$, X has N possible values x_i, and Y has M possible values y_j.

5.1-15. For two random variables X and Y:

$$f_{X,Y}(x, y) = 0.15\delta(x + 1)\delta(y) + 0.1\delta(x)\delta(y) + 0.1\delta(x)\delta(y - 2)$$
$$+ 0.4\delta(x - 1)\delta(y + 2) + 0.2\delta(x - 1)\delta(y - 1) + 0.5\delta(x - 1)\delta(y - 3)$$

Find: (a) the correlation, (b) the covariance, and (c) the correlation coefficient of X and Y. (d) Are X and Y either uncorrelated or orthogonal?

5.1-16. Discrete random variables X and Y have the joint density

$$f_{X,Y}(x, y) = 0.4\delta(x + \alpha)\delta(y - 2) + 0.3\delta(x - \alpha)\delta(y - 2)$$
$$+ 0.1\delta(x - \alpha)\delta(y - \alpha) + 0.2\delta(x - 1)\delta(y - 1)$$

Determine the value of α, if any, that minimizes the correlation between X and Y and find the minimum correlation. Are X and Y orthogonal?

5.1-17. For two discrete random variables X and Y:

$$f_{X,Y}(x, y) = 0.3\delta(x - \alpha)\delta(y - \alpha) + 0.5\delta(x + \alpha)\delta(y - 4) + 0.2\delta(x + 2)\delta(y + 2)$$

Determine the value of α, if any, that minimizes the covariance of X and Y. Find the minimum covariance. Are X and Y uncorrelated?

5.1-18. The density function

$$f_{X,Y}(x, y) = \begin{cases} \dfrac{xy}{9} & 0 < x < 2 \quad \text{and} \quad 0 < y < 3 \\ 0 & \text{elsewhere} \end{cases}$$

applies to two random variables X and Y.
(a) Show, by use of (5.1-6) and (5.1-7), that X and Y are uncorrelated.
(b) Show that X and Y are also statistically independent.

5.1-19. Two random variables X and Y have the density function

$$f_{X,Y}(x, y) = \begin{cases} \dfrac{2}{43}(x + 0.5y)^2 & 0 < x < 2 \quad \text{and} \quad 0 < y < 3 \\ 0 & \text{elsewhere} \end{cases}$$

(a) Find all the first- and second-order moments.
(b) Find the covariance.
(c) Are X and Y uncorrelated?

5.1-20. Define random variables V and W by

$$V = X + aY$$
$$W = X - aY$$

where a is a real number and X and Y are random variables. Determine a in terms of moments of X and Y such that V and W are orthogonal.

*5.1-21. If X and Y in Problem 5.1-20 are gaussian, show that W and V are statistically independent if $a^2 = \sigma_X^2/\sigma_Y^2$, where σ_X^2 and σ_Y^2 are the variances of X and Y, respectively.

*5.1-22. Three uncorrelated random variables X_1, X_2, and X_3 have means $\bar{X}_1 = 1$, $\bar{X}_2 = -3$, and $\bar{X}_3 = 1.5$ and second moments $E[X_1^2] = 2.5$, $E[X_2^2] = 11$, and $E[X_3^2] = 3.5$. Let $Y = X_1 - 2X_2 + 3X_3$ be a new random variable and find:
(a) the mean value, (b) the variance of Y.

5.1-23. Given $W = (aX + 3Y)^2$ where X and Y are zero-mean random variables with variances $\sigma_X^2 = 4$ and $\sigma_Y^2 = 16$. Their correlation coefficient is $\rho = -0.5$.
(a) Find a value for the parameter a that minimizes the mean value of W.
(b) Find the minimum mean value.

5.1-24. Two random variables have a uniform density on a circular region defined by

$$f_{X,Y}(x, y) = \begin{cases} 1/\pi r^2 & x^2 + y^2 \le r^2 \\ 0 & \text{elsewhere} \end{cases}$$

Find the mean value of the function $g(X, Y) = X^2 + Y^2$.

*5.1-25. Define the conditional expected value of a function $g(X, Y)$ of random variables X and Y as

$$E[g(X, Y)|B] = \int_{-\infty}^{\infty}\int_{-\infty}^{\infty} g(x, y) f_{X,Y}(x, y|B)\, dx\, dy$$

(a) If event B is defined as $B = \{y_a < Y \le y_b\}$, where $y_a < y_b$ are constants,
evaluate $E[g(X, Y)|B]$. (*Hint*: Use results of Problem 4.4-8).
(b) If B is defined by $B = \{Y = y\}$ what does the conditional expected value of
part (a) become?

5.1-26. For random variables X and Y having $\bar{X} = 1$, $\bar{Y} = 2$, $\sigma_X^2 = 6$, $\sigma_Y^2 = 9$, and
$\rho = -\frac{2}{3}$, find (a) the covariance of X and Y, (b) the correlation of X and Y,
and (c) the moments m_{20} and m_{02}.

5.1-27. $\bar{X} = \frac{1}{2}$, $\overline{X^2} = \frac{5}{2}$, $\bar{Y} = 2$, $\overline{Y^2} = \frac{19}{2}$, and $C_{XY} = -1/(2\sqrt{3})$ for random variables X
and Y.
(a) Find $\sigma_X^2, \sigma_Y^2, R_{XY}$, and ρ.
(b) What is the mean value of the random variable $W = (X + 3Y)^2 + 2X + 3$?

5.1-28. Let X and Y be statistically independent random variables with $\bar{X} = \frac{3}{4}$,
$\overline{X^2} = 4$, $\bar{Y} = 1$, and $\overline{Y^2} = 5$. For a random variable $W = X - 2Y + 1$ find
(a) R_{XY}, (b) R_{XW}, (c) R_{YW}, and (d) C_{XY}. (e) Are X and Y uncorrelated?

5.1-29. Statistically independent random variables X and Y have moments $m_{10} = 2$,
$m_{20} = 14$, $m_{02} = 12$, and $m_{11} = -6$. Find the moment μ_{22}.

5.1-30. A joint density is given as

$$f_{X,Y}(x, y) = \begin{cases} x(y + 1.5) & 0 < x < 1 \quad \text{and} \quad 0 < y < 1 \\ 0 & \text{elsewhere} \end{cases}$$

Find all the joint moments m_{nk}, n and $k = 0, 1, \ldots$.

5.1-31. Find all the joint central moments μ_{nk}, n and $k = 0, 1, \ldots$, for the density of
Problem 5.1-30.

5.1-32. Random variables X and Y are defined by the joint density of Problem 4.3-19.
Find all first- and second-order joint moments for these random variables. Are
X and Y orthogonal? Are they uncorrelated?

5.1-33. In a control system, a random voltage X is known to have a mean value
$\bar{X} = m_1 = -2\,\text{V}$ and a second moment $\overline{X^2} = m_2 = 9\,\text{V}^2$. If the voltage X
is amplified by an amplifier that gives an output $Y = -1.5X + 2$, find
$\sigma_X^2, \bar{Y}, \overline{Y^2}, \sigma_Y^2$, and R_{XY}.

5.1-34. Two random variables X and Y are defined by $\bar{X} = 0$, $\bar{Y} = -1$, $\overline{X^2} = 2$,
$\overline{Y^2} = 4$, and $R_{XY} = -2$. Two new random variables W and U are:

$$W = 2X + Y$$
$$U = -X - 3Y.$$

Find $\bar{W}, \bar{U}, \overline{W^2}, \overline{U^2}, R_{WU}, \sigma_X^2$, and σ_Y^2.

5.1-35. Statistically independent random variables X and Y have respective means $\bar{X} = 1$ and $\bar{Y} = -1/2$. Their second moments are $\overline{X^2} = 4$ and $\overline{Y^2} = 11/4$. Another random variable is defined as $W = 3X^2 + 2Y + 1$. Find σ_X^2, σ_Y^2, R_{XY}, C_{XY}, $\bar{W}$, and R_{WY}.

5.1-36. Determine the correlation R_{XY} and correlation coefficient for the random variables defined in Problem 4.5-9.

5.1-37. The *cosine inequality*, sometimes called *Schwarz's inequality* for random variables X and Y is

$$[E(XY)]^2 \leq E(X^2)E(Y^2)$$

Show its validity. (*Hint*: Expand the nonnegative quantity $E[(aX - Y)^2]$, where a is a real parameter.)

5.1-38. The *triangle inequality* for random variables X and Y is

$$\{E[(X + Y)^2]\}^{0.5} \leq \{E(X^2)\}^{0.5} + \{E(Y^2)\}^{0.5}$$

Show its validity. (*Hint*: Expand and combine $E[(X + Y)^2]$ and $\{[E(X^2)]^{0.5} + [E(Y^2)]^{0.5}\}^2$ and use the cosine inequality of Problem 5.1-37.)

*5.2-1. Find the joint characteristic function for X and Y defined in Problem 5.1-3.

*5.2-2. Show that the joint characteristic function of N independent random variables X_i, having characteristic functions $\Phi_{X_i}(\omega_i)$ is

$$\Phi_{X_1,\dots,X_N}(\omega_1,\dots,\omega_N) = \prod_{i=1}^{N} \Phi_{X_i}(\omega_i)$$

*5.2-3. For N random variables, show that

$$|\Phi_{X_1,\dots,X_N}(\omega_1,\dots,\omega_N)| \leq \Phi_{X_1,\dots,X_N}(0,\dots,0) = 1$$

*5.2-4. For two zero-mean gaussian random variables X and Y, show that their joint characteristic function is

$$\Phi_{X,Y}(\omega_1,\omega_2) = \exp\{-\tfrac{1}{2}[\sigma_X^2\omega_1^2 + 2\rho\sigma_X\sigma_Y\omega_1\omega_2 + \sigma_Y^2\omega_2^2]\}$$

*5.2-5. Find the joint characteristic function for random variables X and Y defined by

$$f_{X,Y}(x, y) = (1/2\pi) \, \text{rect} \, (x/\pi) \, \text{rect} \, [(x+y)/\pi] \cos(x+y)$$

Use the result to find the marginal characteristic functions of X and Y.

*5.2-6. Random variables X_1 and X_2 have the joint characteristic function

$$\Phi_{X_1,X_2}(\omega_1,\omega_2) = [(1 - j2\omega_1)(1 - j2\omega_2)]^{-N/2}$$

where $N > 0$ is an integer.
(*a*) Find the correlation and moments m_{20} and m_{02}.
(*b*) Determine the means of X_1 and X_2.
(*c*) What is the correlation coefficient?

*5.2-7. The joint probability density of two discrete random variables X and Y consists of impulses located at all lattice points (mb, nd), where $m = 0, 1, \ldots, M$ and $n = 1, 2, \ldots, N$ with $b > 0$ and $d > 0$ being constants. All possible points are equally probable. Determine the joint characterisic function.

*5.2-8. Let $X_k, k = 1, 2, \ldots, K$, be statistically independent Poisson random variables, each with its own variance b_k (Problem 3.2-13). Show that the sum $X = X_1 + X_2 + \cdots + X_K$ is a Poisson random variable. (*Hint*: Use results of Problems 5.2-2 and 3.2-31.)

*5.2-9. Show that the sum X or N statistically independent Poisson random variables X_i, with different means b_i, is also a Poisson random variable but its mean is $b = b_1 + b_2 + \cdots + b_N$. [*Hint*: Use (5.2-7) and the result of Problem 5.2-2.]

*5.2-10. Show that the sum of N identically distributed statistically independent exponential random variables X_i, as given by (2.5-9) with $a = 0$ and b replaced by $1/a$, is an Erlang random variable, as defined in Problem 3.2-32 and in Appendix F. [*Hint*: Use (5.2-7) and the result of Problem 5.2-2.]

*5.2-11. The chi-square random variable with one degree of freedom is defined by the density

$$f_X(x) = \frac{u(x)e^{-x/2}}{\Gamma(1/2)\sqrt{2x}}$$

where $\Gamma(1/2)$ is a constant approximately equal to 1.772. Show that the sum X of N identically distributed statistically independent chi-square random variables, each with one degree of freedom, is a chi-square random variable with N degrees of freedom as defined in Problem 3.2-27 and Appendix F [see (F-35) through (F-39)]. [*Hint*: Use (5.2-7) and the result of Problem 5.2-2.]

*5.3-1. Zero-mean gaussian random variables X and Y have variances $\sigma_X^2 = 3$ and $\sigma_Y^2 = 4$, respectively, and a correlation coefficient $\rho = -\frac{1}{4}$.
(*a*) Write an expression for the joint density function.
(*b*) Show that a rotation of coordinates through the angle given by (5.3-11) will produce new statistically independent random variables.

*5.3-2. Find the conditional density functions $f_X(x|Y = y)$ and $f_Y(y|X = x)$ applicable to two gaussian random variables X and Y defined by (5.3-1) and show that they are also gaussian.

5.3-3. Assume $\sigma_x = \sigma_Y = \sigma$ in (5.3-1) and show that the locus of the maximum of the joint density is a line passing through the point $(\bar{X}, \bar{Y})$ with slope $\pi/4$ (or $-\pi/4$) when $\rho = 1$ (or -1).

5.3-4. Two gaussian random variables X and Y have variances $\sigma_X^2 = 9$ and $\sigma_Y^2 = 4$, respectively, and correlation coefficient ρ. It is known that a coordinate rotation by an angle $-\pi/8$ results in new random variables Y_1 and Y_2 that are uncorrelated. What is ρ?

*5.3-5. Let X and Y be jointly gaussian random variables where $\sigma_X^2 = \sigma_Y^2$ and $\rho = -1$. Find a transformation matrix such that new random variables Y_1 and Y_2 are statistically independent.

5.3-6. Gaussian random variables X and Y have first- and second-order moments $\bar{X} = -1.0$, $\overline{X^2} = 1.16$, $\bar{Y} = 1.5$, $\overline{Y^2} = 2.89$, and $R_{XY} = -1.724$. Find: (a) C_{XY} and (b) ρ. Also find the angle θ of a coordinate rotation that will generate new random variables that are statistically independent.

5.3-7. Suppose the annual snowfalls (accumulated depths in meters) for two nearby alpine ski resorts are adequately represented by jointly gaussian random variables X and Y, for which $\rho = 0.82$, $\sigma_X = 1.5\,\text{m}$, $\sigma_Y = 1.2\,\text{m}$, and $R_{XY} = 81.476\,\text{m}^2$. If the average snowfall at one resort is $10\,\text{m}$, what is the average at the other resort?

5.3-8. Two gaussian random variables X and Y have a correlation coefficient $\rho = 0.25$. The standard deviation of X is 1.9. A linear transformation (coordinate rotation of $\pi/6$) is known to transform X and Y to new random variables that are statistically independent. What is σ_Y^2?

*5.3-9. Gaussian random variables X_1 and X_2, for which $\bar{X}_1 = 2$, $\sigma_{X_1}^2 = 9$, $\bar{X}_2 = -1$, $\sigma_{X_2}^2 = 4$, and $C_{X_1 X_2} = -3$, are transformed to new random variables Y_1 and Y_2 according to

$$Y_1 = -X_1 + X_2$$
$$Y_2 = -2X_1 - 3X_2.$$

Find: (a) $\overline{X_1^2}$, (b) $\overline{X_2^2}$, (c) $\rho_{X_1 X_2}$, (d) $\sigma_{Y_1}^2$, (e) $\sigma_{Y_2}^2$, and (f) $C_{Y_1 Y_2}$.

*5.4-1. Random variables X and Y having the joint density

$$f_{X,Y}(x, y) = (\tfrac{8}{3})u(x - 2)u(y - 1)xy^2 \exp(4 - 2xy)$$

undergo a transformation

$$[T] = \begin{bmatrix} 1 & 1 \\ 1 & -1 \end{bmatrix}$$

to generate new random variables Y_1 and Y_2.
(a) Find the joint density of Y_1 and Y_2.
(b) Show what points in the $y_1 y_2$ plane correspond to a nonzero value of the new density.

*5.4-2. Three random variables X_1, X_2, and X_3 represent samples of a random noise voltage taken at three times. Their covariance matrix is defined by

$$[C_X] = \begin{bmatrix} 3.0 & 1.8 & 1.1 \\ 1.8 & 3.0 & 1.8 \\ 1.1 & 1.8 & 3.0 \end{bmatrix}.$$

A transformation matrix

$$[T] = \begin{bmatrix} 4 & -1 & -2 \\ 2 & 2 & 1 \\ -3 & -1 & 3 \end{bmatrix}$$

converts the variables to new random variables Y_1, Y_2, and Y_3. Find the covariance matrix of the new random variables.

*5.4-3. Determine the density of $Y = (X_1^2 + X_2^2)^{0.5}$ when X_1 and X_2 are jointly gaussian random variables with zero means and the same variance. (*Hint*: Use the results of Example 5.4-2.)

*5.5-1. Zero-mean gaussian random variables X_1, X_2, and X_3 having a covariance matrix

$$[C_X] = \begin{bmatrix} 4 & 2.05 & 1.05 \\ 2.05 & 4 & 2.05 \\ 1.05 & 2.05 & 4 \end{bmatrix}$$

are transformed to new variables

$$Y_1 = 5X_1 + 2X_2 - X_3$$
$$Y_2 = -X_1 + 3X_2 + X_3$$
$$Y_3 = 2X_1 - X_2 + 2X_3$$

(a) Find the covariance matrix of Y_1, Y_2, and Y_3.
(b) Write an expression for the joint density function of Y_1, Y_2, and Y_3.

*5.5-2. Two gaussian random variables X_1 and X_2 are defined by the mean and covariance matrices

$$[\bar{X}] = \begin{bmatrix} 2 \\ -1 \end{bmatrix} \quad [C_X] = \begin{bmatrix} 5 & -2/\sqrt{5} \\ -2/\sqrt{5} & 4 \end{bmatrix}$$

Two new random variables Y_1 and Y_2 are formed using the transformation

$$[T] = \begin{bmatrix} 1 & \frac{1}{2} \\ \frac{1}{2} & 1 \end{bmatrix}$$

Find the matrices (a) $[\bar{Y}]$ and (b) $[C_Y]$. (c) Also find the correlation coefficient of Y_1 and Y_2.

*5.6-1. Show that (5.6-2) results from the transformations of (5.6-1).

*5.6-2. Extend the text and show that (5.6-1) can be replaced by

$$Y_1 = T_1(X_1, X_2) = \bar{Y}_1 + \sqrt{-2\sigma_{Y_1}^2 \ln(X_1)} \cos(2\pi X_2)$$
$$Y_2 = T_2(X_1, X_2) = \bar{Y}_2 + \sqrt{-2\sigma_{Y_2}^2 \ln(X_1)} \sin(2\pi X_2)$$

to generate statistically independent gaussian random variables Y_1 and Y_2, with respective means $\bar{Y}_1$ and $\bar{Y}_2$, and respective variances $\sigma_{Y_1}^2$ and $\sigma_{Y_2}^2$.

*5.6-3. Extend the text that leads to (5.6-7) and find transformations of two statistically independent random variables X_1 and X_2, both uniform on $(0, 1)$, that will directly create two correlated gaussian random variables W_1 and W_2 having correlation coefficient ρ_W, means $\bar{W}_1$ and $\bar{W}_2$, and variances $\sigma_{W_1}^2$ and $\sigma_{W_2}^2$.

*5.6-4. Work Problem 5.6-3, except generate the random variables R and Θ for which (5.6-16) applies.

*5.6-5. Repeat Example 5.6-1 except use $N = 1000$ values of X_1 and 1000 values of X_2. Note the improvement in the accuracy of the estimated quantities for random variables W_1 and W_2 as compared to the example.

5.7-1. Find the mean value of the power estimator of (5.7-7) and give arguments why the estimator is unbiased.

5.7-2. Find the variance of the power estimator of (5.7-7) and show that it approaches zero as N becomes infinite.

5.7-3. If the factor $1/(N-1)$ in the variance estimator of (5.7-8) is replaced by $1/N$, show that the mean of the modified estimator is biased. Determine the amount of bias. How does the bias behave as N becomes very large?

5.7-4. Rework Example 5.7-3 for 1000 values of Y generated by MATLAB. Compare the new values of sample mean, second moment, and variance with those found in the example. Are they more accurate?

*5.8-1. A complex random variable Z is defined by

$$Z = \cos(X) + j\sin(Y)$$

where X and Y are independent real random variables uniformly distributed from $-\pi$ to π.
(a) Find the mean value of Z.
(b) Find the variance of Z.

*5.8-2. Complex random variables Z_1 and Z_2 have zero means. The correlation of the real parts of Z_1 and Z_2 is 4, while the correlation of the imaginary parts is 6. The real part of Z_1 and the imaginary part of Z_2 are statistically independent as a pair, as are the imaginary part of Z_1 and the real part of Z_2.
(a) What is the correlation of Z_1 and Z_2?
(b) Are Z_1 and Z_2 statistically independent?

Random Processes—Temporal Characteristics

6.0
INTRODUCTION

In the real world of engineering and science, it is necessary that we be able to deal with time waveforms. Indeed, we frequently encounter *random* time waveforms in practical systems. More often than not, a *desired* signal in some system is random. For example, the bit stream in a binary communication system is a random message because each bit in the stream occurs randomly. On the other hand, a desired signal is often accompanied by an *undesired* random waveform, noise. The noise interferes with the message and ultimately limits the performance of the system. Thus, any hope we have of determining the performance of systems with random waveforms hinges on our ability to describe and deal with such waveforms. In this chapter we introduce concepts that allow the description of random waveforms in a probabilistic sense.

6.1
THE RANDOM PROCESS CONCEPT

The concept of a random process is based on enlarging the random variable concept to include time. Since a random variable X is, by its definition, a function of the possible outcomes s of an experiment, it now becomes a function of both s and time. In other words, we assign, according to some rule, a time function

$$x(t, s) \qquad (6.1\text{-}1)$$

to every outcome s. The family of all such functions, denoted $X(t, s)$, is called a *random process*. As with random variables where x was denoted as a specific value of the random variable X, we shall often use the convenient short-form notation $x(t)$ to represent a specific waveform of a random process denoted by $X(t)$.

Clearly, a random process $X(t, s)$ represents a family or *ensemble* of time functions when t and s are variables. Figure 6.1-1 illustrates a few members of an ensemble. Each member time function is called a *sample function, ensemble member*, or sometimes a *realization* of the process. Thus, a random process also represents a *single* time function when t is a variable and s is fixed at a specific value (outcome).

A random process also represents a random variable when t is fixed and s is a variable. For example, the random variable $X(t_1, s) = X(t_1)$ is obtained from the process when time is "frozen" at the value t_1. We often use the notation X_1 to denote the random variable associated with the process $X(t)$ at time t_1. X_1 corresponds to a vertical "slice" through the ensemble at time t_1, as illustrated in Figure 6.1-1. The statistical properties of $X_1 = X(t_1)$ describe the statistical properties of the random process at time t_1. The expected value of X_1 is called the *ensemble average* as well as the expected or mean value of

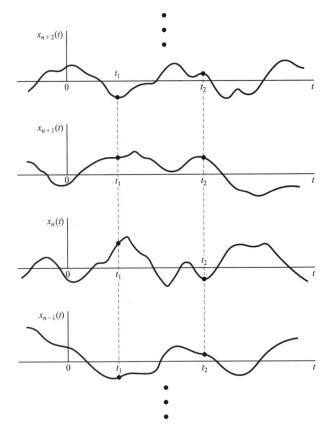

FIGURE 6.1-1
A continuous random process. [*Reproduced from Peebles (1976) with permission of publishers Addison–Wesley, Advanced Book Program.*]

the random process (at time t_1). Since t_1 may have various values, the mean value of a process may not be constant; in general, it may be a function of time. We easily visualize any number of random variables X_i derived from a random process $X(t)$ at times t_i, $i = 1, 2, \ldots$:

$$X_i = X(t_i, s) = X(t_i) \tag{6.1-2}$$

181

CHAPTER 6:
Random
Processes—
Temporal
Characteristics

A random process can also represent a mere number when t and s are both fixed.

Classification of Processes

It is convenient to classify random processes according to the characteristics of t and the random variable $X = X(t)$ at time t. We shall consider only four cases based on t and X having values in the ranges $-\infty < t < \infty$ and $-\infty < x < \infty$.†

If X is continuous and t can have any of a continuum of values, then $X(t)$ is called a *continuous random process*. Figure 6.1-1 is an illustration of this class of process. Thermal noise generated by any realizable network is a practical example of a waveform that is modeled as a sample function of a continuous random process. In this example, the network is the outcome in the underlying random experiment of selecting a network. (The presumption is that many networks are available from which to choose; this may not be the case in the real world, but it should not prevent us from imagining a production line producing any number of similar networks.) Each network establishes a sample function, and all sample functions form the process.‡

A second class of random process, called a *discrete random process*, corresponds to the random variable X having only discrete values while t is continuous. Figure 6.1-2 illustrates such a process derived by heavily limiting the sample functions shown in Figure 6.1-1. The sample functions have only two discrete values: the positive level is generated whenever a sample function in Figure 6.1-1 is positive and the negative level occurs for other times.

A random process for which X is continuous but time has only discrete values is called a *continuous random sequence* (Thomas, 1969, p. 80). One example of such a sequence can be formed by periodically sampling the ensemble members of Figure 6.1-1. The result is illustrated in Figure 6.1-3. Since a continuous random sequence is defined at only discrete (sample) times, it is also frequently called a *discrete-time* (DT) *random process*; its sample functions are often referred to as a *DT random signal*. Technically, a DT random process is a set of random variables denoted by $\{X(t_n)\}$ for sample times $t_n = nT_s$, $n = 0$, ± 1, $\pm 2, \ldots$, with T_s called the *sampling interval*. The *sampling rate* is $1/T_s$

---

†Other cases can be defined based on a definition of random processes on a finite time interval (see for example: Rosenblatt (1974), p. 91; Prabhu (1965), p. 1; Miller (1974), p. 31; Parzen (1962), p. 7; Dubes (1968), p. 320; Ross (1972), p. 56). Other recent texts on random processes are Helstrom (1984), and Gray and Davisson (1986).

‡Note that finding the mean value of the process at any time t is equivalent to finding the average voltage that would be produced by all the various networks at time t.

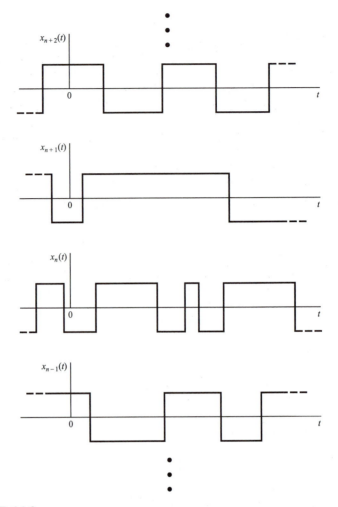

FIGURE 6.1-2
A discrete random process formed by heavily limiting the waveforms of Figure
6.1-1. [*Reproduced from Peebles (1976) with permission of publishers Addison–
Wesley, Advanced Book Program.*]

samples per second. However, for most practical work it is sufficient to refer to
a DT random process as $X(nT_s)$. These types of processes are important in the
analysis of various *digital signal processing* (DSP) systems where the consant T_s
is not important.† The process is then usually referred to as a *discrete-time
sequence*, and notation $X[n]$ is adopted.‡

---

†As subsequently described, T_s becomes important in digital systems mainly when a digital signal is
required to be converted back to its analog form.
‡In most DSP literature brackets are used to imply a DT sequence from which some sampling
interval has been tacitly omitted. In essence, $X[n]$ becomes a DT sequence function of index n.

183

CHAPTER 6:
Random
Processes—
Temporal
Characteristics

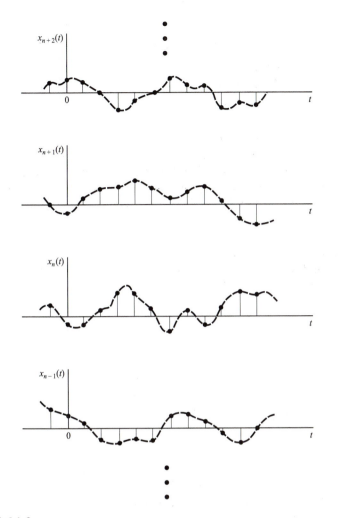

FIGURE 6.1-3
A continuous random sequence (or discrete-time random process) formed by sampling the waveforms of Figure 6.1-1. [*Adapted from Peebles (1976), with permission of publishers Addison–Wesley, Advanced Book Program.*]

A fourth class of random process, called a *discrete random sequence*, corresponds to both time and the random variable being discrete. Figure 6.1-4 illustrates a discrete random sequence developed by sampling the sample functions of Figure 6.1-2. Alternatively, it can derive from rounding off samples of a DT random process (continuous random sequence). This operation is exactly what happens in DSP systems. The rounding consists of choosing a discrete amplitude, from a finite set of discrete amplitudes, that most closely equals each sample value of the DT random process. The operation is called *quantization* and is necessary to convert the process to a form suitable for use in a digital computer. For the DSP application one might refer to the quantized DT random process as a *digital process*. However, as we shall subsequently note, there

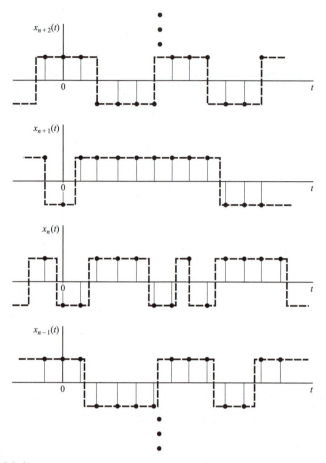

FIGURE 6.1-4
A discrete random sequence formed by sampling the waveforms of Figure 6.1-2.
[*Adapted from Peebles (1976) with permission of publishers Addison–Wesley,
Advanced Book Program.*]

is usually no need to model DSP signals as a digital process because the errors
committed in quantization are typically small enough to be ignored. Analysis
then proceeds based on the nonquantized DT random process.

Deterministic and Nondeterministic Processes

In addition to the classes described above, a random process can be described
by the form of its sample functions. If future values of any sample function
cannot be predicted exactly from observed past values, the process is called
nondeterministic. The process of Figure 6.1-1 is one example.

A process is called *deterministic* if future values of any sample function
can be predicted from past values. An example is the random process defined
by

$$X(t) = A \cos(\omega_0 t + \Theta) \qquad (6.1\text{-}3)$$

185

CHAPTER 6:
Random
Processes—
Temporal
Characteristics

Here A, Θ, or ω_0 (or all) may be random variables. Any one sample function corresponds to (6.1-3) with particular values of these random variables. Therefore, knowledge of the sample function prior to any time instant automatically allows prediction of the sample function's future values because its form is known.

6.2
STATIONARITY AND INDEPENDENCE

As previously stated, a random process becomes a random variable when time is fixed at some particular value. The random variable will possess statistical properties, such as a mean value, moments, variance, etc., that are related to its density function. If *two* random variables are obtained from the process for two time instants, they will have statistical properties (means, variances, joint moments, etc.) related to their joint density function. More generally, N random variables will possess statistical properties related to their N-dimensional joint density function.

Broadly speaking, a random process is said to be *stationary* if all its statistical properties do not change with time. Other processes are called *nonstationary*. These statements are not intended as definitions of stationarity but are meant to convey only a general meaning. More concrete definitions follow. Indeed, there are several "levels" of stationarity, all of which depend on the density functions of the random variables of the process. Our definitions apply to both continuous and discrete processes.

Distribution and Density Functions

To define stationarity, we must first define distribution and density functions as they apply to a random process $X(t)$. For a particular time t_1, the distribution function associated with the random variable $X_1 = X(t_1)$ will be denoted $F_X(x_1; t_1)$. It is defined as†

$$F_X(x_1; t_1) = P\{X(t_1) \le x_1\} \qquad (6.2\text{-}1)$$

for any real number x_1. This is the same definition used all along for the distribution function of one random variable. Only the notation has been altered to reflect the fact that it is possibly now a function of time choice t_1.

For two random variables $X_1 = X(t_1)$ and $X_2 = X(t_2)$, the *second-order joint distribution function* is the two-dimensional extension of (6.2-1):

$$F_X(x_1, x_2; t_1, t_2) = P\{X(t_1) \le x_1, X(t_2) \le x_2\} \qquad (6.2\text{-}2)$$

---

†$F_X(x_1; t_1)$ is known as the *first-order distribution function* of the process $X(t)$.

In a similar manner, for N random variables $X_i = X(t_i), i = 1, 2, \ldots, N$, the *Nth-order joint distribution function* is

$$F_X(x_1, \ldots, x_N; t_1, \ldots, t_N) = P\{X(t_1) \leq x_1, \ldots, X(t_N) \leq x_N\} \qquad (6.2\text{-}3)$$

Joint density functions of interest are found from appropriate derivatives of the above three relationships:†

$$f_X(x_1; t_1) = dF_X(x_1; t_1)/dx_1 \qquad (6.2\text{-}4)$$

$$f_X(x, x_2; t_1, t_2) = \partial^2 F_X(x_1, x_2; t_1, t_2)/(\partial x_1 \, \partial x_2) \qquad (6.2\text{-}5)$$

$$f_X(x_1, \ldots, x_N; t_1, \ldots, t_N) = \partial^N F_X(x_1, \ldots, x_N; t_1, \ldots, t_N)/(\partial x_1 \cdots \partial x_N) \qquad (6.2\text{-}6)$$

Statistical Independence

Two processes $X(t)$ and $Y(t)$ are *statistically independent* if the random variable group $X(t_1), X(t_2), \ldots, X(t_N)$ is independent of the group $Y(t_1'), Y(t_2'),$ $\ldots, Y(t_M')$ for any choice of times $t_1, t_2, \ldots, t_N, t_1', t_2', \ldots, t_M'$. Independence requires that the joint density be factorable by groups:

$$f_{X,Y}(x_1, \ldots, x_N, y_1, \ldots, y_M; t_1, \ldots, t_N, t_1', \ldots, t_M')$$
$$= f_X(x_1, \ldots, x_N; t_1, \ldots, t_N) f_Y(y_1, \ldots, y_M; t_1', \ldots, t_M') \qquad (6.2\text{-}7)$$

First-Order Stationary Processes

A random process is called *stationary to order one* if its first-order density function does not change with a shift in time origin. In other words

$$f_X(x_1; t_1) = f_X(x_1; t_1 + \Delta) \qquad (6.2\text{-}8)$$

must be true for any t_1 and any real number Δ if $X(t)$ is to be a first-order stationary process.

Consequences of (6.2-8) are that $f_X(x_1; t_1)$ is independent of t_1 and the process mean value $E[X(t)]$ is a constant:

$$E[X(t)] = \bar{X} = \text{constant} \qquad (6.2\text{-}9)$$

To prove (6.2-9), we find mean values of the random variables $X_1 = X(t_1)$ and $X_2 = X(t_2)$. For X_1:

$$E[X_1] = E[X(t_1)] = \int_{-\infty}^{\infty} x_1 f_X(x_1; t_1) \, dx_1 \qquad (6.2\text{-}10)$$

---

†Analogous to distribution functions, these are called *first-*, *second*, and *Nth-order density functions*, respectively.

For X_2:

187

CHAPTER 6:
Random
Processes—
Temporal
Characteristics

$$E[X_2] = E[X(t_2)] = \int_{-\infty}^{\infty} x_1 f_X(x_1; t_2) \, dx_1 \dagger \qquad (6.2\text{-}11)$$

Now by letting $t_2 = t_1 + \Delta$ in (6.2-11), substituting (6.2-8), and using (6.2-10), we get

$$E[X(t_1 + \Delta)] = E[X(t_1)] \qquad (6.2\text{-}12)$$

which must be a constant because t_1 and Δ are arbitrary.

Second-Order and Wide-Sense Stationarity

A process is called *stationary to order two* if its second-order density function satisfies

$$f_X(x_1, x_2; t_1, t_2) = f_X(x_1, x_2; t_1 + \Delta, t_2 + \Delta) \qquad (6.2\text{-}13)$$

for all t_1, t_2, and Δ. After some thought, the reader will conclude that (6.2-13) is a function of time differences $t_2 - t_1$ and not absolute time (let arbitrary $\Delta = -t_1$). A second-order stationary process is also first-order stationary because the second-order density function determines the lower, first-order, density.

Now the correlation $E[X_1 X_2] = E[X(t_1)X(t_2)]$ of a random process will, in general, be a function of t_1 and t_2. Let us denote this function by $R_{XX}(t_1, t_2)$ and call it the *autocorrelation function* of the random process $X(t)$:

$$R_{XX}(t_1, t_2) = E[X(t_1)X(t_2)]. \qquad (6.2\text{-}14)$$

A consequence of (6.2-13), however, is that the autocorrelation function of a second-order stationary process is a function only of time differences and not absolute time; that is, if

$$\tau = t_2 - t_1 \qquad (6.2\text{-}15)$$

then (6.2-14) becomes

$$R_{XX}(t_1, t_1 + \tau) = E[X(t_1)X(t_1 + \tau)] = R_{XX}(\tau) \qquad (6.2\text{-}16)$$

Proof of (6.2-16) uses (6.2-13); it is left as a reader exercise (see Problem 6.2-2).

Many practical problems require that we deal with the autocorrelation function and mean value of a random process. Problem solutions are greatly simplified if these quantities are not dependent on absolute time. Of course, second-order stationarity is sufficient to guarantee these characteristics. However, it is often more restrictive than necessary, and a more relaxed form of stationarity is desirable. The most useful form is the *wide-sense stationary process*, defined as that for which two conditions are true:

---

†Note that the variable x_2 of integration has been replaced by the alternative variable x_1 for convenience.

$$E[X(t)] = \bar{X} = \text{constant} \qquad (6.2\text{-}17a)$$

$$E[X(t)X(t + \tau)] = R_{XX}(\tau) \qquad (6.2\text{-}17b)$$

A process stationary to order 2 is clearly wide-sense stationary. However, the converse is not necessarily true.

> EXAMPLE 6.2-1. We show that the random process
>
> $$X(t) = A\cos(\omega_0 t + \Theta)$$
>
> is wide-sense stationary if it is assumed that A and ω_0 are constants and Θ is a uniformly distributed random variable on the interval $(0, 2\pi)$. The mean value is
>
> $$E[X(t)] = \int_0^{2\pi} A\cos(\omega_0 t + \theta)\frac{1}{2\pi}d\theta = 0$$
>
> The autocorrelation function, from (6.2-14) with $t_1 = t$ and $t_2 = t + \tau$, becomes
>
> $$R_{XX}(t, t + \tau) = E[A\cos(\omega_0 t + \Theta)A\cos(\omega_0 t + \omega_0\tau + \Theta)]$$
>
> $$= \frac{A^2}{2}E[\cos(\omega_0\tau) + \cos(2\omega_0 t + \omega_0\tau + 2\Theta)]$$
>
> $$= \frac{A^2}{2}\cos(\omega_0\tau) + \frac{A^2}{2}E[\cos(2\omega_0 t + \omega_0\tau + 2\Theta)]$$
>
> The second term easily evaluates to 0. Thus, the autocorrelation function depends only on τ and the mean value is a constant, so $X(t)$ is wide-sense stationary.

When we are concerned with two random processes $X(t)$ and $Y(t)$, we say they are *jointly* wide-sense stationary if each satisfies (6.2-17) and their *cross-correlation function*, defined in general by

$$R_{XY}(t_1, t_2) = E[X(t_1)Y(t_2)] \qquad (6.2\text{-}18)$$

is a function only of time difference $\tau = t_2 - t_1$ and not absolute time; that is, if

$$R_{XY}(t, t + \tau) = E[X(t)Y(t + \tau)] = R_{XY}(\tau) \qquad (6.2\text{-}19)$$

N-Order and Strict-Sense Stationarity

By extending the above reasoning to N random variables $X_i = X(t_i)$, $i = 1, 2, \ldots, N$, we say a random process is *stationary to order N* if its Nth-order density function is invariant to a time origin shift; that is, if

$$f_X(x_1, \ldots, x_N; t_1, \ldots, t_N) = f_X(x_1, \ldots, x_N; t_1 + \Delta, \ldots, t_N + \Delta) \qquad (6.2\text{-}20)$$

for all $t_1, \ldots, t_N$ and Δ. Stationarity of order N implies stationarity to all orders $k \le N$. A process stationary to *all* orders $N = 1, 2, \ldots$, is called *strict-sense stationary*.

Time Averages and Ergodicity

189

CHAPTER 6:
Random
Processes—
Temporal
Characteristics

The time average of a quantity is defined as

$$A[\cdot] = \lim_{T \to \infty} \frac{1}{2T} \int_{-T}^{T} [\cdot] \, dt \qquad (6.2\text{-}21)$$

Here A is used to denote time average in a manner analogous to E for the statistical average. Time average is taken over all time because, as applied to random processes, sample functions of processes are presumed to exist for all time.

Specific averages of interest are the mean value $\bar{x} = A[x(t)]$ of a sample function (a lowercase letter is used to imply a sample function), and the *time autocorrelation function*, denoted $\mathcal{R}_{xx}(\tau) = A[x(t)x(t + \tau)]$. These functions are defined by

$$\bar{x} = A[x(t)] = \lim_{T \to \infty} \frac{1}{2T} \int_{-T}^{T} x(t) \, dt \qquad (6.2\text{-}22)$$

$$\mathcal{R}_{xx}(\tau) = A[x(t)x(t + \tau)]$$

$$= \lim_{T \to \infty} \frac{1}{2T} \int_{-T}^{T} x(t)x(t + \tau) \, dt \qquad (6.2\text{-}23)$$

For any *one* sample function of the process $X(t)$, these last two integrals simply produce two numbers (for a fixed value of τ). However, when all sample functions are considered, we see that $\bar{x}$ and $\mathcal{R}_{xx}(\tau)$ are actually *random variables*. By taking the expected value on both sides of (6.2-22) and (6.2-23), and assuming the expectation can be brought inside the integrals, we obtain†

$$E[\bar{x}] = \bar{X} \qquad (6.2\text{-}24)$$

$$E[\mathcal{R}_{xx}(\tau)] = R_{XX}(\tau) \qquad (6.2\text{-}25)$$

Now suppose by some theorem the random variables $\bar{x}$ and $\mathcal{R}_{xx}(\tau)$ could be made to have zero variances; that is, $\bar{x}$ and $\mathcal{R}_{xx}(\tau)$ actually become constants. Then we could write

$$\bar{x} = \bar{X} \qquad (6.2\text{-}26)$$

$$\mathcal{R}_{xx}(\tau) = R_{XX}(\tau) \qquad (6.2\text{-}27)$$

In other words, the time averages $\bar{x}$ and $\mathcal{R}_{xx}(\tau)$ equal the statistical averages $\bar{X}$ and $R_{XX}(\tau)$, respectively. The *ergodic theorem* allows the validity of (6.2-26) and (6.2-27). Stated in loose terms, it more generally allows all time averages to equal the corresponding statistical averages. Processes that satisfy the ergodic theorem are called *ergodic processes*.

Ergodicity is a very restrictive form of stationarity, and it may be difficult to prove that it constitutes a reasonable assumption in any physical situation.

---

†We assume also that $X(t)$ is a stationary process so that the mean and the autocorrelation function are not time-dependent.

Nevertheless, we shall often assume a process is ergodic to simplify problems. In the real world, we are usually forced to work with only one sample function of a process and therefore must, like it or not, derive mean value, correlation functions, etc., from the time waveform. By assuming ergodicity, we may infer the similar statistical characteristics of the process. The reader may feel that our theory is on shaky ground on the basis of these comments. However, it must be remembered that all our theory only serves to model real-world conditions. Therefore, what difference do our assumptions really make provided the assumed model does truly reflect real conditions?

Two random processes are called *jointly ergodic* if they are individually ergodic and also have a *time cross-correlation function* that equals the statistical cross-correlation function:

$$\mathscr{R}_{xy}(\tau) = \lim_{T \to \infty} \frac{1}{2T} \int_{-T}^{T} x(t)y(t+\tau)\,dt = R_{XY}(\tau) \qquad (6.2\text{-}28)$$

The above discussion of ergodic processes is meant to convey only a general idea of the meaning of ergodicity; it is not intended to be very rigorous or precise. Indeed, a careful discussion of ergodic processes requires that several levels of ergodicity be defined, somewhat analogous to stationarity of processes. In the following two subsections we give slightly more detailed discussions of two forms of ergodicity. For more rigorous additional developments the reader is referred to the literature [see, for example, Papoulis (1965, 1991), pp. 323–332, 427–442; Gray and Davisson (1986), pp. 170–178; Gardner (1990), Chapter 8; and Viniotis (1998), pp. 413–417].

Mean-Ergodic Processes

A process $X(t)$ with a constant mean value $\bar{X}$ is called *mean-ergodic*, or *ergodic in the mean*, if its statistical average $\bar{X}$ equals the time average $\bar{x}$ of any sample function $x(t)$ with probability 1 for all sample functions; that is, if

$$E[X(t)] = \bar{X} = A[x(t)] = \bar{x} \qquad (6.2\text{-}29)$$

with probability 1 for all $x(t)$.

To develop a proof of the preceding statements, we start by assuming four conditions are true for $X(t)$. First, $X(t)$ has a finite constant mean $\bar{X}$ for all t. Second, $X(t)$ is bounded; that is, $|x(t)| < \infty$ for all t and all $x(t)$ comprising $X(t)$. Third,

$$\lim_{T \to \infty} \frac{1}{2T} \int_{-T}^{T} E[|X(t)|]\,dt < \infty \qquad (6.2\text{-}30)$$

The second and third assumptions are necessary to allow the interchange of operations of expectation and integration subsequently needed. The fourth assumption is that $X(t)$ is a *regular process* [Prabhu (1965), p. 20]; that is,

$$E[|X(t)|^2] = R_{XX}(t, t) < \infty \qquad (6.2\text{-}31)$$

For a real wide-sense stationary process this is equivalent to $E\{[X(t)]^2\} = R_{XX}(0) < \infty$, and since $\bar{X}$ is finite by assumption, it means $C_{XX}(0) = R_{XX}(0) - \bar{X}^2 < \infty$ is also true. [See (6.3-21) for a definition of the *auto-covariance function* $C_{XX}(t, t + \tau)$.]

191

CHAPTER 6:
Random
Processes—
Temporal
Characteristics

Equation (6.2-22) represents a number $\bar{x}$ for a given process sample function $x(t)$. If we imagine the time average is taken of the process, we can define a random variable A_X by

$$A_X = \lim_{T \to \infty} \frac{1}{2T} \int_{-T}^{T} X(t)\, dt \qquad (6.2\text{-}32)$$

to define all values of $\bar{x}$ produced by all process sample functions. The mean value of A_X is

$$\bar{A}_X = E[A_X]$$

$$= E\left\{ \lim_{T \to \infty} \frac{1}{2T} \int_{-T}^{T} X(t)\, dt \right\} = \lim_{T \to \infty} \frac{1}{2T} \int_{-T}^{T} \bar{X}\, dt = \bar{X} \qquad (6.2\text{-}33)$$

On use of Chebychev's inequality of (3.2-10) we have

$$P\{|A_X - \bar{A}_X| < \epsilon\} \geq 1 - (\sigma_{A_X}^2 / \epsilon^2) \qquad \text{any } \epsilon > 0 \qquad (6.2\text{-}34)$$

where $\sigma_{A_X}^2$ is the variance of A_X. Now for any $\epsilon > 0$, no matter how small, if $\sigma_{A_X}^2 = 0$, then $A_X = \bar{A}_X$ with probability 1. In other words, the random variable A_X is a constant.

From (6.2-34) we have found that $\bar{A}_X = \bar{x}$, all $x(t)$, with probability 1 if the variance of A_X is zero. We next show conditions under which the variance is zero. Begin with

$$\sigma_{A_X}^2 = E[(A_X - \bar{A}_X)^2] = E\left\{ \left[\lim_{T \to \infty} \frac{1}{2T} \int_{-T}^{T} [X(t) - \bar{X}]\, dt \right]^2 \right\}$$

$$= E\left\{ \lim_{T \to \infty} \left(\frac{1}{2T}\right)^2 \int_{-T}^{T} \int_{-T}^{T} [X(t) - \bar{X}][X(t_1) - \bar{X}]\, dt\, dt_1 \right\}$$

$$= \lim_{T \to \infty} \left(\frac{1}{2T}\right)^2 \int_{-T}^{T} \int_{-T}^{T} C_{XX}(t, t_1)\, dt\, dt_1 \qquad (6.2\text{-}35)$$

Thus, $\sigma_{A_X}^2 = 0$ if, and only if, the last form of (6.2-35) equals zero. This general condition can be specialized for wide-sense stationary processes where $C_{XX}(t, t_1) = C_{XX}(\tau)$, $\tau = t_1 - t$, $dt_1 = d\tau$:

$$\sigma_{A_X}^2 = \lim_{T \to \infty} \left(\frac{1}{2T}\right)^2 \int_{t=-T}^{T} \int_{\tau=-T-t}^{T-t} C_{XX}(\tau)\, d\tau\, dt \qquad (6.2\text{-}36)$$

This integral represents using horizontal Riemann strips in the region of integration over the τt plane as sketched in Figure 6.2-1. If vertical strips are used instead, and the symmetry $C_{XX}(-\tau) = C_{XX}(\tau)$ is invoked, the integral evaluates readily (Problem 6.2-19)

$$\sigma_{A_X}^2 = \lim_{T \to \infty} \frac{1}{2T} \int_{-2T}^{2T} \left(1 - \frac{|\tau|}{2T}\right) C_{XX}(\tau)\, d\tau < \lim_{T \to \infty} \frac{1}{2T} \int_{-2T}^{2T} |C_{XX}(\tau)|\, d\tau \qquad (6.2\text{-}37)$$

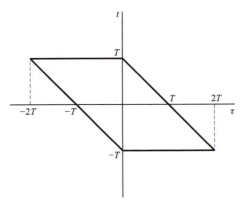

FIGURE 6.2-1
Closed region of integration applicable to (6.2-36).

Finally, we see from (6.2-37) that the variance is zero if

1. $C_{XX}(0) < \infty$ and $C_{XX}(\tau) \to 0$ as $|\tau| \to \infty$, and $\qquad$ (6.2-38a)

2. $\displaystyle\int_{-\infty}^{\infty} |C_{XX}(\tau)| < \infty$ $\qquad$ (6.2-38b)

Consequently, $X(t)$ will be mean-ergodic.

> **EXAMPLE 6.2-2.** A zero-mean wide-sense stationary process $X(t)$ has an autocorrelation function $R_{XX}(\tau) = C_{XX}(\tau) = \exp(-2\alpha|\tau|)$ for $\alpha > 0$ a constant. We determine if $X(t)$ is mean-ergodic. Let I represent the middle form of (6.2-37) and observe that $C_{XX}(-\tau) = C_{XX}(\tau)$, so
>
> $$I = \lim_{T \to \infty} \frac{1}{T} \int_0^{2T} \left(1 - \frac{\tau}{2T}\right) e^{-2\alpha\tau} \, d\tau$$
>
> Simple integrals from Appendix C produce
>
> $$I = \lim_{T \to \infty} \left\{ \frac{1}{2\alpha T}\left(1 - e^{-4\alpha T}\right) - \frac{1}{8\alpha^2 T^2}\left(1 - e^{-4\alpha T} - 4\alpha T e^{-4\alpha T}\right) \right\} = 0$$
>
> Since $C_{XX}(\tau)$ satisfies (6.2-37) and $\sigma_{A_X}^2 = 0$, then $X(t)$ is mean-ergodic.

Let a sequence of values be generated from a wide-sense stationary random process $X(t)$ with constant mean value $\bar{X}$ by sampling at $2N + 1$ times at periodic points in time from $-T$ to $+T$. The samples may be represented by a discrete-time sequence of identically distributed random variables $X[n]$ for sample n. Samples are assumed to be statistically independent, which is approximately true in practice if samples are separated far enough in time. The discrete sequence is called mean-ergodic if the time average of samples equals the statistical average with probability 1.

$$A_X = \lim_{N \to \infty} \frac{1}{2N + 1} \sum_{n=-N}^{N} X[n] = \bar{X}, \qquad \text{probability 1} \qquad (6.2\text{-}39)$$

It can be shown that this result is true if

193

CHAPTER 6:
Random
Processes—
Temporal
Characteristics

$$\sigma_{A_X}^2 = E[(A_X - \bar{A}_X)^2] = \lim_{N \to \infty} \frac{1}{2N+1} \sum_{n=-2N}^{2N} \left(1 - \frac{|n|}{2N+1}\right) C_{XX}[n] = 0$$

$$(6.2\text{-}40)$$

[See Papoulis, 1991, p. 432; Childers, 1997, p. 277; Leon-Garcia, 1989, p. 364.]

Correlation-Ergodic Processes

Analogous to a mean-ergodic process, a stationary continuous process $X(t)$ with autocorrelation function $R_{XX}(\tau)$ is called *autocorrelation-ergodic* or *ergodic in the autocorrelation* if, and only if, for all τ

$$\lim_{T \to \infty} \frac{1}{2T} \int_{-T}^{T} X(t)X(t+\tau)\,dt = R_{XX}(\tau) \qquad (6.2\text{-}41)$$

A necessary and sufficient condition for (6.2-41) to be true follows a similar development leading to (6.2-37) except with $X(t)$ replaced now by

$$W(t) = X(t)X(t+\lambda) \qquad (6.2\text{-}42)$$

where λ is a time offset. We have

$$E[W(t)] = E[X(t)X(t+\lambda)] = R_{XX}(\lambda) \qquad (6.2\text{-}43)$$

$$R_{WW}(\tau) = E[W(t)W(t+\tau)]$$

$$= E[X(t)X(t+\lambda)X(t+\tau)X(t+\tau+\lambda)] \qquad (6.2\text{-}44)$$

$$C_{WW}(\tau) = R_{WW}(\tau) - \{E[W(t)]\}^2$$

$$= R_{WW}(\tau) - R_{XX}^2(\lambda) \qquad (6.2\text{-}45)$$

Thus, $X(t)$ is autocorrelation-ergodic if $C_{XX}(\tau)$ of the middle form of (6.2-37) is replaced by $C_{WW}(\tau)$ of (6.2-45), and the integral is zero. This and other forms of ergodicity are described in Papoulis (1984). We note that autocorrelation-ergodicity requires fourth-order moments of $X(t)$, as defined in (6.2-44). For processes in general this requirement may be severe. However, for gaussian processes, fourth-order moments are known in terms of only moments of order two or less [Thomas (1969), p. 64].

A wide-sense stationary sequence $X[n]$ is autocorrelation-ergodic if, and only if, for all k

$$\lim_{N \to \infty} \frac{1}{2N+1} \sum_{n=-N}^{N} X[n]X[n+k] = R_{XX}[k] \qquad (6.2\text{-}46)$$

[Childers (1997), p. 278].

Finally, we mention that two processes $X(t)$ and $Y(t)$ may be called *cross-correlation-ergodic*, or *ergodic in the correlation*, if the time cross-correlation function is equal to the statistical cross-correlation function.

6.3
CORRELATION FUNCTIONS

The autocorrelation and cross-correlation functions were introduced in the previous section. These functions are examined further in this section, along with their properties. In addition, other correlation-type functions are introduced that are important to the study of random processes.

Autocorrelation Function and Its Properties

Recall that the autocorrelation function of a random process $X(t)$ is the correlation $E[X_1 X_2]$ of two random variables $X_1 = X(t_1)$ and $X_2 = X(t_2)$ defined by the process at times t_1 and t_2. Mathematically,

$$R_{XX}(t_1, t_2) = E[X(t_1)X(t_2)] \tag{6.3-1}$$

For time assignments, $t_1 = t$ and $t_2 = t_1 + \tau$, with τ a real number, (6.3-1) assumes the convenient form

$$R_{XX}(t, t + \tau) = E[X(t)X(t + \tau)] \tag{6.3-2}$$

If $X(t)$ is at least wide-sense stationary, it was noted in Section 6.2 that $R_{XX}(t, t + \tau)$ must be a function only of time difference $\tau = t_2 - t_1$. Thus, for wide-sense stationary processes

$$R_{XX}(\tau) = E[X(t)X(t + \tau)] \tag{6.3-3}$$

For such processes the autocorrelation function exhibits the following properties:

(1) $|R_{XX}(\tau)| \leq R_{XX}(0)$ $\qquad\qquad$ (6.3-4)

(2) $R_{XX}(-\tau) = R_{XX}(\tau)$ $\qquad\qquad$ (6.3-5)

(3) $R_{XX}(0) = E[X^2(t)]$ $\qquad\qquad$ (6.3-6)

The first property shows that $R_{XX}(\tau)$ is bounded by its value at the origin, while the third property states that this bound is equal to the mean-squared value called the *power* in the process. The second property indicates that an autocorrelation function has even symmetry.

Other properties of stationary processes may also be stated [see Cooper and McGillem (1971), pp. 112–114, and (1986), pp. 196–199, Melsa and Sage (1973), pp. 207–208, and Leon-Garcia (1989), pp. 357–358]:

(4) If $E[X(t)] = \bar{X} \neq 0$ and $X(t)$ is ergodic with no periodic components then

$$\lim_{|\tau| \to \infty} R_{XX}(\tau) = \bar{X}^2 \tag{6.3-7}$$

(5) If $X(t)$ has a periodic component, then $R_{XX}(\tau)$ will have a periodic component with the same period. $\qquad\qquad$ (6.3-8)

(6) If $X(t)$ is ergodic, zero-mean, and has no periodic component, then

195

CHAPTER 6:
Random
Processes—
Temporal
Characteristics

$$\lim_{|\tau| \to \infty} R_{XX}(\tau) = 0 \qquad (6.3\text{-}9)$$

(7) $R_{XX}(\tau)$ cannot have an arbitrary shape. $\qquad (6.3\text{-}10)$

Properties 4 through 6 are more or less self-explanatory. Property 7 simply says that any arbitrary function cannot be an autocorrelation function. This fact will be more apparent when the *power density spectrum* is introduced in Chapter 7. It will be shown there that $R_{XX}(\tau)$ is related to the power density spectrum through the Fourier transform and the form of the spectrum is not arbitrary.

EXAMPLE 6.3-1. Given the autocorrelation function, for a stationary ergodic process with no periodic components, is

$$R_{XX}(\tau) = 25 + \frac{4}{1 + 6\tau^2}$$

we shall find the mean value and variance of the process $X(t)$. From property 4, the mean value is $E[X(t)] = \bar{X} = \sqrt{25} = \pm 5$. Note that property 4 yields only the magnitude of $\bar{X}$; it cannot reveal its sign. The variance is given by (3.2-6), so

$$\sigma_X^2 = E[X^2(t)] - (E[X(t)])^2$$

But $E[X^2(t)] = R_{XX}(0) = 25 + 4 = 29$ from property 3, so

$$\sigma_X^2 = 29 - 25 = 4$$

EXAMPLE 6.3-2. Let $X(t)$ be a wide-sense stationary random process with autocorrelation function

$$R_{XX}(\tau) = e^{-a|\tau|}$$

where $a > 0$ is a constant. We assume $X(t)$ "amplitude modulates" a "carrier" $\cos(\omega_0 t + \Theta)$ as shown in Figure 6.3-1a, where ω_0 is a constant and Θ is a random variable uniform on $(-\pi, \pi)$ that is statistically independent of $X(t)$. We determine the autocorrelation function of $Y(t)$.

$$
\begin{aligned}
R_{YY}(t, t + \tau) &= E[Y(t)Y(t + \tau)] \\
&= E[X(t)X(t + \tau)\cos(\omega_0 t + \Theta)\cos(\omega_0 t + \omega_0 \tau + \Theta)] \\
&= R_{XX}(\tau)\tfrac{1}{2}E[\cos(\omega_0 \tau) + \cos(2\omega_0 t + \omega_0 \tau + 2\Theta)]
\end{aligned}
$$

Since

$$E[\cos(2\omega_0 t + \omega_0 \tau + 2\Theta)] = \int_{-\pi}^{\pi} \frac{1}{2\pi}[\cos(2\omega_0 t + \omega_0 \tau + 2\Theta)]\, d\theta$$

$$= 0$$

we have

$$R_{YY}(t, t + \tau) = \tfrac{1}{2}R_{XX}(\tau)\cos(\omega_0 \tau) = R_{YY}(\tau)$$

which is sketched in Figure 6.3-1b.

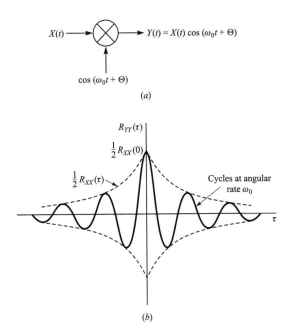

FIGURE 6.3-1
(*a*) Amplitude modulation of a carrier with random phase by a random process $X(t)$. (*b*) A plot of the autocorrelation function of $Y(t)$.

Cross-Correlation Function and Its Properties

The cross-correlation function of two random processes $X(t)$ and $Y(t)$ was defined in (6.2-18). Setting $t_1 = t$ and $\tau = t_2 - t_1$, we may write (6.2-18) as

$$R_{XY}(t, t + \tau) = E[X(t)Y(t + \tau)] \tag{6.3-11}$$

If $X(t)$ and $Y(t)$ are at least jointly wide-sense stationary, $R_{XY}(t, t + \tau)$ is independent of absolute time and we can write

$$R_{XY}(\tau) = E[X(t)Y(t + \tau)] \tag{6.3-12}$$

If

$$R_{XY}(t, t + \tau) = 0 \tag{6.3-13}$$

then $X(t)$ and $Y(t)$ are called *orthogonal processes*. If the two processes are statistically independent, the cross-correlation function becomes

$$R_{XY}(t, t + \tau) = E[X(t)]E[Y(t + \tau)] \tag{6.3-14}$$

If, in addition to being independent, $X(t)$ and $Y(t)$ are at least wide-sense stationary, (6.3-14) becomes

$$R_{XY}(\tau) = \bar{X}\bar{Y} \tag{6.3-15}$$

which is a constant.

We may list some properties of the cross-correlation function applicable to processes that are at least wide-sense stationary:

197

CHAPTER 6:
Random
Processes—
Temporal
Characteristics

$$(1) \quad R_{XY}(-\tau) = R_{YX}(\tau) \tag{6.3-16}$$

$$(2) \quad |R_{XY}(\tau)| \leq \sqrt{R_{XX}(0)R_{YY}(0)} \tag{6.3-17}$$

$$(3) \quad |R_{XY}(\tau)| \leq \tfrac{1}{2}[R_{XX}(0) + R_{YY}(0)] \tag{6.3-18}$$

Property 1 follows from the definition (6.3-12). It describes the symmetry of $R_{XY}(\tau)$. Property 2 can be proven by expanding the inequality

$$E[\{Y(t+\tau) + \alpha X(t)\}^2] \geq 0 \tag{6.3-19}$$

where α is a real number (see Problem 6.3-18). Properties 2 and 3 both constitute bounds on the magnitude of $R_{XY}(\tau)$. Equation (6.3-17) represents a tighter bound than that of (6.3-18), because the geometric mean of two positive numbers cannot exceed their arithmetic mean; that is

$$\sqrt{R_{XX}(0)R_{YY}(0)} \leq \tfrac{1}{2}[R_{XX}(0) + R_{YY}(0)] \tag{6.3-20}$$

EXAMPLE 6.3-3. Let two random processes $X(t)$ and $Y(t)$ be defined by

$$X(t) = A\cos(\omega_0 t) + B\sin(\omega_0 t)$$

$$Y(t) = B\cos(\omega_0 t) - A\sin(\omega_0 t)$$

where A and B are random variables and ω_0 is a constant. It can be shown (Problem 6.3-3) that $X(t)$ is wide-sense stationary if A and B are uncorrelated, zero-mean random variables with the same variance (they may have different density functions, however). With these same constraints on A and B, $Y(t)$ is also wide-sense stationary. We shall now find the cross-correlation function $R_{XY}(t, t + \tau)$ and show that $X(t)$ and $Y(t)$ are *jointly* wide-sense stationary. By use of (6.3-11) we have

$$
\begin{aligned}
R_{XY}(t, t+\tau) &= E[X(t)Y(t+\tau)] \\
&= E[AB\cos(\omega_0 t)\cos(\omega_0 t + \omega_0 \tau) \\
&\quad + B^2 \sin(\omega_0 t)\cos(\omega_0 t + \omega_0 \tau) \\
&\quad - A^2 \cos(\omega_0 t)\sin(\omega_0 t + \omega_0 \tau) \\
&\quad - AB\sin(\omega_0 t)\sin(\omega_0 t + \omega_0 \tau)] \\
&= E[AB]\cos(2\omega_0 t + \omega_0 \tau) \\
&\quad + E[B^2]\sin(\omega_0 t)\cos(\omega_0 t + \omega_0 \tau) \\
&\quad - E[A^2]\cos(\omega_0 t)\sin(\omega_0 t + \omega_0 \tau)
\end{aligned}
$$

Since A and B are assumed to be zero-mean, uncorrelated random variables, $E[AB] = 0$. Also, since A and B are assumed to have equal variances, $E[A^2] = E[B^2] = \sigma^2$ and we obtain

$$R_{XY}(t, t+\tau) = -\sigma^2 \sin(\omega_0 \tau)$$

Thus, $X(t)$ and $Y(t)$ are jointly wide-sense stationary because $R_{XY}(t, t + \tau)$ depends only on τ.

Note from the above results that cross-correlation functions are not necessarily even functions of τ with the maximum at $\tau = 0$, as is the case with autocorrelation functions.

Covariance Functions

The concept of the covariance of two random variables, as defined by (5.1-13), can be extended to random processes. The *autocovariance function* is defined by

$$C_{XX}(t, t + \tau) = E[\{X(t) - E[X(t)]\}\{X(t + \tau) - E[X(t + \tau)]\}] \qquad (6.3\text{-}21)$$

which can also be put in the form

$$C_{XX}(t, t + \tau) = R_{XX}(t, t + \tau) - E[X(t)]E[X(t + \tau)] \qquad (6.3\text{-}22)$$

The *cross-covariance function* for two processes $X(t)$ and $Y(t)$ is defined by

$$C_{XY}(t, t + \tau) = E[\{X(t) - E[X(t)]\}\{Y(t + \tau) - E[Y(t + \tau)]\}] \qquad (6.3\text{-}23)$$

or, alternatively,

$$C_{XY}(t, t + \tau) = R_{XY}(t, t + \tau) - E[X(t)]E[Y(t + \tau)] \qquad (6.3\text{-}24)$$

For processes that are at least jointly wide-sense stationary, (6.3-22) and (6.3-24) reduce to

$$C_{XX}(\tau) = R_{XX}(\tau) - \bar{X}^2 \qquad (6.3\text{-}25)$$

and

$$C_{XY}(\tau) = R_{XY}(\tau) - \bar{X}\bar{Y} \qquad (6.3\text{-}26)$$

The *variance* of a random process is given in general by (6.3-21) with $\tau = 0$. For a wide-sense stationary process, variance does not depend on time and is given by (6.3-25) with $\tau = 0$:

$$\sigma_X^2 = E[\{X(t) - E[X(t)]\}^2] = R_{XX}(0) - \bar{X}^2 \qquad (6.3\text{-}27)$$

For two random processes, if

$$C_{XY}(t, t + \tau) = 0 \qquad (6.3\text{-}28)$$

they are called *uncorrelated*. From (6.3-24) this means that

$$R_{XY}(t, t + \tau) = E[X(t)]E[Y(t + \tau)] \qquad (6.3\text{-}29)$$

Since this result is the same as (6.3-14), which applies to independent processes, we conclude that independent processes are uncorrelated. The converse case is not necessarily true, although it is true for *joint gaussian processes*, which we consider in Section 6.5.

Discrete-Time Processes and Sequences

199

CHAPTER 6:
Random
Processes—
Temporal
Characteristics

The preceding results for various correlation functions hold for DT processes and DT sequences. However, such processes and sequences are defined only at the "sample" times nT_s, as discussed in Section 6.1. Therefore, time offsets, denoted above generally by τ, can now have only discrete values, say, kT_s. For a DT process $X(nT_s)$ we summarize the specific definitions for the process mean, autocorrelation, and autocovariance functions as follows:

$$\text{Mean} = E[X(nT_s)] \tag{6.3-30}$$

$$R_{XX}(nT_s, nT_s + kT_s) = E[X(nT_s)X(nT_s + kT_s)] \tag{6.3-31}$$

$$
\begin{aligned}
&C_{XX}(nT_s, nT_s + kT_s) \\
&= E(\{X(nT_s) - E[X(nT_s)]\}\{X(nT_s + kT_s) - E[X(nT_s + kT_s)]\}) \\
&= R_{XX}(nT_s, nT_s + kT_s) - E[X(nT_s)]E[X(nT_s + kT_s)]
\end{aligned} \tag{6.3-32}
$$

Similar statements hold for a DT process $Y(nT_s)$. For cross-correlation and cross-covariance functions we have

$$R_{XY}(nT_s, nT_s + kT_s) = E[X(nT_s)Y(nT_s + kT_s)] \tag{6.3-33}$$

$$
\begin{aligned}
&C_{XY}(nT_s, nT_s + kT_s) \\
&= E(\{X(nT_s) - E[X(nT_s)]\}\{Y(nT_s + kT_s) - E[Y(nT_s + kT_s)]\}) \\
&= R_{XY}(nT_s, nT_s + kT_s) - E[X(nT_s)]E[Y(nT_s + kT_s)]
\end{aligned} \tag{6.3-34}
$$

For the functions $R_{YX}(\cdot, \cdot)$ and $C_{YX}(\cdot, \cdot)$ simply interchange X and Y in (6.3-33) and (6.3-34).

For processes that are at least jointly wide-sense stationary, our expressions reduce to constants ($\bar{X}$ or $\bar{Y}$) and functions of only kT_s.

$$E[X(nT_s)] = \bar{X} \qquad E[Y(nT_s)] = \bar{Y} \tag{6.3-35}$$

$$R_{XX}(nT_s, nT_s + kT_s) = R_{XX}(kT_s) \tag{6.3-36}$$

$$R_{YY}(nT_s, nT_s + kT_s) = R_{YY}(kT_s) \tag{6.3-37}$$

$$C_{XX}(nT_s, nT_s + kT_s) = R_{XX}(kT_s) - \bar{X}^2 \tag{6.3-38}$$

$$C_{YY}(nT_s, nT_s + kT_s) = R_{YY}(kT_s) - \bar{Y}^2 \tag{6.3-39}$$

$$R_{XY}(nT_s, nT_s + kT_s) = R_{XY}(kT_s) \tag{6.3-40}$$

$$R_{YX}(nT_s, nT_s + kT_s) = R_{YX}(kT_s) \tag{6.3-41}$$

$$C_{XY}(nT_s, nT_s + kT_s) = R_{XY}(kT_s) - \bar{X}\bar{Y} \tag{6.3-42}$$

$$C_{YX}(nT_s, nT_s + kT_s) = R_{YX}(kT_s) - \bar{Y}\bar{X} \tag{6.3-43}$$

Finally, we note that for DT sequences, where the dependence on T_s is dropped (only implied) as discussed in Section 6.1, all the above results apply if T_s is everywhere omitted and the functional notation is everywhere replaced by brackets. As examples: $R_{XY}(nT_s, nT_s + kT_s)$ becomes $R_{XY}[n, n + k]$ while $E[X(nT_s)]$ becomes $E\{X[n]\}$.

6.4
MEASUREMENT OF CORRELATION FUNCTIONS

In the real world, we can never measure the true correlation functions of two random processes $X(t)$ and $Y(t)$ because we never have *all* sample functions of the ensemble at our disposal. Indeed, we may typically have available for measurements only a portion of one sample function from each process. Thus, our only recourse is to determine time averages based on finite time portions of single sample functions, taken large enough to approximate true results for ergodic processes. Because we are able to work only with time functions, we are forced, like it or not, to presume that given processes are ergodic. This fact should not prove too disconcerting, however, if we remember that assumptions only reflect the details of our mathematical model of a real-world situation. Provided that the model gives consistent agreement with the real situation, it is of little importance whether ergodicity is assumed or not.

Figure 6.4-1 illustrates the block diagram of a possible system for measuring the approximate time cross-correlation function of two jointly ergodic random processes $X(t)$ and $Y(t)$. Sample functions $x(t)$ and $y(t)$ are delayed by amounts T and $T - \tau$, respectively, and the product of the delayed waveforms is formed. This product is then integrated to form the output which equals the integral at time $t_1 + 2T$, where t_1 is arbitrary and $2T$ is the integration period. The integrator can be of the integrate-and-dump variety described by Peebles (1976, p. 361).

If we assume $x(t)$ and $y(t)$ exist at least during the interval $-T < t$ and t_1 is an arbitrary time except $0 \leq t_1$, then the output is easily found to be

$$R_o(t_1 + 2T) = \frac{1}{2T} \int_{t_1 - T}^{t_1 + T} x(t) y(t + \tau) \, dt \tag{6.4-1}$$

Now if we choose $t_1 = 0$† and assume T is large, then we have

$$R_o(2T) = \frac{1}{2T} \int_{-T}^{T} x(t) y(t + \tau) dt \approx \mathscr{R}_{xy}(\tau) = R_{XY}(\tau) \tag{6.4-2}$$

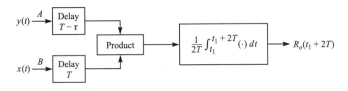

FIGURE 6.4-1
A time cross-correlation function measurement system. Autocorrelation function measurement is possible by connecting points A and B and applying either $x(t)$ or $y(t)$.

---

†Since the processes are assumed jointly ergodic and therefore jointly stationary, the integral (6.4-1) will tend to be independent of t_1 if T is large enough.

Thus, for jointly ergodic processes, the system of Figure 6.4-1 can approximately measure their cross-correlation function (τ is varied to obtain the complete function).

201

CHAPTER 6:
Random
Processes—
Temporal
Characteristics

Clearly, by connecting points A and B and applying either $x(t)$ or $y(t)$ to the system, we can also measure the autocorrelation functions $R_{XX}(\tau)$ and $R_{YY}(\tau)$.

EXAMPLE 6.4-1. We connect points A and B together in Figure 6.4-1 and use the system to measure the autocorrelation function of the process $X(t)$ of Example 6.2-1. From (6.4-2)

$$R_o(2T) = \frac{1}{2T} \int_{-T}^{T} A^2 \cos(\omega_0 t + \theta) \cos(\omega_0 t + \theta + \omega_0 \tau)\, dt$$

$$= \frac{A^2}{4T} \int_{-T}^{T} [\cos(\omega_0 \tau) + \cos(2\omega_0 t + 2\theta + \omega_0 \tau)]\, dt$$

In writing this result θ represents a specific value of the random variable Θ, the value that corresponds to the specific ensemble member being used in (6.4-2). On straightforward reduction of the above integral we obtain

$$R_o(2T) = R_{XX}(\tau) + \varepsilon(T)$$

where

$$R_{XX}(\tau) = (A^2/2) \cos(\omega_0 \tau)$$

is the true autocorrelation function of $X(t)$, and

$$\varepsilon(t) = (A^2/2) \cos(\omega_0 \tau + 2\theta) \frac{\sin(2\omega_0 T)}{2\omega_0 T}$$

is an error term. If we require the error term's magnitude to be at least 20 times smaller than the largest value of the true autocorrelation function then $|\varepsilon(T)| < 0.05 R_{XX}(0)$ is necessary. Thus, we must have $1/2\omega_0 T \le 0.05$ or

$$T \ge 10/\omega_0$$

In other words, if $T \ge 10/\omega_0$ the error in using Figure 6.4-1 to measure the autocorrelation function of the process $X(t) = A \cos(\omega_0 t + \Theta)$ will be 5% or less of the largest value of the true autocorrelation function.

6.5
GAUSSIAN RANDOM PROCESSES

A number of random processes are important enough to have been given names. In this section we shall discuss the most important of these, the *gaussian random process*.

Consider a continuous random process such as illustrated in Figure 6.1-1 and define N random variables $X_1 = X(t_1), \ldots, X_i = X(t_i), \ldots, X_N = X(t_N)$ corresponding to N time instants $t_1, \ldots, t_i, \ldots, t_N$. If, for any $N = 1, 2, \ldots$

and any times $t_1, \ldots, t_N$, these random variables are jointly gaussian, that is, they have a joint density as given by (5.3-12), the process is called gaussian. Equation (5.3-12) can be written in the form

$$f_X(x_1, \ldots, x_N; t_1, \ldots, t_N) = \frac{\exp\{-(1/2)[x - \bar{X}]^t[C_X]^{-1}[x - \bar{X}]\}}{\sqrt{(2\pi)^N |[C_X]|}} \tag{6.5-1}$$

where matrices $[x - \bar{X}]$ and $[C_X]$ are defined in (5.3-13) and (5.3-15), respectively. The mean values $\bar{X}_i$ of $X(t_i)$ are

$$\bar{X}_i = E[X_i] = E[X(t_i)] \tag{6.5-2}$$

The elements of the covariance matrix $[C_X]$ are

$$\begin{aligned} C_{ik} = C_{X_i X_k} &= E[(X_i - \bar{X}_i)(X_k - \bar{X}_k)] \\ &= E[\{X(t_i) - E[X(t_i)]\}\{X(t_k) - E[X(t_k)]\}] \\ &= C_{XX}(t_i, t_k) \end{aligned} \tag{6.5-3}$$

which is the autocovariance of $X(t_i)$ and $X(t_k)$ from (6.3-21).

From (6.5-2) and (6.5-3), when used in (6.5-1), we see that the mean and autocovariance functions are all that are needed to completely specify a gaussian random process. By expanding (6.5-3) to get

$$C_{XX}(t_i, t_k) = R_{XX}(t_i, t_k) - E[X(t_i)]E[X(t_k)] \tag{6.5-4}$$

we see that an alternative specification using only the mean and autocorrelation function $R_{XX}(t_i, t_k)$ is possible.

If the gaussian process is not stationary the mean and autocovariance functions will, in general, depend on absolute time. However, for the important case where the process is wide-sense stationary, the mean will be constant,

$$\bar{X}_i = E[X(t_i)] = \bar{X} \qquad \text{(constant)} \tag{6.5-5}$$

while the autocovariance and autocorrelation functions will depend only on time differences and not absolute time,

$$C_{XX}(t_i, t_k) = C_{XX}(t_k - t_i) \tag{6.5-6}$$

$$R_{XX}(t_i, t_k) = R_{XX}(t_k - t_i) \tag{6.5-7}$$

It follows from the preceding discussions that a wide-sense stationary gaussian process is also strictly stationary.

We illustrate some of the above remarks with an example.

EXAMPLE 6.5-1. A gaussian random process is known to be wide-sense stationary with a mean of $\bar{X} = 4$ and autocorrelation function

$$R_{XX}(\tau) = 25e^{-3|\tau|}$$

We seek to specify the joint density function for three random variables $X(t_i)$, $i = 1, 2, 3$, defined at times $t_i = t_0 + [(i - 1)/2]$, with t_0 a constant. Here $t_k - t_i = (k - i)/2$, i and $k = 1, 2, 3$, so

$$R_{XX}(t_k - t_i) = 25e^{-3|k-i|/2}$$

and

203

CHAPTER 6:
Random
Processes—
Temporal
Characteristics

$$C_{XX}(t_k - t_i) = 25e^{-3|k-i|/2} - 16$$

from (6.5-4) through (6.5-7). Elements of the covariance matrix are found from (6.5-3). Thus

$$[C_X] = \begin{bmatrix} (25 - 16) & (25e^{-3/2} - 16) & (25e^{-6/2} - 16) \\ (25e^{-3/2} - 16) & (25 - 16) & (25e^{-3/2} - 16) \\ (25e^{-6/2} - 16) & (25e^{-3/2} - 16) & (25 - 16) \end{bmatrix}$$

and $\bar{X}_i = 4$ completely determine (6.5-1) for this case where $N = 3$.

Two random processes $X(t)$ and $Y(t)$ are said to be *jointly gaussian* if the random variables $X(t_1), \ldots, X(t_N)$, $Y(t'_1), \ldots, Y(t'_M)$ defined at times $t_1, \ldots, t_N$ for $X(t)$ and times $t'_1, \ldots, t'_M$ for $Y(t)$, are jointly gaussian for any N, $t_1, \ldots, t_N$, M, $t'_1, \ldots, t'_M$.

6.6
POISSON RANDOM PROCESS

In this section we consider an important example of a discrete random process known as the *Poisson process*.† It describes the number of times that some event has occurred as a function of time, where the events occur at random times. The event might be the arrival of a customer at a bank or supermarket check-out register, the occurrence of a lightning strike within some prescribed area, the failure of some component in a system, or the emission of an electron from the surface of a light-sensitive material (photodetector). In each of these examples a single event occurs at a random time and the process amounts to counting the number of such occurrences with time. For this reason, the process is also known as the *Poisson counting process*.

To visualize the Poisson process, let $X(t)$ represent the number of event occurrences with time (the process); then $X(t)$ has integer-valued, nondecreasing sample functions, as illustrated in Figure 6.6-1a for the random occurrence times of Figure 6.6-1b. For convenience, we take $X(t) = 0$ at $t = 0$; for $t > 0$, $X(t)$ is the number of occurrences in the interval $(0, t)$; for $t < 0$, $X(t)$ is the *negative* of the number of occurrences in the interval $(t, 0)$. (See Shanmugan and Briepohl, 1988, p. 296.) In many situations only the process' behavior for $t > 0$ is of interest, and in the remainder of this section we shall assume the process is defined only for $t > 0$ (and is zero for $t < 0$).

To define the Poisson process we shall require two conditions. First, we require that the event occur only once in any vanishingly small interval of time. In essence, we require that only one event can occur at a time. This condition does not prevent the times of occurrence of events from being very

† For additional reading, some recent books that also cover this topic are Shanmugan and Briepohl (1988) and Gardner (1990).

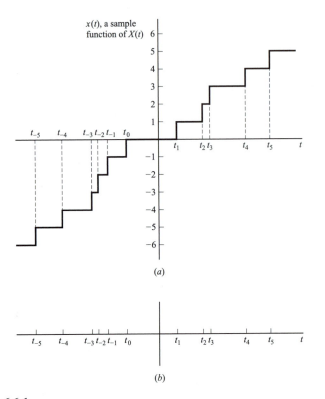

FIGURE 6.6-1
(a) A sample function of a Poisson discrete random process, and (b) the random times of occurrence of events being counted to form the process.

close together, only that they do not coincide. Second, we require that occurrence times be statistically independent so that the number that occurs in any given time interval is independent of the number in any other nonoverlapping time interval; this independence is required to apply regardless of the number of time intervals of interest. A consequence of the two conditions is that the number of event occurrences in any finite interval of time is described by the Poisson distribution where the average rate of occurrences is denoted by λ [see Section 2.5, equations (2.5-4)–(2.5-6)]. [For more detail, see Gray and Davisson (1986), pp. 282–284.]

Probability Density Function

Thus, from (2.5-4) with $b = \lambda t$, the probability of exactly k occurrences over a time interval $(0, t)$ is

$$P[X(t) = k] = \frac{(\lambda t)^k e^{-\lambda t}}{k!} \qquad k = 0, 1, 2, \ldots \qquad (6.6\text{-}1)$$

and the probability density of the number of occurrences is

$$f_X(x) = \sum_{k=0}^{\infty} \frac{(\lambda t)^k e^{-\lambda t}}{k!} \delta(x - k) \tag{6.6-2}$$

205

CHAPTER 6:
Random
Processes—
Temporal
Characteristics

From Problem 3.2-13 we know that the mean and variance of a Poisson random variable are each equal to λt. Thus, from (3.2-6) we also know the second moment, which is $E[X^2(t)] = \sigma_X^2 + \{E[X(t)]\}^2 = \lambda t + \lambda^2 t^2$. These facts are used to establish useful equations by formally computing the mean and second moment:

$$E[X(t)] = \int_{-\infty}^{\infty} x f_X(x) \, dx = \int_{-\infty}^{\infty} x \sum_{k=0}^{\infty} \frac{(\lambda t)^k e^{-\lambda t}}{k!} \delta(x - k) \, dx$$

$$= \sum_{k=0}^{\infty} \frac{k(\lambda t)^k e^{-\lambda t}}{k!} = \lambda t \tag{6.6-3}$$

$$E[X^2(t)] = \int_{-\infty}^{\infty} x^2 f_X(x) \, dx = \sum_{k=0}^{\infty} \frac{k^2(\lambda t)^k e^{-\lambda t}}{k!} = \lambda t[1 + \lambda t] \tag{6.6-4}$$

Joint Probability Density

To determine the joint probability density function for the Poisson process at times $0 < t_1 < t_2$, first observe that the probability of k_1 event occurrences over $(0, t_1)$ is

$$P[X(t_1) = k_1] = \frac{(\lambda t_1)^{k_1} e^{-\lambda t_1}}{k_1!} \qquad k_1, = 0, 1, 2, \ldots \tag{6.6-5}$$

from (6.6-1). Next, the conditional probability of k_2 occurrences over $(0, t_2)$, given that k_1 events occurred over $(0, t_1)$, is just the probability that $k_2 - k_1$ events occurred over (t_1, t_2), which is

$$P[X(t_2) = k_2 | X(t_1) = k_1] = \frac{[\lambda(t_2 - t_1)]^{k_2 - k_1} e^{-\lambda(t_2 - t_1)}}{(k_2 - k_1)!} \tag{6.6-6}$$

for $k_2 \geq k_1$. The joint probability of k_2 occurrences at time t_2 and k_1 occurrences at time t_1 is the product of (6.6-5) and (6.6-6):

$$P(k_1, k_2) = P[X(t_2) = k_2 | X(t_1) = k_1] \cdot P[X(t_1) = k_1]$$

$$= \frac{(\lambda t_1)^{k_1} [\lambda(t_2 - t_1)]^{k_2 - k_1} e^{-\lambda t_2}}{k_1!(k_2 - k_1)!} \qquad k_2 \geq k_1 \tag{6.6-7}$$

The joint density now becomes

$$f_X(x_1, x_2) = \sum_{k_1=0}^{\infty} \sum_{k_2=k_1}^{\infty} P(k_1, k_2) \delta(x_1 - k_1) \delta(x_2 - k_2) \tag{6.6-8}$$

for the process' random variables $X(t_1) = X_1$, and $X(t_2) = X_2$.

A principal reason for developing (6.6-8) is that the autocorrelation function of the process can be determined. This is left as an exercise for the reader (see Problem 6.6-5).

EXAMPLE 6.6-1. By example, we illustrate how the higher-dimensional probability density of the Poisson process can be derived. We take only the case of three random variables defined at times $0 < t_1 < t_2 < t_3$. For $k_1 \le k_2 \le k_3$ occurrences at the respective times, we have

$$P(k_1, k_2, k_3) = P[X(t_3) = k_3 | X(t_2) = k_2, X(t_1) = k_1]$$
$$\cdot P[X(t_2) = k_2 | X(t_1) = k_1] P[X(t_1) = k_1]$$
$$= \frac{[\lambda(t_3 - t_2)]^{k_3-k_2} e^{-\lambda(t_3-t_2)}}{(k_3 - k_2)!} \cdot \frac{[\lambda(t_2 - t_1)]^{k_2-k_1} e^{-\lambda(t_2-t_1)}}{(k_2 - k_1)!}$$
$$\cdot \frac{(\lambda t_1)^{k_1} e^{-\lambda t_1}}{k_1!} = \frac{(\lambda t_1)^{k_1} [\lambda(t_2 - t_1)]^{k_2-k_1} [\lambda(t_3 - t_2)]^{k_3-k_2} e^{-\lambda t_3}}{k_1!(k_2 - k_1)!(k_3 - k_2)!}$$

and

$$f_X(x_1, x_2, x_3) = \sum_{k_1=0}^{\infty} \sum_{k_2=k_1}^{\infty} \sum_{k_3=k_2}^{\infty} P(k_1, k_2, k_3)\delta(x_1 - k_1)\delta(x_2 - k_2)\delta(x_3 - k_3)$$

*6.7
COMPLEX RANDOM PROCESSES

If the complex random variable of Section 5.8 is generalized to include time, the result is a *complex random process Z(t) given by*

$$Z(t) = X(t) + jY(t) \tag{6.7-1}$$

where $X(t)$ and $Y(t)$ are real processes. $Z(t)$ is called stationary if $X(t)$ and $Y(t)$ are jointly stationary. If $X(t)$ and $Y(t)$ are jointly wide-sense stationary, then $Z(t)$ is said to be wide-sense stationary.

Two complex processes $Z_i(t)$ and $Z_j(t)$ are jointly wide-sense stationary if each is wide-sense stationary and their cross-correlation function (defined below) is a function of time differences only and not absolute time.

We may extend the operations involving process mean value, autocorrelation function, and autocovariance function to include complex processes. The *mean value* of $Z(t)$ is

$$E[Z(t)] = E[X(t)] + jE[Y(t)] \tag{6.7-2}$$

Autocorrelation function is defined by

$$R_{ZZ}(t, t + \tau) = E[Z^*(t)Z(t + \tau)] \tag{6.7-3}$$

where the asterisk * denotes the complex conjugate. *Autocovariance function* is defined by

$$C_{ZZ}(t, t + \tau) = E[\{Z(t) - E[Z(t)]\}^* \{Z(t + \tau) - E[Z(t + \tau)]\}] \tag{6.7-4}$$

If $Z(t)$ is at least wide-sense stationary, the mean value becomes a constant

$$\bar{Z} = \bar{X} + j\bar{Y} \tag{6.7-5}$$

and the correlation functions are independent of absolute time:

$$R_{ZZ}(t, t + \tau) = R_{ZZ}(\tau) \tag{6.7-6}$$

$$C_{ZZ}(t, t + \tau) = C_{ZZ}(\tau) \tag{6.7-7}$$

For two complex processes $Z_i(t)$ and $Z_j(t)$, *cross-correlation* and *cross-covariance functions* are defined by

$$R_{Z_i Z_j}(t, t + \tau) = E[Z_i^*(t) Z_j(t + \tau)] \qquad i \neq j \tag{6.7-8}$$

and

$$C_{Z_i Z_j}(t, t + \tau) = E[\{Z_i(t) - E[Z_i(t)]\}^* \{Z_j(t + \tau) - E[Z_j(t + \tau)]\}] \qquad i \neq j \tag{6.7-9}$$

respectively. If the two processes are at least jointly wide-sense stationary, we obtain

$$R_{Z_i Z_j}(t, t + \tau) = R_{Z_i Z_j}(\tau) \qquad i \neq j \tag{6.7-10}$$

$$C_{Z_i Z_j}(t, t + \tau) = C_{Z_i Z_j}(\tau) \qquad i \neq j \tag{6.7-11}$$

$Z_i(t)$ and $Z_j(t)$ are said to be *uncorrelated processes* if $C_{Z_i Z_j}(t, t + \tau) = 0, i \neq j$. They are called *orthogonal processes* if $R_{Z_i Z_j}(t, t + \tau) = 0, i \neq j$.

EXAMPLE 6.7-1. A complex random process $V(t)$ is comprised of a sum of N complex signals:

$$V(t) = \sum_{n=1}^{N} A_n e^{j\omega_0 t + j\Theta_n}$$

Here $\omega_0/2\pi$ is the (constant) frequency of each signal. A_n is a random variable representing the random amplitude of the nth signal. Similarly, Θ_n is a random variable representing a random phase angle. We assume all the variables A_n and Θ_n, for $n = 1, 2, \ldots, N$, are statistically independent and the Θ_n are uniformly distributed on $(0, 2\pi)$. We find the auto-correlation function of $V(t)$.

From (6.7-3):

$$R_{VV}(t, t + \tau) = E[V^*(t) V(t + \tau)]$$

$$= E\left[\sum_{n=1}^{N} A_n e^{-j\omega_0 t - j\Theta_n} \sum_{m=1}^{N} A_m e^{j\omega_0 t + j\omega_0 \tau + j\Theta_m} \right]$$

$$= \sum_{n=1}^{N} \sum_{m=1}^{N} e^{j\omega_0 \tau} E[A_n A_m e^{j(\Theta_m - \Theta_n)}] = R_{VV}(\tau)$$

207

CHAPTER 6:
Random
Processes—
Temporal
Characteristics

From statistical independence:

$$R_{VV}(\tau) = e^{j\omega_0 \tau} \sum_{n=1}^{N} \sum_{m=1}^{N} E[A_n A_m] E[\exp\{j(\Theta_m - \Theta_n)\}]$$

However,

$$E[\exp\{j(\Theta_m - \Theta_n)\}] = E[\cos(\theta_m - \theta_n)] + jE[\sin(\Theta_m - \Theta_n)]$$

$$= \int_0^{2\pi} \int_0^{2\pi} \frac{1}{(2\pi)^2} [\cos(\theta_m - \theta_n) + j\sin(\theta_m - \theta_n)] \, d\theta_n \, d\theta_m$$

$$= \begin{cases} 0 & m \neq n \\ 1 & m = n \end{cases}$$

so

$$R_{VV}(\tau) = e^{j\omega_0 \tau} \sum_{n=1}^{N} \overline{A_n^2}$$

6.8
SUMMARY

This chapter represents a major change in the book's direction. All prior material was extended to include *time* so that random processes could be defined in order to model random time signals. In essence, the random variable concept was extended to include time, which results in a random process. Many details were covered regarding random processes:

- The random process was defined and both deterministic and nondeterministic types were discussed.
- Stationarity and statistical independence of processes were defined, including various forms of ergodic processes.
- Autocorrelation, cross-correlation, autocovariance, and cross-covariance functions were defined for continuous-time, as well as discrete-time, processes and sequences.
- Measurement of correlation functions was included.
- Both gaussian and Poisson processes were developed in some detail.
- Finally, the more advanced concepts of complex random processes and their characteristics were included.

PROBLEMS

6.1-1. A random experiment consists of selecting a point on some city street that has two-way automobile traffic. Define and classify a random process for this experiment that is related to traffic flow.

6.1-2. A 10-meter section of a busy downtown sidewalk is actually the platform of a scale that produces a voltage proportional to the total weight of people on the scale at any time.

(a) Sketch a typical sample function for this process.
(b) What is the underlying random experiment for the process?
(c) Classify the process.

*6.1-3. An experiment consists of measuring the weight W of some person each 10 minutes. The person is randomly male or female (which is not known though) with equal probability. A two-level discrete process $X(t)$ is generated where

$$X(t) = \pm 10$$

The level -10 is generated in the period following a measurement if the measured weight does not exceed W_0 (some constant). Level $+10$ is generated if weight exceeds W_0. Let the weight of men in kg be a random variable having the gaussian density

$$f_W(w|\text{male}) = \frac{1}{\sqrt{2\pi}11.3} \exp[-(w - 77.1)^2/2(11.3)^2]$$

Similarly, for women

$$f_W(w|\text{female}) = \frac{1}{\sqrt{2\pi}6.8} \exp[-(w - 54.4)^2/2(6.8)^2]$$

(a) Find W_0 so that $P\{W > W_0|\text{male}\}$ is equal to $P\{W \leq W_0|\text{female}\}$.
(b) If the levels ± 10 are interpreted as "decisions" about whether the weight measurement of a person corresponds to a male or female, give a physical significance of their generation.
(c) Sketch a possible sample function.

6.1-4. The two-level *semirandom binary process* is defined by

$$X(t) = A \text{ or } -A \qquad (n - 1)T < t < nT$$

where the levels A and $-A$ occur with equal probability, T is a positive constant, and $n = 0, \pm 1, \pm 2, \ldots$.
(a) Sketch a typical sample function.
(b) Classify the process.
(c) Is the process deterministic?

6.1-5. For a random process $X(t)$ it is known that $f_X(x_1, x_2, x_3; t_1, t_2, t_3) = f_X(x_1, x_2, x_3; t_1 + \Delta, t_2 + \Delta, t_3 + \Delta)$ for any t_1, t_2, t_3 and Δ. Indicate which of the following statements are unequivocally true: $X(t)$ is (a) stationary to order 1, (b) stationary to order 2, (c) stationary to order 3, (d) strictly stationary, (e) wide-sense stationary, (f) not stationary in any sense, and (g) ergodic.

6.1-6. A random process is defined by $X(t) = A$, where A is a continuous random variable uniformly distributed on $(0, 1)$.
(a) Determine the form of the sample functions.
(b) Classify the process.
(c) Is it deterministic?

6.1-7. Work Problem 6.1-6 except assume $X(t) = At$ (t represents time).

6.2-1. Sample functions in a discrete random process are constants; that is

$$X(t) = C = \text{constant}$$

where C is a discrete random variable having possible values $c_1 = 1, c_2 = 2$, and $c_3 = 3$ occurring with probabilities 0.6, 0.3, and 0.1, respectively.
(a) Is $X(t)$ deterministic?
(b) Find the first-order density function of $X(t)$ at any time t.

6.2-2. Utilize (6.2-13) to prove (6.2-16).

***6.2-3.** A random process $X(t)$ has periodic sample functions as shown in Figure P6.2-3 where B, T, and $4t_0 \leq T$ are constants but ε is a random variable uniformly distributed on the interval $(0, T)$.
(a) Find the first-order distribution function of $X(t)$.
(b) Find the first-order density function.
(c) Find $E[X(t)]$, $E[X^2(t)]$, and σ_X^2.

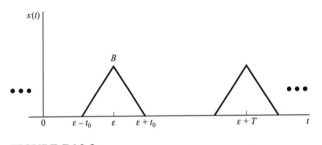

FIGURE P6.2-3

6.2-4. Work Problem 6.2-3 for the waveform of Figure P6.2-4. Assume $2t_0 < T$.

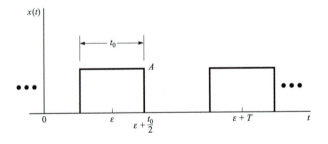

FIGURE P6.2-4

***6.2-5.** Work Problem 6.2-3 for the waveform of Figure P6.2-5. Assume $4t_0 \leq T$.

6.2-6. A random process is defined by $X(t) = X_0 + Vt$ where X_0 and V are statistically independent random variables uniformly distributed on intervals $[X_{01}, X_{02}]$ and $[V_1, V_2]$, respectively. Find (a) the mean, (b) the autocorrelation, and (c) the autocovariance functions of $X(t)$. (d) Is $X(t)$ stationary in any sense? If so, state the type.

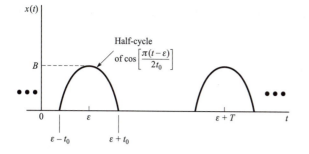

FIGURE P6.2-5

211

CHAPTER 6:
Random
Processes—
Temporal
Characteristics

*6.2-7. (*a*) Find the first-order density of the random process of Problem 6.2-6.
   (*b*) Plot the density for $t = k(X_{02} - X_{01})/(V_2 - V_1)$ with $k = 0, \frac{1}{2}, 1,$ and 2. Assume $V_2 = 3V_1$ in all plots.

6.2-8. Assume a wide-sense stationary process $X(t)$ has a known mean $\bar{X}$ and a known autocorrelation function $R_{XX}(\tau)$. Now suppose the process is observed at time t_1 and we wish to *estimate*, that is, *predict*, what the process will be at time $t_1 + \tau$ with $t > 0$. We assume the estimate has the form

$$\hat{X}(t_1 + \tau) = \alpha X(t_1) + \beta$$

where α and β are constants.
   (*a*) Find α and β so that the mean-squared prediction error

$$\overline{\varepsilon^2} = E[\{X(t_1 + \tau) - \hat{X}(t_1 + \tau)\}^2]$$

   is minimum.
   (*b*) Find the minimum mean-squared error in terms of $R_{XX}(\tau)$. Develop an alternative form in terms of the autocovariance function.

6.2-9. Find the time average and time autocorrelation function of the random process of Example 6.2-1. Compare these results with the statistical mean and autocorrelation found in the example.

6.2-10. Assume that an ergodic random process $X(t)$ has an autocorrelation function

$$R_{XX}(\tau) = 18 + \frac{2}{6 + \tau^2}[1 + 4\cos(12\tau)]$$

   (*a*) Find $|\bar{X}|$.
   (*b*) Does this process have a periodic component?
   (*c*) What is the average power in $X(t)$?

6.2-11. Define a random process $X(t)$ as follows: (1) $X(t)$ assumes only one of two possible levels 1 or -1 at any time, (2) $X(t)$ switches back and forth between its two levels randomly with time, (3) the number of level transitions in any time interval τ is a Poisson random variable, that is, the probability of exactly k transitions, when the average rate of transitions is λ, is given by $[(\lambda\tau)^k/k!]\exp(-\lambda\tau)$, (4) transitions occurring in any time interval are statistically independent of transitions in any other interval, and (5) the levels at the

start of any interval are equally probable. $X(t)$ is usually called the *random telegraph process*. It is an example of a discrete random process.
(a) Find the autocorrelation function of the process.
(b) Find probabilities $P\{X(t) = 1\}$ and $P\{X(t) = -1\}$ for any t.
(c) What is $E[X(t)]$?
(d) Discuss the stationarity of $X(t)$.

6.2-12. Work Problem 6.2-11 assuming the random telegraph signal has levels 0 and 1.

6.2-13. $\bar{X} = 6$ and $R_{XX}(t, t + \tau) = 36 + 25\exp(-|\tau|)$ for a random process $X(t)$. Indicate which of the following statements are true based on what is known with certainty. $X(t)$ (a) is first-order stationary, (b) has total average power of 61 W, (c) is ergodic, (d) is wide-sense stationary, (e) has a periodic component, and (f) has an ac power of 36 W.

6.2-14. (a) Determine whether the process of Problem 6.1-6 is first-order stationary.
(b) Also determine whether it is wide-sense stationary.
(c) How would the results of (a) and (b) change if the random variable A had a nonuniform density?

6.2-15. Find the first- and second-order density functions for the process of Problem 6.1-6.

6.2-16. Find the autocorrelation function and mean of the process of Problem 6.1-6.

6.2-17. Determine if the constant process $X(t) = A$, where A is a random variable with mean $\bar{A}$ and variance σ_A^2, is mean-ergodic.

6.2-18. Let $N(t)$ be a zero-mean wide-sense stationary noise process for which $R_{NN}(\tau) = (\mathcal{N}_0/2)\delta(\tau)$, where $\mathcal{N}_0 > 0$ is a finite constant. Determine if $N(t)$ is mean-ergodic.

6.2-19. Evaluate the integral of (6.2-36) as indicated in the text and show that (6.2-37) is true.

6.3-1. Given the random process

$$X(t) = A\sin(\omega_0 t + \Theta)$$

where A and ω_0 are constants and Θ is a random variable uniformly distributed on the interval $(-\pi, \pi)$. Define a new random process $Y(t) = X^2(t)$.
(a) Find the autocorrelation function of $Y(t)$.
(b) Find the cross-correlation function of $X(t)$ and $Y(t)$.
(c) Are $X(t)$ and $Y(t)$ wide-sense stationary?
(d) Are $X(t)$ and $Y(t)$ jointly wide-sense stationary?

6.3-2. A random process is defined by

$$Y(t) = X(t)\cos(\omega_0 t + \Theta)$$

where $X(t)$ is a wide-sense stationary random process that amplitude-modulates a carrier of constant angular frequency ω_0 with a random phase Θ independent of $X(t)$ and uniformly distributed on $(-\pi, \pi)$.
(a) Find $E[Y(t)]$.

(b) Find the autocorrelation function of $Y(t)$.
(c) Is $Y(t)$ wide-sense stationary?

213

CHAPTER 6:
Random
Processes—
Temporal
Characteristics

6.3-3. Given the random process

$$X(t) = A\cos(\omega_0 t) + B\sin(\omega_0 t)$$

where ω_0 is a constant, and A and B are uncorrelated zero-mean random variables having different density functions but the same variances σ^2, show that $X(t)$ is wide-sense stationary but not strictly stationary.

6.3-4. If $X(t)$ is a stationary random process having a mean value $E[X(t)] = 3$ and autocorrelation function $R_{XX}(\tau) = 9 + 2e^{-|\tau|}$, find:
(a) the mean value and
(b) the variance of the random variable

$$Y = \int_0^2 X(t)\,dt$$

(*Hint*: Assume expectation and integration operations are interchangeable.)

6.3-5. Define a random process by

$$X(t) = A\cos(\pi t)$$

where A is a gaussian random variable with zero mean and variance σ_A^2.
(a) Find the density functions of $X(0)$ and $X(1)$.
(b) Is $X(t)$ stationary in any sense?

6.3-6. For the random process of Problem 6.1-4, calculate:
(a) the mean value $E[X(t)]$ (b) $R_{XX}(t_1 = 0.5T, t_2 = 0.7T)$
(c) $R_{XX}(t_1 = 0.2T, t_2 = 1.2T)$.

6.3-7. A random process consists of three sample functions $X(t, s_1) = 2$, $X(t, s_2) = 2\cos(t)$, and $X(t, s_3) = 3\sin(t)$, each occurring with equal probability. Is the process stationary in any sense?

6.3-8. Statistically independent, zero-mean random processes $X(t)$ and $Y(t)$ have autocorrelation functions

$$R_{XX}(\tau) = e^{-|\tau|}$$

and

$$R_{YY}(\tau) = \cos(2\pi\tau)$$

respectively.
(a) Find the autocorrelation function of the sum $W_1(t) = X(t) + Y(t)$.
(b) Find the autocorrelation function of the difference $W_2(t) = X(t) - Y(t)$.
(c) Find the cross-correlation function of $W_1(t)$ and $W_2(t)$.

6.3-9. Define a random process as $X(t) = p(t + \varepsilon)$, where $p(t)$ is any periodic waveform with period T and ε is a random variable uniformly distributed on the interval $(0, T)$. Show that

$$E[X(t)X(t+\tau)] = \frac{1}{T}\int_0^T p(\xi)p(\xi+\tau)\,d\xi = R_{XX}(\tau)$$

*6.3-10. Use the result of Problem 6.3-9 to find the autocorrelation function of random processes having periodic sample function waveforms $p(t)$ defined
(a) by Figure P6.2-3 with $\varepsilon = 0$ and $4t_0 \le T$, and
(b) by Figure P6.2-4 with $\varepsilon = 0$ and $2t_0 \le T$.

6.3-11. Define two random processes by $X(t) = p_1(t+\varepsilon)$ and $Y(t) = p_2(t+\varepsilon)$ when $p_1(t)$ and $p_2(t)$ are both periodic waveforms with period T and ε is a random variable uniformly distributed on the interval $(0, T)$. Find an expression for the cross-correlation function $E[X(t)Y(t+\tau)]$.

6.3-12. Prove:
(a) (6.3-4) and (b) (6.3-5).

6.3-13. Give arguments to justify (6.3-9).

6.3-14. For a stationary ergodic random process having the autocorrelation function shown in Figure P6.3-14, find:
(a) $E[X(t)]$ (b) $E[X^2(t)]$ and (c) σ_X^2.

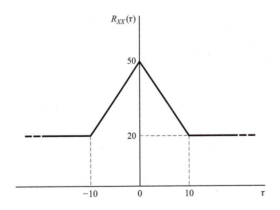

FIGURE P6.3-14

6.3-15. A random process $Y(t) = X(t) - X(t+\tau)$ is defined in terms of a process $X(t)$ that is at least wide-sense stationary.
(a) Show that mean value of $Y(t)$ is 0 even if $X(t)$ has a nonzero mean value.
(b) Show that

$$\sigma_Y^2 = 2[R_{XX}(0) - R_{XX}(\tau)]$$

(c) If $Y(t) = X(t) + X(t+\tau)$, find $E[Y(t)]$ and σ_Y^2. How do these results compare to those of parts (a) and (b).

6.3-16. For two zero-mean, jointly wide-sense stationary random processes $X(t)$ and $Y(t)$, it is known that $\sigma_X^2 = 5$ and $\sigma_Y^2 = 10$. Explain why each of the following functions cannot apply to the processes if they have no periodic components.

215

CHAPTER 6:
Random
Processes—
Temporal
Characteristics

(a) $R_{XX}(\tau) = 6u(\tau)\exp(-3\tau)$ (b) $R_{XX}(\tau) = 5\sin(5\tau)$

(c) $R_{XY}(\tau) = 9(1 + 2\tau^2)^{-1}$ (d) $R_{YY}(\tau) = -\cos(6\tau)\exp(-|\tau|)$

(e) $R_{YY}(\tau) = 5\left[\dfrac{\sin(3\tau)}{3\tau}\right]^2$ (f) $R_{YY}(\tau) = 6 + 4\left[\dfrac{\sin(10\tau)}{10\tau}\right]$

6.3-17. Given two random processes $X(t)$ and $Y(t)$. Find expressions for autocorrelation function of $W(t) = X(t) + Y(t)$ if:

(a) $X(t)$ and $Y(t)$ are correlated.
(b) They are uncorrelated.
(c) They are uncorrelated with zero means.

6.3-18. Use (6.3-19) to prove (6.3-17).

6.3-19. Let $X(t)$ be a stationary continuous random process that is differentiable. Denote its time-derivative by $\dot{X}(t)$.

(a) show that $E[\dot{X}(t)] = 0$.
(b) Find $R_{X\dot{X}}(\tau)$ in terms of $R_{XX}(\tau)$.
(c) Find $R_{\dot{X}\dot{X}}(\tau)$ in terms of $R_{XX}(\tau)$. (*Hint*: Use the definition of the derivative

$$\dot{X}(t) = \lim_{\varepsilon \to 0} \frac{X(t + \varepsilon) - X(t)}{\varepsilon}$$

and assume the order of the limit and expectation operations can be interchanged.)

6.3-20. A stationary zero-mean random process $X(t)$ is ergodic, has average power of 24 W, and has no periodic components. Which of the following can be a valid autocorrelation function? If one cannot, state at least one reason why. (a) $16 + 18\cos(3\tau)$, (b) $24\mathrm{Sa}^2(2\tau)$, (c) $[1 + 3\tau^2]^{-1}\exp(-6\tau)$, and (d) $24\delta(t - \tau)$.

6.3-21. Use the result of Problem 6.3-9 to find the autocorrelation function of a random process with periodic sample function waveform $p(t)$ defined by

$$p(t) = A\cos^2(2\pi t/T)$$

where A and $T > 0$ are constants.

6.3-22. For the random process of Problem 6.1-7, find

(a) $E[X(t)]$.
(b) $R_{XX}(t, t + \tau)$.
(c) Is the process stationary in any sense?

6.3-23. A number of practical systems have "square-law" detectors that produce an output $W(t)$ that is the square of its input $Y(t)$. Let the detector's output be defined by

$$W(t) = Y^2(t) = X^2(t)\cos^2(\omega_0 t + \Theta)$$

where ω_0 is a constant, $X(t)$ is second-order stationary, and Θ is a random variable independent of $X(t)$ and uniform on $(0, 2\pi)$. Find (a) $E[W(t)]$, (b) $R_{WW}(t, t + \tau)$, and (c) whether or not $W(t)$ is wide-sense stationary.

6.3-24. A random process $X(t)$ is known to be wide-sense stationary with $E[X^2(t)] = 11$. Give one or more reasons why each of the following expressions cannot be the autocorrelation function of the process.

(a) $R_{XX}(t, t + \tau) = \cos(8t)e^{-(t+\tau)^2}$

(b) $R_{XX}(t, t + \tau) = \dfrac{\sin(2\tau)}{1 + \tau^2}$

(c) $R_{XX}(t, t + \tau) = \dfrac{11 \sin[5(\tau - 2)]}{5(\tau - 2)}$

(d) $R_{XX}(t, t + \tau) = -11e^{-|\tau|}$

(e) $R_{XX}(t, t + \tau) = \dfrac{11\tau}{1 + 3\tau^2 + 4\tau^4}$

6.3-25. A wide-sense stationary random process $Y(t)$ has a power of $E[Y^2(t)] = 4$. Give at least one reason why each of the following expressions cannot be its autocorrelation function.

(a) $R_{YY}(t, t + \tau) = 4 \tan^{-1}(\tau)$

(b) $R_{YY}(t, t + \tau) = 6 \exp[-2\tau^2 - |\tau|]$

(c) $R_{YY}(t, t + \tau) = \frac{1}{2}u(\tau)$

(d) $R_{YY}(t, t + \tau) = \dfrac{-0.5 + u(\tau)}{1 + 8\tau^4}$

(e) $R_{YY}(t, t + \tau) = \dfrac{\cos^2(6\tau)}{2 + \cos^2(4t)}$

6.3-26. Two random processes are defined by

$$Y_1(t) = X(t)\cos(\omega_0 t)$$
$$Y_2(t) = Y(t)\cos(\omega_0 t + \Theta)$$

where $X(t)$ and $Y(t)$ are jointly wide-sense stationary processes.
(a) If Θ is a constant (nonrandom), is there any value of Θ that will make $Y_1(t)$ and $Y_2(t)$ orthogonal?
(b) If Θ is a uniform random variable, statistically independent of $X(t)$ and $Y(t)$, are there any conditions on Θ that will make $Y_1(t)$ and $Y_2(t)$ orthogonal?

6.3-27. Determine the largest constant K such that the function

$$R_{XY}(\tau) = Ke^{-\tau^2}\sin(\pi\tau)$$

can possibly be a valid cross-correlation function of two jointly wide-sense stationary processes $X(t)$ and $Y(t)$ for which $E[X^2(t)] = 6$ and $E[Y^2(t)] = 4$.

6.4-1. An engineer wants to measure the mean value of a noise signal that can be well-modeled as a sample function of a gaussian process. He uses the sampling estimator of (5.7-4). After 100 samples he wishes his estimate to be within ± 0.1 V of the true mean with probability 0.9606. What is the largest variance the process can have such that his wishes will be true?

217

CHAPTER 6:
Random
Processes—
Temporal
Characteristics

6.4-2. Let $X(t)$ be the sum of a deterministic signal $s(t)$ and a wide-sense stationary noise process $N(t)$. Find the mean value, and autocorrelation and autocovariance functions of $X(t)$. Discuss the stationarity of $X(t)$.

6.4-3. Random processes $X(t)$ and $Y(t)$ are defined by

$$X(t) = A\cos(\omega_0 t + \Theta)$$
$$Y(t) = B\sin(\omega_0 t + \Theta)$$

where A, B, and ω_0 are constants while Θ is a random variable uniform on $(0, 2\pi)$. By the procedures of Example 6.2-1 it is easy to find that $X(t)$ and $Y(t)$ are zero-mean, wide-sense stationary with autocorrelation functions

$$R_{XX}(\tau) = (A^2/2)\cos(\omega_0 \tau)$$
$$R_{YY}(\tau) = (B^2/2)\cos(\omega_0 \tau)$$

(a) Find the cross-correlation function $R_{XY}(t, t + \tau)$ and show that $X(t)$ and $Y(t)$ are jointly wide-sense stationary.
(b) Solve (6.4-2) and show that the response of the system of Figure 6.4-1 equals the true cross-correlation function plus an error term $\varepsilon(T)$ that decreases as T increases.
(c) Sketch $|\varepsilon(T)|$ versus T to show its behavior. How large must T be to make $|\varepsilon(T)|$ less than 1% of the largest value the correct cross-correlation function can have?

6.4-4. Consider random processes

$$X(t) = A\cos(\omega_0 t + \Theta)$$
$$Y(t) = B\cos(\omega_1 t + \Phi)$$

where A, B, ω_1, and ω_0 are constants, while Θ and Φ are statistically independent random variables each uniform on $(0, 2\pi)$.
(a) Show that $X(t)$ and $Y(t)$ are jointly wide-sense stationary.
(b) If $\Theta = \Phi$ show that $X(t)$ and $Y(t)$ are not jointly wide-sense stationary unless $\omega_1 = \omega_0$.

6.4-5. A zero-mean gaussian random process has an autocorrelation function

$$R_{XX}(\tau) = \begin{cases} 13[1 - (|\tau|/6)] & |\tau| \le 6 \\ 0 & \text{elsewhere} \end{cases}$$

Find the covariance function necessary to specify the joint density of random variables defined at times $t_i = 2(i - 1)$, $i = 1, 2, \ldots, 5$. Give the covariance matrix for the $X_i = X(t_i)$.

6.4-6. A random process, as defined by $Y(t) = X(t)\cos(\omega_0 t + \Theta)$ in Problem 6.3-23, is applied to both inputs of the measurement system of Figure 6.4-1. Determine $R_0(2T)$ if ω_0 is large enough so that $\cos(\omega_0 t + \Theta)$ cycles rapidly compared to $X(t)$.

6.5-1. A gaussian random process has an autocorrelation function

$$R_{XX}(\tau) = 6\exp(-|\tau|/2)$$

Determine a covariance matrix for the random variables $X(t)$, $X(t+1)$, $X(t+2)$, and $X(t+3)$.

6.5-2. Work Problem 6.5-1 if

$$R_{XX}(\tau) = 6\frac{\sin(\pi\tau)}{\pi\tau}$$

6.5-3. If the gaussian process of Problem 6.4-5 is shifted to have a constant mean $\bar{X} = -2$ but all else is unchanged, discuss how the autocorrelation function and covariance matrix change. What is the effect on the joint density of the five random variables?

6.6-1. Aircraft arrive at an airport according to a Poisson process at a rate of 12 per hour. All aircraft are handled by one air-traffic controller. If the controller takes a 2-minute coffee break, what is the probability that he will miss one or more arriving aircraft?

6.6-2. Telephone calls are initiated through an exchange at the average rate of 75 per minute and are described by a Poisson process. Find the probability that more than 3 calls are initiated in any 5-second period.

6.6-3. A small store has two check-out lanes that develop waiting lines if more than two customers arrive in any one minute interval. Assume that a Poisson process describes the number of customers that arrive for check-out. Find the probability of a waiting line if the average rate of customer arrivals is (a) 2 per minutes, (b) 1 per minute, and (c) $\frac{1}{2}$ per minute.

6.6-4. A particular commercial system for controlling a petroleum distillation plant has failures (resulting in plant down time) that occur at the average rate of two per 30 days. Assume that the number of failures is a Poisson process and find the probability that one failure will occur during the first 30 days and no other failures will occur for the next 30 days.

6.6-5. Show that the autocorrelation function of the Poisson process is

$$R_{XX}(t_1, t_2) = \begin{cases} \lambda t_1[1 + \lambda t_2] & t_1 < t_2 \\ \lambda t_2[1 + \lambda t_1] & t_1 > t_2 \end{cases}$$

[*Hint*: Make use of the sums in (6.6-3) and (6.6-4).]

6.6-6. Determine the autocovariance function of the Poisson process. (*Hint*: Make use of the result of Problem 6.6-5.)

*6.7-1. A complex random process $Z(t) = X(t) + jY(t)$ is defined by jointly stationary real processes $X(t)$ and $Y(t)$. Show that

$$E[|Z(t)|^2] = R_{XX}(0) + R_{YY}(0)$$

*6.7-2. Let $X_1(t)$, $X_2(t)$, $Y_1(t)$, and $Y_2(t)$ be real random processes and define

$$Z_1(t) = X_1(t) + jY_1(t) \qquad Z_2(t) = X_2(t) - jY_2(t)$$

Find the expressions for the cross-correlation function of $Z_1(t)$ and $Z_2(t)$ if:
(a) All the real processes are correlated.

(b) They are uncorrelated.
(c) They are uncorrelated with zero means.

219

CHAPTER 6:
Random
Processes—
Temporal
Characteristics

*6.7-3. Let $Z(t)$ be a stationary complex random process with an autocorrelation function $R_{ZZ}(\tau)$. Define the random variable

$$W = \int_a^{a+T} Z(t)\, dt$$

where $T > 0$ and a are real numbers. Show that

$$E[|W|^2] = \int_{-T}^T (T - |\tau|) R_{ZZ}(\tau)\, d\tau$$

*6.7-4. Extend Example 6.7-1 to allow the sum of complex-amplitude unequal-frequency phasors. Let Z_i, $i = 1, 2, \ldots, N$ be N complex zero-mean, uncorrelated random variables with variances $\sigma_{Z_i}^2$. Form a random process

$$Z(t) = \sum_{i=1}^N Z_i e^{j\omega_i t}$$

where ω_i are the angular frequencies of the phasors.
(a) Show that $E[Z(t)] = 0$.
(b) Derive the autocorrelation function and show that $Z(t)$ is wide-sense stationary.

*6.7-5. A complex random process is defined by

$$Z(t) = \exp(j\Omega t)$$

where Ω is a zero-mean random variable uniformly distributed on the interval from $\omega_0 - \Delta\omega$ to $\omega_0 + \Delta\omega$, where ω_0 and $\Delta\omega$ are positive constants. Find:
(a) The mean value, and (b) the autocorrelation function of $Z(t)$.
(c) Is $Z(t)$ wide-sense stationary?

*6.7-6. Work Problem 6.7-5 except assume the process

$$Z(t) = e^{j\Omega t} + e^{-j\Omega t} = 2\cos(\Omega t)$$

*6.7-7. Let $X(t)$ and $Y(t)$ be statistically independent wide-sense stationary real processes having the same autocorrelation function $R(\tau)$. Define the complex process

$$Z(t) = X(t)\cos(\omega_0 t) + jY(t)\sin(\omega_0 t)$$

where ω_0 is a positive constant. Find the autocorrelation function of $Z(t)$. Is $Z(t)$ wide-sense stationary?

Random Processes—Spectral Characteristics

7.0
INTRODUCTION

All of the foregoing discussions concerning random processes have involved the time domain. That is, we have characterized processes by means of auto-correlation, cross-correlation, and covariance functions without any consideration of spectral properties. As is well known, both time domain *and* frequency domain analysis methods exist for analyzing linear systems and deterministic waveforms. But what about random waveforms? Is there some way to describe random processes in the frequency domain? The answer is yes, and it is the purpose of this chapter to introduce the most important concepts that apply to characterizing random processes in the frequency domain.

The spectral description of a deterministic waveform is obtained by Fourier transforming the waveform, and the reader would be correct in concluding that Fourier transforms play an important role in the spectral characterization of random waveforms. However, the direct transformation approach is not attractive for random waveforms because the transform may not exist. Thus, spectral analysis of random processes requires a bit more subtlety than do deterministic signals.

An appropriate spectrum to be associated with a random process is introduced in the following section. The concepts rely heavily on theory of Fourier transforms. Readers wishing to refresh their background on Fourier theory are referred to Appendix D, where a short review is given.

7.1
POWER DENSITY SPECTRUM AND ITS PROPERTIES

The spectral properties of a *deterministic* signal $x(t)$ are contained in its *Fourier transform* $X(\omega)$ given by

$$X(\omega) = \int_{-\infty}^{\infty} x(t)e^{-j\omega t}\, dt \qquad (7.1\text{-}1)$$

221

CHAPTER 7:
Random
Processes—
Spectral
Characteristics

The function $X(\omega)$, sometimes called simply the *spectrum* of $x(t)$, has the unit of volts per hertz when $x(t)$ is a voltage and describes the way in which relative signal voltage is distributed with frequency. The Fourier transform can, therefore, be considered to be a *voltage density spectrum* applicable to $x(t)$. Both the amplitudes and the phases of the frequencies present in $x(t)$ are described by $X(\omega)$. For this reason, if $X(\omega)$ is known then $x(t)$ can be recovered by means of the *inverse Fourier transform*

$$x(t) = \frac{1}{2\pi} \int_{-\infty}^{\infty} X(\omega)e^{j\omega t}\, d\omega \qquad (7.1\text{-}2)$$

In other words, $X(\omega)$ forms a complete description of $x(t)$ and vice versa.

In attempting to apply (7.1-1) to a random process, we immediately encounter problems. The principal problem is the fact that $X(\omega)$ may not exist for most sample functions of the process. Thus, we conclude that a spectral description of a random process utilizing a voltage density spectrum (Fourier transform) is not feasible because such a spectrum may not exist. Other problems arise if Laplace transforms are considered (Cooper and McGillem, 1971, p. 132).

On the other hand, if we turn our attention to the description of the *power* in the random process as a function of frequency, instead of voltage, it results that such a function does exist. We next proceed to develop this function, called the *power density spectrum*† of the random process.

The Power Density Spectrum

For a random process $X(t)$, let $x_T(t)$ be defined as that portion of a sample function $x(t)$ that exists between $-T$ and T; that is

$$x_T(t) = \begin{cases} x(t) & -T < t < T \\ 0 & \text{elsewhere} \end{cases} \qquad (7.1\text{-}3)$$

Now so long as T is finite, we presume that $x_T(t)$ has bounded variation, will satisfy

$$\int_{-T}^{T} |x_T(t)|\, dt < \infty \qquad (7.1\text{-}4)$$

and will have a Fourier transform (see Appendix D for conditions sufficient for the existence of Fourier transforms), which we denote $X_T(\omega)$, given by

$$X_T(\omega) = \int_{-T}^{T} x_T(t)e^{-j\omega t}\, dt = \int_{-T}^{T} x(t)e^{-j\omega t}\, dt \qquad (7.1\text{-}5)$$

---

†Many books call this function a *power spectral density*. We shall occasionally use also the names *power density* or *power spectrum*.

The energy contained in $x(t)$ in the interval $(-T, T)$ is†

$$E(T) = \int_{-T}^{T} x_T^2(t)\,dt = \int_{-T}^{T} x^2(t)\,dt \tag{7.1-6}$$

Since $x_T(t)$ is Fourier transformable, its energy must also be related to $X_T(\omega)$ by Parseval's theorem. Thus, from (7.1-6) and (D-21) of Appendix D

$$E(T) = \int_{-T}^{T} x^2(t)\,dt = \frac{1}{2\pi} \int_{-\infty}^{\infty} |X_T(\omega)|^2\,d\omega \tag{7.1-7}$$

By dividing the expressions in (7.1-7) by $2T$, we obtain the average power $P(T)$ in $x(t)$ over the interval $(-T, T)$:

$$P(T) = \frac{1}{2T} \int_{-T}^{T} x^2(t)\,dt = \frac{1}{2\pi} \int_{-\infty}^{\infty} \frac{|X_T(\omega)|^2}{2T}\,d\omega \tag{7.1-8}$$

At this point we observe that $|X_T(\omega)|^2/2T$ is a power density spectrum because power results through its integration. However, it is not the function that we seek, for two reasons. One is the fact that (7.1-8) does not represent the power in an entire sample function. There remains the step of letting T become arbitrarily large so as to include all power in the ensemble member. The second reason is that (7.1-8) is only the power in one sample function and does not represent the process. In other words, $P(T)$ is actually a random variable with respect to the random process. By taking the expected value in (7.1-8), we can obtain an average power P_{XX} for the random process.‡

From the above discussion it is clear that we must still form the limit as $T \to \infty$ and take the expected value of (7.1-8) to obtain a suitable power density spectrum for the random process. It is important that the limiting operation be done last (Thomas, 1969, p. 98, or Cooper and McGillem, 1971, p. 134, and 1986, p. 233). After these operations are performed, (7.1-8) can be written

$$P_{XX} = \lim_{T\to\infty} \frac{1}{2T} \int_{-T}^{T} E[X^2(t)]\,dt = \frac{1}{2\pi} \int_{-\infty}^{\infty} \lim_{T\to\infty} \frac{E[|X_T(\omega)|^2]}{2T}\,d\omega \tag{7.1-9}$$

Equation (7.1-9) establishes two important facts. First, average power P_{XX} in a random process $X(t)$ is given by the time average of its second moment:

$$P_{XX} = \lim_{T\to\infty} \frac{1}{2T} \int_{-T}^{T} E[X^2(t)]\,dt = A\{E[X^2(t)]\} \tag{7.1-10}$$

---

†We assume a real process $X(t)$ and interpret $x(t)$ as either the voltage across a 1-Ω impedance or the current through 1 Ω. In other words, we shall assume a 1-Ω real impedance whenever we discuss energy or power in subsequent work, unless specifically stated otherwise.
‡In taking the expected value we replace $x(t)$ by $X(t)$ in (7.1-8) since the integral of $x^2(t)$ is an operation performed on all sample functions of $X(t)$.

For a process that is at least wide-sense stationary, $E[X^2(t)] = \overline{X^2}$, a constant, and $P_{XX} = \overline{X^2}$. Second, P_{XX} can be obtained by a frequency domain integration. If we define the *power density spectrum* for the random process by

223

CHAPTER 7:
Random
Processes—
Spectral
Characteristics

$$\mathcal{S}_{XX}(\omega) = \lim_{T \to \infty} \frac{E[|X_T(\omega)|^2]}{2T} \qquad (7.1\text{-}11)$$

the applicable integral, which we call the *power formula*, is

$$P_{XX} = \frac{1}{2\pi} \int_{-\infty}^{\infty} \mathcal{S}_{XX}(\omega) \, d\omega \qquad (7.1\text{-}12)$$

from (7.1-9). Two examples will illustrate the above concepts.

EXAMPLE 7.1-1. Consider the random process

$$X(t) = A_0 \cos(\omega_0 t + \Theta)$$

where A_0 and ω_0 are real constants and Θ is a random variable uniformly distributed on the interval $(0, \pi/2)$. We shall find the average power P_{XX} in $X(t)$ by use of (7.1-10). Mean-squared value is

$$E[X^2(t)] = E[A_0^2 \cos^2(\omega_0 t + \Theta)] = E\left[\frac{A_0^2}{2} + \frac{A_0^2}{2}\cos(2\omega_0 t + 2\Theta)\right]$$

$$= \frac{A_0^2}{2} + \frac{A_0^2}{2} \int_0^{\pi/2} \frac{2}{\pi}\cos(2\omega_0 t + 2\theta) \, d\theta$$

$$= \frac{A_0^2}{2} - \frac{A_0^2}{\pi}\sin(2\omega_0 t)$$

This process is not even wide-sense stationary, since the above function is time-dependent. The time average of the above expression is

$$A\{E[X^2(t)]\} = \lim_{T \to \infty} \frac{1}{2T} \int_{-T}^{T} \left[\frac{A_0^2}{2} - \frac{A_0^2}{\pi}\sin(2\omega_0 t)\right] dt$$

which easily evaluates to

$$P_{XX} = A\{E[X^2(t)]\} = A_0^2/2$$

EXAMPLE 7.1-2. We reconsider the process of the above example to find $\mathcal{S}_{XX}(\omega)$ and average power P_{XX} by use of (7.1-11) and (7.1-12), respectively. First we find $X_T(\omega)$:

$$X_T(\omega) = \int_{-T}^{T} A_0 \cos(\omega_0 t + \Theta) \exp(-j\omega t) \, dt$$

$$= \frac{A_0}{2} \exp(j\Theta) \int_{-T}^{T} \exp[j(\omega_0 - \omega)t] \, dt$$

$$+ \frac{A_0}{2} \exp(-j\Theta) \int_{-T}^{T} \exp[-j(\omega_0 + \omega)t] \, dt$$

$$= A_0 T \exp(j\Theta) \frac{\sin[(\omega - \omega_0)T]}{(\omega - \omega_0)T}$$

$$+ A_0 T \exp(-j\Theta) \frac{\sin[(\omega + \omega_0)T]}{(\omega + \omega_0)T}$$

Next we determine $|X_T(\omega)|^2 = X_T(\omega)X_T^*(\omega)$ and find its expected value. After some simple algebraic reduction we obtain

$$\frac{E[|X_T(\omega)|^2]}{2T} = \frac{A_0^2 \pi}{2} \left\{ \frac{T}{\pi} \frac{\sin^2[(\omega - \omega_0)T]}{[(\omega - \omega_0)T]^2} + \frac{T}{\pi} \frac{\sin^2[(\omega + \omega_0)T]}{[(\omega + \omega_0)T]^2} \right\}$$

In obtaining this result we have neglected cross terms each having factors at widely separate frequencies ω_0 and $-\omega_0$ such that their product is small. Now it is known that

$$\lim_{T \to \infty} \frac{T}{\pi} \left[\frac{\sin(\alpha T)}{\alpha T} \right]^2 = \delta(\alpha)$$

(Lathi, 1968, p. 24), so (7.1-11) and the above result give

$$\mathcal{S}_{XX}(\omega) = \frac{A_0^2 \pi}{2}[\delta(\omega - \omega_0) + \delta(\omega + \omega_0)]$$

Finally, we use this result to obtain average power from (7.1-12):

$$P_{XX} = \frac{1}{2\pi} \int_{-\infty}^{\infty} \frac{A_0^2 \pi}{2}[\delta(\omega - \omega_0) + \delta(\omega + \omega_0)] \, d\omega = \frac{A_0^2}{2}$$

Thus, P_{XX} found here agrees with that of the earlier Example 7.1-1.

Properties of the Power Density Spectrum

The power density spectrum possesses a number of important properties:

$$(1) \quad \mathcal{S}_{XX}(\omega) \geq 0 \qquad\qquad\qquad (7.1\text{-}13)$$

$$(2) \quad \mathcal{S}_{XX}(-\omega) = \mathcal{S}_{XX}(\omega) \qquad X(t) \text{ real} \qquad (7.1\text{-}14)$$

$$(3) \quad \mathcal{S}_{XX}(\omega) \text{ is real} \qquad\qquad\qquad (7.1\text{-}15)$$

$$(4) \quad \frac{1}{2\pi} \int_{-\infty}^{\infty} \mathcal{S}_{XX}(\omega) \, d\omega = A\{E[X^2(t)]\} \qquad (7.1\text{-}16)$$

Property 1 follows from the definition (7.1-11) and the fact that the expected value of a nonnegative function is nonnegative. Similarly, property 3 is true from (7.1-11) since $|X_T(\omega)|^2$ is real. Some reflection on the properties of Fourier transforms of real functions will verify property 2 (see Problem 7.1-9). Property 4 is just another statement of (7.1-9).

Sometimes another property is included in a list of properties:

$$(5) \quad \mathcal{S}_{\dot{X}\dot{X}}(\omega) = \omega^2 \mathcal{S}_{XX}(\omega) \qquad\qquad (7.1\text{-}17)$$

It says that the power density spectrum of the derivative $\dot{X}(t) = dX(t)/dt$ is ω^2 times the power spectrum of $X(t)$. Proof of this property is left as a reader exercise (Problem 7.1-10).

A final property we list is

225

CHAPTER 7:
Random
Processes—
Spectral
Characteristics

$$(6) \quad \frac{1}{2\pi} \int_{-\infty}^{\infty} \mathscr{S}_{XX}(\omega) e^{j\omega\tau} \, d\omega = A[R_{XX}(t, t + \tau)] \quad (7.1\text{-}18)$$

$$\mathscr{S}_{XX}(\omega) = \int_{-\infty}^{\infty} A[R_{XX}(t, t + \tau)] e^{-j\omega\tau} \, d\tau \quad (7.1\text{-}19)$$

It states that the power density spectrum and the time average of the auto-correlation function form a Fourier transform pair. We prove this very important property in Section 7.2. Of course, if $X(t)$ is at least wide-sense stationary, $A[R_{XX}(t, t + \tau)] = R_{XX}(\tau)$, and property 6 indicates that the power spectrum and the autocorrelation function form a Fourier transform pair. Thus

$$\mathscr{S}_{XX}(\omega) = \int_{-\infty}^{\infty} R_{XX}(\tau) e^{-j\omega\tau} \, d\tau \quad (7.1\text{-}20)$$

$$R_{XX}(\tau) = \frac{1}{2\pi} \int_{-\infty}^{\infty} \mathscr{S}_{XX}(\omega) e^{j\omega\tau} \, d\omega \quad (7.1\text{-}21)$$

for a wide-sense stationary process.

Bandwidth of the Power Density Spectrum

Assume that $X(t)$ is a lowpass process; that is, its spectral components are clustered near $\omega = 0$ and have decreasing magnitudes at higher frequencies. Such processes are also called *baseband*. Except for the fact that the area of $\mathscr{S}_{XX}(\omega)$ is not necessarily unity, $\mathscr{S}_{XX}(\omega)$ has characteristics similar to a probability density function (it is nonnegative and real). Indeed, by dividing $\mathscr{S}_{XX}(\omega)$ by its area, a new function is formed with area of unity that is analogous to a density function.

Recall that standard deviation is a measure of the spread in a density function. The analogous quantity for the normalized power spectrum is a measure of its spread that we call *rms bandwidth*,† which we denote W_{rms} (rad/s). Now since $\mathscr{S}_{XX}(\omega)$ is an even function for a real process, its "mean value" is zero and its "standard deviation" is the square root of its second moment. Thus, upon normalization, the rms bandwidth is given by

$$W_{rms}^2 = \frac{\displaystyle\int_{-\infty}^{\infty} \omega^2 \mathscr{S}_{XX}(\omega) \, d\omega}{\displaystyle\int_{-\infty}^{\infty} \mathscr{S}_{XX}(\omega) \, d\omega} \quad (7.1\text{-}22)$$

†The notation rms bandwidth stands for *root-mean-squared* bandwidth.

EXAMPLE 7.1-3. Given the power spectrum

$$\mathcal{S}_{XX}(\omega) = \frac{10}{[1 + (\omega/10)^2]^2}$$

where the 6-dB bandwidth is 10 radians per second, we find W_{rms}. First, using (C-28) from Appendix C,

$$\int_{-\infty}^{\infty} \frac{10 \, d\omega}{[1 + (\omega/10)^2]^2} = 10^5 \int_{-\infty}^{\infty} \frac{d\omega}{(100 + \omega^2)^2}$$

$$= 10^5 \left\{ \frac{\omega}{200(100 + \omega^2)} \bigg|_{-\infty}^{\infty} + \frac{1}{2000} \tan^{-1}\left(\frac{\omega}{10}\right) \bigg|_{-\infty}^{\infty} \right\}$$

$$= 50\pi$$

Next, from (C-30) of Appendix C:

$$\int_{-\infty}^{\infty} \frac{10\omega^2 \, d\omega}{[1 + (\omega/10)^2]^2} = 10^5 \int_{-\infty}^{\infty} \frac{\omega^2 \, d\omega}{(100 + \omega^2)^2}$$

$$= 10^5 \left\{ \frac{-\omega}{2(100 + \omega^2)} \bigg|_{-\infty}^{\infty} + \frac{1}{20} \tan^{-1}\left(\frac{\omega}{10}\right) \bigg|_{-\infty}^{\infty} \right\}$$

$$= 5000\pi$$

Thus

$$W_{\text{rms}} = \sqrt{\frac{5000\pi}{50\pi}} = 10 \, \text{rad/s}$$

Although W_{rms} and the 6-dB bandwidth of $\mathcal{S}_{XX}(\omega)$ are equal in this case, they are not equal in general.

The above concept is readily extended to a process that has a bandpass form of power spectrum; that is, its significant spectral components cluster near some frequencies $\bar{\omega}_0$ and $-\bar{\omega}_0$. If we assume that the process $X(t)$ is real, $\mathcal{S}_{XX}(\omega)$ will be real and have even symmetry about $\omega = 0$. With this assumption we define a *mean* frequency $\bar{\omega}_0$ by

$$\bar{\omega}_0 = \frac{\displaystyle\int_0^\infty \omega \mathcal{S}_{XX}(\omega) \, d\omega}{\displaystyle\int_0^\infty \mathcal{S}_{XX}(\omega) \, d\omega} \tag{7.1-23}$$

and rms bandwidth by

$$W_{\text{rms}}^2 = \frac{4\displaystyle\int_0^\infty (\omega - \bar{\omega}_0)^2 \mathcal{S}_{XX}(\omega) \, d\omega}{\displaystyle\int_0^\infty \mathcal{S}_{XX}(\omega) \, d\omega} \tag{7.1-24}$$

The reader is encouraged to sketch a few lowpass and bandpass power spectrums and justify for himself why the factor of 4 appears in (7.1-24).

7.2

227

CHAPTER 7:
Random
Processes—
Spectral
Characteristics

RELATIONSHIP BETWEEN POWER SPECTRUM AND AUTOCORRELATION FUNCTION

In Section 7.1 it was stated that the inverse Fourier transform of the power density spectrum is the time average of the autocorrelation function; that is

$$\frac{1}{2\pi} \int_{-\infty}^{\infty} \mathscr{S}_{XX}(\omega) e^{j\omega\tau}\, d\omega = A[R_{XX}(t, t+\tau)] \qquad (7.2\text{-}1)$$

This expression will now be proved.

If we use (7.1-5), which is the definition of $X_T(\omega)$, in the defining equation (7.1-11) for the power spectrum we have†

$$\mathscr{S}_{XX}(\omega) = \lim_{T \to \infty} E\left[\frac{1}{2T} \int_{-T}^{T} X(t_1) e^{j\omega t_1}\, dt_1 \int_{-T}^{T} X(t_2) e^{-j\omega t_2}\, dt_2\right]$$

$$= \lim_{T \to \infty} \frac{1}{2T} \int_{-T}^{T} \int_{-T}^{T} E[X(t_1)X(t_2)] e^{-j\omega(t_2 - t_1)}\, dt_2\, dt_1 \qquad (7.2\text{-}2)$$

The expectation in the integrand of (7.2-2) is identified as the autocorrelation function of $X(t)$:

$$E[X(t_1)X(t_2)] = R_{XX}(t_1, t_2) \qquad -T < (t_1 \text{ and } t_2) < T \qquad (7.2\text{-}3)$$

Thus, (7.2-2) becomes

$$\mathscr{S}_{XX}(\omega) = \lim_{T \to \infty} \frac{1}{2T} \int_{-T}^{T} \int_{-T}^{T} R_{XX}(t_1, t_2) e^{-j\omega(t_2 - t_1)}\, dt_1\, dt_2 \qquad (7.2\text{-}4)$$

Next, we seek the inverse Fourier transform of (7.2-4).‡ The transform will exist since the power spectrum is real, nonnegative, and is absolutely integrable from (7.1-12) because we presume $P_{XX} < \infty$. The inverse transform is

$$\frac{1}{2\pi} \int_{-\infty}^{\infty} \mathscr{S}_{XX}(\omega)\, e^{j\omega\tau}\, d\omega$$

$$= \frac{1}{2\pi} \int_{-\infty}^{\infty} \lim_{T \to \infty} \frac{1}{2T} \int_{-T}^{T} \int_{-T}^{T} R_{XX}(t, t_1)\, e^{-j\omega(t_1 - t)}\, dt\, dt_1\, e^{j\omega\tau}\, d\omega \qquad (7.2\text{-}5)$$

$$= \lim_{T \to \infty} \frac{1}{2T} \int_{-T}^{T} \int_{-T}^{T} R_{XX}(t, t_1) \frac{1}{2\pi} \int_{-\infty}^{\infty} e^{j\omega(\tau - t_1 + t)}\, d\omega\, dt_1\, dt$$

From (A-23) and the event symmetry of impulses, we recognize the inner integral of (7.2-5) as the impulse $2\pi\delta(\tau - t_1 + t) = 2\pi\delta(t_1 - t - \tau)$, so

---

†We use $X(t)$ in (7.1-5), rather than $x(t)$, to imply that the operations performed take place on the process as opposed to one sample function.
‡Our development follows that of Middleton (1960), pp. 142, 194.

$$\frac{1}{2\pi} \int_{-\infty}^{\infty} \mathscr{S}_{XX}(\omega) e^{j\omega\tau} d\omega$$

$$= \lim_{T\to\infty} \frac{1}{2T} \int_{-T}^{T} \int_{-T}^{T} R_{XX}(t, t_1)\delta(t_1 - t - \tau) dt_1 \, dt \qquad (7.2\text{-}6)$$

The definition of the impulse given in (A-16) allows the immediate evaluation of the integral over t_1:

$$\frac{1}{2\pi} \int_{-\infty}^{\infty} \mathscr{S}_{XX}(\omega) e^{j\omega\tau} d\omega$$

$$= \lim_{T\to\infty} \frac{1}{2T} \int_{-T}^{T} R_{XX}(t, t+\tau) dt \qquad -T < t+\tau < T \qquad (7.2\text{-}7)$$

In the limit as $T \to \infty$ the condition $-T < t + \tau < T$ has little meaning. The quantity on the right side of (7.2-7) is recognized as the time average of the process' autocorrelation function

$$A[R_{XX}(t, t+\tau)] = \lim_{T\to\infty} \frac{1}{2T} \int_{-T}^{T} R_{XX}(t, t+\tau) dt \qquad (7.2\text{-}8)$$

Thus, (7.2-7) says that the inverse Fourier transform of the power density spectrum is the time average of the process' autocorrelation function, which proves (7.2-1). The direct transform must also hold, giving

$$\mathscr{S}_{XX}(\omega) = \int_{-\infty}^{\infty} A[R_{XX}(t, t+\tau)]e^{-j\omega\tau} d\tau \qquad (7.2\text{-}9)$$

which shows that $\mathscr{S}_{XX}(\omega)$ and $A[R_{XX}(t, t+\tau)]$ form a Fourier transform pair:

$$A[R_{XX}(t, t+\tau)] \leftrightarrow \mathscr{S}_{XX}(\omega) \qquad (7.2\text{-}10)$$

For the important case where $X(t)$ is at least wide-sense stationary, $A[R_{XX}(t, t+\tau)] = R_{XX}(\tau)$ and we get

$$\mathscr{S}_{XX}(\omega) = \int_{-\infty}^{\infty} R_{XX}(\tau)e^{-j\omega\tau} d\tau \qquad (7.2\text{-}11)$$

$$R_{XX}(\tau) = \frac{1}{2\pi} \int_{-\infty}^{\infty} \mathscr{S}_{XX}(\omega)e^{j\omega\tau} d\omega \qquad (7.2\text{-}12)$$

or

$$R_{XX}(\tau) \leftrightarrow \mathscr{S}_{XX}(\omega) \qquad (7.2\text{-}13)$$

The expressions (7.2-11) and (7.2-12) are usually called the *Wiener-Khintchine relations* after the great American mathematician Norbert Wiener (1894–1964) and the Russian mathematician A. I. Khintchine (1894–1959). They form the basic link between the time domain description (correlation functions) of processes and their description in the frequency domain (power spectrum).

From (7.2-13), it is clear that knowledge of the power spectrum of a process allows complete recovery of the autocorrelation function when $X(t)$ is at least wide-sense stationary; for a nonstationary process, only the time average of the autocorrelation function is recoverable from (7.2-10).

EXAMPLE 7.2-1. The power spectrum will be found for the random process of Example 6.2-1 that has the autocorrelation function

229

CHAPTER 7:
Random
Processes—
Spectral
Characteristics

$$R_{XX}(\tau) = (A_0^2/2)\cos(\omega_0 \tau)$$

where A_0 and ω_0 are constants. This equation can be written in the form

$$R_{XX}(\tau) = \frac{A_0^2}{4}(e^{j\omega_0\tau} + e^{-j\omega_0\tau})$$

Now we note that the inverse transform of a frequency domain impulse function is

$$\frac{1}{2\pi}\int_{-\infty}^{\infty}\delta(\omega)e^{j\omega\tau}\,d\omega = \frac{1}{2\pi}$$

from (A-2) of Appendix A. Thus

$$1 \leftrightarrow 2\pi\delta(\omega)$$

and, from the frequency-shifting property of Fourier transforms given by (D-7) of Appendix D, we get

$$e^{j\omega_0\tau} \leftrightarrow 2\pi\delta(\omega - \omega_0)$$

By using this last result, the Fourier transform of $R_{XX}(\tau)$ becomes

$$\mathscr{S}_{XX}(\omega) = \frac{A_0^2\pi}{2}[(\omega - \omega_0) + \delta(\omega + \omega_0)]$$

This function and $R_{XX}(\tau)$ are illustrated in Figure 7.2-1.

EXAMPLE 7.2-2. As another example, assume a wide-sense stationary process $X(t)$ has an autocorrelation function

$$R_{XX}(\tau) = \begin{cases} A_0[1 - (|\tau|/T) & -T \le t \le T \\ 0 & \text{elsewhere} \end{cases}$$

where $T > 0$ and A_0 are constants. $R_{XX}(\tau)$ is sketched in Figure 7.2-2a for $A_0 > 0$. The power spectrum is found from the Fourier transform

$$\mathscr{S}_{XX}(\omega) = \int_{-\infty}^{\infty} R_{XX}(\tau)e^{-j\omega\tau}\,d\tau$$

$$= A_0\int_{-T}^{0}[1 + (\tau/T)]e^{-j\omega\tau}\,d\tau + A_0\int_{0}^{T}[1 - (\tau/T)]e^{-j\omega\tau}\,d\tau$$

These integrals evaluate using (C-45) and (C-46) to give

$$\mathscr{S}_{XX}(\omega) = A_0 T\,\text{Sa}^2(\omega T/2)$$

which is sketched in Figure 7.2-2b.

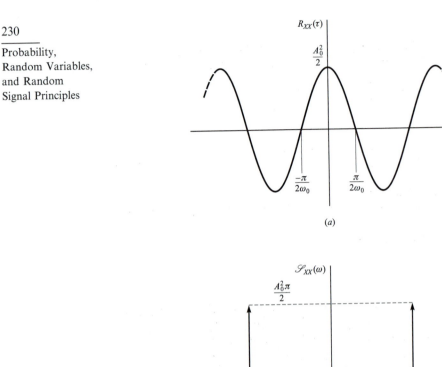

FIGURE 7.2-1
The autocorrelation function (a) and power density spectrum (b) of the wide-sense stationary random process of Example 7.2-1.

7.3
CROSS-POWER DENSITY SPECTRUM AND ITS PROPERTIES

Consider a real random process $W(t)$ given by the sum of two other real processes $X(t)$ and $Y(t)$:

$$W(t) = X(t) + Y(t) \tag{7.3-1}$$

The autocorrelation function of $W(t)$ is

$$\begin{aligned}
R_{WW}(t, t + \tau) &= E[W(t)W(t + \tau)] \\
&= E\{[X(t) + Y(t)][X(t + \tau) + Y(t + \tau)]\} \\
&= R_{XX}(t, t + \tau) + R_{YY}(t, t + \tau) \\
&\quad + R_{XY}(t, t + \tau) + R_{YX}(t, t + \tau)
\end{aligned} \tag{7.3-2}$$

Now if we take the time average of both sides of (7.3-2) and Fourier transform the resulting expression by applying (7.2-9), we have

$$\mathscr{S}_{WW}(\omega) = \mathscr{S}_{XX}(\omega) + \mathscr{S}_{YY}(\omega) + \mathscr{F}\{A[R_{XY}(t, t + \tau)]\} + \mathscr{F}\{A[R_{YX}(t, t + \tau)]\} \tag{7.3-3}$$

231

CHAPTER 7:
Random
Processes—
Spectral
Characteristics

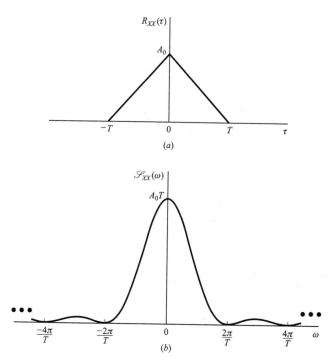

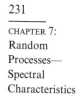

FIGURE 7.2-2
The autocorrelation function (*a*), and power spectrum (*b*), for the wide-sense
stationary process of Example 7.2-2.

where $\mathscr{F}\{\cdot\}$ represents the Fourier transform. It is clear that the left side of
(7.3-3) is just the power spectrum of $W(t)$. Similarly, the first two right-side
terms are the power spectrums of $X(t)$ and $Y(t)$, respectively. The second two
right-side terms are new quantities that are the subjects of this section. It will
be shown that they are *cross-power density spectrums* defined by (7.3-12) and
(7.3-14) below.

The Cross-Power Density Spectrum

For two real random processes $X(t)$ and $Y(t)$, we define $x_T(t)$ and $y_T(t)$ as
truncated ensemble members; that is

$$x_T(t) = \begin{cases} x(t) & -T < t < T \\ 0 & \text{elsewhere} \end{cases} \qquad (7.3\text{-}4)$$

and

$$y_T(t) = \begin{cases} y(t) & -T < t < T \\ 0 & \text{elsewhere} \end{cases} \qquad (7.3\text{-}5)$$

Both $x_T(t)$ and $y_T(t)$ are assumed to have bounded variation and to be magnitude integrable over the interval $(-T, T)$ as indicated by (7.1-4). As a con-

sequence, they will possess Fourier transforms that we denote by $X_T(\omega)$ and $Y_T(\omega)$, respectively:

$$x_T(t) \leftrightarrow X_T(\omega) \qquad (7.3\text{-}6)$$

$$y_T(t) \leftrightarrow Y_T(\omega) \qquad (7.3\text{-}7)$$

We next define the *cross power* $P_{XY}(T)$ in the two processes within the interval $(-T, T)$ by

$$P_{XY}(T) = \frac{1}{2T} \int_{-T}^{T} x_T(t) y_T(t)\, dt = \frac{1}{2T} \int_{-T}^{T} x(t) y(t)\, dt \qquad (7.3\text{-}8)$$

Since $x_T(t)$ and $y_T(t)$ are Fourier transformable, Parseval's theorem (D-20) applies; its left side is the same as (7.3-8). Thus, we may write

$$P_{XY}(T) = \frac{1}{2T} \int_{-T}^{T} x(t) y(t)\, dt = \frac{1}{2\pi} \int_{-\infty}^{\infty} \frac{X_T^*(\omega) Y_T(\omega)}{2T}\, d\omega \qquad (7.3\text{-}9)$$

This cross power is a random quantity since its value will vary depending on which ensemble member is considered. We form the average cross power, denoted $\bar{P}_{XY}(T)$, by taking the expected value in (7.3-9). The result is

$$\bar{P}_{XY}(T) = \frac{1}{2T} \int_{-T}^{T} R_{XY}(t, t)\, dt = \frac{1}{2\pi} \int_{-\infty}^{\infty} \frac{E[X_T^*(\omega) Y_T(\omega)]}{2T}\, d\omega \qquad (7.3\text{-}10)$$

Finally, we form the total average cross power P_{XY} by letting $T \to \infty$:

$$P_{XY} = \lim_{T \to \infty} \frac{1}{2T} \int_{-T}^{T} R_{XY}(t, t)\, dt = \frac{1}{2\pi} \int_{-\infty}^{\infty} \lim_{T \to \infty} \frac{E[X_T^*(\omega) Y_T(\omega)]}{2T}\, d\omega \qquad (7.3\text{-}11)$$

It is clear that the integrand involving ω can be defined as a *cross-power density spectrum*; it is a function of ω which we denote

$$\mathscr{S}_{XX}(\omega) = \lim_{T \to \infty} \frac{E[X_T^*(\omega) Y_T(\omega)]}{2T} \qquad (7.3\text{-}12)$$

Thus, we obtain what we call the *cross-power formula*

$$P_{XY} = \frac{1}{2\pi} \int_{-\infty}^{\infty} \mathscr{S}_{XY}(\omega)\, d\omega \qquad (7.3\text{-}13)$$

By repeating the above procedure, we can also define another cross-power density spectrum by

$$\mathscr{S}_{YX}(\omega) = \lim_{T \to \infty} \frac{E[Y_T^*(\omega) X_T(\omega)]}{2T} \qquad (7.3\text{-}14)$$

Cross power is given by

$$P_{YX} = \frac{1}{2\pi} \int_{-\infty}^{\infty} \mathscr{S}_{YX}(\omega)\, d\omega = P_{XY}^* \qquad (7.3\text{-}15)$$

Total cross power $P_{XY} + P_{YX}$ can be interpreted as the additional power two processes are capable of generating, over and above their individual powers, due to the fact that they are not orthogonal.

Properties of the Cross-Power Density Spectrum 233

CHAPTER 7:
Random
Processes—
Spectral
Characteristics

Some properties of the cross-power spectrum of real random processes $X(t)$ and $Y(t)$ are listed below without formal proofs.

(1) $\mathscr{S}_{XY}(\omega) = \mathscr{S}_{YX}(-\omega) = \mathscr{S}^*_{YX}(\omega)$ (7.3-16)

(2) Re $[\mathscr{S}_{XY}(\omega)]$ and Re $[\mathscr{S}_{YX}(\omega)]$ are even functions of ω (see Problem 7.3-4)

       (7.3-17)

(3) Im $[\mathscr{S}_{XY}(\omega)]$ and Im $[\mathscr{S}_{YX}(\omega)]$ are odd functions of ω (see Problem 7.3-4).

       (7.3-18)

(4) $\mathscr{S}_{XY}(\omega) = 0$ and $\mathscr{S}_{YX}(\omega) = 0$ if $X(t)$ and $Y(t)$ are orthogonal.   (7.3-19)

(5) If $X(t)$ and $Y(t)$ are uncorrelated and have constant means $\bar{X}$ and $\bar{Y}$

$$\mathscr{S}_{XY}(\omega) = \mathscr{S}_{YX}(\omega) = 2\pi \bar{X} \bar{Y} \delta(\omega) \qquad (7.3\text{-}20)$$

(6)
$$A[R_{XY}(t, t+\tau)] \leftrightarrow \mathscr{S}_{XY}(\omega) \qquad (7.3\text{-}21)$$
$$A[R_{YX}(t, t+\tau)] \leftrightarrow \mathscr{S}_{YX}(\omega) \qquad (7.3\text{-}22)$$

In the above properties, Re $[\cdot]$ and Im $[\cdot]$ represent the real and imaginary parts, respectively, and $A[\cdot]$ represents the time average, as usual, defined by (6.2-21).

Property 1 follows from (7.3-12) and (7.3-14). Properties 2 and 3 are proved by considering the symmetry that $X_T(\omega)$ and $Y_T(\omega)$ must possess for real processes. Properties 4 and 5 may be proved by substituting the integral (Fourier transform) forms for $X_T(\omega)$ and $Y_T(\omega)$ into $E[X^*_T(\omega)Y_T(\omega)]$ and showing that the function has the necessary behavior under the stated assumptions.

Property 6 states that the cross-power density spectrum and the time average of the cross-correlation function are a Fourier transform pair; its development is given in Section 7.4. For the case of jointly wide-sense stationary processes, (7.3-21) and (7.3-22) reduce to the especially useful forms

$$\mathscr{S}_{XY}(\omega) = \int_{-\infty}^{\infty} R_{XY}(\tau)e^{-j\omega\tau} \, d\tau \qquad (7.3\text{-}23)$$

$$\mathscr{S}_{YX}(\omega) = \int_{-\infty}^{\infty} R_{YX}(\tau)e^{-j\omega\tau} \, d\tau \qquad (7.3\text{-}24)$$

$$R_{XY}(\tau) = \frac{1}{2\pi} \int_{-\infty}^{\infty} \mathscr{S}_{XY}(\omega)e^{j\omega\tau} \, d\omega \qquad (7.3\text{-}25)$$

$$R_{YX}(\tau) = \frac{1}{2\pi} \int_{-\infty}^{\infty} \mathscr{S}_{YX}(\omega)e^{j\omega\tau} \, d\omega \qquad (7.3\text{-}26)$$

EXAMPLE 7.3-1. Suppose we are given a cross-power spectrum defined by

$$\mathscr{S}_{XY}(\omega) = \begin{cases} a + jb\omega/W & -W < \omega < W \\ 0 & \text{elsewhere} \end{cases}$$

where $W > 0$, a and b are real constants. We use (7.3-25) to find the cross-correlation function. It is

$$R_{XY}(\tau) = \frac{1}{2\pi} \int_{-W}^{W} \left(a + j\frac{b\omega}{W} \right) e^{j\omega\tau} d\omega$$

$$= \frac{a}{2\pi} \int_{-W}^{W} e^{j\omega\tau} d\omega + j\frac{b}{2\pi W} \int_{-W}^{W} \omega e^{j\omega\tau} d\omega$$

On using (C-45) and (C-46) this expression will readily reduce to

$$R_{XY}(\tau) = \frac{a}{2\pi} \left[\frac{e^{j\omega\tau}}{j\tau} \Big|_{-W}^{W} \right] + j\frac{b}{2\pi W} \left\{ e^{j\omega\tau} \left[\frac{\omega}{j\tau} - \frac{1}{(j\tau)^2} \right] \Big|_{-W}^{W} \right\}$$

$$= \frac{1}{\pi W \tau^2} [(aW\tau - b)\sin(W\tau) + bW\tau\cos(W\tau)]$$

EXAMPLE 7.3-2. We determine the cross-correlation function corresponding to the cross-power density spectrum

$$\mathcal{S}_{XY}(\omega) = \frac{8}{(\alpha + j\omega)^3}$$

where $\alpha > 0$ is a constant. We write $\mathcal{S}_{XY}(\omega) = 4G(\omega)$ where $G(\omega) = 2/(\alpha + j\omega)^3$. From pair 17 of Appendix E, we have

$$g(\tau) = u(\tau)\tau^2 e^{-\alpha\tau} \leftrightarrow G(\omega)$$

From the linearity property of Fourier transforms and pair 17, we get

$$R_{XY}(\tau) = 4u(\tau)\tau^2 e^{-\alpha\tau}$$

$|\mathcal{S}_{XY}(\omega)|$ and $R_{XY}(\tau)$ are sketched in Figure 7.3-1.

*7.4
RELATIONSHIP BETWEEN CROSS-POWER SPECTRUM AND CROSS-CORRELATION FUNCTION

In the following discussion we show that

$$\mathcal{S}_{XY}(\omega) = \int_{-\infty}^{\infty} \left\{ \lim_{T \to \infty} \frac{1}{2T} \int_{-T}^{T} R_{XY}(t, t + \tau) dt \right\} e^{-j\omega\tau} d\tau \qquad (7.4-1)$$

as indicated in (7.3-21).

The development consists of using the transforms of the truncated processes given by

$$X_T(\omega) = \int_{-T}^{T} X(t)e^{-j\omega t} dt \qquad (7.4-2)$$

$$Y_T(\omega) = \int_{-T}^{T} Y(t_1)e^{-j\omega t_1} dt_1 \qquad (7.4-3)$$

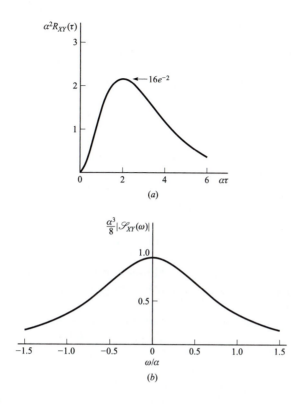

235

CHAPTER 7:
Random
Processes—
Spectral
Characteristics

FIGURE 7.3-1
Cross-correlation function (a) and magnitude of cross-power spectrum (b) for the
process of Example 7.3-2.

in (7.3-12). First, we use (7.4-2) and (7.4-3) to form

$$X_T^*(\omega)Y_T(\omega) = \int_{-T}^{T} X(t)e^{j\omega t}\,dt \int_{-T}^{T} Y(t_1)e^{-j\omega t_1}\,dt_1 \qquad (7.4\text{-}4)$$

This result is used in (7.3-12) to form the cross-power density spectrum.

$$\begin{aligned}
\mathscr{S}_{XY}(\omega) &= \lim_{T\to\infty} \frac{E[X_T^*(\omega)Y_T(\omega)]}{2T} \\
&= \lim_{T\to\infty} \frac{1}{2T} E\left[\int_{-T}^{T} X(t)e^{j\omega t}\,dt \int_{-T}^{T} Y(t_1)e^{-j\omega t_1}\,dt_1\right] \\
&= \lim_{T\to\infty} \frac{1}{2T} \int_{-T}^{T}\int_{-T}^{T} R_{XY}(t,t_1)e^{-j\omega(t_1-t)}\,dt\,dt_1 \qquad (7.4\text{-}5)
\end{aligned}$$

Next, we use a procedure similar to that developed in Section 7.2. First, we
inverse Fourier transform both sides of (7.4-5) and identify one integral as an
impulse function

$$\frac{1}{2\pi} \int_{-\infty}^{\infty} \mathscr{S}_{XY}(\omega) e^{j\omega\tau} d\omega$$

$$= \frac{1}{2\pi} \int_{-\infty}^{\infty} \lim_{T\to\infty} \frac{1}{2T} \int_{-T}^{T} \int_{-T}^{T} R_{XY}(t, t_1) e^{-j\omega(t_1-t)} dt\, dt_1 e^{j\omega\tau} d\omega$$

$$= \lim_{T\to\infty} \frac{1}{2T} \int_{-T}^{T} \int_{-T}^{T} R_{XY}(t, t_1) \frac{1}{2\pi} \int_{-\infty}^{\infty} e^{j\omega(\tau-t_1+t)} d\omega\, dt_1\, dt$$

$$= \lim_{T\to\infty} \frac{1}{2T} \int_{-T}^{T} \int_{-T}^{T} R_{XY}(t, t_1)\delta(t_1 - \tau - t) dt_1\, dt \qquad (7.4\text{-}6)$$

Further reduction is possible on use of (A-16). The definition of the impulse allows the immediate solution for the integral over t_1.

$$\frac{1}{2\pi} \int_{-\infty}^{\infty} \mathscr{S}_{XY}(\omega) e^{j\omega\tau} d\omega = \lim_{T\to\infty} \frac{1}{2T} \int_{-T}^{T} R_{XY}(t, t+\tau) dt \qquad (7.4\text{-}7)$$

which is valid for $-T < t + \tau < T$. This condition arises from the requirement that the impulse in (7.4-6) be within the range of integration. The condition can be ignored as $T \to \infty$ in the limit. Equation (7.4-7) indicates that the cross-power spectrum and the time-average cross-correlation function form a Fourier transform pair. The result proves (7.4-1) because it is the direct transform part of the pair, while (7.2-7) is the inverse part.

It should be noted from (7.4-7) that given the cross-power spectrum, the cross-correlation function cannot in general be recovered; only its time average can. For jointly wide-sense stationary processes, however, the cross-correlation function $R_{XY}(\tau)$ *can* be found from $\mathscr{S}_{XY}(\omega)$ since its time average is just $R_{XY}(\tau)$.

Although we shall not give the proof, a development similar to the above shows that (7.3-22) is true.

EXAMPLE 7.4-1. Let the cross-correlation function of two processes $X(t)$ and $Y(t)$ be

$$R_{XY}(t, t+\tau) = \frac{AB}{2} \{\sin(\omega_0 \tau) + \cos[\omega_0(2t + \tau)]\}$$

where A, B, and ω_0 are constants. We find the cross-power spectrum by use of (7.4-1). First, the time average is formed

$$\lim_{T\to\infty} \frac{1}{2T} \int_{-T}^{T} R_{XY}(t, t+\tau) dt$$

$$= \frac{AB}{2} \sin(\omega_0 \tau) + \frac{AB}{2} \lim_{T\to\infty} \frac{1}{2T} \int_{-T}^{T} \cos[\omega_0(2t + \tau)] dt$$

The integral is readily evaluated and is found to be zero. Finally we Fourier transform the time-averaged cross-correlation function with the aid of pair 12 of Appendix E:

$$\mathscr{S}_{XY}(\omega) = \mathscr{F}\left\{ \frac{AB}{2} \sin(\omega_0 \tau) \right\}$$

$$= \frac{-j\pi AB}{2} [\delta(\omega - \omega_0) - \delta(\omega + \omega_0)]$$

237

CHAPTER 7:
Random
Processes—
Spectral
Characteristics

7.5
POWER SPECTRUMS FOR DISCRETE-TIME PROCESSES AND SEQUENCES

As with continuous time processes, power spectrums for discrete-time (DT) processes and sequences, as defined in Section 6.1, result from Fourier transformation of appropriate correlation functions. To fully appreciate these topics, we need material to be developed in Chapter 8. The reason is that the developments depend, in part, on sampling theory and the response of a network to a random signal (the material of Chapter 8). However, for completeness here, it is reasonable to just summarize the important results, without proofs. We shall consider only baseband processes and sequences that are at least jointly wide-sense stationary.

Discrete-Time Processes

Let $X(t)$ be a band-limited† random process for which samples are taken at times $nT_s, n = 0, \pm 1, \pm 2, \ldots$, to form a discrete-time process $X(nT_s)$. When sampling is the result of multiplying $X(t)$ by a periodic "sampling" pluse train consisting of rectangular pulses of short duration T_p and amplitude $1/T_p$ occurring each T_s seconds, the sampling theory of Chapter 8 shows that the process' autocorrelation function $R_{XX}(\tau)$ has a sampled representation, denoted by $R_{X_s X_s}(\tau)$, that is given by

$$R_{X_s X_s}(\tau) = R_{XX}(\tau) \frac{1}{T_p} \sum_{n=-\infty}^{\infty} \text{rect}\left(\frac{\tau - nT_s}{T_p} \right)$$

$$\approx R_{XX}(\tau) \sum_{n=-\infty}^{\infty} \delta(\tau - nT_s)$$

$$= \sum_{n=-\infty}^{\infty} R_{XX}(nT_s)\delta(\tau - nT_s) \tag{7.5-1}$$

The second sum is an approximation to the first sum that results from use of narrow sample pulses (T_p small). The third sum derives from the second by use

---

†A band-limited process, for our present purposes, is one with a power density spectrum that is zero at all frequencies except over a finite band where $|\omega| < \omega_s/2$, with $\omega_s = 2\pi/T_s$ and T_s is the constant time between any adjacent pairs of samples. T_s is called the *sampling interval* or *sampling period*.

of (A-29). Sampling with very narrow pulses (impulses in the limit) is called *ideal sampling.*

Direct Fourier transformation of (7.5-1) defines the power spectrum of the discrete-time (DT) random process, which we denote by $\mathscr{S}_{X_sX_s}(\omega)$:

$$\mathscr{S}_{X_sX_s}(\omega) = \sum_{n=-\infty}^{\infty} R_{XX}(nT_s)e^{-jn\omega T_s} \qquad (7.5\text{-}2)$$

It is to be noted that (7.5-2) is periodic in the variable ω with period $\omega_s = 2\pi/T_s$.

It is of interest to show how the power spectrum of the DT random process is related to the power spectrum of the process $X(t)$. First, we substitute the relationship (Problem 7.5-2)

$$\sum_{n=-\infty}^{\infty} \delta(\tau - nT_s) = \frac{1}{T_s} \sum_{n=-\infty}^{\infty} e^{jn\omega_s\tau} \qquad (7.5\text{-}3)$$

into the middle form of (7.5-1) and then Fourier transform the result to get (Problem 7.5-3)

$$\mathscr{S}_{X_sX_s}(\omega) = \frac{1}{T_s} \sum_{n=-\infty}^{\infty} \mathscr{S}_{XX}(\omega - n\omega_s) \qquad (7.5\text{-}4)$$

Now $\mathscr{S}_{XX}(\omega)$ is the central term of (7.5-4) where $n = 0$. If we write a rectangular function $\text{rect}(\omega/\omega_s)$ on both sides to select out $\mathscr{S}_{XX}(\omega)$, we have

$$\mathscr{S}_{XX}(\omega) = T_s \, \text{rect}(\omega/\omega_s)\mathscr{S}_{X_sX_s}(\omega) \qquad (7.5\text{-}5)$$

Our result, (7.5-5), shows that the process' power spectrum is T_s times the central period portion of the periodic power spectrum of the DT process. Thus, (7.5-2) in its central period yields the process' power spectrum within a constant factor.

Alternatively, (7.5-5) can be obtained another way. In Chapter 8 it is shown that the autocorrelation function has the following valid representation:

$$R_{XX}(\tau) = \sum_{n=-\infty}^{\infty} R_{XX}(nT_s) \, \text{Sa}[\omega_s(\tau - nT_s)/2] \qquad (7.5\text{-}6)$$

based on sampling theory. The direct Fourier transformation of (7.5-6) produces

$$\mathscr{S}_{XX}(\omega) = T_s \, \text{rect}(\omega/\omega_s) \sum_{n=-\infty}^{\infty} R_{XX}(nT_s) \, e^{-jn\omega T_s}$$

$$= T_s \, \text{rect}(\omega/\omega_s)\mathscr{S}_{X_sX_s}(\omega) \qquad (7.5\text{-}7)$$

Which is the same as (7.5-5).

The above procedures and results apply to any other wide-sense stationary, band-limited, baseband process, say, $Y(t)$. One only needs to revise subscripts. Similarly, for $X(t)$ and $Y(t)$ jointly wide-sense stationary, appropriate cross-correlation functions and cross-power spectrums result from subscripts XY and YX.

239

CHAPTER 7:
Random
Processes—
Spectral
Characteristics

It is emphasized that the above results require processes to be band-limited and the sample rate ω_s to be large enough.† For processes that are not perfectly band-limited, there is usually a practical value of ω above which the power spectrum can be considered negligible. If the sample rate exceeds at least twice this practical value, the power density expressions can still be used, but give the desired density with a small (negligible) error. The error is due to an effect called *aliasing*, and it can be reduced by faster sampling.

Discrete-Time Sequences

As noted in Section 6.1, the DT sequence is essentially a DT process, but some notation is changed to be consistent with that in the DSP (digital signal processing) literature. Since computers treat samples as simply a sequence of numbers, they care not about the time separation (T_s) used between samples. In recognition of this fact, the explicit dependence on T_s is dropped and $X(nT_s)$ is written as $X[n]$, a function of index n denoted by brackets rather than parentheses. Two other changes must also be recognized.

First, a new "variable" Ω, called the *discrete frequency*, is defined, even though it is actually an angle, as

$$\Omega = \omega T_s \qquad (7.5\text{-}8)$$

Second, it results that z-transforms are extensively used in DSP. For our purposes it can be recognized that the right side of (7.5-2) is a z-transform of the sequence of samples of the autocorrelation function. For example, the *bilateral* (or *two-sided*) *z-transform*, denoted by $\mathscr{S}_{X_s X_s}(z)$ (boldface type used for the z-transform), of the sequence $R_{XX}[n]$ is

$$\mathscr{S}_{X_s X_s}(z) = \sum_{n=-\infty}^{\infty} R_{XX}[n]\, z^{-n} \qquad (7.5\text{-}9)$$

By comparing (7.5-2) with (7.5-9) we have

$$\mathscr{S}_{X_s X_s}(\omega) = \sum_{n=-\infty}^{\infty} R_{XX}[n] e^{-jn\omega T_s} = \sum_{n=-\infty}^{\infty} R_{XX}[n]\, e^{-jn\Omega}$$

$$= \mathscr{S}_{X_s X_s}(z)\Big|_{z=e^{j\Omega}} = \mathscr{S}_{X_s X_s}(e^{j\Omega}) \qquad (7.5\text{-}10)$$

In general, z can be any complex number. However, (7.5-10) shows that the power density spectrum is the z-transform evaluated on the unit circle [where $z = \exp(j\Omega)$, $|z| = 1$, and the angle of z is Ω]. The second right-side sum of (7.5-10) is called the *discrete-time Fourier transform* (DTFT) of the sequence $R_{XX}[n]$.

---

†If the frequency of the highest nonzero spectral term in $\mathscr{S}_{XX}(\omega)$ is W_X, called the *spectral extent*, the sampling rate must be $\omega_s > 2W_X$, where $2W_X$ is known as the *Nyquist rate* (see Section 8.7).

EXAMPLE 7.5-1. Suppose the autocorrelation sequence of a DT sequence $X[n]$ is

$$R_{XX}[n] = a^{|n|} \qquad |a| < 1$$

We use (7.5-10) to find the DTFT of $X[n]$.

$$\mathscr{S}_{X_s X_s}(\omega) = \sum_{n=-\infty}^{\infty} R_{XX}[n] e^{-jn\Omega} = \sum_{n=-\infty}^{\infty} a^{|n|} e^{-jn\Omega}$$

$$= \sum_{n=-\infty}^{-1} a^{-n} e^{-jn\Omega} + \sum_{n=0}^{\infty} a^n e^{-jn\Omega} \qquad (1)$$

We use the sum of (C-62) by identifying $w = [a \exp(j\Omega)]^{-1}$ and $w = a \exp(-j\Omega)$, respectively, for the two sums in the last form of (1). We have

$$\mathscr{S}_{X_s X_s}(\omega) = -\frac{a}{a - e^{-j\Omega}} + \frac{1}{1 - ae^{-j\Omega}}$$

$$= \frac{(1 - a^2)}{(1 + a^2) - 2a\cos(\Omega)} \qquad \Omega = \omega T_s$$

Note that this power spectrum is periodic in the variable Ω, as are all power spectrums of all DT sequences.

Developments similar to those leading to (7.5-10) produce expressions for the power spectrum of $Y[n]$ for a process $Y(t)$ and for the applicable cross-power spectrums for $X[n]$ and $Y[n]$. These are left as reader exercises.

In (7.5-10) the DTFT was defined as the second right-side form where we say $\mathscr{S}_{X_s X_s}(e^{j\Omega})$ is the DTFT of $R_{XX}[n]$. There is also an *inverse DTFT* (or IDTFT) where $R_{XX}[n]$ can be recovered from $\mathscr{S}_{X_s X_s}(e^{j\Omega})$. The two form a DTFT pair:

$$\mathscr{S}_{X_s X_s}(e^{j\Omega}) = \sum_{n=-\infty}^{\infty} R_{XX}[n] e^{-jn\Omega} \qquad \text{(DTFT)} \qquad (7.5\text{-}11)$$

$$R_{XX}[n] = \frac{1}{2\pi} \int_{-\pi}^{\pi} \mathscr{S}_{X_s X_s}(e^{j\Omega}) e^{jn\Omega} \, d\Omega \qquad \text{(IDTFT)} \qquad (7.5\text{-}12)$$

EXAMPLE 7.5-2. We show that (7.5-12) is valid by proving that the right side results in $R_{XX}[n]$.

$$\frac{1}{2\pi} \int_{-\pi}^{\pi} \mathscr{S}_{X_s X_s}(e^{j\Omega}) e^{jn\Omega} \, d\Omega = \frac{1}{2\pi} \int_{-\pi}^{\pi} \sum_{m=-\infty}^{\infty} R_{XX}[m] e^{-jm\Omega} e^{jn\Omega} \, d\Omega$$

$$= \sum_{m=-\infty}^{\infty} R_{XX}[m] \frac{1}{2\pi} \int_{-\pi}^{\pi} e^{j(n-m)\Omega} \, d\Omega$$

$$= \sum_{m=-\infty}^{\infty} R_{XX}[m] \frac{\sin[(n-m)\pi]}{(n-m)\pi}$$

Since $\sin[(n-m)\pi]/[(n-m)\pi] = 0$ for all $m \neq n$ and equals unity for $m = n$, the sum reduces to $R_{XX}[n]$. This result equals the left side of (7.5-12), proving the IDTFT is valid.

EXAMPLE 7.5-3. Let $X(t)$ be a (white†) gaussian noise random process having a sequence of samples $X[n]$. Assume the sequence of autocorrelation values is such that $R_{XX}[k] = \sigma_X^2 = 4$ when $k = 0$ and is zero for all other values of k. If a new sequence $Y[n] = X[n] + 0.5X[n-1] + 0.25X[n-2]$ is formed, then it can be shown (Problem 7.5-13) that

$$R_{YY}[k] = \begin{cases} (21/16)\sigma_x^2 & k = 0 \\ (5/8)\sigma_X^2 & k = +1 \text{ and } -1 \\ (1/4)\sigma_X^2 & k = +2 \text{ and } -2 \\ 0 & \text{all other } k \end{cases} \tag{1}$$

It can also be shown (Problem 7.5-14) that the power spectrum of $Y[n]$ is

$$\mathscr{S}_{Y_sY_s}(\omega) = \mathscr{S}_{Y_sY_s}(e^{j\Omega}) = \sum_{k=-2}^{2} R_{YY}[k]e^{-jk\Omega}$$

$$= \frac{\sigma_X^2}{16}[21 + 20\cos(\Omega) + 8\cos(2\Omega)] \tag{2}$$

from (7.5-11).

Our example consists of using MATLAB for simulating the sequence $Y[n]$, computing estimates of its autocorrelation function and power spectrum, and comparing the results to the true quantities. The MATLAB code used is given in Figure 7.5-1. Specifically, we first use the procedures of Example 5.6-1 to generate 1002 values of zero-mean, gaussian random variable X with a variance of 4. These values are used to generate 1000 values of $Y[n]$ as defined above. Next, we estimate the autocorrelation sequence $R_{YY}[k]$ using the formula (Childers, 1997, p. 297)

$$\hat{R}_{YY}[k] = \frac{1}{N}\sum_{n=0}^{N-1-|k|} Y[n]Y[n+|k|] \qquad |k| < N$$

where $N = 1000$. Results are shown as the stem plot of Figure 7.5-2, where the exact function of (1) is shown as the piecewise continuous curve for comparison. We see that some significant errors remain, even with $N = 1000$ data points. Finally, we compute an estimate of the power spectrum using the formula (Childers, 1997, p. 303)

$$\hat{\mathscr{S}}_{YY}(\omega) = \sum_{k=-(N-1)}^{N-1} \hat{R}_{YY}[k]e^{-jk\Omega} \qquad \Omega = \omega T_s$$

Direct results shown in Figure 7.5-3 for $N = 1000$ prove to be very noisy, so the code that was used has averaged adjacent intervals of 100 frequencies to smooth the plot to that shown in Figure 7.5-4 [the lower curve is the exact result of (2)]. Clearly, even for 1000 data values our estimate shows some significant error.

---

†White noise is properly defined in Section 7.6.

```
%%%%%%%%%%%%%%% Example 7.5-3 %%%%%%%%%%%%%%%%%%

clear

N = 1000;
xvar = 4; % variance of x
lag = 10;

randn('state',140);
x = sqrt(xvar)*randn(1,N+2); % i.i.d. Gaussian random variable

xn = x(3:N+2); % x(n)
xn1 = x(2:N+1); % x(n-1)
xn2 = xn1(1:N); % x(n-2)

y = xn + 0.5*xn1 + 0.25*xn2;

Ryy = zeros(1,2*N-1); % initialize
for k = -N+1:N-1
   ndx1 = max([1 1+k]) :min([N+k N]);
   ndx2 = max([1 1-k]) :min([N-k N]);
   Ryy(N+k) = sum(y(ndx1).*y(ndx2))./N; % autocorrelation
end

Rtrue = xvar*[0 0 1/4 5/8 21/16 5/8 1/4 0 0]; % true value of Ryy

M = 2*N - 1;
w = -pi+pi/M:2*pi/M:pi; % frequency vector

Strue = xvar/16*(21+20*cos(w) + 8*cos(2*w)); % true value of Syy
Syy_noisy = abs(fftshift(fft(Ryy))); % power spectral estimate

stp = 100; % number of points to average (stp must be evenly divisible
% by M+1)
Syy = zeros(1, (M+1)/stp);
for i = 1:stp:M % smooth the power spectrum estimate
    if i < (M+1)/2
      Syy((i-1)/stp+1) = mean(Syy_noisy(i:i+stp-1));
    else
      Syy((i-1)/stp+1) = mean(Syy_noisy(i-1:i+stp-2));
    end
end

clf
plot([-lag -3:3 lag],Rtrue,'k')
hold
stem(-lag:lag,Ryy(N-lag:N+lag),'k')

xlabel('Lag')
ylabel('Magnitude')
title('Autocorrelation')

figure
plot(w,Syy_noisy(1:M),'k')
xlabel('Normalized Frequency (rad)')
ylabel('Magnitude')
title('Power Spectrum')

figure
plot(w(round(stp/2):stp:length(w)),Syy,'--k',w,Strue,'k')

xlabel('Normalized Frequency (rad)')
ylabel('Magnitude')
title('Power Spectrum')
```

FIGURE 7.5-1
The MATLAB code used in Example 7.5-3.

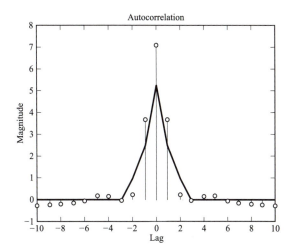

243

CHAPTER 7:
Random
Processes—
Spectral
Characteristics

FIGURE 7.5-2
True autocorrelation function and estimated autocorrelation sequence applicable to
Example 7.5-3.

Discrete Fourier Transform

Equations (7.5-11) and (7.5-12) apply to a band-limited signal, be it deterministic or random. The principal points to be made are: (1) there is theoretically
an infinite number of values of autocorrelation function needed to complete
the sampled signal's power spectrum, and (2) this computed power spectrum is
a *continuous* function of Ω that consists of nonoverlapping replicas of the
power spectrum of the continuous signal that was sampled. When a digital

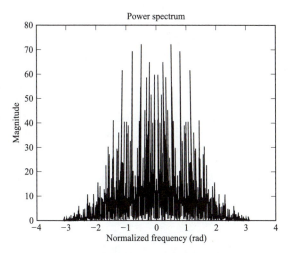

FIGURE 7.5-3
Unsmoothed power spectrum estimate applicable to Example 7.5-3.

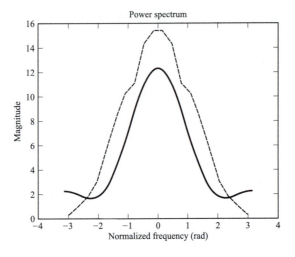

FIGURE 7.5-4
Smoothed estimated and exact power spectrums applicable to Example 7.5-3.

computer is used to compute either (7.5-11) or (7.5-12) some problems immediately arise.

In regard to point (1), the computer cannot use an infinite number of
values $R_{XX}[n]$, even if they were available, since we cannot wait forever for
the final result. Furthermore, such a procedure could involve an infinite
amount of memory and/or an infinite number of calculations. Thus, from a
practical standpoint, the computer typically works with a finite number of
values of $R_{XX}[n]$, say, N. For truly band-limited signals this limitation requires
truncation of $R_{XX}[n]$ to N values, and some error will result in computing the
power spectrum. Truncation results in some spectral distortion and aliasing
(the overlap of spectral replicas due to distortion-caused spectral spread—see
Section 8.7 for more detail). Because even truncated waveforms have spectrums that decrease with increasing frequency, there is always some practical
frequency beyond which the spectrum's magnitude becomes negligible; this
fact allows truncation errors to be reduced by sampling at a rate higher than
twice the practical frequency.

In regard to point (2), even if a computer could determine the continuous
power spectrum (the DTFT), it must still produce results for discrete values of
Ω (or ω, since $\Omega = \omega T_s$). Thus, it is common practice to discretize the variable
Ω (or ω) to a finite number of values. Usually the number is chosen the same
as the number of sample values (denoted by N above) to facilitate practical
computations (by using the *fast Fourier transform* or FFT).

From the above comments it is clear that the practical use of the DTFT of
(7.5-11) and the IDTFT of (7.5-12) involves sequences of finite length N. Let
us define the extent of the samples as NT_s, and the discrete values ω_k and Ω_k of
ω and Ω, respectively, by

$$\omega_k = 2\pi k/(NT_s) \qquad k = 0, 1, \ldots, (N-1) \qquad (7.5\text{-}13)$$

$$\Omega_k = \omega T_s|_{\omega=\omega_k} = 2\pi k/N \qquad k = 0, 1, \ldots, (N-1) \qquad (7.5\text{-}14)$$

Next, we define the discrete values of the power spectrum by

245

CHAPTER 7:
Random
Processes—
Spectral
Characteristics

$$\mathscr{S}_{X_sX_s}(\omega_k) = \mathscr{S}_{X_sX_s}(e^{j\Omega_k}) = \mathscr{S}_{X_sX_s}(e^{j\omega_k T_s}) \tag{7.5-15}$$

$$\mathscr{S}_{X_sX_s}[k] = \mathscr{S}_{X_sX_s}(e^{j\Omega_k}) = \mathscr{S}_{X_sX_s}(e^{j2\pi k/N}) \tag{7.5-16}$$

These definitions allow us to write the truncated version of (7.5-11) and the discretized version of (7.5-12), respectively, as

$$\mathscr{S}_{X_sX_s}[k] = \sum_{n=0}^{N-1} R_{XX}[n] e^{-j2\pi nk/N} \qquad k = 0, 1, \ldots, (N-1) \tag{7.5-17}$$

$$R_{XX}[n] = \frac{1}{N}\sum_{k=0}^{N-1} \mathscr{S}_{X_sX_s}[k] e^{j2\pi nk/N} \qquad n = 0, 1, \ldots, (N-1) \tag{7.5-18}$$

Equation (7.5-17) is called the *discrete Fourier transform* (DFT) of the finite autocorrelation sequence, while (7.5-18) is known as the *inverse discrete Fourier transform* (IDFT) of the discretized power spectrum. Together the two results form a DFT *pair*.

An alternative version of the DFT pair results from use of the DT sequence $R_{XX}(nT_s)$:

$$\mathscr{S}_{X_sX_s}(\omega_k) = \sum_{n=0}^{N-1} R_{XX}(nT_s) e^{-jn\omega_k T_s} \qquad k = 0, 1, \ldots, (N-1) \tag{7.5-19}$$

$$R_{XX}(nT_s) = \frac{1}{N}\sum_{k=0}^{N-1} \mathscr{S}_{X_sX_s}(\omega_k) e^{jn\omega_k T_s} \qquad n = 0, 1, \ldots, (N-1) \tag{7.5-20}$$

EXAMPLE 7.5-4. To set the stage for the following example we determine the autocorrelation function $R_{XX}(\tau)$ for a continuous time process $X(t)$ that we assume has the power spectrum

$$\mathscr{S}_{XX}(\omega) = K \cos\left(\frac{\pi\omega}{2W_X}\right) \text{rect}\left(\frac{\omega}{2W_X}\right)$$

Here $K > 0$ is a real constant and W_X is the spectral extent of the process. On inverse Fourier transformation we get

$$R_{XX}(\tau) = \frac{KW_X}{2\pi}\{\text{Sa}[W_X\tau - (\pi/2)] + \text{Sa}[W_X\tau + (\pi/2)]\}$$

$$= \frac{(KW_X/2)\cos(W_X\tau)}{(\pi/2)^2 - (W_X\tau)^2} \tag{1}$$

In computing (1) we have used (C-7) and (C-45) to easily evaluate the inverse transform's integral.

EXAMPLE 7.5-5. We demonstrate the solution of (7.5-19) by use of MATLAB software. First, we assume values of the power spectrum are to be found at $N = 5$ frequencies. Next, we choose to calculate the power spectrum of Example 7.5-4 by using five true values of the autocorrelation function of (1) when the sampling rate is $\omega_s = 2\pi/T_s = 2W_X$. These values are found to be

$$R_{XX}[0] = K2/(\pi T_s) \qquad R_{XX}[1] = K2/(3\pi T_s)$$
$$R_{XX}[2] = -K2/(15\pi T_s) \qquad R_{XX}[3] = K2/(35\pi T_s)$$
$$R_{XX}[4] = -K2/(63\pi T_s)$$

The MATLAB code to solve (7.5-19) when these values are used is given in Figure 7.5-5.

In Figure 7.5-6 the solid line curve is the amplitude-normalized plot of the true power spectrum of (1) in Example 7.5-4 versus normalized frequency $[-\pi < (\Omega = \omega T_s) < \pi]$. The computed frequency values are shown as the stem plots at five points. Clearly, the program is producing the correct values of the power spectrum at the limited number of points used. This example is revisited in Problem 7.5-15, where the number of frequencies is increased to show that accuracy is preserved while calculating the function at more discrete points.

7.6
SOME NOISE DEFINITIONS AND OTHER TOPICS

In many practical problems it is helpful to sometimes characterize noise through its power density spectrum. Indeed, in the following discussions we *define* two forms of noise on the basis of their power spectrums. We also consider the response of a product device when one of its input waveforms is a random signal or noise.

White and Colored Noise

A sample function $n(t)$ of a wide-sense stationary noise random process $N(t)$ is called *white noise* if the power density spectrum of $N(t)$ is a constant at all frequencies. Thus, we define

$$\mathscr{S}_{NN}(\omega) = \mathscr{N}_0/2 \qquad (7.6-1)$$

for white noise, where $\mathscr{N}_0$ is a real positive constant. By inverse Fourier transformation of (7.6-1), the autocorrelation function of $N(t)$ is found to be

$$R_{NN}(\tau) = (\mathscr{N}_0/2)\,\delta(\tau) \qquad (7.6-2)$$

The above two functions are illustrated in Figure 7.6-1. White noise derives its name by analogy with "white" light, which contains all visible light frequencies in its spectrum.

White noise is unrealizable, as can be seen by the fact that it possesses infinite average power:

$$\frac{1}{2\pi}\int_{-\infty}^{\infty} \mathscr{S}_{NN}(\omega)\,d\omega = \infty \qquad (7.6-3)$$

%%%%%%%%%%%%%% **Example 7.5-5** %%%%%%%%%%%%%%%%%%%

247

CHAPTER 7:
Random
Processes—
Spectral
Characteristics

```
clear

N = 5;
k = 1;
Ts = 1;

wx = pi/Ts;

if rem(N,2) == 0
    w = -wx : 2*wx/N : wx - 2*wx/N; % frequency vector
else
    w = -wx + wx/N : 2*wx/N : wx;
end

tau = -0.5*N:(0.5*N-1);

sinc1 = zeros(1,length(tau));
sinc2 = zeros(1,length(tau));

f = find(pi*tau - pi/2 ~= 0); % avoid sin(0)/0
sinc1(f) = sin(pi*tau(f) - pi/2)./(pi*tau(f) - pi/2);
f = find(pi*tau - pi/2 == 0); % sin(0)/0 = 1
sinc1(f) = ones(1,length(f));

f = find(pi*tau + pi/2 ~= 0);
sinc2(f) = sin(pi*tau(f) + pi/2)./(pi*tau(f) + pi/2);
f = find(pi*tau + pi/2 == 0);
sinc2(f) = ones(1,length(f));

Rtrue = Ts*k*wx/(2*pi)*(sinc1 + sinc2); % true
autocorrelation

w2 = -wx : 2*wx/128 : wx - 2*wx/128; % frequency vector
Strue = k*cos(pi*w2/(2*wx)); % true power spectrum
Sxx = abs(fftshift(fft(Rtrue,N))); % estimated power spectrum

clf
plot(w2,Strue,'k')
hold
stem(w,Sxx,'k')

xlabel('Normalized Frequency (rad)')
ylabel('Magnitude')
title('Power Spectrum')
```

FIGURE 7.5-5
MATLAB code used in Example 7.5-5.

However, one type of real-world noise closely approximates white noise. *Thermal noise* generated by thermal agitation of electrons in any electrical conductor has a power spectrum that is constant up to very high frequencies and then decreases. For example, a resistor at temperature T in kelvin

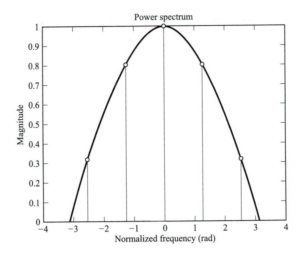

FIGURE 7.5-6
True power spectrum (solid curve) and MATLAB-calculated power spectrum (stem plot) applicable to Example 7.5-5.

produces a noise voltage across its open-circuited terminals having the power spectrum† (Carlson, 1975, p. 118)

$$\mathscr{S}_{NN}(\omega) = \frac{(\mathscr{N}_0/2)(\alpha|\omega|/T)}{e^{\alpha|\omega|/T} - 1} \tag{7.6-4}$$

where $\alpha = 7.64(10^{-12})$ kelvin-seconds is a constant. At a temperature of $T = 290 \, \text{K}$ (usually called *room temprature* although it corresponds to a rather cool room at 63°F), this function remains above 0.9 ($\mathscr{N}_0/2$) for frequencies up

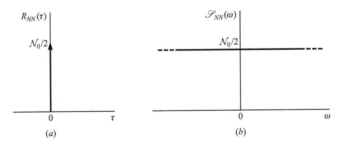

FIGURE 7.6-1
(*a*) The autocorrelation function and (*b*) the power density spectrum of white noise. [*Adapted from Peebles (1976) with permission of publishers Addison-Wesley, Advanced Book Program.*]

---

†The unit of $\mathscr{S}_{NN}(\omega)$ is actually volts squared per hertz. According to our convention, we obtain watts per hertz by presuming the voltage exists across a 1-Ω resistor.

to 10^{12} Hz or 1000 GHz. Thus, thermal noise has a nearly flat spectrum at all frequencies that are likely to ever be used in radio, microwave, or millimeter-wave systems.†

249

CHAPTER 7:
Random
Processes—
Spectral
Characteristics

Noise having a nonzero and constant power spectrum over a *finite* frequency band and zero everywhere else is called *band-limited white noise*. Figure 7.6-2a depicts such a power spectrum that is lowpass. Here

$$\mathscr{S}_{NN}(\omega) = \begin{cases} \dfrac{P\pi}{W} & -W < \omega < W \\ 0 & \text{elsewhere} \end{cases} \qquad (7.6\text{-}5)$$

Inverse transformation of (7.6-5) gives the autocorrelation function shown in Figure 7.6-2b:

$$R_{NN}(\tau) = P\frac{\sin(W\tau)}{W\tau} \qquad (7.6\text{-}6)$$

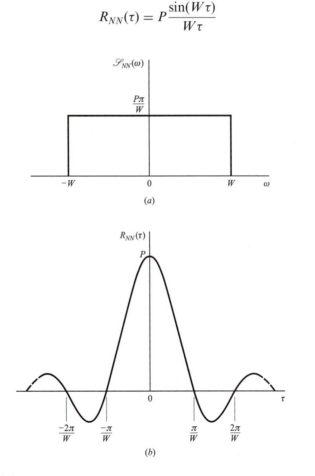

(a)

(b)

FIGURE 7.6-2
Power density spectrum (*a*) and autocorrelation function (*b*) of lowpass band-limited white noise.

---

†This statement must be reexamined for $T < 290\,\text{K}$, such as in some superconducting systems or other low-temperature devices (masers).

The constant P equals the power in the noise.

Band-limited white noise can also be bandpass as illustrated in Figure 7.6-3. The applicable power spectrum and autocorrelation function are:

$$\mathscr{S}_{NN}(\omega) = \begin{cases} P\pi/W & \omega_0 - (W/2) < |\omega| < \omega_0 + (W/2) \\ 0 & \text{elsewhere} \end{cases} \qquad (7.6\text{-}7)$$

and

$$R_{NN}(\tau) = P\frac{\sin(W\tau/2)}{(W\tau/2)}\cos(\omega_0\tau) \qquad (7.6\text{-}8)$$

where ω_0 and W are constants and P is the power in the noise.

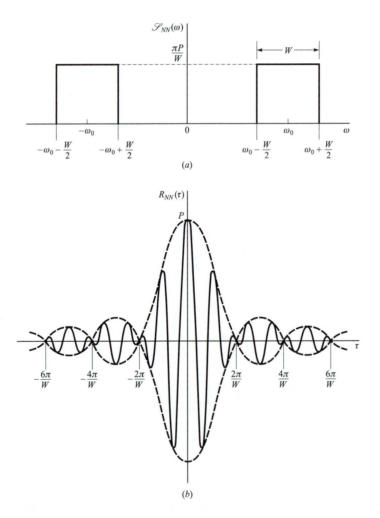

(a)

(b)

FIGURE 7.6-3

Power density spectrum (*a*) and autocorrelation function (*b*) for bandpass band-limited white noise. [*Adapted from Peebles (1976) with permission of publishers Addison-Wesley, Advanced Book Program.*]

Again, by analogy with colored light that has only a portion of the visible light frequencies in its spectrum, we define *colored noise* as any noise that is not white. An example serves to illustrate colored noise.

251

CHAPTER 7:
Random
Processes—
Spectral
Characteristics

EXAMPLE 7.6-1. A wide-sense stationary noise process $N(t)$ has an auto-correlation function

$$R_{NN}(\tau) = Pe^{-3|\tau|}$$

where P is a constant. We find its power spectrum. It is

$$\mathscr{S}_{NN}(\omega) = \int_{-\infty}^{\infty} Pe^{-3|\tau|}e^{-j\omega\tau} \, d\tau$$

$$= P\int_{0}^{\infty} e^{-(3+j\omega)\tau} \, d\tau + P\int_{-\infty}^{0} e^{(3-j\omega)\tau} \, d\tau$$

These integrals easily evaluate using (C-45) to give

$$\mathscr{S}_{NN}(\omega) = \frac{P}{3+j\omega} + \frac{P}{3-j\omega} = \frac{6P}{9+\omega^2}$$

This power spectrum is sketched in Figure 7.6-4 along with the preceding autocorrelation function.

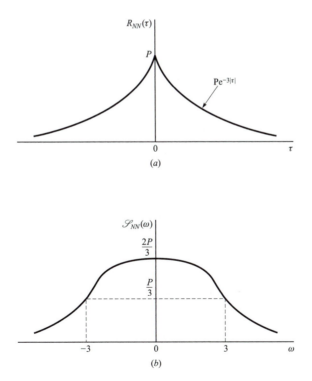

FIGURE 7.6-4
The autocorrelation function (*a*) and power spectrum (*b*) of the colored noise of Example 7.6-1. [*Adapted from Peebles (1976) with permission of publishers Addison-Wesley, Advanced Book Program.*]

Product Device Response to a Random Signal

Product devices are frequently encountered in electrical systems. Often they involve the product of a random waveform $X(t)$ (either signal or noise or the sum of signal and noise) with a cosine (or sine) "carrier" wave as illustrated in Figure 7.6-5. The response is the new process

$$Y(t) = X(t)A_0 \cos(\omega_0 t) \qquad (7.6\text{-}9)$$

where A_0 and ω_0 are constants. We seek to find the power spectrum $\mathcal{S}_{YY}(\omega)$ of $Y(t)$ in terms of the power spectrum $\mathcal{S}_{XX}(\omega)$ of $X(t)$.

The autocorrelation function of $Y(t)$ is

$$R_{YY}(t, t + \tau) = E[Y(t)Y(t + \tau)]$$

$$= E[A_0^2 X(t)X(t + \tau) \cos(\omega_0 t) \cos(\omega_0 t + \omega_0 \tau)]$$

$$= \frac{A_0^2}{2} R_{XX}(t, t + \tau)[\cos(\omega_0 \tau) + \cos(2\omega_0 t + \omega_0 \tau)] \qquad (7.6\text{-}10)$$

Even if $X(t)$ is wide-sense stationary $Y(t)$ is not since $R_{YY}(t, t + \tau)$ depends on t. Thus, we apply (7.1-19) to obtain $\mathcal{S}_{YY}(\omega)$ after we take the time average of $R_{YY}(t, t + \tau)$. Let $X(t)$ be assumed wide-sense stationary. Then (7.6-10) becomes

$$A[R_{YY}(t, t + \tau)] = \frac{A_0^2}{2} R_{XX}(\tau) \cos(\omega_0 \tau) \qquad (7.6\text{-}11)$$

On Fourier transforming (7.6-11) we have

$$\mathcal{S}_{YY}(\omega) = \frac{A_0^2}{4}[\mathcal{S}_{XX}(\omega - \omega_0) + \mathcal{S}_{XX}(\omega + \omega_0)] \qquad (7.6\text{-}12)$$

A possible power density spectrum of $X(t)$ and that given by (7.6-12) are illustrated in Figure 7.6-6. It presumes that $X(t)$ is a lowpass process, although this is not a constraint in applying (7.6-12).

> **EXAMPLE 7.6-2.** One important use of the product device is in recovery (demodulation) of the information signal (music, speech, etc.) conveyed in the wave transmitted from a conventional broadcast radio station that uses AM (*amplitude modulation*). The wave received by a receiver tuned to a station with frequency $\omega_0/2\pi$ is one input to the product device. The

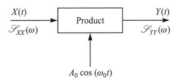

FIGURE 7.6-5
A product of interest in electrical systems. [*Adapted from Peebles (1976) with permission of publishers Addison-Wesley, Advanced Book Program.*]

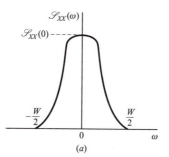

(a)

253

CHAPTER 7:
Random
Processes—
Spectral
Characteristics

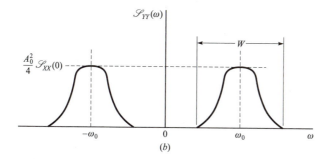

(b)

FIGURE 7.6-6
Power density spectrums applicable to Figure 7.6-5; (a) at the input and (b) at the
output. [*Adapted from Peebles (1976) with permission of publishers Addison-Wesley,
Advanced Book Program.*]

other is a "local oscillator" signal $A_0 \cos(\omega_0 t)$ generated within the receiver. The product device output passes through a lowpass filter which has
as its output the desired information signal. Unfortunately, this signal
also contains noise because noise is also present at the input to the product device; the input noise is added to the received radio wave. We shall
calculate the power in the output noise of the product demodulator.

Let the power spectrum of the input noise, denoted $X(t)$, be approximated by an idealized (rectangular) function with bandwidth W_{RF} centered at $\pm\omega_0$. Thus,

$$\mathscr{S}_{XX}(\omega) = \begin{cases} \mathscr{N}_0/2 & -\omega_0 - (W_{RF}/2) < \omega < -\omega_0 + (W_{RF}/2) \\ \mathscr{N}_0/2 & \omega_0 - (W_{RF}/2) < \omega < \omega_0 + (W_{RF}/2) \\ 0 & \text{elsewhere} \end{cases}$$

where $\mathscr{N}_0/2$ is the power density within the noise band. By applying
(7.6-12) the power density spectrum of the output noise $Y(t)$ of the product device is readily found (by sketch) to be

$$\mathscr{S}_{YY}(\omega) = \begin{cases} \mathscr{N}_0 A_0^2/8 & -2\omega_0 - (W_{RF}/2) < \omega < -2\omega_0 + (W_{RF}/2) \\ \mathscr{N}_0 A_0^2/4 & -W_{RF}/2 < \omega < W_{RF}/2 \\ \mathscr{N}_0 A_0^2/8 & 2\omega_0 - (W_{RF}/2) < \omega < 2\omega_0 + (W_{RF}/2) \\ 0 & \text{elsewhere} \end{cases}$$

Now only the noise in the band $-W_{RF}/2 < \omega < W_{RF}/2$ cannot be removed by a lowpass filter (which usually follows the product device to remove unwanted noise and other undesired outputs) because the desired signal is in the same band. This remaining component of $\mathscr{S}_{YY}(\omega)$ gives rise to the final output noise power, denoted N_o,

$$N_o = \frac{1}{2\pi} \int_{-W_{RF}/2}^{W_{RF}/2} \frac{\mathscr{N}_0 A_0^2}{4} \, d\omega = \frac{\mathscr{N}_0 A_0^2 W_{RF}}{8\pi}$$

*7.7
POWER SPECTRUMS OF COMPLEX PROCESSES

Power spectrums may readily be defined for complex processes. We consider only those processes that are at least wide-sense stationary. In terms of the autocorrelation function $R_{ZZ}(\tau)$ of a complex random process $Z(t)$, the power density spectrum is defined as its Fourier transform

$$\mathscr{S}_{ZZ}(\omega) = \int_{-\infty}^{\infty} R_{ZZ}(\tau)e^{-j\omega\tau} \, d\tau \tag{7.7-1}$$

The inverse transform applies, so

$$R_{ZZ}(\tau) = \frac{1}{2\pi} \int_{-\infty}^{\infty} \mathscr{S}_{ZZ}(\omega)e^{j\omega\tau} \, d\omega \tag{7.7-2}$$

For two jointly wide-sense stationary complex processes $Z_m(t)$ and $Z_n(t)$, their cross-power density spectrum and cross-correlation function are a Fourier transform pair:

$$\mathscr{S}_{Z_m Z_n}(\omega) = \int_{-\infty}^{\infty} R_{Z_m Z_n}(\tau)e^{-j\omega\tau} \, d\tau \tag{7.7-3}$$

$$R_{Z_m Z_n}(\tau) = \frac{1}{2\pi} \int_{-\infty}^{\infty} \mathscr{S}_{Z_m Z_n}(\omega)e^{j\omega\tau} \, d\omega \tag{7.7-4}$$

An equivalent statement is:

$$R_{Z_m Z_n}(\tau) \leftrightarrow \mathscr{S}_{Z_m Z_n}(\omega) \tag{7.6-5}$$

EXAMPLE 7.7-1. We reconsider the complex process $V(t)$ of Example 6.7-1 and find its power spectrum. From the previous example

$$R_{VV}(\tau) = e^{j\omega_0\tau} \sum_{n=1}^{N} \overline{A_n^2}$$

On Fourier transforming this autocorrelation function we obtain

255

CHAPTER 7:
Random
Processes—
Spectral
Characteristics

$$\mathcal{S}_{VV}(\omega) = \mathcal{F}\left\{ e^{j\omega_0\tau} \sum_{n=1}^{N} \overline{A_n^2} \right\}$$

$$= \sum_{n=1}^{N} \overline{A_n^2} \mathcal{F}\{e^{j\omega_0\tau}\}$$

$$= 2\pi\delta(\omega - \omega_0) \sum_{n=1}^{N} \overline{A_n^2}$$

after using pair 9 of Appendix E.

7.8
SUMMARY

The material of Chapter 6 developed random processes through the use of time domain concepts. This chapter developed processes from the standpoint of the frequency domain. Central to the development is the concept of a spectrum describing the manner in which *power* is distributed with frequency in a random process. This power density spectrum defines the process' spectral properties, and is somewhat analogous to the *voltage* density with frequency for deterministic waveforms. However, the analogy is principally due to the fact that in both cases, the time domain characteristics are related to those of the frequency domain through Fourier transforms. Important topics discussed herein were:

- The power density spectrum and its properties were developed for any (stationary or nonstationary) random process.
- It was proven that the time average of a process' autocorrelation function and the power density spectrum form a Fourier transform pair.
- Similarly, a cross-power density spectrum and its properties were presented for two random processes.
- The power density spectrums of discrete-time processes and sequences were developed and related to the power spectrums of the continuous-time process from which they were derived via sampling. The relationship of the discrete-time Fourier transform (DTFT) and the inverse DTFT (IDTFT) were related to the bilateral z-transform and methods used in digital signal processing (DSP).
- Computer examples (and chapter-end problems) were included to support (using MATLAB) the materials on discrete-time processes and sequences.
- Some practical noise definitions of white and colored noise were included. The response of a product device was analyzed; it is valuable as a model for some mixers and phase detectors used in the practical world.
- The chapter closed with some more advanced material on power spectrums of complex random processes.

PROBLEMS

7.1-1. We are given the random process

$$X(t) = A_0 \cos(\omega_0 t + \Theta)$$

where A_0 and ω_0 are constants and Θ is a random variable uniformly distributed on the interval $(0, \pi)$.
(a) Is $X(t)$ wide-sense stationary?
(b) Find the power in $X(t)$ by using (7.1-10).
(c) Find the power spectrum of $X(t)$ by using (7.1-11) and calculate power from (7.1-12). Do your two powers agree?

7.1-2. Work Problem 7.1-1 if the process is defined by

$$X(t) = u(t)A_0 \cos(\omega_0 t + \Theta)$$

where $u(t)$ is the unit-step function.

*7.1-3. Work Problem 7.1-2 assuming Θ is uniform on the interval $(0, \pi/2)$.

7.1-4. Work Problem 7.1-1 if the random process is given by $X(t) = A_0 \sin(\omega_0 t + \Theta)$.

*7.1-5. Work Problem 7.1-1 if the random process is

$$X(t) = A_0^2 \cos^2(\omega_0 t + \Theta)$$

7.1-6. Let A_0 and B_0 be random variables. We form the random process

$$X(t) = A_0 \cos(\omega_0 t) + B_0 \sin(\omega_0 t)$$

where ω_0 is a real constant.
(a) Show that if A_0 and B_0 are uncorrelated with zero means and equal variances, then $X(t)$ is wide-sense stationary.
(b) Find the autocorrelation function of $X(t)$.
(c) Find the power density spectrum.

7.1-7. A limiting form for the impulse function was given in Example 7.1-2. Give arguments to show that the following are also true:

(a) $\lim_{T \to \infty} T \exp[-\pi\alpha^2 T^2] = \delta(\alpha)$

(b) $\lim_{T \to \infty} \frac{T}{2} \exp[-|\alpha|T] = \delta(\alpha)$

7.1-8. Work Problem 7.1-7 for the following cases:

(a) $\lim_{T \to \infty} \frac{T \sin(\alpha T)}{\alpha T} = \delta(\alpha)$

(b) $\lim_{\substack{T \to \infty \\ |\alpha| < 1/T}} T[1 - |\alpha|T] = \delta(\alpha)$

7.1-9. Show that (7.1-14) is true.

7.1-10. Prove (7.1-7). [*Hint*: Use (D-6) of Appendix D and the definition of the derivative.]

7.1-11. A random process is defined by

$$Y(t) = X(t)\cos(\omega_0 t + \Theta)$$

257

CHAPTER 7:
Random
Processes—
Spectral
Characteristics

where $X(t)$ is a lowpass wide-sense stationary process, ω_0 is a real constant, and Θ is a random variable uniformly distributed on the interval $(0, 2\pi)$. Find and sketch the power density spectrum of $Y(t)$ in terms of that of $X(t)$. Assume Θ is independent of $X(t)$.

7.1-12. Determine which of the following functions can and cannot be valid power density spectrums. For those that are not, explain why.

(a) $\dfrac{\omega^2}{\omega^6 + 3\omega^2 + 3}$ (b) $\exp[-(\omega - 1)^2]$

(c) $\dfrac{\omega^2}{\omega^4 + 1} - \delta(\omega)$ (d) $\dfrac{\omega^4}{1 + \omega^2 + j\omega^6}$

7.1-13. Work Problem 7.1-12 for the following functions.

(a) $\dfrac{\cos(3\omega)}{1 + \omega^2}$ (b) $\dfrac{1}{(1 + \omega^2)^2}$

(c) $\dfrac{|\omega|}{1 + 2\omega + \omega^2}$ (d) $\dfrac{1}{\sqrt{1 - 3\omega^2}}$

7.1-14. Given that $X(t) = \sum_{i=1}^{N} \alpha_i X_i(t)$ where $\{\alpha_i\}$ is a set of real constants and the processes $X_i(t)$ are stationary and orthogonal, show that

$$\mathscr{S}_{XX}(\omega) = \sum_{i=1}^{N} \alpha_i^2 \mathscr{S}_{X_i X_i}(\omega)$$

7.1-15. A random process is given by

$$X(t) = A_0 \cos(\Omega t + \Theta)$$

where A_0 is a real constant, Ω is a random variable with density function $f_\Omega(\cdot)$, and Θ is a random variable uniformly distributed on the interval $(0, 2\pi)$ independent of Ω. Show that the power spectrum of $X(t)$ is

$$\mathscr{S}_{XX}(\omega) = \frac{\pi A_0^2}{2}[f_\Omega(\omega) + f_\Omega(-\omega)]$$

7.1-16. If $X(t)$ is a stationary process, find the power spectrum of

$$Y(t) = A_0 + B_0 X(t)$$

in terms of the power spectrum of $X(t)$ if A_0 and B_0 are real constants.

7.1-17. Find the rms bandwidth of the power spectrum

$$\mathscr{S}_{XX}(\omega) = \begin{cases} \dfrac{P}{1 + (\omega/W)^2} & |\omega| < KW \\ 0 & |\omega| > KW \end{cases}$$

where P, W, and K are real positive constants. If $K \to \infty$, what happens?

7.1-18. Find the rms bandwidth of the power spectrum

$$\mathscr{S}_{XX}(\omega) = \begin{cases} P\cos(\pi\omega/2W) & |\omega| \leq W \\ 0 & |\omega| > W \end{cases}$$

where $W > 0$ and $P > 0$ are constants.

7.1-19. Determine the rms bandwidths of the power spectrums given by:

(a) $\mathscr{S}_{XX}(\omega) = \begin{cases} P & |\omega| < W \\ 0 & |\omega| > W \end{cases}$

(b) $\mathscr{S}_{XX}(\omega) = \begin{cases} P[1 - |\omega/W|] & |\omega| \leq W \\ 0 & |\omega| > W \end{cases}$

where P and W are real positive constants.

*7.1-20. Given the power spectrum

$$\mathscr{S}_{XX}(\omega) = \frac{P}{\left[1 + \left(\frac{\omega - \alpha}{W}\right)^2\right]^2} + \frac{P}{\left[1 + \left(\frac{\omega + \alpha}{W}\right)^2\right]^2}$$

where P, α, and W are real positive constants, find the mean frequency and rms bandwidth.

7.1-21. Show that the rms bandwidth of the power spectrum of a real bandpass process $X(t)$ is given by

$$W_{\text{rms}}^2 = 4[\overline{W^2} - \bar{\omega}_0^2]$$

where $\bar{\omega}_0$ is given by (7.1-23) and $\overline{W^2}$ is given by the right side of (7.1-22).

7.1-22. The autocorrelation function of a random process $X(t)$ is

$$R_{XX}(\tau) = 3 + 2\exp(-4\tau^2)$$

(a) Find the power spectrum of $X(t)$.
(b) What is the average power in $X(t)$?
(c) What fraction of the power lies in the frequency band $-1/\sqrt{2} \leq \omega \leq 1/\sqrt{2}$?

7.1-23. State whether or not each of the following functions can be a valid power density spectrum. For those that cannot, explain why.

(a) $\dfrac{|\omega|\exp(-4\omega^2)}{1 + j\omega}$

(b) $\cos(3\omega)\exp(-\omega^2 + j2\omega)$

(c) $\dfrac{\omega^6}{(12 + \omega^2)^6}$

(d) $6\tan[12\omega/(1 + \omega^2)]$

(e) $\cos^2(\omega)\exp(-8\omega^2)$

(f) $(-j\omega)(j\omega)/(3 - j\omega)^2(3 + j\omega)^2$

7.1-24. If $\mathscr{S}_{XX}(\omega)$ is a valid power spectrum of a random process $X(t)$, discuss whether the functions $d\mathscr{S}_{XX}(\omega)/d\omega$ and $d^2\mathscr{S}_{XX}(\omega)/d\omega^2$ can be valid power spectrums.

7.1-25. (*a*) Rework Problem 7.1-15 and show that even if Θ is a constant (not random) the power spectrum is still given by

259

CHAPTER 7:
Random
Processes—
Spectral
Characteristics

$$\mathcal{S}_{XX}(\omega) = (\pi A_0^2/2)[f_\Omega(\omega) + f_\Omega(-\omega)]$$

[*Hint*: Time-average autocorrelation function before Fourier transforming to obtain $\mathcal{S}_{XX}(\omega)$.]

(*b*) Find the total power in $x(t)$ and show that it is independent of the form of the density function $f_\Omega(\omega)$.

7.1-26. Find the rms bandwidth of the power spectrum

$$\mathcal{S}_{XX}(\omega) = 1/[1 + (\omega/W)^2]^3$$

where $W > 0$ is a constant.

7.1-27. Work Problem 7.1-26 for the power spectrum

$$\mathcal{S}_{XX}(\omega) = \omega^2/[1 + (\omega/W)^2]^3$$

7.1-28. Work Problem 7.1-26 for the power spectrum

$$\mathcal{S}_{XX}(\omega) = 1/[1 + (\omega/W)^2]^4$$

7.1-29. Work Problem 7.1-26 for the power spectrum

$$\mathcal{S}_{XX}(\omega) = \omega^2/[1 + (\omega/W)^2]^4$$

*7.1-30. Generalize Problems 7.1-26 and 7.1-28 by finding the rms bandwidth of the power spectrum

$$\mathcal{S}_{XX}(\omega) = 1/[1 + (\omega/W)^2]^N$$

where $N \geq 2$ is an integer.

*7.1-31. Generalize Problems 7.1-27 and 7.1-29 by finding the rms bandwidth of the power spectrum

$$\mathcal{S}_{XX}(\omega) = \omega^2/[1 + (\omega/W)^2]^N$$

where $N \geq 3$ is an integer.

7.1-32. Assume a random process has a power spectrum

$$\mathcal{S}_{XX}(\omega) = \begin{cases} 4 - (\omega^2/9) & |\omega| \leq 6 \\ 0 & \text{elsewhere} \end{cases}$$

Find (*a*) the average power, (*b*) the rms bandwidth, and (*c*) the autocorrelation function of the process.

7.1-33. Show that rms bandwidth of a lowpass random process $X(t)$, as given by (7.1-22), can also be obtained from

$$W_{\text{rms}}^2 = \frac{-1}{R_{XX}(0)} \left. \frac{d^2 R_{XX}(\tau)}{d\tau^2} \right|_{\tau=0}$$

where $R_{XX}(\tau)$ is the autocorrelation function of $X(t)$.

7.1-34. Assume a random process $X(t)$ has a power spectrum

$$\mathscr{S}_{XX}(\omega) = \frac{6|\omega|}{(W^2 + \omega^2)^2}$$

where $W > 0$ is a constant.
(a) Sketch $\mathscr{S}_{XX}(\omega)$.
(b) At what positive value of ω, denoted by ω_{max}, does $\mathscr{S}_{XX}(\omega)$ reach a maximum value?

7.1-35. Treat the power spectrum of Problem 7.1-34 as bandpass and find its mean frequency $\bar{\omega}_0$ and rms bandwidth W_{rms}.

7.1-36. Work Problem 7.1-35, except assume the power spectrum

$$\mathscr{S}_{XX}(\omega) = \frac{6|\omega|}{(W^2 + \omega^2)^3}$$

7.1-37. Determine which of the following functions can be a valid power density of some random process. For those that cannot, give at least one reason why.

(a) $\displaystyle\sum_{n=-3}^{3} (-2)^n \delta(\omega - 3n)$

(b) $\exp[-4\sin(6\omega)]$

(c) $\delta(\omega - 4) + \delta(\omega + 4) + e^{-|\omega|}\cos^2(10\omega)$

(d) $\dfrac{-5\omega}{(10 + \omega^4)^2}$

7.1-38. Find $\bar{\omega}_0$ and W_{rms} for the power spectrum shown in Figure P7.6-6.

7.2-1. Find the power density spectrum of the random process for which

$$R_{XX}(\tau) = P\cos^4(\omega_0\tau)$$

if P and ω_0 are constants. Determine the power in the process by use of (7.1-12).

7.2-2. A random process has the power density spectrum

$$\mathscr{S}_{XX}(\omega) = \frac{6\omega^2}{1 + \omega^4}$$

Find the average power in the process.

7.2-3. Work Problem 7.2-2 for the power spectrum

$$\mathscr{S}_{XX}(\omega) = \frac{6\omega^2}{[1 + \omega^2]^3}$$

7.2-4. Work Problem 7.2-2 for the power spectrum

$$\mathscr{S}_{XX}(\omega) = \frac{6\omega^2}{(1 + \omega^2)^4}$$

261

CHAPTER 7:
Random
Processes—
Spectral
Characteristics

7.2-5. Assume $X(t)$ is a wide-sense stationary process with nonzero mean value $\bar{X} \neq 0$. Show that

$$\mathscr{S}_{XX}(\omega) = 2\pi \bar{X}^2 \delta(\omega) + \int_{-\infty}^{\infty} C_{XX}(\tau) e^{-j\omega\tau} \, d\tau$$

where $C_{XX}(\tau)$ is the autocovariance function of $X(t)$.

7.2-6. For a random process $X(t)$, assume that

$$R_{XX}(\tau) = Pe^{-\tau^2/2a^2}$$

where $P > 0$ and $a > 0$ are constants. Find the power density spectrum of $X(t)$. [*Hint*: Use Appendix E to evaluate the Fourier transform of $R_{XX}(\tau)$.]

7.2-7. A random process has an autocorrelation function

$$R_{XX}(\tau) = \begin{cases} P[1-(2\tau/T)] & 0 < \tau \le T/2 \\ P[1+(2\tau/T)] & -T/2 \le \tau \le 0 \\ 0 & \tau < -T/2 \quad \text{and} \quad \tau > T/2 \end{cases}$$

Find and sketch its power density spectrum. (*Hint*: Use Appendix E.)

*7.2-8. A random process $X(t)$ has a periodic autocorrelation function where the function of Problem 7.2-7 forms the central period of duration T. Find and sketch the power spectrum.

7.2-9. If the random processes of Problem 7.1-14 are stationary, zero-mean, statistically independent processes, show that the power spectrum of the sum is the same as for orthogonal processes. For stationary independent processes with nonzero means, what is $\mathscr{S}_{XX}(\omega)$?

7.2-10. Given that a process $X(t)$ has the autocorrelation function

$$R_{XX}(\tau) = Ae^{-\alpha|\tau|} \cos(\omega_0 \tau)$$

where $A > 0$, $\alpha > 0$, and ω_0 are real constants, find the power spectrum of $X(t)$.

7.2-11. A random process $X(t)$ having the power spectrum of Problem 7.2-3 is applied to an ideal differentiator.
(*a*) Find the power spectrum of the differentiator's output.
(*b*) What is the power in the derivative?

7.2-12. Work Problem 7.2-11 for the power spectrum of Problem 7.2-4.

7.2-13. A wide-sense stationary random process $X(t)$ is used to define another process by

$$Y(t) = \int_{-\infty}^{\infty} h(\xi)X(t-\xi) \, d\xi$$

where $h(t)$ is some real function having a Fourier transform $H(\omega)$. Show that the power spectrum of $Y(t)$ is given by

$$\mathscr{S}_{YY}(\omega) = \mathscr{S}_{XX}(\omega)|H(\omega)|^2$$

7.2-14. A deterministic signal $A \cos(\omega_0 t)$, where A and ω_0 are real constants, is added to a noise process $N(t)$ for which

$$\mathscr{S}_{NN}(\omega) = \frac{W^2}{W^2 + \omega^2}$$

and $W > 0$ is a constant.
(a) Find the ratio of average signal power to average noise power.
(b) What value of W maximizes the signal-to-noise ratio? What is the consequence of choosing this value of W?

7.1-15. A random process has the autocorrelation function

$$R_{XX}(\tau) = B \cos^2(\omega_0 \tau) \exp(-W|\tau|)$$

where B, ω_0, and W are postiive constants.
(a) Find and sketch the power spectrum of $X(t)$ when ω_0 is at least several times larger than W.
(b) Compute the average power in the lowpass part of the power spectrum. Repeat for the bandpass part. In each case assume $\omega_0 \gg W$.

*7.2-16. Generalize Problem 7.2-15 by replacing $\cos^2(\omega_0 \tau)$ with $\cos^N(\omega_0 \tau)$ where $N \geq 0$ is an integer. What is the resulting power spectrum when N is (a) odd, and (b) even?

*7.2-17. The product of a wide-sense stationary gaussian random process $X(t)$ with itself delayed by T seconds forms a new process $Y(t) = X(t)X(t-T)$. Determine (a) the autocorrelation function, and (b) the power spectrum of $Y(t)$. {*Hint:* Use the fact that $E[X_1 X_2 X_3 X_4] = E[X_1 X_2]E[X_3 X_4] + E[X_1 X_3] E[X_2 X_4] + E[X_1 X_4] \ E[X_2 X_3] - 2E[X_1]E[X_2]E[X_3]E[X_4]$ for gaussian random variables X_1, X_2, X_3, and X_4. (Thomas, 1969, p. 64.)}

7.2-18. For a random process $X(t)$, assume its autocorrelation function is

$$R_{XX}(t, t+\tau) = 12 e^{-4\tau^2} \cos^2(24t)$$

(a) Is $X(t)$ wide-sense stationary?
(b) Find $R_{XX}(\tau)$.
(c) Find the power spectrum of $X(t)$.

7.2-19. A random process is defined by $Y(t) = X(t) - X(t-a)$, where $X(t)$ is a wide-sense stationary process and $a > 0$ is a constant. Find the autocorrelation function and power density spectrum of $Y(t)$ in terms of the corresponding quantities for $X(t)$.

7.2-20. Find the autocorrelation function corresponding to the power density spectrum

$$\mathscr{S}_{XX}(\omega) = \frac{157 + 12\omega^2}{(16 + \omega^2)(9 + \omega^2)}$$

[*Hint:* Use a partial fraction expansion (Peebles and Giuma, 1991, pp. 149–156) and Table E-1.]

7.2-21. Find the autocorrelation function corresponding to the power spectrum

$$\mathscr{S}_{XX}(\omega) = \frac{8}{(9 + \omega^2)^2}$$

[*Hint*: Use the convolution property of Fourier transforms given by (D-16).]

263

CHAPTER 7:
Random
Processes—
Spectral
Characteristics

*7.2-22. Find the power spectrum corresponding to the autocorrelation function

$$R_{XX}(\tau) = [\cos(\alpha\tau) + \sin(\alpha|\tau|)]e^{-\alpha|\tau|}$$

where $\alpha > 0$ is a constant.

*7.3-1. Joint wide-sense stationary random processes $X(t)$ and $Y(t)$ define a process $W(t)$ by

$$W(t) = X(t)\cos(\omega_0 t) + Y(t)\sin(\omega_0 t)$$

where ω_0 is a real positive constant.
(a) Develop some conditions on the mean values and correlation functions of $X(t)$ and $Y(t)$ such that $W(t)$ is wide-sense stationary.
(b) With the conditions of part (a) applied to $W(t)$, find its power spectrum in terms of power spectrums of $X(t)$ and $Y(t)$.
(c) If $X(t)$ and $Y(t)$ are also uncorrelated, what is the power spectrum of $W(t)$?

7.3-2. A random process is given by

$$W(t) = AX(t) + BY(t)$$

where A and B are real constants and $X(t)$ and $Y(t)$ are jointly wide-sense stationary processes.
(a) Find the power spectrum $\mathscr{S}_{WW}(\omega)$ of $W(t)$.
(b) Find $\mathscr{S}_{WW}(\omega)$ if $X(t)$ and $Y(t)$ are uncorrelated.
(c) Find the cross-power spectrums $\mathscr{S}_{XW}(\omega)$ and $\mathscr{S}_{YW}(\omega)$.

*7.3-3. Define two random processes by

$$X(t) = A\cos(\omega_0 t + \Theta)$$
$$Y(t) = W(t)\cos(\omega_0 t + \Theta)$$

where A and ω_0 are real positive constants, Θ is a random variable independent of $W(t)$, and $W(t)$ is a random process with a constant mean value $\bar{W}$. By using (7.3-12), show that

$$\mathscr{S}_{XY}(\omega) = \frac{A\bar{W}\pi}{2}[\delta(\omega - \omega_0) + \delta(\omega + \omega_0)]$$

regardless of the form of the probability density function of Θ.

7.3-4. Decompose the cross-power spectrums into real and imaginary parts according to

$$\mathscr{S}_{XY}(\omega) = R_{XY}(\omega) + jI_{XY}(\omega)$$
$$\mathscr{S}_{YX}(\omega) = R_{YX}(\omega) + jI_{YX}(\omega)$$

and prove that

$$R_{XY}(\omega) = R_{YX}(-\omega) = -R_{YX}(\omega)$$
$$I_{XY}(\omega) = I_{YX}(-\omega) = -I_{YX}(\omega)$$

7.3-5. From the results of Problem 7.3-4, prove (7.3-16).

7.3-6. Show that (7.3-19) and (7.3-20) are true.

7.3-7. Find the cross-correlation function $R_{XY}(t, t+\tau)$ and cross-power spectrum $\mathscr{S}_{XY}(\omega)$ for the delay-and-multiply device of Problem 7.2-17. {*Hint*: Use the fact that $E[X_1 X_2 X_3] = E[X_1]E[X_2 X_3] + E[X_2]E[X_3 X_1] + E[X_3]E[X_1 X_2] - 2E[X_1]E[X_2]E[X_3]$ for three gaussian random variables X_1, X_2, and X_3. (Thomas, 1969, p. 64.)}

7.3-8. If $X(t)$ and $Y(t)$ are real random processes, determine which of the following functions can be valid. For those that are not, state at least one reason why.

(a) $R_{XX}(\tau) = \exp(-|\tau|)$ (b) $|R_{XY}(\tau)| \leq j\sqrt{R_{XX}(0)R_{YY}(0)}$

(c) $R_{XX}(\tau) = 2\sin(3\tau)$ (d) $\mathscr{S}_{XX}(\omega) = 6/(6 + 7\omega^3)$

(e) $\mathscr{S}_{XX}(\omega) = \dfrac{4\exp(-3|\tau|)}{1 + \omega^2}$ (f) $\mathscr{S}_{XY}(\omega) = 3 + j\omega^2$

(g) $\mathscr{S}_{XY}(\omega) = 18\delta(\omega)$

7.3-9. Form the product of two statistically independent jointly wide-sense stationary random processes $X(t)$ and $Y(t)$ as

$$W(t) = X(t)Y(t)$$

Find general expressions for the following correlation functions and power spectrums in terms of those of $X(t)$ and $Y(t)$: (a) $R_{WW}(t, t+\tau)$ and $\mathscr{S}_{WW}(\omega)$, (b) $R_{XW}(t, t+\tau)$ and $\mathscr{S}_{XW}(\omega)$, and (c) $R_{WX}(t, t+\tau)$ and $\mathscr{S}_{WX}(\omega)$. (d) If

$$R_{XX}(\tau) = (W_1/\pi)\mathrm{Sa}(W_1\tau)$$

and

$$R_{YY}(\tau) = (W_2/\pi)\mathrm{Sa}(W_2\tau)$$

with constants $W_2 > W_1$, find explicit functions for $R_{WW}(t, t+\tau)$ and $\mathscr{S}_{WW}(\omega)$.

7.3-10. An engineer is working with the function

$$R_{XY}(\tau) = P(1+\tau)\exp(-W^2\tau^2)$$

where $P > 0$ and $W > 0$ are constants. He suspects that the function may not be a valid cross-correlation for two jointly stationary processes $X(t)$ and $Y(t)$, as he has been told. Determine if his suspicions are true. [*Hint*: Find the cross-power spectrum and see if it satisfies properties (7.3-16) through (7.3-18).]

7.3-11. A wide-sense stationary process $X(t)$ is applied to an ideal differentiator having the response $Y(t) = dX(t)/dt$. The cross-correlation of the input-output processes is known to be

$$R_{XY}(\tau) = dR_{XX}(\tau)/d\tau$$

265

CHAPTER 7:
Random
Processes—
Spectral
Characteristics

(a) Determine $\mathscr{S}_{XY}(\omega)$ and $\mathscr{S}_{YX}(\omega)$ in terms of the power spectrum $\mathscr{S}_{XX}(\omega)$ of $X(t)$.

(b) Since $\mathscr{S}_{XX}(\omega)$ must be real, nonnegative, and have even symmetry, what are the properties of $\mathscr{S}_{XY}(\omega)$?

7.3-12. The cross-correlation of jointly wide-sense stationary processes $X(t)$ and $Y(t)$ is assumed to be

$$R_{XY}(\tau) = Bu(\tau)\exp(-W\tau)$$

where $B > 0$ and $W > 0$ are constants.
(a) Find $R_{YX}(\tau)$.
(b) Find $\mathscr{S}_{XY}(\omega)$ and $\mathscr{S}_{YX}(\omega)$.

7.3-13. Work Problem 7.3-12 for the function

$$R_{XY}(\tau) = Bu(\tau)\tau\exp(-W\tau)$$

7.3-14. The cross-power spectrum for random processes $X(t)$ and $Y(t)$ can be written as

$$\mathscr{S}_{XY}(\omega) = \mathscr{S}_{XX}(\omega)H(\omega)$$

where $\mathscr{S}_{XX}(\omega)$ is the power spectrum of $X(t)$ and $H(\omega)$ is a function with an inverse Fourier transform $h(\tau)$. Derive expressions for $R_{XY}(\tau)$ and $R_{YX}(\tau)$ in terms of $R_{XX}(\tau)$ and $h(\tau)$.

7.3-15. Determine the cross-power density spectrum corresponding to the cross-correlation function

$$R_{XY}(\tau) = u(-\tau)\frac{e^{b\tau}}{a+b} + \frac{u(\tau)e^{-b\tau}}{a^2 - b^2}[a+b-2be^{-(a-b)\tau}]$$

where $a > 0$ and $b > 0$ are constants.

7.3-16. Find the cross-correlation function corresponding to the cross-power spectrum

$$\mathscr{S}_{XY}(\omega) = \frac{6}{(9+\omega^2)(3+j\omega)^2}$$

*7.4-1. Again consider the random processes of Problem 7.3-3.
(a) Use (6.3-11) to show that the cross-correlation function is given by

$$R_{XY}(t, t+\tau) = \frac{A\bar{W}}{2}\{\cos(\omega_0\tau) + E[\cos(2\Theta)]\cos(2\omega_0 t + \omega_0\tau)$$
$$- E[\sin(2\Theta)]\sin(2\omega_0 t + \omega_0\tau)\}$$

where the expectation is with respect to Θ only.
(b) Find the time average of $R_{XY}(t, t+\tau)$ and determine the cross-power density spectrum $\mathscr{S}_{XY}(\omega)$.

7.5-1. Fourier transform the first right-side form of (7.5-1), assuming T_p is small enough that $R_{XX}(\tau)$ does not change appreciably over any time interval of length T_p. Does your result become (7.5-2) when $T_p \to 0$?

7.5-2. Show that (7.5-3) is true.

7.5-3. Substitute (7.5-3) into the middle form of (7.5-1) and prove that (7.5-4) is true.

7.5-4. Fourier transform (7.5-6) and show that (7.5-7) is true. (*Hint*: Make use of the table of transforms in Appendix E.)

7.5-5. Periodic samples of the autocorrelation function of *white noise* $N(t)$ with period T_s are defined by

$$R_{NN}(kT_s) = \begin{cases} \sigma_N^2 & k = 0 \\ 0 & k \neq 0 \end{cases}$$

Find the power spectrum of the DT random process.

7.5-6. A discrete-time random sequence $X[n]$ has a DTFT given by

$$\mathcal{S}_{X_sX_s}(e^{j\Omega}) = A_0 \cos(\Omega) \sum_{n=-\infty}^{\infty} \text{rect}\left(\frac{\Omega - n2\pi}{\pi}\right)$$

where A_0 is a real positive constant. Find the sequence $R_{XX}[n]$ by use of the IDTFT of (7.5-12).

7.5-7. A random sequence $Y[n]$ is formed by adding the white noise sequence of Problem 7.5-5 to a one-unit delayed white noise sequence according to

$$Y[n] = N[n] + b_1 N[n-1]$$

where b_1 is a real constant. Show that

$$R_{YY}[k] = \begin{cases} (1 + b_1^2)\sigma_N^2 & k = 0 \\ b_1\sigma_n^2 & k = +1 \text{ and } -1 \\ 0 & \text{all other } k \end{cases}$$

7.5-8. Find the power density spectrum of the sequence $Y[n]$ of Problem 7.5-7 by use of (7.5-11).

7.5-9. Work Problem 7.5-7 except for the sequence $Y[n] = N[n] + b_m N[n-m]$, where b_m is a real constant and m is a positive integer. Use (7.5-11) to find the power spectrum of $Y[n]$.

7.5-10. Extend Problem 7.5-7 to consider the sequence

$$Y[n] = N[n] + b_1 N[n-1] + b_2 N[n-2]$$

where b_2 is a real constant. Find $R_{YY}[k]$ and the applicable power spectrum from (7.5-11).

7.5-11. The white noise of Problem 7.5-5 is added to a "signal" X that is a random variable statistically independent of the noise and having a mean value of zero and a variance σ_X^2. Assume X is a constant with time. Denote the DT sequence of samples of signal-plus-noise by $W[n] = X[n] + N[n]$ where, of course, $X[n] = X$, a constant with n. Find the autocorrelation function of this sequence and then determine the DTFT of the sequence. [*Hint*: Use the known sum

$$\sum_{n=-\infty}^{\infty} \exp(-jm\Omega) = 2\pi \sum_{n=-\infty}^{\infty} \delta(\Omega - n2\pi).]$$

267

CHAPTER 7:
Random
Processes—
Spectral
Characteristics

7.5-12. A DT random sequence $X[n]$ has an autocorrelation function defined by the sequence $R_{XX}[n] = (1 - |n|/N)$ for $|n| \leq N$ and is zero for $|n| > N$, with N a positive integer. Find the DTFT of the sequence.

7.5-13. For the sequence $Y[n]$ defined in Example 7.5-3, show that its autocorrelation sequence is given by (1).

7.5-14. Show that the autocorrelation sequence defined in Equation (1) of Example 7.5-3 has the power spectrum of equation (2).

7.5-15. Rework Example 7.5-5 except use $N = 10$ and observe the effect of increasing N.

7.5-16. Extend Example 7.5-4 by showing that the autocorrelation function for the continuous-time process having the power spectrum

$$\mathcal{S}_{XX}(\omega) = K \cos^2\left(\frac{\pi\omega}{2W_X}\right) \text{rect}\left(\frac{\omega}{2W_X}\right)$$

is

$$R_{XX}(\tau) = \frac{KW_X}{4\pi}[2\text{Sa}(W_X\tau) + \text{Sa}(W_X\tau - \pi) + \text{Sa}(W_X\tau + \pi)]$$
$$= \frac{(KW_X\pi/2)\sin(W_X\tau)}{W_X\tau[\pi^2 - (W_X\tau)^2]}$$

7.5-17. Rework Example 7.5-5 except assume the random process defined in Problem 7.5-16. Rework your results for $N = 10$ and compare to the results for $N = 5$.

7.5-18. Show that (7.5-12) can be written as

$$R_{XX}[n] = \frac{T_s}{2\pi}\int_{-\pi/T_s}^{\pi/T_s} \mathcal{S}_{X_sX_s}(\omega) e^{jn\omega T_s} d\omega$$

7.6-1. (a) Sketch the power spectrum of (7.6-4) as a function of $\alpha\omega/T$.
(b) For what values of ω will $\mathcal{S}_{NN}(\omega)$ remain above $0.5(\mathcal{N}_0/2)$ when $T = 4.2\,\text{K}$ (the value of liquid helium at one atmosphere of pressure)? These values form the region where thermal noise is approximately white in some amplifiers operated at very low temperatures, such as a *maser*.

7.6-2. For the power spectrum given in Figure 7.6-2a, show that (7.6-6) defines the corresponding band-limited noise autocorrelation function.

7.6-3. Show that (7.6-8) gives the autocorrelation function of the bandpass band-limited noise defined by Figure 7.6-3a.

7.6-4. A lowpass random process $X(t)$ has a continuous power spectrum $\mathcal{S}_{XX}(\omega)$ and $\mathcal{S}_{XX}(0) \neq 0$. Find the bandwidth W of a lowpass band-limited white-noise power spectrum having a density $\mathcal{S}_{XX}(0)$ and the same total power as in $X(t)$.

7.6-5. Work Problem 7.6-4 for a bandpass process assuming $\mathcal{S}_{XX}(\omega_0) \neq 0$, where ω_0 is some convenient frequency about which the spectral components of $X(t)$ cluster.

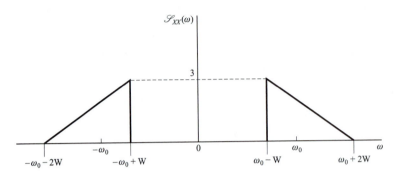

FIGURE P7.6-6

7.6-6. The power spectrum of a bandpass process $X(t)$ is shown in Figure P7.6-6. $X(t)$ is applied to a product device where the second multiplying input is $3\cos(\omega_0 t)$. Plot the power spectrum of the device's output $3X(t)\cos(\omega_0 t)$.

7.6-7. Let the "carrier" $A_0\cos(\omega_0 t)$ in Figure 7.6-5 be modified to add a phase random variable Θ so that $Y(t) = A_0 X(t)\cos(\omega_0 t + \Theta)$. If Θ is uniformly distributed on $(0, 2\pi)$ and is independent of $X(t)$, find $R_{YY}(t, t + \tau)$ and $\mathcal{S}_{YY}(\omega)$ when $X(t)$ is wide-sense stationary.

7.6-8. Assume a stationary bandpass process $X(t)$ is adequately approximated by the power spectrum

$$\mathcal{S}_{XX}(\omega) = Pu(\omega - \omega_0)(\omega - \omega_0)\exp[-(\omega - \omega_0)^2/b]$$
$$+ Pu(-\omega - \omega_0)(-\omega - \omega_0)\exp[-(\omega + \omega_0)^2/b]$$

where ω_0, $P > 0$, and $b > 0$ are constants. The product $Y(t) = X(t)\cos(\omega_0 t)$ is formed.
(a) Find and sketch the power spectrum of $Y(t)$.
(b) Determine the average power in $X(t)$ and $Y(t)$.

7.6-9. Approximate (7.6-4) as a rectangular lowpass power spectrum with a constant amplitude $\mathcal{N}_0/2$ for $|\omega| < W$ where W is the angular frequency at which (7.6-4) drops to $\mathcal{N}_0/4$ when $T = 2\,\text{K}$. What average noise power exists in the approximate power spectrum if $\mathcal{N}_0/2 = 5.5(10^{-19})$? (*Hint:* Assume the exponential is adequately approximated by the first three terms in its series representation.)

7.6-10. A signal $s_i(t) = 2.3\cos(1000t)$ plus an input noise process $N_i(t)$ having the power spectrum shown in Figure P7.6-10a are applied to the product device shown in (b). The ideal lowpass filter (LPF) acts only to remove all spectral

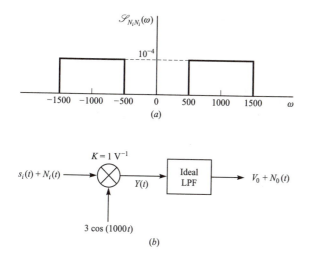

269

CHAPTER 7:
Random
Processes—
Spectral
Characteristics

FIGURE P7.6-10

components (signal and noise) that are outside the band $|\omega| < 500\,\text{rad/s}$ and does not affect components inside the band.

(a) Find the output dc level V_0.

(b) Sketch the power density spectrum of $N_0(t)$.

(c) What signal power to average noise power ratio, $V_0^2/E[N_0^2(t)]$, occurs at the output? Note that this circuit acts as a detector of the signal's amplitude in the presence of noise.

*7.7-1. A complex random process is given by

$$Z(t) = Ae^{j\Omega t}$$

where Ω is a random variable with probability density function $f_\Omega(\cdot)$ and A is a complex constant. Show that the power spectrum of $Z(t)$ is

$$\mathscr{S}_{ZZ}(\omega) = 2\pi|A|^2 f_\Omega(\omega)$$

*7.7-2. Compute the power spectrum of the complex process of Problem 6.7-4.

*7.7-3. Let $X(t)$ and $Y(t)$ be statistically independent processes with power spectrums

$$\mathscr{S}_{XX}(\omega) = 2\delta(\omega) + 1/[1 + (\omega/10)^2]$$

and

$$\mathscr{S}_{YY}(\omega) = 4/[1 + (\omega/2)^2]$$

A complex process

$$Z(t) = [X(t) + jY(t)]\exp(j\omega_0 t)$$

is formed where ω_0 is a constant much larger than 10.

(a) Determine the autocorrelation function of $Z(t)$.

(b) Find and sketch the power spectrum of $Z(t)$.

Linear Systems with Random Inputs

8.0
INTRODUCTION

A large part of our preceding work has been aimed at describing a random signal by modeling it as a sample function of a random process. We have found that time domain methods based on correlation functions, and frequency domain techniques based on power spectrums, constitute powerful ways of defining the behavior of random signals. Our work must not stop here, however, because one of the most important aspects of random signals is how they interact with linear systems. The knowledge of how to describe a random waveform would be of little value to a communication or control system engineer, for example, unless he was also able to determine how such a waveform would alter the desired output of his system.

In this chapter, we explore methods of describing the response of a linear system when the applied waveform is random. We begin by discussing some basic aspects of linear systems in the following section. Those readers well-versed in linear system theory can proceed directly to Section 8.2 without loss. For others, the topic of Section 8.1 should serve as a brief review and summary.

8.1
LINEAR SYSTEM FUNDAMENTALS

In this section, a brief summary of the basic aspects of linear systems is given. Attention will be limited to a system having only one input and one output, or response, as illustrated in Figure 8.1-1. It is assumed that the input signal $x(t)$ and the response $y(t)$ are deterministic signals, even though some of the topics

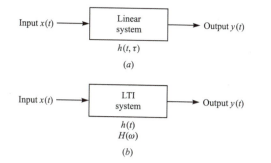

(a)

(b)

FIGURE 8.1-1

(a) A general single-input single-output linear system, and (b) a similar linear, time-invariant (LTI) system.

discussed apply to random waveforms. Which topics are applicable to random signals will be made clear when they are used in later sections.

The General Linear System

Clearly, the linear system (Figure 8.1-1a) will, in general, cause the response $y(t)$ to be different from the input signal $x(t)$. We think of the system as *operating* on $x(t)$ to cause $y(t)$ and write

$$y(t) = L[x(t)] \qquad (8.1\text{-}1)$$

Here L is an *operator* representing the action of the system on $x(t)$.

A system is said to be linear if its response to a sum of inputs $x_n(t)$, $n = 1, 2, \ldots, N$, is equal to the sum of responses taken separately. Thus, if $x_n(t)$ causes a response $y_n(t), n = 1, 2, \ldots, N$, then for a linear system

$$y(t) = L\left[\sum_{n=1}^{N} \alpha_n x_n(t)\right] = \sum_{n=1}^{N} \alpha_n L[x_n(t)] = \sum_{n=1}^{N} \alpha_n y_n(t) \qquad (8.1\text{-}2)$$

must hold, where the α_n are arbitrary constants and N may be infinite.

From the definition (2.3-2) and properties of the impulse function we may write

$$x(t) = \int_{-\infty}^{\infty} x(\xi)\delta(t - \xi) \, d\xi \qquad (8.1\text{-}3)$$

By substituting (8.1-3) into (8.1-1) and observing that the operator operates on the time function, we obtain

$$y(t) = L[x(t)] = L\left[\int_{-\infty}^{\infty} x(\xi)\delta(t - \xi) \, d\xi\right] = \int_{-\infty}^{\infty} x(\xi)L[\delta(t - \xi)] \, d\xi \qquad (8.1\text{-}4)$$

We now *define* a new function $h(t, \xi)$ as the *impulse response* of the linear system; that is,

$$L[\delta(t - \xi)] = h(t, \xi) \qquad (8.1\text{-}5)$$

Equation (8.1-4) becomes

$$y(t) = \int_{-\infty}^{\infty} x(\xi)h(t, \xi)\, d\xi \qquad (8.1\text{-}6)$$

which shows that the response of a general linear system is completely determined by its impulse response through (8.1-6).

Linear Time-Invariant Systems

A general linear system is said to be also time-invariant if the *form* of its impulse response $h(t, \xi)$ does not depend on the time that the impulse is applied. Thus, if an impulse $\delta(t)$, occurring at $t = 0$, causes the response $h(t)$, then an impulse $\delta(t - \xi)$, occurring at $t = \xi$, must cause the response $h(t - \xi)$ if the system is time-invariant. This fact means that

$$h(t, \xi) = h(t - \xi) \qquad (8.1\text{-}7)$$

for a linear-time invariant system, so (8.1-6) becomes

$$y(t) = \int_{-\infty}^{\infty} x(\xi)h(t - \xi)\, d\xi \qquad (8.1\text{-}8)$$

Equation (8.1-8) is known as the *convolution integral* of $x(t)$ and $h(t)$; it is sometimes written in the short form

$$y(t) = x(t) * h(t) \qquad (8.1\text{-}9)$$

By a suitable change of variables, (8.1-8) can be put in the alternative form

$$y(t) = \int_{-\infty}^{\infty} h(\xi)x(t - \xi)\, d\xi \qquad (8.1\text{-}10)$$

Time-Invariant System Transfer Function

Either (8.1-8) or (8.1-10) shows that a linear time-invariant system is completely characterized by its impulse response, which is a temporal characterization. By Fourier transformation of $y(t)$, we may derive an equivalent characterization in the frequency domain. Hence, if $X(\omega)$, $Y(\omega)$, and $H(\omega)$ are the respective Fourier transforms of $x(t)$, $y(t)$, and $h(t)$, then

$$Y(\omega) = \int_{-\infty}^{\infty} y(t)e^{-j\omega t}\, dt = \int_{-\infty}^{\infty}\left[\int_{-\infty}^{\infty} x(\xi)h(t - \xi)\, d\xi\right]e^{-j\omega t}\, dt$$

$$= \int_{-\infty}^{\infty} x(\xi)\left[\int_{-\infty}^{\infty} h(t - \xi)e^{-j\omega(t-\xi)}\, dt\right]e^{-j\omega\xi}\, d\xi$$

$$= \int_{-\infty}^{\infty} x(\xi)H(\omega)e^{-j\omega\xi}\, d\xi = X(\omega)h(\omega) \qquad (8.1\text{-}11)$$

The function $H(\omega)$ is called the *transfer function* of the system. Equation (8.1-11) shows that the Fourier transform of the response of any linear time-invariant system is equal to the product of the transform of the input signal and the transform of the network impulse response.

In the actual calculation of a transfer function for a given network, an alternative definition based on the response of the system to an exponential signal

$$x(t) = e^{j\omega t} \tag{8.1-12}$$

may be more convenient. It can be shown (Thomas, 1969, p. 142), or Papoulis, 1962, p. 83) that†

$$H(\omega) = \frac{L[e^{j\omega t}]}{e^{j\omega t}} = \frac{y(t)}{x(t)} \tag{8.1-13}$$

where

$$y(t) = L[e^{j\omega t}] \tag{8.1-14}$$

An example serves to illustrate the determination of $H(\omega)$ by means of (8.1-13).

EXAMPLE 8.1-1. We find $H(\omega)$ for the network shown in Figure 8.1-2. By assuming a clockwise current i (and no loading in the output circuit), we have‡

$$x(t) = L\frac{di}{dt} + y(t)$$

But $y(t) = iR$ so

$$\frac{di}{dt} = \frac{1}{R}\frac{dy(t)}{dt}$$

and

$$x(t) = \frac{L}{R}\frac{dy(t)}{dt} + y(t)$$

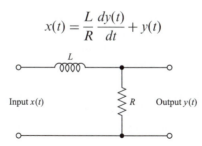

FIGURE 8.1-2
A linear time-invariant network. [*Reproduced from Peebles (1976) with permission of publishers Addison-Wesley, Advanced Book Program.*]

---

†It should be carefully observed that (8.1-13) holds *only* for $x(t)$ given by (8.1-12); that is, for an exponential waveform.
‡L in the network is an inductance and should not be confused with L above, which stands for a linear system operator.

With $x(t) = \exp(j\omega t)$ as the input we must have an output $y(t) = H(\omega)x(t)$ from (8.1-13). Hence, $dy(t)/dt = H(\omega)j\omega x(t)$ and

$$x(t) = \frac{L}{R} H(\omega) j\omega x(t) + H(\omega)x(t)$$

Finally, we solve for $H(\omega)$:

$$H(\omega) = \frac{1}{1 + (j\omega L/R)}$$

EXAMPLE 8.1-2. As a second example we prove (8.1-13) by direct use of (8.1-10). For $x(t) = \exp(j\omega t)$ we have

$$y(t) = \int_{-\infty}^{\infty} h(\xi)e^{j\omega(t-\xi)}\, d\xi = x(t)\int_{-\infty}^{\infty} h(\xi)e^{-j\omega\xi}\, d\xi$$

But the integral is $H(\omega)$, the Fourier transform of $h(t)$, so

$$y(t) = x(t)H(\omega)$$

which gives (8.1-13).

Idealized Systems

To simplify the analysis of many complex systems, it is often convenient to *approximate* the system's transfer function $H(\omega)$ by an idealized one. Idealized transfer functions are illustrated in Figure 8.1-3a for a lowpass system; (b) applies to a highpass system and (c) applies to a bandpass system. In every case the *idealized system* has a transfer function magnitude that is flat within its passband and zero outside this band; its midband gain is unity and its phase $\theta(\omega)$ is defined to be a linear function of frequency.

In replacing an actual system with an idealized one, the latter would be assigned a midband gain and phase slope that approximate the actual values. The bandwidth W (in lowpass and bandpass cases) is chosen according to some convenient basis. For example, W could be made equal to the 3-dB bandwidth of the actual system, or alternatively, it could be chosen to satisfy a specific requirement. An example of the latter case is considered in Section 8.5 where W, called *noise bandwidth*, is selected to cause the actual and ideal systems to produce the same output noise power when each is excited by the same noise source.

Causal and Stable Systems

To complete our summary of basic topics in linear system theory, we consider two final items.

A linear time-invariant system is said to be *causal* if it does not respond prior to the application of an input signal. Mathematically, this implies that

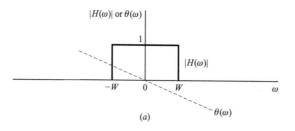

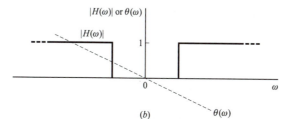

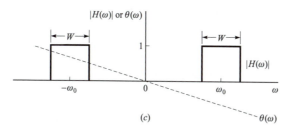

FIGURE 8.1-3
Ideal system transfer function. (*a*) Lowpass, (*b*) highpass, and (*c*) bandpass systems.
[*Reproduced from Peebles (1976) with permission of publishers Addison-Wesley, Advanced Book Program.*]

$y(t) = 0$ for $t < t_0$ if $x(t) = 0$ for $t < t_0$, where t_0 is any real constant. From (8.1-10), this condition requires that

$$h(t) = 0 \qquad \text{for} \qquad t < 0 \qquad (8.1\text{-}15)$$

All passive, linear time-invariant networks that can be constructed will satisfy (8.1-15). As a consequence, a system satisfying (8.1-15) is often called *physically realizable*.

A linear time-invariant system is said to be *stable* if its response to any bounded input is bounded; that is, if $|x(t)| < M$, where M is some constant, then $|y(t)| < MI$ for a stable system where I is another constant independent of the input. By considering (8.1-10), it is readily shown that

$$I = \int_{-\infty}^{\infty} |h(t)| \, dt < \infty \qquad (8.1\text{-}16)$$

will ensure that a system having the impulse response $h(t)$ will be stable.

8.2
RANDOM SIGNAL RESPONSE OF LINEAR SYSTEMS

With the preceding summary of linear systems theory in mind, we proceed now to determine characteristics of the response of a stable, linear, time-invariant system as illustrated in Figure 8.1-1b when the applied waveform is an ensemble member $x(t)$ of a random process $X(t)$. We assume in all work that the system's impulse response $h(t)$ is a real function.† In this section we restrict our attention to temporal characteristics such as mean value and mean-squared value of the response, its autocorrelation function, and applicable cross-correlation functions. Spectral characteristics are developed in Section 8.4.

System Response—Convolution

Even when $x(t)$ is a random signal, the network's response $y(t)$ is given by the convolution integral:

$$y(t) = \int_{-\infty}^{\infty} x(\xi)h(t - \xi)\,d\xi \tag{8.2-1}$$

or

$$y(t) = \int_{-\infty}^{\infty} h(\xi)x(t - \xi)\,d\xi \tag{8.2-2}$$

where $h(t)$ is the network's impulse response.

We may view (8.2-2) as an operation on an ensemble member $x(t)$ of the random process $X(t)$ that produces an ensemble member of a new process $Y(t)$. With this viewpoint, we may think of (8.2-2) as defining the process $Y(t)$ in terms of the process $X(t)$:

$$Y(t) = \int_{-\infty}^{\infty} h(\xi)X(t - \xi)\,d\xi \tag{8.2-3}$$

Thus, we may envision the system as accepting the random process $X(t)$ as its input and responding with the new process $Y(t)$ according to (8.2-3).

Mean and Mean-Squared Value of System Response

We may readily apply (8.2-3) to find the mean value of the system's response. By assuming $X(t)$ is wide-sense stationary, we have‡

---

†All real-world networks have real impulse responses.
‡We shall assume that expectation and integration operations are interchangeable whenever needed. Some justification can be found in Cooper and McGillem (1986), p. 288, who state that the operation

$$E[Y(t)] = E\left[\int_{-\infty}^{\infty} h(\xi)X(t-\xi)\,d\xi\right]$$

$$= \int_{-\infty}^{\infty} h(\xi)E[X(t-\xi)]\,d\xi$$

$$= \bar{X}\int_{-\infty}^{\infty} h(\xi)\,d\xi = \bar{Y} \quad \text{(constant)} \tag{8.2-4}$$

This expression indicates that the mean value of $Y(t)$ equals the mean value of $X(t)$ times the area under the impulse response if $X(t)$ is wide-sense stationary.

For the mean-squared value of $Y(t)$, we calculate

$$E[Y^2(t)] = E\left[\int_{-\infty}^{\infty} h(\xi_1)X(t-\xi_1)\,d\xi_1 \int_{-\infty}^{\infty} h(\xi_2)X(t-\xi_2)\,d\xi_2\right]$$

$$= \int_{-\infty}^{\infty}\int_{-\infty}^{\infty} E[X(t-\xi_1)X(t-\xi_2)]h(\xi_1)h(\xi_2)\,d\xi_1\,d\xi_2 \tag{8.2-5}$$

If we assume the input is wide-sense stationary then

$$E[X(t-\xi_1)X(t-\xi_2)] = R_{XX}(\xi_1 - \xi_2) \tag{8.2-6}$$

and (8.2-5) becomes independent of t:

$$\overline{Y^2} = E[Y^2(t)] = \int_{-\infty}^{\infty}\int_{-\infty}^{\infty} R_{XX}(\xi_1 - \xi_2)h(\xi_1)h(\xi_2)\,d\xi_1\,d\xi_2 \tag{8.2-7}$$

Although this expression gives the power in $Y(t)$, it may be tedious to calculate in most cases. We develop an example of its solution for a simple case.

EXAMPLE 8.2-1. We find $\overline{Y^2}$ for a system having white noise at its input. Here

$$R_{XX}(\xi_1 - \xi_2) = (\mathcal{N}_0/2)\delta(\xi_1 - \xi_2)$$

where $\mathcal{N}_0$ is a positive real constant. From (8.2-7):

$$\overline{Y^2} = \int_{-\infty}^{\infty}\int_{-\infty}^{\infty} (\mathcal{N}_0/2)\delta(\xi_1 - \xi_2)h(\xi_1)\,d\xi_1 h(\xi_2)\,d\xi_2$$

$$= (\mathcal{N}_0/2)\int_{-\infty}^{\infty} h^2(\xi_2)\,d\xi_2$$

Output power becomes proportional to the area under the square of $h(t)$ in this case.

$$E\left[\int_{t_1}^{t_2} W(t)h(t)\,dt\right] = \int_{t_1}^{t_2} E[W(t)]h(t)\,dt$$

is valid, where $W(t)$ is some bounded function of a random process [on the interval (t_1, t_2)] and $h(t)$ is a nonrandom time function, if

$$\int_{t_1}^{t_2} E[|W(t)|]|h(t)|\,dt < \infty$$

where t_1 and t_2 are real constants that may be infinite. This condition is satisfied in all *physical* cases if $W(t)$ is wide-sense stationary because $W(t)$ will be bounded and the systems are stable [see (8.1-16)]. See also Appendix G.

Autocorrelation Function of Response

Let $X(t)$ be wide-sense stationary. The autocorrelation function of $Y(t)$ is

$$R_{YY}(t, t + \tau) = E[Y(t)Y(t + \tau)]$$

$$= E\left[\int_{-\infty}^{\infty} h(\xi_1)X(t - \xi_1)\,d\xi_1 \int_{-\infty}^{\infty} h(\xi_2)X(t + \tau - \xi_2)\,d\xi_2\right]$$

$$= \int_{-\infty}^{\infty}\int_{-\infty}^{\infty} E[X(t - \xi_1)X(t + \tau - \xi_2)]h(\xi_1)h(\xi_2)\,d\xi_1\,d\xi_2 \qquad (8.2\text{-}8)$$

which reduces to

$$R_{YY}(\tau) = \int_{-\infty}^{\infty}\int_{-\infty}^{\infty} R_{XX}(\tau + \xi_1 - \xi_2)h(\xi_1)h(\xi_2)\,d\xi_1\,d\xi_2 \qquad (8.2\text{-}9)$$

because $X(t)$ is assumed wide-sense stationary.

Two facts result from (8.2-9). First, $Y(t)$ is wide-sense stationary if $X(t)$ is wide-sense stationary because $R_{YY}(\tau)$ does not depend on t and $E[Y(t)]$ is a constant from (8.2-4). Second, the form of (8.2-9) shows that $R_{YY}(\tau)$ is the twofold convolution of the input autocorrelation function with the network's impulse response; that is

$$R_{YY}(\tau) = R_{XX}(\tau) * h(-\tau) * h(\tau) \qquad (8.2\text{-}10)$$

Cross-Correlation Functions of Input and Output

The cross-correlation function of $X(t)$ and $Y(t)$ is

$$R_{XY}(t, t + \tau) = E[X(t)Y(t + \tau)] = E\left[X(t)\int_{-\infty}^{\infty} h(\xi)X(t + \tau - \xi)\,d\xi\right]$$

$$= \int_{-\infty}^{\infty} E[X(t)X(t + \tau - \xi)]h(\xi)\,d\xi \qquad (8.2\text{-}11)$$

If $X(t)$ is wide-sense stationary, (8.2-11) reduces to

$$R_{XY}(\tau) = \int_{-\infty}^{\infty} R_{XX}(\tau - \xi)h(\xi)\,d\xi \qquad (8.2\text{-}12)$$

which is the convolution $R_{XX}(\tau)$ with $h(\tau)$:

$$R_{XY}(\tau) = R_{XX}(\tau) * h(\tau) \qquad (8.2\text{-}13)$$

A similar development shows that

$$R_{YX}(\tau) = \int_{-\infty}^{\infty} R_{XX}(\tau - \xi)h(-\xi)\,d\xi \qquad (8.2\text{-}14)$$

or

$$R_{YX}(\tau) = R_{XX}(\tau) * h(-\tau) \qquad (8.2\text{-}15)$$

From (8.2-12) and (8.2-14), it is clear that the cross-correlation functions depend on τ and not on absolute time t. As a consequence of this fact $X(t)$ and

$Y(t)$ are *jointly* wide-sense stationary if $X(t)$ is wide-sense stationary, because we have already shown $Y(t)$ to be wide-sense stationary.

By substituting (8.2-12) into (8.2-9), autocorrelation function and cross-correlation functions are seen to be related by

$$R_{YY}(\tau) = \int_{-\infty}^{\infty} R_{XY}(\tau + \xi_1)h(\xi_1)\, d\xi_1 \qquad (8.2\text{-}16)$$

or

$$R_{YY}(\tau) = R_{XY}(\tau) * h(-\tau) \qquad (8.2\text{-}17)$$

A similar substitute of (8.2-14) into (8.2-9) gives

$$R_{YY}(\tau) = \int_{-\infty}^{\infty} R_{YX}(\tau - \xi_2)h(\xi_2)\, d\xi_2 \qquad (8.2\text{-}18)$$

or

$$R_{YY}(\tau) = R_{YX}(\tau) * h(\tau) \qquad (8.2\text{-}19)$$

> **EXAMPLE 8.2-2.** We shall continue Example 8.2-1 by finding the cross-correlation functions $R_{XY}(\tau)$ and $R_{YX}(\tau)$. From (8.2-12)
>
> $$R_{XY}(\tau) = \int_{-\infty}^{\infty} (\mathcal{N}_0/2)\delta(\tau - \xi)h(\xi)\, d\xi$$
> $$= (\mathcal{N}_0/2)h(\tau)$$
>
> From (8.2-14)
>
> $$R_{YX}(\tau) = \int_{-\infty}^{\infty} (\mathcal{N}_0/2)\delta(\tau - \xi)h(-\xi)\, d\xi$$
> $$= (\mathcal{N}_0/2)h(-\tau) = R_{XY}(-\tau)$$
>
> These two results are seen to satisfy (6.3-16), as they should.

8.3
SYSTEM EVALUATION USING RANDOM NOISE

A practical application of the foregoing theory can be immediately developed; it is based on the cross-correlation function of (8.2-12). Suppose we desire to find the impulse response of some linear time-invariant system. If we have available a broadband (relative to the system) noise source having a flat power spectrum, and a cross-correlation measurement device, such as shown in Figure 6.4-1, $h(t)$ can easily be determined.

For the approximately white noise source

$$R_{XX}(\tau) \approx \left(\frac{\mathcal{N}_0}{2}\right)\delta(\tau) \qquad (8.3\text{-}1)$$

With this noise applied to the system, the cross-correlation function from (8.2-12) or Example 8.2-2 becomes

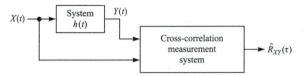

FIGURE 8.3-1
A method for finding a system's impulse response. [*Reproduced from Peebles (1976) with permission of publishers Addison-Wesley, Advanced Book Program.*]

$$R_{XY}(\tau) \approx \int_{-\infty}^{\infty} \left(\frac{\mathcal{N}_0}{2}\right) \delta(\tau - \xi) h(\xi)\, d\xi$$

$$= \left(\frac{\mathcal{N}_0}{2}\right) h(\tau) \tag{8.3-2}$$

or

$$h(\tau) \approx \left(\frac{2}{\mathcal{N}_0}\right) R_{XY}(\tau) \tag{8.3-3}$$

Since a measurement $\hat{R}_{XY}(\tau)$ of $R_{XY}(\tau)$ can be obtained from the cross-correlation measurement device, (8.3-3) gives us a measurement $\hat{h}(\tau)$ of $h(\tau)$

$$\hat{h}(\tau) = \left(\frac{2}{\mathcal{N}_0}\right) \hat{R}_{XY}(\tau) \approx h(\tau) \tag{8.3-4}$$

Figure 8.3-1 illustrates the concepts described here.

8.4
SPECTRAL CHARACTERISTICS OF SYSTEM RESPONSE

Because the Fourier transform of a correlation function (autocorrelation or cross-correlation) is a power spectrum for wide-sense stationary processes, it would seem that if $R_{XX}(\tau)$ is known for the input process, one can find $R_{YY}(\tau)$, $R_{XY}(\tau)$, and $R_{YX}(\tau)$ as described in Section 8.2 and therefore obtain power spectrums by transformation. Indeed, this approach is conceptually valid. However, from a practical standpoint the integrals involved may be difficult to evaluate.

In this section an alternative approach is taken where the desired power spectrum involving the system's reponse is related to the power spectrum of the input. In every case, the input process $X(t)$ is assumed to be wide-sense stationary, which, as previously proved, means that $Y(t)$ and $X(t)$ are jointly wide-sense stationary.

Power Density Spectrum of Response

We show now that the power density spectrum $\mathcal{S}_{YY}(\omega)$ of the response of a linear time-invariant system having a transfer function $H(\omega)$ is given by

$$\mathcal{S}_{YY}(\omega) = \mathcal{S}_{XX}(\omega)|H(\omega)|^2 \qquad (8.4\text{-}1)$$

where $\mathcal{S}_{XX}(\omega)$ is the power spectrum of the input process $X(t)$. We call $|H(\omega)|^2$ the *power transfer function* of the system.

The proof of (8.4-1) begins by writing $\mathcal{S}_{YY}(\omega)$ as the Fourier transform of the output autocorrelation function

$$\mathcal{S}_{YY}(\omega) = \int_{-\infty}^{\infty} R_{YY}(\tau)e^{-j\omega\tau}\,d\tau \qquad (8.4\text{-}2)$$

On substitution of (8.2-9), (8.4-2) becomes

$$\mathcal{S}_{YY}(\omega) = \int_{-\infty}^{\infty} h(\xi_1)\int_{-\infty}^{\infty} h(\xi_2)\int_{-\infty}^{\infty} R_{XX}(\tau + \xi_1 - \xi_2)e^{-j\omega\tau}\,d\tau\,d\xi_2\,d\xi_1 \qquad (8.4\text{-}3)$$

The change of variable $\xi = \tau + \xi_1 - \xi_2$, $d\xi = d\tau$, produces

$$\mathcal{S}_{YY}(\omega) = \int_{-\infty}^{\infty} h(\xi_1)e^{j\omega\xi_1}\,d\xi_1\int_{-\infty}^{\infty} h(\xi_2)e^{-j\omega\xi_2}\,d\xi_2\int_{-\infty}^{\infty} R_{XX}(\xi)e^{-j\omega\xi}\,d\xi \qquad (8.4\text{-}4)$$

These three integrals are recognized as $H^*(\omega)$, $H(\omega)$, and $\mathcal{S}_{XX}(\omega)$, respectively. Hence

$$\mathcal{S}_{YY}(\omega) = H^*(\omega)H(\omega)\mathcal{S}_{XX}(\omega) = \mathcal{S}_{XX}(\omega)|H(\omega)|^2 \qquad (8.4\text{-}5)$$

and (8.4-1) is proved.

The average power, denoted P_{YY}, in the system's response is readily found by using (8.4-5):

$$P_{YY} = \frac{1}{2\pi}\int_{-\infty}^{\infty} \mathcal{S}_{XX}(\omega)|H\omega)|^2\,d\omega \qquad (8.4\text{-}6)$$

EXAMPLE 8.4-1. The power spectrum and average power of the response of the network of Example 8.1-1 will be found when $X(t)$ is white noise for which

$$\mathcal{S}_{XX}(\omega) = \frac{\mathcal{N}_0}{2}$$

Here $H(\omega) = 1 + (j\omega L/R)]^{-1}$ so

$$|H(\omega)|^2 = \frac{1}{1 + (\omega L/R)^2}$$

and

$$\mathcal{S}_{YY}(\omega) = \mathcal{S}_{XX}(\omega)|H(\omega)|^2 = \frac{\mathcal{N}_0/2}{1 + (\omega L/R)^2}$$

Average power in $Y(t)$, from (8.4-6), is

$$P_{YY} = \frac{1}{2\pi}\int_{-\infty}^{\infty} \mathcal{S}_{YY}(\omega)\,d\omega = \frac{\mathcal{N}_0}{4\pi}\int_{-\infty}^{\infty} \frac{d\omega}{1 + (\omega L/R)^2} = \frac{\mathcal{N}_0 R}{4L}$$

after an integral from Appendix C is used.

As a check on the calculation of P_{YY}, we note that (pair 15, Appendix E)

$$h(t) = (R/L)u(t)e^{-Rt/L} \leftrightarrow H(\omega) = \frac{1}{1 + (j\omega L/R)}$$

for this network, and, using the result of Example 8.2-1, we get

$$P_{YY} = \overline{Y^2} = \left(\frac{\mathcal{N}_0}{2}\right)\int_{-\infty}^{\infty} \left(\frac{R}{L}\right)^2 e^{-2Rt/L} \, dt = \frac{\mathcal{N}_0 R}{4L}$$

The two powers are in agreement.

Cross-Power Density Spectrums of Input and Output

It is easily shown (see Problem 8.4-5) that the Fourier transforms of the cross-correlation functions of (8.2-12) and (8.2-14) may be written as

$$\mathcal{S}_{XY}(\omega) = \mathcal{S}_{XX}(\omega)H(\omega) \tag{8.4-7}$$

$$\mathcal{S}_{YX}(\omega) = \mathcal{S}_{XX}(\omega)H(-\omega) \tag{8.4-8}$$

respectively.

Measurement of Power Density Spectrums

The practical measurement of a power density spectrum is usually discussed in books as an "estimation" of the power spectrum. Although the theory behind spectral estimation is extensive and detailed,† a simple discussion can be given that provides insight and a plausible basis for measuring power spectrums.

To measure the power spectrum of a lowpass process $X(t)$, consider the system of Figure 8.4-1a. $X(t)$, having the power spectrum of (b), is applied to a real linear filter with a very narrowband, bandpass transfer function as illustrated in (c). The center frequency, ω_f, or the filter's transfer function is presumed adjustable from near $\omega = 0$ out to angular frequency W, the spectral extent of $X(t)$. The filter's output, $Y(t)$, is applied to a power meter that measures the average power in $Y(t)$. Both $X(t)$ and $Y(t)$ are assumed stationary, ergodic processes. We also assume that the power meter averages $Y^2(t)$ over a very long time such that any fluctuations in its reading are small relative to the measured power.

By assuming the spectral extent, W_f, of the filter is very small relative to W, and using the facts that $|H(\omega)|^2$ and $\mathcal{S}_{XX}(\omega)$ are even functions of ω for real filters and real $X(t)$, we can expand the power, $P_{YY}(\omega_f)$, in $Y(t)$ as

---

†For additional detail the reader is referred to some of the literature, such as Bendat and Piersol (1986), Kay (1986), and Blackman and Tukey (1958).

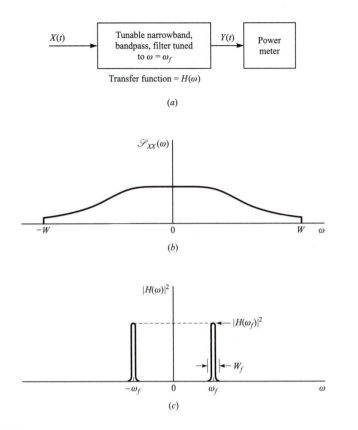

FIGURE 8.4-1
(a) A system for the measurement of a lowpass power density spectrum as in (b), and (c) the squared magnitude of the filter's transfer function.

$$P_{YY}(\omega_f) = \frac{1}{2\pi} \int_{-\infty}^{\infty} \mathscr{S}_{XX}(\omega)|H(\omega)|^2 \, d\omega$$

$$= \frac{1}{\pi} \int_{0}^{\infty} \mathscr{S}_{XX}(\omega)|H(\omega)|^2 \, d\omega$$

$$\approx \frac{1}{\pi} \mathscr{S}_{XX}(\omega_f) \int_{0}^{\infty} |H(\omega)|^2 \, d\omega$$

$$= \frac{\mathscr{S}_{XX}(\omega_f)|H(\omega_f)|^2 W_N}{\pi} \tag{8.4-9}$$

The last form in (8.4-9) uses a quantity called *noise bandwidth*, defined in the next section, as given by

$$W_N = \frac{\int_{0}^{\infty} |H(\omega)|^2 \, d\omega}{|H(\omega_f)|^2} \tag{8.4-10}$$

Finally, we write (8.4-9) as

$$\mathscr{S}_{XX}(\omega_f) \approx \frac{\pi P_{YY}(\omega_f)}{W_N |H(\omega_f)|^2} \tag{8.4-11}$$

In words, the power $P_{YY}(\omega_f)$, measured when the filter is tuned to $\omega = \omega_f$, is multiplied by the known constant $\pi/[W_N |H(\omega_f)|^2]$, and the result is an approximation to $\mathscr{S}_{XX}(\omega)$ at $\omega = \omega_f$. By varying ω_f, the system can measure the power spectrum for various ω.

When $X(t)$ is a bandpass process, the system of Figure 8.4-2a is more convenient to use. Here $X(t)$ has a power spectrum centered at some high angular frequency ω_0, as sketched in (b). From (7.5-12), the effect of the product device is to scale the power spectrum of $X(t)$ by $A_0^2/4$ and shift it both higher and lower in angular frequency by an amount $\omega_0 + \omega_{IF} + \omega_f$, the frequency of the local oscillator. Four spectral components are created, as shown in (c), for the product signal $P(t)$. The narrowband filter has a transfer function centered at a fixed angular frequency ω_{IF}, as sketched in (d). Its spectral extent W_f is assumed to be much smaller than the spectral extent W of the process $X(t)$.

For the system of Figure 8.4-2, the power in $Y(t)$ becomes

$$\begin{aligned} P_{YY}(\omega_0 + \omega_f) &= \frac{1}{2\pi} \int_{-\infty}^{\infty} \frac{A_0^2}{4} [\mathscr{S}_{XX}(\omega - \omega_0 - \omega_{IF} - \omega_f) \\ &\quad + \mathscr{S}_{XX}(\omega + \omega_0 + \omega_{IF} + \omega_f)] |H(\omega)|^2 \, d\omega \\ &= 2 \frac{1}{2\pi} \int_{-\infty}^{\infty} \frac{A_0^2}{4} \mathscr{S}_{XX}(\omega - \omega_0 - \omega_{IF} - \omega_f) |H(\omega)|^2 \, d\omega \\ &= \frac{A_0^2}{2} \frac{1}{2\pi} \int_{0}^{\infty} \mathscr{S}_{XX}(\omega - \omega_0 - \omega_{IF} - \omega_f) |H(\omega)|^2 \, d\omega \\ &\approx \frac{A_0^2}{4\pi} \mathscr{S}_{XX}(-\omega_0 - \omega_f) \int_{0}^{\infty} |H(\omega)|^2 \, d\omega \\ &= \frac{A_0^2}{4\pi} \mathscr{S}_{XX}(\omega_0 + \omega_f) W_N |H(\omega_{IF})|^2 \end{aligned} \tag{8.4-12}$$

where W_N is the filter's noise bandwidth according to (8.4-10). Thus,

$$\mathscr{S}_{XX}(\omega_0 + \omega_f) \approx \frac{4\pi P_{YY}(\omega_0 + \omega_f)}{A_0^2 W_N |H(\omega_{IF})|^2} \tag{8.4-13}$$

In words, an approximation (measurement) of $\mathscr{S}_{XX}(\omega)$ at angular frequency $\omega_0 + \omega_f$ is equal to the known constant $4\pi/[A_0^2 W_N |H(\omega_{IF})|^2]$ multiplied by the average power $P_{YY}(\omega_0 + \omega_f)$, measured in $Y(t)$ when the local oscillator is tuned to an angular frequency *larger* than the center frequency ω_{IF} of the filter by an amount $\omega_0 + \omega_f$. By varying the frequency of the local oscillator (changing ω_f), $\mathscr{S}_{XX}(\omega)$ can be measured for all frequencies around ω_0.

For proper performance of the system of Figure 8.4-2, $W_f \ll W$ is required so that $\mathscr{S}_{PP}(\omega) \approx \mathscr{S}_{PP}(\omega_{IF})$ for all frequencies near ω_{IF} that are in the passband of the filter. Furthermore ω_{IF} must not be chosen too small; it should satisfy $\omega_{IF} > W + (W_f/2)$ if the spectral terms in Figure 8.4-2c are not

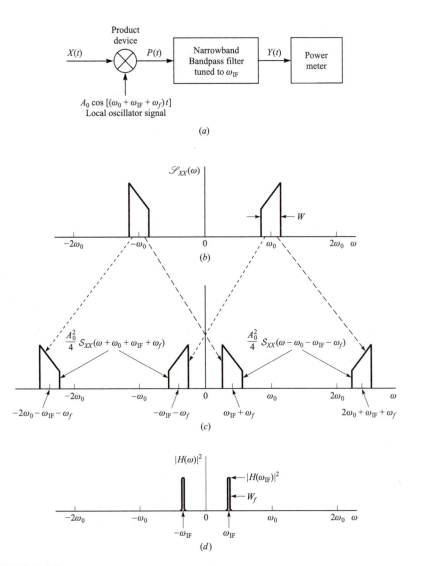

FIGURE 8.4-2
(*a*) System for the measurement of a bandpass power spectrum as in (*b*). (*c*) The power density spectrum of $P(t)$. (*d*) The squared magnitude of the filter's transfer function.

to overlap when changes take place in ω_f. We next discuss an upper bound for ω_{IF} by means of an example.

> **EXAMPLE 8.4-2.** When a real, practical device is used to form the product in Figure 8.4-2, the output $P(t)$ always contains a term proportional to the input $X(t)$ because of practical "leakage," sometimes called "feed-through." This leakage causes two additional terms in the spectrum of (*c*) centered at ω_0 and $-\omega_0$. As the other spectral terms change their

positions in frequency in response to the oscillator's frequency adjustments, the leakage terms do not move. When ω_f has its largest value of $(W + W_f)/2$, the highest frequency in the spectral component being measured is at $\omega_{IF} + (W + W_f)/2 + (W/2) = \omega_{IF} + W + (W_f/2)$. If the frequency is to be lower than the lowest frequency in the leakage spectrum, which is $\omega_0 - (W/2)$, then we require

$$\omega_{IF} < \omega_0 - [(3W + W_f)/2]$$

This expression provides an upper bound on the choice of ω_{IF} in design.

8.5
NOISE BANDWIDTH

Consider a system having a lowpass transfer function $H(\omega)$. Assume white noise is applied at the input. The power density of this white noise is $\mathcal{N}_0/2$ where $\mathcal{N}_0$ is a real positive constant. The total average power emerging from the network is [from (8.4-6)]

$$P_{YY} = \frac{1}{2\pi} \int_{-\infty}^{\infty} \left(\frac{\mathcal{N}_0}{2}\right) |H(\omega)|^2 \, d\omega \tag{8.5-1}$$

By assuming the system impulse response is real,† $|H(\omega)|^2$ will be an even function of ω and (8.5-1) can be written

$$P_{YY} = \frac{\mathcal{N}_0}{2\pi} \int_{0}^{\infty} |H(\omega)|^2 \, d\omega \tag{8.5-2}$$

Now consider an idealized system that is equivalent to the actual system in the sense that both produce the same output average power when they both are excited by the same white noise source, and both have the same value of power transfer function at midband; that is, $|H(0)|^2$ is the same in both systems. The principal difference between the two systems is that the idealized one has a rectangularly shaped power transfer function $|H_I(\omega)|^2$ defined by

$$|H_I(\omega)|^2 = \begin{cases} |H(0)|^2 & |\omega| < W_N \\ 0 & |\omega| > W_N \end{cases} \tag{8.5-3}$$

where W_N is a positive constant selected to make output powers in the two systems equal. The output power in the idealized system is

$$\frac{1}{2\pi} \int_{-\infty}^{\infty} \left(\frac{\mathcal{N}_0}{2}\right) |H_I(\omega)|^2 \, d\omega = \frac{\mathcal{N}_0}{2\pi} \int_{0}^{W_N} |H(0)|^2 \, d\omega = \frac{\mathcal{N}_0 |H(0)|^2 W_N}{2\pi} \tag{8.5-4}$$

By equating (8.5-2) and (8.5-4), we require that W_N be given by

---

†The impulse response of any physical system is always real.

$$W_N = \frac{\int_0^\infty |H(\omega)|^2 \, d\omega}{|H(0)|^2} \qquad (8.5\text{-}5)$$

W_N is called the *noise bandwidth* of the system.

EXAMPLE 8.5-1. The noise bandwidth is found for a system having the power transfer function

$$|H(\omega)|^2 = \frac{1}{1 + (\omega/W)^2}$$

where W is the 3-dB bandwidth in radians per second. Here $|H(0)|^2 = 1$, so

$$W_N = \int_0^\infty \frac{W^2 \, d\omega}{W^2 + \omega^2} = W \tan^{-1}\left(\frac{\omega}{W}\right)\Bigg|_0^\infty = \frac{W\pi}{2}$$

This expression shows that W_N is larger than the system 3-dB bandwidth by a factor of about 1.57.

If we repeat the above development for a bandpass transfer function with a centerband frequency ω_0 it will be found that

$$W_N = \frac{\int_0^\infty |H(\omega)|^2 \, d\omega}{|H(\omega_0)|^2} \qquad (8.5\text{-}6)$$

Proof of this result is left as a reader exercise (see Problem 8.5-1). The development also provides a simple expression for output noise power in terms of noise bandwidth:

$$P_{YY} = \frac{\mathcal{N}_0}{2\pi} |H(\omega_0)|^2 W_N \qquad (8.5\text{-}7)$$

For a lowpass filter, (8.5-7) applies by letting $\omega_0 = 0$.

8.6
BANDPASS, BAND-LIMITED, AND NARROWBAND PROCESSES

A random process $N(t)$ will be called *bandpass* if its power density spectrum $\mathcal{S}_{NN}(\omega)$ has its significant components clustered in a band of width W (rad/s) that does not include $\omega = 0$. Such a power spectrum is illustrated in Figure 8.6-1a.† Our definition does not prevent the power spectrum from being non-zero at $\omega = 0$; it only requires that $\mathcal{S}_{NN}(0)$ be small in relation to more

---

†Power spectrums arising in physical systems will always decrease as frequency becomes sufficiently large, so a suitable value of W can always be found. For example, W could be chosen to include all frequencies for which $\mathcal{S}_{NN}(\omega) \geq 0.1\mathcal{S}_{NN}(\omega_0)$ where ω_0 is some convenient frequency near where $\mathcal{S}_{NN}(\omega)$ has its largest magnitude (see Figure 8.6-1).

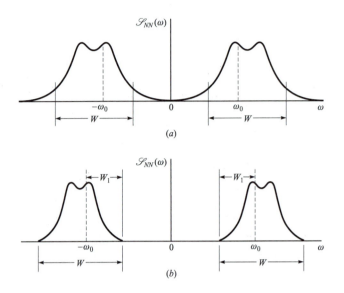

FIGURE 8.6-1
Power density spectrums (*a*) for a bandpass random process and (*b*) for a band-limited bandpass process.

significant values, so as to distinguish the bandpass case from a lowpass power spectrum with significant peaking at higher frequencies.

All subsequent discussions in this section will relate to special forms of bandpass processes.

Band-Limited Processes

If the power spectrum of a bandpass random process is *zero* outside some frequency band of width W (rad/s) that does not include $\omega = 0$, the process is called *band-limited*. The concept of a band-limited process forms a convenient approximation for physical processes that often allows analytical problem solutions that otherwise might not be possible. A band-limited bandpass process power spectrum is illustrated in Figure 8.6-1*b*.

Narrowband Processes

A band-limited random process is said to be *narrowband* if $W \ll \omega_0$, where ω_0 is some conveniently chosen frequency near band-center or near where the power spectrum is at its maximum. A power spectrum of a narrowband process is sketched in Figure 8.6-2*a*. A typical sample function, if viewed on an oscilloscope, might look as shown in (*b*). The appearance of *n(t)* suggests that the process might be represented by a cosine function with angular frequency ω_0 and slowly varying amplitude and phase; that is, by

$$N(t) = A(t)\cos[\omega_0 t + \Theta(t)] \tag{8.6-1}$$

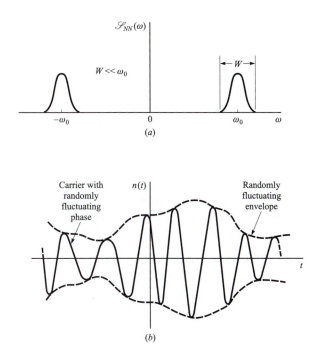

FIGURE 8.6-2
(*a*) A power spectrum of a narrowband random process $N(t)$ and (*b*) a typical ensemble member $n(t)$. [*Reproduced from Peebles (1976) with permission of publishers Addison-Wesley, Advanced Book Program.*]

where $A(t)$ is a random process representing the slowly varying amplitude and $\Theta(t)$ is a process representing the slowly varying phase. Indeed this is the case, and, for the important practical case where $N(t)$ is gaussian noise, it is known that $A(t)$ and $\Theta(t)$ have Rayleigh and uniform (over 2π) first-order probability density functions, respectively. The processes $A(t)$ and $\Theta(t)$ are not statistically independent when $N(t)$ is gaussian (Davenport, 1970, p. 522, or Davenport and Root, 1958, pp. 161–165), but for any one instant in time the process *random variables* are independent.

In some problems, (8.6-1) is a preferred representation for $N(t)$. For others, it is convenient to use the equivalent form

$$N(t) = X(t)\cos(\omega_0 t) - Y(t)\sin(\omega_0 t) \qquad (8.6\text{-}2)$$

where the processes $X(t)$ and $Y(t)$ are given by

$$X(t) = A(t)\cos[\Theta(t)] \qquad (8.6\text{-}3)$$

$$Y(t) = A(t)\sin[\Theta(t)] \qquad (8.6\text{-}4)$$

Expressions relating $A(t)$ and $\Theta(t)$ to $X(t)$ and $Y(t)$ are

$$A(t) = \sqrt{X^2(t) + Y^2(t)} \qquad (8.6\text{-}5)$$

$$\Theta(t) = \tan^{-1}[Y(t)/X(t)] \qquad (8.6\text{-}6)$$

Properties of Band-Limited Processes

The representations (8.6-1) and (8.6-2) are actually more general than implied above; they can also be applied to any band-limited random process. For the remainder of this section we concern ourselves only with (8.6-2).

Let $N(t)$ be any band-limited wide-sense stationary real random process with a mean value of zero and a power density spectrum that satisfies

$$\mathcal{S}_{NN}(\omega) \neq 0 \qquad 0 < \omega_0 - W_1 < |\omega| < \omega_0 - W_1 + W$$

$$\mathcal{S}_{NN}(\omega) = 0 \qquad \text{elsewhere} \tag{8.6-7}$$

where W_1 and W are real positive constants. Then $N(t)$ can be represented by the right side of (8.6-2),† where the random processes $X(t)$ and $Y(t)$ have the following properties:

(1) $X(t)$ and $Y(t)$ are jointly wide-sense stationary   (8.6-8)

(2) $E[X(t)] = 0 \qquad E[Y(t)] = 0$ (8.6-9)

(3) $E[X^2(t)] = E[Y^2(t)] = E[N^2(t)]$ (8.6-10)

(4) $R_{XX}(\tau) = \dfrac{1}{\pi} \displaystyle\int_0^\infty \mathcal{S}_{NN}(\omega) \cos[(\omega - \omega_0)\tau]\, d\omega$ (8.6-11)

(5) $R_{YY}(\tau) = R_{XX}(\tau)$ (8.6-12)

(6) $R_{XY}(\tau) = \dfrac{1}{\pi} \displaystyle\int_0^\infty \mathcal{S}_{NN}(\omega) \sin[(\omega - \omega_0)\tau]\, d\omega$ (8.6-13)

(7) $R_{YX}(\tau) = -R_{XY}(\tau) \qquad R_{XY}(\tau) = -R_{XY}(-\tau)$ (8.6-14)

(8) $R_{XY}(0) = E[X(t)Y(t)] = 0 \qquad R_{YX}(0) = 0$ (8.6-15)

(9) $\mathcal{S}_{XX}(\omega) = L_p[\mathcal{S}_{NN}(\omega - \omega_0) + \mathcal{S}_{NN}(\omega + \omega_0)]$ (8.6-16)

(10) $\mathcal{S}_{YY}(\omega) = \mathcal{S}_{XX}(\omega)$ (8.6-17)

(11) $\mathcal{S}_{XY}(\omega) = jL_p[\mathcal{S}_{NN}(\omega - \omega_0) - \mathcal{S}_{NN}(\omega + \omega_0)]$ (8.6-18)

(12) $\mathcal{S}_{YX}(\omega) = -\mathcal{S}_{XY}(\omega)$ (8.6-19)

In the preceding 12 results, ω_0 is any convenient frequency within the band of $\mathcal{S}_{NN}(\omega)$; $R_{XX}(\tau)$, $R_{YY}(\tau)$, $R_{XY}(\tau)$, and $R_{YX}(\tau)$ are autocorrelation and cross-correlation functions of $X(t)$ and $Y(t)$ while $\mathcal{S}_{XX}(\omega)$, $\mathcal{S}_{YY}(\omega)$, $\mathcal{S}_{XY}(\omega)$, and $\mathcal{S}_{YX}(\omega)$ are the corresponding power spectrums; and $L_p[\cdot]$ denotes preserving only the lowpass part of the quantity within the brackets.

We outline the proofs of the above properties in the next subsection. Here we discuss their meaning and develop an example. We see that in addition to being zero-mean (property 2) wide-sense stationary (property 1) processes, $X(t)$ and $Y(t)$ also have equal powers (property 3), the same autocorrelation function (property 5), and therefore the same power spectrum (property 10).

---

†If we denote the right side of (8.6-2) by $\hat{N}(t)$, the equality in (8.6-2) must be interpreted in the sense of zero mean-squared error; that is, $N(t)$ equals $\hat{N}(t)$ in the sense that

$$E[\{N(t) - \hat{N}(t)\}^2] = 0$$

(Ziemer and Tranter, 1976, p. 241).

Random variables defined for the processes $X(t)$ and $Y(t)$ at any one time are orthogonal (property 8). If $N(t)$ has a power spectrum with components having even symmetry about $\omega = \pm\omega_0$, then $X(t)$ and $Y(t)$ will be orthogonal processes (property 6). A consequence of this last point is that the cross-power spectrums of $X(t)$ and $Y(t)$ are zero (properties 11 and 12).

EXAMPLE 8.6-1. Consider the bandpass process having the power density spectrum shown in Figure 8.6-3a. We shall find $\mathcal{S}_{XX}(\omega)$, $\mathcal{S}_{XY}(\omega)$, and $R_{XY}(\tau)$. By shifting $\mathcal{S}_{NN}(\omega)$ by $+\omega_0$ and $-\omega_0$, as shown in (b), we may construct $\mathcal{S}_{XX}(\omega)$ according to (8.6-16) as the lowpass portion of $\mathcal{S}_{NN}(\omega - \omega_0) + \mathcal{S}_{NN}(\omega + \omega_0)$, as illustrated in ($c$). This function also equals $\mathcal{S}_{YY}(\omega)$ by (8.6-17). Similarly, we form the difference of the spectrums in (b) to obtain $\mathcal{S}_{XY}(\omega)$ according to (8.6-18) as shown in (d). This function also gives $\mathcal{S}_{YX}(\omega)$ from (8.6-19) as shown.

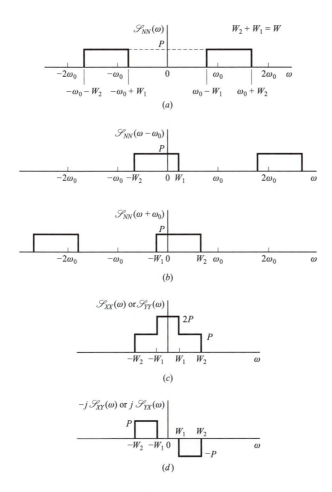

FIGURE 8.6-3
Power spectrums applicable to Example 8.6-1.

To find $R_{XY}(\tau)$ we apply (8.6-13):

$$R_{XY}(\tau) = \frac{1}{\pi} \int_{\omega_0 - W_1}^{\omega_0 + W_2} P \sin[(\omega - \omega_0)\tau] \, d\omega = \frac{P}{\pi\tau} \int_{-W_1\tau}^{W_2\tau} \sin(x) \, dx$$

$$= \frac{P}{\pi\tau}[\cos(W_1\tau) - \cos(W_2\tau)]$$

$$= \frac{P}{\pi\tau}\left\{ \cos\left[\frac{(W_2 + W_1)\tau}{2} - \frac{(W_2 - W_1)\tau}{2}\right] \right.$$

$$\left. - \cos\left[\frac{(W_2 + W_1)\tau}{2} + \frac{(W_2 - W_1)\tau}{2}\right] \right\}$$

$$= \frac{2P}{\pi\tau} \sin\left[\frac{(W_2 + W_1)\tau}{2}\right] \sin\left[\frac{(W_2 - W_1)\tau}{2}\right]$$

Now since $W_1 + W_2 = W$, we may write this result as

$$R_{XY}(\tau) = \frac{WP}{\pi} \frac{\sin(W\tau/2)}{(W\tau/2)} \sin[(W - 2W_1)\tau/2]$$

which is an odd function of τ as (8.6-14) indicates it should be. Figure 8.6-4 illustrates a plot of $R_{XY}(\tau)$ for the special case $W_1 = W/6$.

It should be noted that if $W_1 = W/2$, corresponding to $\mathscr{S}_{NN}(\omega)$ having even components about $\omega = \pm\omega_0$, we get $R_{XY}(\tau) = 0$ for all τ. In this case, $X(t)$ and $Y(t)$ are orthogonal processes; they are also independent if $N(t)$ is gaussian.

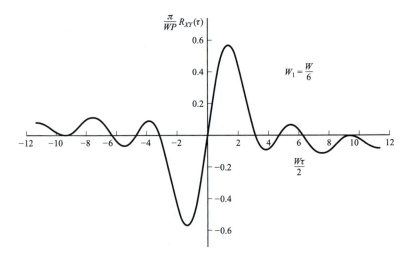

FIGURE 8.6-4
Cross-correlation function of Example 8.6-1.

*Proof of Properties of Band-Limited Processes

293

CHAPTER 8:
Linear Systems
with Random
Inputs

It is a quite long and involved task to prove all 12 properties of band-limited processes in detail. Therefore, we shall outline most of the proofs and give the details on only a few.

Property 2 is proved by taking the expected value on both sides of (8.6-2). Since $N(t)$ is assumed wide-sense stationary with a mean value of zero, then $E[X(t)] = 0$ and $E[Y(t)] = 0$ are necessary and property 2 follows.

The sequence of developments leading to the proofs of properties 9 and 4 will now be given. We begin by assuming the usual case $W_1 = W/2$ (see Figure 8.6-1b) and observing that the network of Figure 8.6-5a gives $X(t)$ at its output if the ideal lowpass filter has a bandwidth $W/2$ and if $\omega_0 > W/2$.† We shall assume these conditions true. Thus

$$
\begin{aligned}
V_1(t) &= 2N(t)\cos(\omega_0 t) \\
&= 2[X(t)\cos^2(\omega_0 t) - Y(t)\sin(\omega_0 t)\cos(\omega_0 t)] \\
&= X(t) + [X(t)\cos(2\omega_0 t) - Y(t)\sin(2\omega_0 t)] \qquad (8.6\text{-}20)
\end{aligned}
$$

The filter will remove the bandpass process contained within the brackets so that only $X(t)$ appears in the output. Next, we develop an expression for $R_{XX}(t, t + \tau)$:

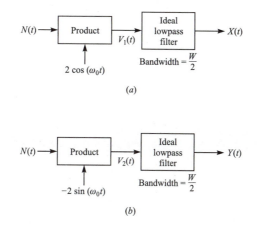

(a)

(b)

FIGURE 8.6-5

Block diagrams of networks that realize (a) $X(t)$ and (b) $Y(t)$ from a random process $N(t) = X(t)\cos(\omega_0 t) - Y(t)\sin(\omega_0 t)$. [*Reproduced from Peebles (1976) with permission of publishers Addison-Wesley, Advanced Book Program.*]

---

†These are idealized values based on an ideal product device. Practical values of bandwidth and ω_0 may be considerably different. The assumption $W_1 = W/2$ is for simple definition of filter bandwidth and is not a constraint in the proofs of properties 9 or 4.

$$R_{XX}(t, t + \tau) = E[X(t)X(t + \tau)]$$

$$= E\left[\int_{-\infty}^{\infty} h(u)V_1(t - u) \, du \int_{-\infty}^{\infty} h(v)V_1(t + \tau - v) \, dv\right]$$

$$= \int_{-\infty}^{\infty}\int_{-\infty}^{\infty} h(u)h(v)R_{NN}(\tau + u - v)4\cos[\omega_0(t - u)]$$

$$\cdot \cos[\omega_0(t + \tau - v)] \, du \, dv \qquad (8.6\text{-}21)$$

In developing (8.6-21), we have written $X(t)$ and $X(t + \tau)$ in terms of the convolution integral involving $h(t)$, the impulse reponse of the lowpass filter, substituted $V_1(t)$ from (8.6-20), and used the fact that $N(t)$ is assumed wide-sense stationary. The further reduction of (8.6-21) is lengthy (Peebles, 1976, p. 157) and will only be outlined. If the cosine factors are replaced by their exponential forms and if $R_{NN}(\tau + u - v)$ is replaced by its equivalent, the inverse transform of the power spectrum $\mathcal{S}_{NN}(\omega)$, (8.6-21) becomes the sum of four integrals. It can be shown that two of these integrals, the only two involving t, are zero. Thus, $R_{XX}(t, t + \tau)$ becomes a function of τ only and $X(t)$ is therefore wide-sense stationary, proving part of property 1. The two remaining integrals are used to prove properties 9 and 4.

A procedure exactly the same as discussed in the last paragraph can be used to prove first that $Y(t)$ is wide-sense stationary, thereby providing the proof of another part of property 1. The development also proves properties 10 and 5; it is based on the fact that $Y(t)$ is produced by the operations shown in Figure 8.6-5b.

Property 3 next results from use of property 5 with $\tau = 0$ and the integration of $\mathcal{S}_{XX}(\omega)$ using property 9.

Properties 11, 6, 8, and the balance of property 1 are proved by considering the cross-correlation function

$$R_{XY}(t, t + \tau) = E[X(t)Y(t + \tau)]$$

$$= E\left[\int_{-\infty}^{\infty} h(u)V_1(t - u) \, du \int_{-\infty}^{\infty} h(v)V_2(t + \tau - v) \, dv\right]$$

$$= -\int_{-\infty}^{\infty}\int_{-\infty}^{\infty} h(u)h(v)R_{NN}(\tau + u - v)4\cos[\omega_0(t - u)]$$

$$\cdot \sin[\omega_0(t + \tau - v)] \, dv \, du \qquad (8.6\text{-}22)$$

which is developed in a manner analogous to (8.6-21). Reduction of (8.6-22) as discussed earlier shows that $R_{XY}(t, t + \tau)$ depends only on τ, so that $X(t)$ and $Y(t)$ are jointly wide-sense stationary (proving property 1); it also proves properties 11 and 6. Property 8 results from property 6 with $\tau = 0$.

Proofs of the remaining properties, 7 and 12, follow from consideration of the autocorrelation function of $N(t)$. It is readily found by using (8.6-2) that

$$R_{NN}(t, t + \tau) = E[N(t)N(t + \tau)]$$

$$= [R_{XX}(\tau) + R_{YY}(\tau)]\tfrac{1}{2}\cos(\omega_0\tau)$$

$$+ [R_{XX}(\tau) - R_{YY}(\tau)]\tfrac{1}{2}\cos(2\omega_0 t + \omega_0\tau)$$

$$- [R_{XY}(\tau) - R_{YX}(\tau)]\tfrac{1}{2}\sin(\omega_0\tau)$$

$$- [R_{XY}(\tau) + R_{YX}(\tau)]\tfrac{1}{2}\sin(2\omega_0 t + \omega_0\tau) \qquad (8.6\text{-}23)$$

Since $N(t)$ is wide-sense stationary by original assumption, its autocorrelation function cannot be a function of t. Thus, we require

$$R_{XX}(\tau) = R_{YY}(\tau) \qquad (8.6\text{-}24)$$

and

$$R_{XY}(\tau) = -R_{YX}(\tau) \qquad (8.6\text{-}25)$$

in (8.6-23); these results prove property 12 and the first part of property 7. Finally, recognizing that $R_{XY}(\tau) = R_{YX}(-\tau)$ for a cross-correlation function, we obtain the second part of property 7, which says that $R_{XY}(\tau)$ is an odd function of τ.

8.7
SAMPLING OF PROCESSES

The concept of a band-limited function is exceptionally important. It forms a basis on which digital systems become possible. The reason is that digital systems (computers) cannot work with a complete waveform. They must rely on samples of waveforms taken (usually) at periodic points in time. If somehow these samples are to be useful, they must be able to represent (reconstruct) a waveform without error for *all* time. The existence of a band-limited waveform allows these results to be true (as we prove below).

The reader may be aware that there are *no* truly band-limited waveforms available to us in the real world for two reasons. First, we never have an infinite amount of time over which to take and process samples, as needed for a truly band-limited signal (an example is any periodic waveform with a finite number of frequencies in its Fourier series). Second, real waveforms are always time-limited and cannot be truly band-limited.

So, how can we even consider band-limited signals? The answer makes use of the fact that the spectrum of any practical signal decreases as frequency gets farther away from center band. For a baseband function, this means the spectral terms approach zero in amplitude as $|\omega| \to \infty$. For a bandpass function with principal spectral terms clustered near frequencies $\pm\omega_0$, it means amplitudes become small at frequencies far removed from $-\omega_0$ or $+\omega_0$. These practical observations allow us to say that there is usually some bandwidth of important frequencies in any real signal such that outside this bandwidth the spectral terms can be considered negligible and approximated as zero. We then say the practical waveform is band-limited to this bandwidth and accept the negligible error that can result.

On having justified that a band-limited assumption is always possible for real waveforms, it remains to show that the assumption always allows a signal to be completely specified at all times by use of only its samples. The proof resides in the famous sampling theorem.

Sampling theory for deterministic waveforms is well developed and understood. A review of the various methods, with historical references, is given by Peebles (1987). Here we are interested more in sampling *random* waveforms. Our principal interests are in the most-needed theorems, those applying to baseband and bandpass random waveforms. Of course, sampling of a given random signal is modeled as sampling of the corresponding random process from which the signal is taken. However, to best understand sampling of random processes, it is first necessary to consider nonrandom signals, as we develop below.

Our work is especially appropriate to the modern world because sampling produces a discrete-time (DT) signal which is modeled as a sample function of a DT random process. Because DT signals are used in modern digital signal processing (DSP), it is important to define ways of modeling such signals. We consider the baseband case first.

Baseband Sampling Theorem

THEOREM. Let $g(t)$ be a real or complex baseband nonrandom waveform having a Fourier transform $G(\omega)$ that is band-limited such that it is nonzero only over a band $-W_g \leq \omega \leq W_g$, where $W_g > 0$ is a constant called the *spectral extent* of $g(t)$; it is the highest spectral frequency in $g(t)$ having non-zero amplitude, as illustrated in Figure 8.7-1. The signal $g(t)$ can be completely specified (recovered) without error from its periodic samples taken at times nT_s, $n = 0, \pm1, \pm2, \ldots$, provided the sample rate ω_s satisfies

$$\omega_s = 2\pi/T_s > 2W_g \qquad \text{(rad/s)} \qquad (8.7\text{-}1)$$

where T_s is the sampling interval and the lower bound $2W_g$ is called the *Nyquist rate*.

To demonstrate the theorem's proof we enlist the help of Figure 8.7-2*a*, which illustrates *natural sampling*. For convenience we assume $p(t)$ has rectangular pulses of duration T_p and amplitude $1/T_p$. More general natural sampling can use pulses of arbitrary shape and amplitude [Peebles (1987)]; however, the general developments are more difficult to visualize, while the final results differ from our case by only a constant of proportionality. The

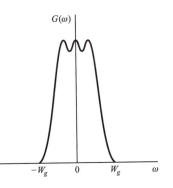

FIGURE 8.7-1
Sketch of $G(\omega)$, the band-limited spectrum of a nonrandom waveform $g(t)$.

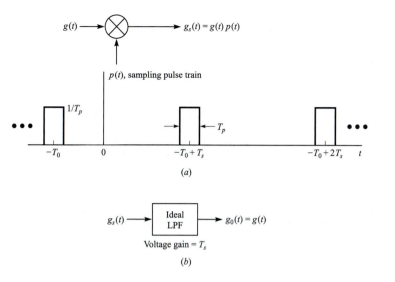

FIGURE 8.7-2
(a) Natural sampling of a waveform $g(t)$ to generate its sampled version $g_s(t)$. (b) Recovery of $g(t)$ by filtering.

pulse amplitude choice of $1/T_p$ allows natural sampling to become *ideal sampling* if $T_p \to 0$, that is, pulses in $p(t)$ become impulses when $T_p \to 0$.

The constant T_0 in the pulse train $p(t)$ is present to allow arbitrary timing of the train. It is reasonably obvious that

$$p(t) = \frac{1}{T_p} \sum_{n=-\infty}^{\infty} \text{rect}\left(\frac{t + T_0 - nT_s}{T_p}\right) \tag{8.7-2}$$

which has the Fourier transform (Problem 8.7-1)

$$P(\omega) = \frac{2\pi}{T_s} e^{j\omega T_0} \text{Sa}(\omega T_p/2) \sum_{n=-\infty}^{\infty} \delta(\omega - n\omega_s) \tag{8.7-3}$$

Since the sampled version of $g(t)$ is $g_s(t) = g(t)p(t)$, its transform, denoted by $G_s(\omega)$, is the convolution of $P(\omega)$ and $G(\omega)$, the transform of $g(t)$ [see (D-17)]. We have (Problem 8.7-2)

$$G_s(\omega) = \frac{1}{2\pi} \int_{-\infty}^{\infty} G(\xi)P(\omega - \xi)\, d\xi$$

$$= \frac{1}{T_s} \sum_{n=-\infty}^{\infty} \text{Sa}(n\omega_s T_p/2)\, e^{jn\omega_s T_0} G(\omega - n\omega_s) \tag{8.7-4}$$

A sketch of $G_s(\omega)$, neglecting the phase term involving T_0, is shown in Figure 8.7-3. Note the replicas of $G(\omega)$ displaced to frequencies $\pm n\omega_s$, $n = 1, 2, \ldots$. There is a spectral taper due to the sampling pulse's shape that scales the replicas by a factor $(1/T_s)\text{Sa}(n\omega_s T_p/2)$. If T_p is small, that is, if $T_p \to 0$, then $\text{Sa}(n\omega_s T_p/2) \to 1$ for all values of n. For $T_p \to 0$ we have the sampled signal's spectrum for ideal sampling

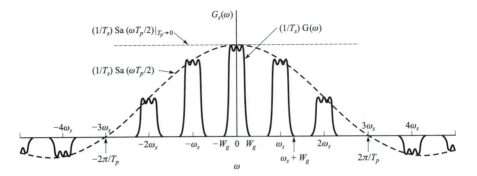

FIGURE 8.7-3
Sketch of the spectrum $G_s(\omega)$ of the sampled signal $g_s(t)$ showing spectral replicas
of $G(\omega)$ displaced to multiples of ω_s about $\omega = 0$.

$$G_s(\omega)|_{T_p \to 0} = \frac{1}{T_s} \sum_{n=-\infty}^{\infty} G(\omega - n\omega_s)e^{jn\omega_s T_0} \qquad (8.7-5)$$

Because later work is little affected by letting $T_p \to 0$, we use (8.7-5) in the
following developments.

On inverse transformation of (8.7-5) an especially useful form is achieved
for $g_s(t)$ (see Problem 8.7-3)

$$g_s(t) = g(t) \sum_{n=-\infty}^{\infty} \delta(t - nT_s + T_0)$$

$$= \sum_{n=-\infty}^{\infty} g(nT_s - T_0)\delta(t - nT_s + T_0) \qquad (8.7-6)$$

This representation is used to show how $g(t)$ can be represented at *all times*
through its periodic samples.

Imagine $g_s(t)$ is applied to an ideal lowpass filter with a transfer function
$H(\omega)$ and impulse response $h(t)$ defined by

$$h(t) = \text{Sa}(\omega_s t/2) \leftrightarrow H(\omega) = T_s \, \text{rect}(\omega/\omega_s) \qquad (8.7-7)$$

This operation is shown in Figure 8.7-2b. The response $g_0(t)$ is

$$g_0(t) = \sum_{n=-\infty}^{\infty} g(nT_s - T_0)\text{Sa}[\omega_s(t - nT_s + T_0)/2] = g(t) \qquad (8.7-8)$$

since each impulse in (8.7-6) evokes an impulse response in the output. The
last equality in (8.7-8) must be true because the filter passes only the central
term of (8.7-5) for $n = 0$. Observe that for the filter to pass an undistorted
spectrum $G(\omega)$ for $n = 0$, it is necessary that $\omega_s > 2W_g$ be true so that spectral
replicas do not overlap, a condition called *aliasing*. Also note that the gain of
the filter was chosen to give $g(t)$ as its response rather than a response proportional to $g(t)$.

If we let $T_0 = 0$ in (8.7-8), then

$$g(t) = \sum_{n=-\infty}^{\infty} g(nT_s)\, \text{Sa}[\omega_s(t - nT_s)/2] \qquad (8.7\text{-}9)$$

which is the statement typically seen for the baseband sampling theorem. It states that $g(t)$ is represented (known) for all time without error by an infinite sum of terms. Each term has an amplitude equal to a sample value and a form given by the function $\text{Sa}(\omega_s t/2)$ displaced to the time of the sample. $\text{Sa}(\cdot)$ is called a *sampling function*. At any given sample time all these functions pass through nulls except the one associated with the sample at the sample time. Thus, at any sample time $g(t)$ equals its sample value. At all times between samples the sum of the sampling functions provides error-free interpolation to give $g(t)$ exactly. Because of this action the sampling function is also known as the *interpolating function*. Inherent in the validity of (8.7-9) is that $g(t)$ be band-limited to W_g and sampling is at a rate $\omega_s > 2W_g$ to avoid aliasing.

> **EXAMPLE 8.7-1.** As an example related to sampling functions, we show their orthogonality. Since
>
> $$\text{Sa}[\omega_s(t - mT_s)/2] \leftrightarrow T_s e^{-jm\omega T_s}\, \text{rect}(\omega/\omega_s) \qquad (1)$$
>
> where $\omega_s = 2\pi/T_s$, we use Parseval's theorem
>
> $$\int_{-\infty}^{\infty} x(t)y^*(t)\, dt = \frac{1}{2\pi}\int_{-\infty}^{\infty} X(\omega)Y^*(\omega)\, d\omega \qquad (2)$$
>
> where $x(t) \leftrightarrow X(\omega)$ and $y(t) \leftrightarrow Y(\omega)$. On substituting (1) into (2):
>
> $$\int_{-\infty}^{\infty} \text{Sa}[\omega_s(t - mT_s)/2]\text{Sa}[\omega_s(t - kT_s)/2]\, dt$$
>
> $$= \frac{1}{2\pi}\int_{-\infty}^{\infty} T_s\, \text{rect}(\omega/\omega_s)e^{-jm\omega T_s} T_s\, \text{rect}(\omega/\omega_s)e^{jk\omega T_s}\, d\omega$$
>
> $$= \frac{T_s^2}{2\pi}\int_{-\omega_s/2}^{\omega_s/2} e^{j(k-m)\omega T_s}\, d\omega = T_s \text{Sa}[(k - m)\pi]$$
>
> $$= \begin{cases} 0 & k \neq m \\ T_s & k = m \end{cases} \qquad (3)$$
>
> When any two sampling functions satisfy (3), they are called orthogonal.

Baseband Sampling Theorem for Random Processes

The above sampling theorem is easily applied to the sampling of band-limited, baseband, wide-sense stationary (WSS) random processes. First, we state the theorem and then discuss and show its validity.

THEOREM. A wide-sense stationary baseband random process $X(t)$ with autocorrelation function $R_{XX}(\tau)$ and power spectrum $\mathscr{S}_{XX}(\omega)$ that is band-limited such that $\mathscr{S}_{XX}(\omega) = 0$ for all $|\omega| > W_x$ can be represented by

$$\hat{X}(t) = \sum_{n=-\infty}^{\infty} X(nT_s)\mathrm{Sa}[\omega_s(t - nT_s)/2] \qquad (8.7\text{-}10)$$

where $\hat{X}(t)$ converges to $X(t)$ in the sense of zero mean-squared error $\overline{\epsilon^2}$. That is,

$$\overline{\epsilon^2} = E\left\{ [\hat{X}(t) - X(t)]^2 \right\} = 0 \qquad (8.7\text{-}11)$$

Proof of (8.7-10) begins by noting that $R_{XX}(\tau)$ is analogous to $g(t)$ in the development of the baseband sampling theorem where (8.7-8) holds. Thus, for the present problem

$$R_{XX}(\tau) = \sum_{n=-\infty}^{\infty} R_{XX}(nT_s - T_0)\,\mathrm{Sa}[\omega_s(\tau - nT_s + T_0)/2] \qquad (8.7\text{-}12)$$

must be true. Since (8.7-12) is the sampling theorem representation of $R_{XX}(\tau)$ over all time τ, the result must also hold for a delay, or time shift, of T_0. After the shift we have

$$R_{XX}(\tau - T_0) = \sum_{n=-\infty}^{\infty} R_{XX}(nT_s - T_0)\,\mathrm{Sa}[\omega_s(\tau - nT_s)/2] \qquad (8.7\text{-}13)$$

These last two results are needed in the proof of (8.7-11).
On expanding (8.7-11)

$$\overline{\epsilon^2} = E\left\{ [X(t)]^2 - 2X(t)\hat{X}(t) + [\hat{X}(t)]^2 \right\}$$
$$= R_{XX}(0) - 2E[X(t)\hat{X}(t)] + E[\hat{X}^2(t)] \qquad (8.7\text{-}14)$$

We now expand the middle right-side term as follows:

$$E[X(t)\hat{X}(t)] = E\left\{ X(t) \sum_{n=-\infty}^{\infty} X(nT_s)\,\mathrm{Sa}[\omega_s(t - nT_s)/2] \right\}$$
$$= \sum_{n=-\infty}^{\infty} R_{XX}(nT_s - t)\mathrm{Sa}[\omega_s(t - nT_s)/2] \qquad (8.7\text{-}15)$$

where (8.7-10) has been used. Since T_0 can have any value in (8.7-12), we set $T_0 = t$ and $\tau = 0$ and find that (8.7-15) equals $R_{XX}(0)$. As a consequence, (8.7-14) reduces to

$$\overline{\epsilon^2} = -R_{XX}(0) + E[\hat{X}^2(t)] \qquad (8.7\text{-}16)$$

Again using (8.7-10) we expand the last term in (8.7-16)

$$E[\hat{X}^2(t)] = E\left\{ \sum_{n=-\infty}^{\infty} X(nT_s)\text{Sa}[\omega_s(t - nT_s)/2] \right.$$

$$\left. \cdot \sum_{m=-\infty}^{\infty} X(mT_s)\text{Sa}[\omega_s(t - mT_s)/2] \right\}$$

$$= \sum_{n=-\infty}^{\infty} \left\{ \sum_{m=-\infty}^{\infty} R_{XX}(mT_s - nT_s)\text{Sa}[\omega_s(t - mT_s)/2] \right\}$$

$$\cdot \text{Sa}[\omega_s(t - nT_s)/2] \qquad (8.7\text{-}17)$$

From (8.7-13) with n, τ, and T_0 replaced, respectively, by m, t, and nT_s, the inner sum in (8.7-17) evaluates to $R_{XX}(nT_s - t)$, so

$$E[\hat{X}^2(t)] = \sum_{n=-\infty}^{\infty} R_{XX}(nT_s - t)\text{Sa}[\omega_s(t - nT_s)/2] = R_{XX}(0) \qquad (8.7\text{-}18)$$

The last form derives from (8.7-12) with T_0 and τ equal, respectively, to t and zero. Finally, when (8.7-18) is used in (8.7-16), we have $\epsilon^2 = 0$ and (8.7-11) is proved, which means (8.7-10) is a valid representation of $X(t)$ and equal to it in the sense of zero mean-squared error in their difference.

Bandpass Sampling of Random Processes

Consider a bandpass random process $X(t)$ that is at least wide-sense stationary and has a power density spectrum $\mathscr{S}_{XX}(\omega)$ that is band-limited and centered about some "carrier" frequency $\omega_0 + \omega_d$,† as shown in Figure 8.7-4. The spectral width of the power spectrum is W_X. Since the highest nonzero spectral term is at frequency $\omega_0 + \omega_d + (W_X/2)$, if we tried to represent $X(t)$ by the *baseband* sampling theorem, one would expect to sample at over twice this

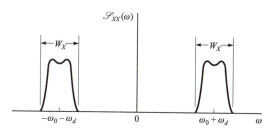

FIGURE 8.7-4
Power density spectrum of a band-limited bandpass random process.

†The quantity ω_d represents an offset, or slight difference, between the nominal frequency of $X(t)$ and the frequency of the I and Q processor to be developed below.

highest frequency. Such a rate can be excessive if $\omega_0 + \omega_d$ is large and is, in fact, not necessary. There are "bandpass" sampling theorems where sampling can occur at a rate near $2W_X$ if the baseband sampling function in (8.7-10) is replaced by a suitable bandpass sampling function [see Peebles (1987) for a summary and discussion, and Kohlenberg (1953), for the original theory]. However, such direct bandpass sampling is not usually done in practice. Instead, a method called *I and Q sampling* is used to achieve certain practical advantages.

I and Q sampling is illustrated in Figure 8.7-5a. Let $X(t)$ be represented in the form of (8.6-1)

$$X(t) = A(t)\cos[(\omega_0 + \omega_d)t + \Theta(t)] \tag{8.7-19}$$

where $A(t)$ and $\Theta(t)$ are baseband processes representing the amplitude and phase of $X(t)$. Because the lowpass filters in Figure 8.7-5a remove terms at frequency $(2\omega_0 + \omega_d)$, simple analysis shows that

$$X_I(t) = A(t)\cos[\omega_d t + \Theta(t)] \tag{8.7-20}$$
$$X_Q(t) = A(t)\sin[\omega_d t + \Theta(t)] \tag{8.7-21}$$

For ω_d small each of these responses is baseband and it is easy to show they have spectral extent $\omega_d + (W_X/2)$. Each can be sampled at a rate of at least

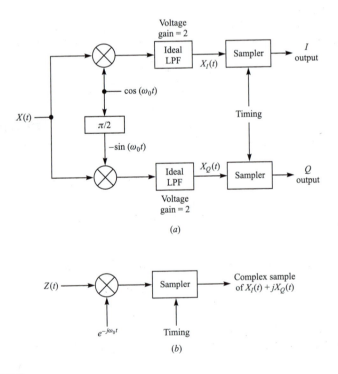

FIGURE 8.7-5
(a) Functions performed in I and Q sampling of bandpass process $X(t)$. (b) An equivalent interpretation of sampling by use of the complex representation $Z(t)$ for $X(t)$.

$2\omega_d + W_X$ and completely recovered, as guaranteed by the baseband sampling theorem.

If recovery of $X(t)$ is desired, first $X_I(t)$ and $X_Q(t)$ are recovered using a baseband recovery method† and then $X(t)$ is recovered by the method of Figure 8.7-6 (see Problem 8.7-4).

An interesting interpretation of I and Q sampling is derived by thinking of samples of $X_I(t)$ and $X_Q(t)$ as components of a complex number treated as a sample of $X_I(t) + jX_Q(t)$. To visualize this idea, suppose $X(t)$ of (8.7-19) is written in its complex representation as

$$Z(t) = A(t)e^{j(\omega_0 + \omega_d)t + j\Theta(t)} \tag{8.7-22}$$

The product of Figure 8.7-5b gives the complex response

$$\begin{aligned} Z(t)e^{-j\omega_0 t} &= A(t)e^{j\omega_d t + j\Theta(t)} \\ &= A(t)\cos[\omega_d t + \Theta(t)] + jA(t)\sin[\omega_d t + \Theta(t)] \\ &= X_I(t) + jX_Q(t) \end{aligned} \tag{8.7-23}$$

A sample of this response is a complex sample of $X_I(t) + jX_Q(t)$. Thus, if samples in the I and Q processor are used to form a complex sample, this sample is the same as the sample derived from the equivalent complex processor of Figure 8.7-5b.

Discrete-Time Power Spectrums

The power spectrums of DT processes and DT sequences were stated without justification in Section 7.5 because of the need to develop the above topics in sampling theory. We may now show some details on the development of these power spectrums.

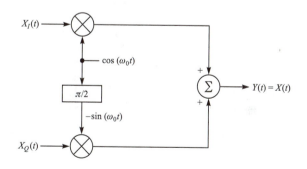

FIGURE 8.7-6
Recovery of a bandpass random process $X(t)$ after recovery of $X_I(t)$ and $X_Q(t)$ from their I and Q samples.

---

†We have described only the filter method of recovery given in Figure 8.7-2b. Other methods are described by Peebles (1987).

We have shown above that (8.7-10) is a valid representation of a band-limited, wide-sense stationary baseband random process in terms of its periodic samples. The proof evolved around the use of Figure 8.7-2 as a representation of the sampling process with $g(t)$ representing the autocorrelation function $R_{XX}(\tau)$ of the process prior to sampling. The output of the product device, which is $R_{XX}(\tau)p(\tau)$, must represent the sampled form of the process' autocorrelation function, which we denote by $R_{X_sX_s}(\tau)$. Thus, if sample pulse duration T_p is small

$$R_{X_sX_s}(\tau) = R_{XX}(\tau)p(\tau) = R_{XX}(\tau)\frac{1}{T_p}\sum_{n=-\infty}^{\infty}\text{rect}\left(\frac{\tau - nT_s}{T_p}\right)$$

$$\approx \sum_{n=-\infty}^{\infty}R_{XX}(nT_s)\frac{1}{T_p}\text{rect}\left(\frac{\tau - nT_s}{T_p}\right) \tag{8.7-24}$$

where we have let $T_0 = 0$ in using the pulse train of (8.7-2).

The power spectrum of the DT process is defined as the Fourier transform of its autocorrelation function of (8.7-24) (Viniotis, 1998, p. 479; Leon-Garcia, 1989, p. 386). Denote the power spectrum by $\mathscr{S}_{X_sX_s}(\omega)$. Then

$$\mathscr{S}_{X_sX_s}(\omega) = \sum_{n=-\infty}^{\infty}R_{XX}(nT_s)\,\text{Sa}(\omega T_p/2)\,e^{-jn\omega T_s} \tag{8.7-25}$$

For almost instantaneous samples, as is approached in practice, we can assume $T_p \to 0$ so that $\text{Sa}(\omega T_p/2) \to 1$ for all ω. Finally,

$$\mathscr{S}_{X_sX_s}(\omega) = \sum_{n=-\infty}^{\infty}R_{XX}(nT_s)\,e^{-jn\omega T_s} \tag{8.7-26}$$

Equation (8.7-26) is the same as (7.5-2) given earlier with only minimal development.

The logic showing how the power spectrum of a DT sequence is derived from (8.7-26) was previously given in Section 7.5, with the result being the discrete-time Fourier transform (DTFT) of (7.5-11). Also given was the inverse DTFT (or IDTFT) of (7.5-12) that allows recovery of $R_{XX}[n]$ from the power spectrum of the DT sequence.

EXAMPLE 8.7-2. Consider a DT random process for which the autocorrelation sequence

$$R_{XX}(nT_s) = (\alpha/2)e^{-\alpha|n|T_s} \tag{1}$$

applies with $\alpha > 0$ a constant. For the continuous random process that produced (1), α is the 3-dB bandwidth of its power spectrum in rad/s.

We use MATLAB to approximate the power spectrum of (8.7-26) by using $2N + 1$ samples at times nT_s, $-N \le n \le N$. We also assume $T_s = \pi/(5\alpha)$, or $\alpha T_s = \pi/5$, which allows our range of discrete frequencies to extend to five times the bandwidth. The MATLAB code is shown in Figure 8.7-7. Figure 8.7-8 plots $(2/\alpha)\mathscr{S}_{X_sX_s}(\omega)$, for $N = 10$, as linear line segments connecting calculated points at normalized frequencies $k = 2\omega_k/\alpha$, $-10 \le k \le 10$. When ω_k reaches the 3-dB bandwidth of the power spectrum $k = 2\alpha/\alpha = 2$.

%%%%%%%%%%%%%%% **Example 8.7-2** %%%%%%%%%%%%%%%%%%%%

305

CHAPTER 8:
Linear Systems
with Random
Inputs

```
clear

N = 10; % number of samples
alpha = 1;
Ts = 2*pi/(11*alpha); % sample period

k = -N:N;
M = 2*N+1;

Rxx = (alpha/2)*exp(-alpha*abs(k)*Ts); % auto-correlation
Sxx = abs(fftshift(fft(Rxx))); % estimated power spectrum

clf
plot(k,2/alpha*Sxx,'k')

xlabel('Normalized Frequency (rad), normalized to alpha')
ylabel('Magnitude')
title('Power Spectrum')
```

FIGURE 8.7-7
MATLAB code applicable to Example 8.7-2.

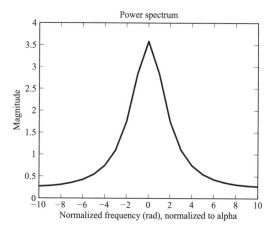

FIGURE 8.7-8
Normalized power spectrum applicable to Example 8.7-2 as found from a finite-term approximation of (8.7-26).

8.8
DISCRETE-TIME SYSTEMS

The most common and important problem solved in modern digital signal processing (DSP) systems is to convert an analog information signal to digital form, process the signal to achieve some modified signal, and then (perhaps) convert the modified signal back to analog form for further use. Clearly, three major operations are involved. We label these analog-to-digital (A/D) conversion, digital-to-digital (D/D) processing, and digital-to-analog (D/A) conversion, respectively.

For example, a radar's received signal (a pulse) might be converted to a digital form, processed in a computer to impart special filtering to the pulse to maximize performance, and then regenerate the filtered pulse waveform in analog form for use by other operations. With modern, fast computers, the advantages gained are smaller size, less weight and power consumption, greater flexibility (filters can be adaptable or changed easily), and absolute control over the processing, as compared to construction of a lumped-element filter.

In this section we give brief overviews of A/D and D/A conversions so that the overall DSP operations can be placed in focus. However, most of our efforts will be centered on D/D processing where the computer (called here a discrete-time system) processes a discrete-time sequence representing a discrete-time process formed from samples of a continuous-time process $X(t)$ that is assumed to be at least wide-sense stationary.

A/D Conversion

A/D conversion usually consists of three things. First is sampling of the analog signal. Second is a procedure called *quantization*, where the sample amplitudes are each rounded off to the closest one of a finite number of amplitudes. The third item is to encode each of the quantized samples into a suitable digital codeword compatible with the computer being used. This codeword is nearly always a binary word. In some definitions of digital systems encoding might be considered part of the D/D processor. Either choice is, of course, valid.

Since our interest is in random processes, let $X(t)$ represent some information source and assume the process to be at least wide-sense stationary. Sampling produces the DT random process $X(nT_s)$ for samples at times nT_s, $n = 0, \pm 1, \pm 2, \ldots$, where T_s is the sampling interval. As noted earlier (Section 6.1), the D/D processor has little need for the value of T_s. Its value becomes important mainly when D/A conversion occurs. Thus, the computer (DT system) views the stream of samples as a sequence function of the integer index n, denoted by $X[n]$. Since most DSP literature uses brackets rather than parentheses to imply a DT sequence, we also use this notation.

The digital processor (DT system) cannot work with the continuous-amplitude values of the sequence $X[n]$ and they must be rounded off, or quantized, to the nearest of a finite number, say, L, of values in a device called a *quantizer*. Clearly, some information is lost in quantization that can never be recovered. The loss is called a *quantization error*. This error can be minimized, or reduced to an acceptable level, in practical systems by choosing L large. The L levels must also span the entire variation of the sequence. In other words, if $|X[n]| < |X|_{max}$, then all L levels should at least span amplitudes from $-|X|_{max}$ to $+|X|_{max}$. If some sample amplitudes exceed the quantizer's amplitude range it is said to suffer *amplitude overload*. Overload can cause serious loss in performance over and above the effect of quantization error. There are many forms of quantizers, and most of these are discussed by Peebles (1987).

EXAMPLE 8.8-1. A *uniform quantizer* places its L levels uniformly separated by a constant amount Δ. If N_b is a positive interger such that $L = 2^{N_b}$ and the extreme quantizer levels are to be set at $\pm 6\,\text{V}$, we find Δ. Since L is an even number, half the levels are positive and half are negative. A simple sketch indicates levels at

$$L_i = -\left(\frac{L-1}{2}\right)\Delta + i\Delta, \quad i = 0, 1, 2, \ldots, (L-1)$$

so $L_{L-1} = (L-1)\Delta/2 = 6\,\text{V}$ and $\Delta = 12/(L-1)$. For a numerical value, assume $L = 128$, so $\Delta = 12/127 = 94.488\,\text{mV}$. In this case, $N_b = 7$.

EXAMPLE 8.8-2. We continue Example 8.8-1 by noting that the power in the error represents a performance degradation in the recovery of the original signal represented by the quantized version of the DT sequence $X[n]$. One common performance measure is the signal power-to-quantization noise power ratio, denoted by S_0/N_q. For a uniform quantizer operated at its limit before amplitude overload, and using L levels, it is given by (Peebles, 1998, p. 632)

$$\left(\frac{S_0}{N_q}\right) = 3L^2/K_{cr}^2$$

where K_{cr} is called a *crest factor*. K_{cr} is related to the form (shape) of the original signal. K_{cr} is typically larger than 1. For our case, with $K_{cr} = 4$ assumed, $(S_0/N_q) = 3(128^2)/16 = 3072.0$, or about $34.87\,\text{dB}$ (a fairly decent, but not an excellent, value).

Coding of samples is important to the actual digital processor (its hardware). However, the processor only performs operations of addition, subtraction, multiplication, division, storage, and delay, which are just mathematical operations. Furthermore, all these operations are done only on command from a suitable mathematical algorithm (programming). For purposes here it is not necessary to study these hardware principles. It is sufficient to describe the mathematical operations that the processor must carry out. Because of these facts, we may ignore coding and quantization (with L chosen large enough), and only describe how a digital system is structured and how it behaves with the DT sequence $X[n]$ at its input.

D/A Conversion

This operation is essentially the inverse of A/D conversion. First, the codewords from the computer are converted back to discrete amplitudes. Strictly, there is no inverse operation to quantization. For large enough L, the discrete levels are a good representation of samples of the desired response and recovery is generated using procedures dictated by the sampling theorem.

In many applications D/A conversion is not used at all. For example, an aircraft that measures its altitude may output results directly in digital form (perhaps to base 10), as needed in a display.

The Discrete-Time System

For present purposes, a discrete-time (DT) system will refer to the digital operations involved in accepting an "input" DT sequence $X[n]$ and generating an "output" sequence $Y[n]$ having some desired properties typically different from those of the input.

A DT system is called *finite-dimensional* if, for some positive integer N and nonnegative integer M, $Y[n]$ is some function only of the N most recent past output values, n, and the current and M most recent past values of the input [Kamen and Heck (1997), p. 42]. The integer N is called the *order* or *dimension* of the system. Any other system is *infinite-dimensional*. The function forms an input–output *difference equation*. The system is called *linear* if, and only if, the difference equation is linear, that is, if it can be written in the form

$$Y[n] = -\sum_{i=1}^{N} a_i(n) Y[n - i] + \sum_{i=0}^{M} b_i(n) X[n - i] \qquad (8.8\text{-}1)$$

where $a_i(n)$ and $b_i(n)$ are coefficients that may, in general, depend on n.

Analogous to a continuous-time, linear, time-invariant system, the linear system of (8.8-1) is time-invariant if, and only if, the coefficients are constants, that is, $a_i(n) = a_i$ and $b_i(n) = b_i$ are independent of n for all i. For a linear time-invariant (LTI) DT system†

$$Y[n] = -\sum_{i=1}^{N} a_i Y[n - i] + \sum_{i=0}^{M} b_i X[n - i] \qquad (8.8\text{-}2)$$

If the LTI DT system is initially at rest, (8.8-2) shows that there is no response prior to application of a nonzero input, a condition that defines the system as *causal* when $N \geq M$ [Taylor (1994), p. 442]. In statistical literature $Y[n]$ of (8.8-2) is called an *autoregressive moving average* (ARMA) process when $X[n]$ corresponds to white noise (see Section 8.9).

In the subsequent discussions we shall only discuss DT systems defined by (8.8-2). If we label the second sum in (8.8-2) as $Y_1[n]$, the structure of the system can be drawn as in Figure 8.8-1a, where D represents one unit of delay. This structure is called *direct form* I, since it is a direct realization of (8.8-2). It results that two DT systems in cascade are invariant to order, which means the two portions of Figure 8.8-1a can be reversed. The result allows some commonality of delay units as given in (b). This form is called the *canonic form* or

---

†LTI DT systems are sometimes called linear *shift-invariant* (LSI).

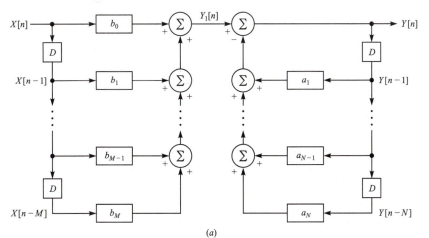

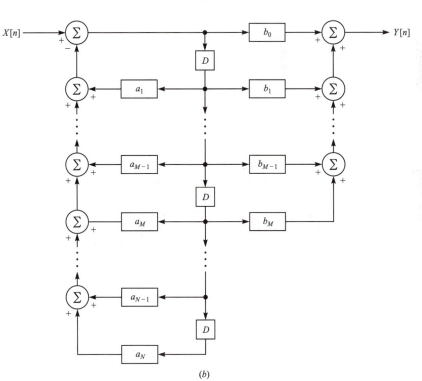

FIGURE 8.8-1

Block diagram of a linear, time-invariant, discrete-time system. (*a*) Direct form I, and (*b*) direct form II, also called the canonic form.

direct form II [Oppenheim and Schafer (1989), p. 296]; it minimizes the required number of delay units for given N and M.

The specification of an LTI DT system is complete if the coefficients in (8.8-2) are determined. The two most important analysis/synthesis methods involve sequence domain and transform domain techniques. These are analogous to time and frequency domain methods of CT systems, where impulse responses and transfer functions are important. DT systems also have impulse responses and transfer functions that we subsequently define.

Sequence Domain Methods for DT Systems

In DT systems we define a *unit-impulse* (also called a *unit-pulse sequence, unit sample*, or *Kronecker delta function*) by a single value of amplitude 1 at $n = 0$:

$$\delta[n] = \begin{cases} 1 & n = 0 \\ 0 & n \neq 0 \end{cases} \tag{8.8-3}$$

It behaves as an even function, that is, $\delta[-n] = \delta[n]$, and satisfies the property

$$\sum_{n=-\infty}^{\infty} \delta[n] = 1 \tag{8.8-4}$$

If a unit-impulse is applied to an LTI system it evokes a response, denoted by $h[n]$ and called the *impulse response*. Because of time invariance, a shifted impulse $\delta[n - k]$ evokes a shifted response $h[n - k]$. Since any sequence can be written as

$$X[n] = \sum_{m=-\infty}^{\infty} X[m]\delta[n - m] \tag{8.8-5}$$

This general sequence evokes a general response

$$Y[n] = \sum_{m=-\infty}^{\infty} X[m]h[n - m] = \sum_{k=-\infty}^{\infty} X[n - k]h[k] \tag{8.8-6}$$

The last form of (8.8-6) results from a simple index change.

Equation (8.8-6) is called a *convolution sum*, or just a *convolution*. On using the short-form notation of (8.1-9) to represent convolution, (8.8-6) becomes

$$Y[n] = \sum_{m=-\infty}^{\infty} X[m]h[n - m] = X[n] * h[n] \tag{8.8-7}$$

We mention some properties of LTI DT systems without proof. Order of cascade of two systems with impulse responses $h_1[n]$ and $h_2[n]$ is unimportant; this means $h[n]$ for the cascade is

$$h[n] = h_1[n] * h_2[n] = h_2[n] * h_1[n] \tag{8.8-8}$$

The impulse response of two systems placed in parallel is the sum of their impulse responses: $h[n] = h_1[n] + h_2[n]$. LTI DT systems at rest are inherently linear such that if $Y_1[n]$ and $Y_2[n]$ are the respective responses of a system to inputs $X_1[n]$ and $X_2[n]$, then the response to a linear combination of the inputs is the linear combination of responses. That is, for input $X[n] = a_1 X_1[n] + a_2 X_2[n]$ the response is

$$Y[n] = a_1 Y_1[n] + a_2 Y_2[n] \tag{8.8-9}$$

for any inputs and any constants a_1 and a_2. Finally, if an LTI DT system is excited by any bounded input sequence, the response is called *bounded-input-bounded-output* (BIBO) stable if it is bounded for every bounded input. A system is BIBO stable if, and only if,

$$\sum_{n=-\infty}^{\infty} |h[n]| < \infty \qquad (8.8\text{-}10)$$

A DT system falls into one of two categories. If it has a purely feed-forward structure where $a_i = 0$, all i, in (8.8-2), it is called *finite impulse response* (FIR) and

$$Y[n] = \sum_{i=0}^{M} b_i X[n - i] \qquad (8.8\text{-}11)$$

For an input impulse, $X[n] = \delta[n]$, the FIR system's impulse response reduces to

$$h[n] = \sum_{i=0}^{M} b_i \delta[n - i] = \begin{cases} b_n & 0 \le n \le M \\ 0 & \text{all other } n \end{cases} \qquad (8.8\text{-}12)$$

In one type of problem an impulse response $h[n]$ is specified. The system coefficients then derive directly from $h[n]$ according to (8.8-12). FIR systems are simple but usually require high complexity (large M) for modern-day applications [Taylor (1994), p. 424]. The responses of FIR systems are sometimes known as *moving average* processes when the input is a white noise sequence.

> **EXAMPLE 8.8-3.** Suppose an LTI discrete-time system is defined by (8.8-2) with $a_i = 0$, all i, and $b_i = 1/4$, $i = 0$, 1, 2, 3. We discuss the system's behavior.
>
> From (8.8-2) $Y[n] = \{X[n] + X[n - 1] + X[n - 2] + X[n - 3]\}(1/4)$. This system uniformly averages four input values: the most current input $X[n]$, and the three most recent past input values. It is, therefore, a uniform sliding (moving) averager over the four most recent input values. The average "slides" because it includes only the most recent four input values. The system is uniform because the b_i all have the same (uniform) values.
>
> A *nonuniform* sliding averager would correspond to b_i being different with index i. For example, if $|b_i|$ decrease for increasing i, the averager would place less "weight," or importance, on increasingly distant past values of the input.

The second category of DT system is called an *infinite impulse response* (IIR) *system*. It is one with both feed-forward and feed-back connections, as in Figure 8.8-1. IIR systems may be less complex than an FIR system for a given requirement, but can suffer stability problems. IIR systems are sometimes called *recursive* because output values can be recursively computed from input and past output values.

Transform Domain Methods for DT Systems

The concept of a transfer function is critical to the transform domain analysis/ synthesis of a DT system. We previously derived (7.5-11) as the discrete-time Fourier transform (DTFT) of an autocorrelation sequence to be

$$\mathcal{S}_{X_s X_s}(\omega) = \mathcal{S}_{X_s X_s}(e^{j\Omega}) = \sum_{n=-\infty}^{\infty} R_{XX}[n] e^{-jn\Omega} \tag{8.8-13}$$

where

$$\Omega = \omega T_s \tag{8.8-14}$$

and $\mathcal{S}_{X_s X_s}(e^{j\Omega})$ is defined by the z-transform of (7.5-9). Since $h[n]$ is a DT sequence that we consider applicable to a stable system, it has a DTFT, denoted by $H(e^{j\Omega})$, given by

$$H(e^{j\Omega}) = \sum_{n=-\infty}^{\infty} h[n] e^{-jn\Omega} \tag{8.8-15}$$

This function is called the *transfer function* (or sometimes the *frequency response function*) of the DT system.

The transfer function can be derived for a DT system for which (8.8-2) applies. For an impulse at the input, $X[n] = \delta[n]$, we have

$$h[n] = -\sum_{r=1}^{N} a_r h[n-r] + \sum_{r=0}^{M} b_r \delta[n-r] \tag{8.8-16}$$

It can be shown that (see Problems 8.8-1 through 8.8-3)

$$\delta[n] \leftrightarrow 1 \tag{8.8-17}$$

$$\delta[n-r] \leftrightarrow e^{-jr\Omega} \tag{8.8-18}$$

$$h[n-r] \leftrightarrow H(e^{j\Omega}) e^{-jr\Omega} \tag{8.8-19}$$

so the DTFT of (8.8-16) becomes

$$H(e^{j\Omega}) = -\sum_{r=1}^{N} a_r H(e^{j\Omega}) e^{-jr\Omega} + \sum_{r=0}^{M} b_r e^{-jr\Omega} \tag{8.8-20}$$

or

$$H(e^{j\Omega}) = \frac{\displaystyle\sum_{r=0}^{M} b_r e^{-jr\Omega}}{1 + \displaystyle\sum_{r=1}^{N} a_r e^{-jr\Omega}} \tag{8.8-21}$$

Equation (8.8-21) defines the transfer function (transform domain) in terms of sequence domain coefficients a_r and b_r. For a desired transfer function it allows the solution for these coefficients, a procedure equivalent to digital filter design. In essence, by choice of pole and zero locations in the right-side function, a synthesis of some desired transfer function results.

EXAMPLE 8.8-4. Assume a DT system is defined by (8.8-16) with $a_r = 0$, all r, and $b_r = 1$, all r. We find the FIR system's transfer function from (8.8-21):

$$H(e^{j\Omega}) = \sum_{r=0}^{M} b_r e^{-jr\Omega}$$

$$= \sum_{r=0}^{M} e^{-jr\Omega} = \frac{\sin[(M+1)\Omega/2]}{\sin(\Omega/2)} e^{-jM\Omega/2} \qquad (1)$$

from use of (C-60). This example is an extension of the uniform sliding averaging system of Example 8.8-3 to $M+1$ terms. For M large so that the principal responses of $|H(e^{j\Omega})|$ occur for relatively small Ω, we have

$$H(e^{j\Omega}) \approx (M+1)\frac{\sin[(M+1)\Omega/2]}{[(M+1)\Omega/2]} \qquad (2)$$

if the delay term (exponential factor) is ignored in (1). In terms of ω, (2) becomes

$$H(e^{j\omega T_s}) \approx (M+1)\frac{\sin[(M+1)\omega T_s/2]}{[(M+1)\omega T_s/2]}$$

Thus, the DT system behaves as a lowpass filter with a transfer function having the shape of a sampling function, and a delay of $MT_s/2$.

To emphasize the utility of (8.8-15), we develop the autocorrelation function and power spectrum of the response $Y[n]$ of a DT system. By analogy with (8.8-13), the power spectrum is the DTFT of $R_{YY}[n]$. From (8.8-6) with $X[n]$ assumed wide-sense stationary:

$$R_{YY}[n] = E\{Y[k]Y[k+n]\}$$

$$= E\left\{ \sum_{m=-\infty}^{\infty} X[k-m]h[m] \sum_{p=-\infty}^{\infty} X[k+n-p]h[p] \right\}$$

$$= \sum_{m=-\infty}^{\infty} h[m] \sum_{p=-\infty}^{\infty} h[p] R_{XX}[n+m-p] \qquad (8.8-22)$$

The DTFT becomes

$$\mathscr{S}_{Y_s Y_s}(\omega) = \mathscr{S}_{Y_s Y_s}(e^{j\Omega}) = \sum_{n=-\infty}^{\infty} R_{YY}[n]e^{-jn\Omega}$$

$$= \sum_{m=-\infty}^{\infty} h[m] \sum_{p=-\infty}^{\infty} h[p] \sum_{n=-\infty}^{\infty} R_{XX}[n+m-p]e^{-jn\Omega}$$

$$= \sum_{m=-\infty}^{\infty} h[m]e^{jm\Omega} \sum_{p=-\infty}^{\infty} h[p]e^{-jp\Omega} \sum_{r=-\infty}^{\infty} R_{XX}[r]e^{-jr\Omega} \qquad (8.8-23)$$

where the last form results from the index change $r = n+m-p$. The left and middle terms in the last form are recognized as $H^*(e^{j\Omega})$ and $H(e^{j\Omega})$, respectively, from (8.8-15), while the right sum is the power spectrum of the sampled version of $X(t)$. Hence,

$$\mathscr{S}_{Y_s Y_s}(\omega) = \mathscr{S}_{Y_s Y_s}(e^{j\Omega}) = \mathscr{S}_{X_s X_s}(e^{j\Omega})|H(e^{j\Omega})|^2 \qquad (8.8\text{-}24)$$

The dependence on ω is more obvious if Ω is replaced by ωT_s from (8.8-14).

8.9
MODELING OF NOISE SOURCES

All our work in this chapter so far has related to finding the response of a linear system when a random waveform (desired signal or undesired noise) was applied at its input. In every case, the system was assumed to not contain any internal sources. In particular, the system was assumed to be free of any internally generated *noise*. In the real world, such an assumption is never justified because all networks (systems) generate one or more types of noise internally. For example, all conductors or semiconductors in a circuit are known to generate *thermal noise* (see Section 7.6) because of thermal agitation of free electrons.† The question naturally arises: How can we handle practical networks that produce internally generated noise? The remainder of this chapter is concerned with answering this question.

We shall find that, by suitable modeling techniques for both the network and for the external source that drives the network, all the internally generated network noise can be thought of as having been caused *by the external source*. In effect, we shall replace the noisy practical network with a noise-free identical network that is driven by a "more noisy" source.

Our work begins by developing models for noise sources.

Resistive (Thermal) Noise Source

Suppose we have an ideal (noise-free, infinite input impedance) voltmeter that responds to voltages that fall in a small ideal (rectangular) frequency band $d\omega/2\pi$ centered at angular frequency ω. If such a voltmeter is used to measure the voltage across a resistor of resistance R (ohms), it is found, both in practice and theoretically, that a noise voltage $e_n(t)$ would exist having a mean-squared value given by

$$\overline{e_n^2(t)} = \frac{2kTR\,d\omega}{\pi} \qquad (8.9\text{-}1)$$

Here $k = 1.38(10^{-23})$ joule per Kelvin is *Boltzmann's constant*,‡ and T is temperature in Kelvin. This result is independent of the value of ω up to

---

†There are many other types of internally generated noise such as *shot noise, partition noise, induced grid noise, flicker noise, secondary emission noise*, etc. The reader is referred to the literature for more detail (Mumford and Scheibe, 1968; van der Ziel, 1970).
‡Ludwig Boltzmann (1844–1906) was an Austrian physicist.

extremely high frequencies. (See Section 7.6 where $\mathcal{N}_0/2$ equals $2kTR$ here. The reader should justify this fact as an exercise.)

Now because the voltmeter does not load the resistor, $\overline{e_n^2(t)}$ is the mean-squared open-circuit voltage of the resistor which can be treated as a voltage source with internal impedance R. In other words, the noisy resistor can be modeled as a Thevenin† voltage source as shown in Figure 8.9-1a. An equivalent current source is shown in (b) where

$$\overline{i_n^2(t)} = \overline{e_n^2(t)}/R^2 = \frac{2kT\,d\omega}{\pi R} \tag{8.9-2}$$

is the short-circuit mean-squared current.

From Figure 8.9-1a it is found that the *incremental noise power* dN_L delivered to the load in the incremental band $d\omega$ by the noisy resistor as a source is

$$dN_L = \frac{\overline{e_n^2(t)}R_L}{(R+R_L)^2} = \frac{2kTRR_L\,d\omega}{\pi(R+R_L)^2} \tag{8.9-3}$$

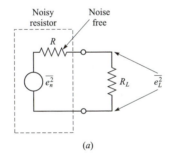

(a)

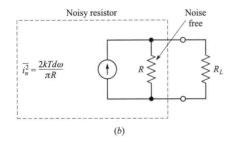

(b)

FIGURE 8.9-1
Equivalent circuit models of a noisy resistor: (a) voltage model and (b) current model. [*Adapted from Peebles (1976) with permission of publishers Addison-Wesley, Advanced Book Program.*]

---

†Named for the French physicist Léon Thevenin (1857–1926).

The maximum delivered power occurs when $R_L = R$. We call this maximum power the *incremental available power* of the source and denote it by dN_{as}; it is given by

$$dN_{as} = \overline{e_n^2(t)}/4R = \frac{kT\,d\omega}{2\pi} \tag{8.9-4}$$

We see from (8.9-4) that the incremental power *available* from a resistor source is *independent of the resistance of the source* and depends only on its physical temperature T. These facts may be used as a basis for modeling arbitrary sources.

Arbitrary Noise Sources, Effective Noise Temperature

Suppose an actual noise source has an incremental available noise power dN_{as}, open-circuit output mean-squared voltage $\overline{e_n^2(t)}$, and impedance as measured between its output terminals of $Z_o(\omega) = R_o(\omega) + jX_o(\omega)$. The available noise power is easily found to be

$$dN_{as} = \frac{\overline{e_n^2(t)}}{4R_o(\omega)} \tag{8.9-5}$$

If we now ascribe all the source's noise to the resistive part $R_o(\omega)$ of its output impedance by *defining* an *effective noise temperature* T_s such that (8.9-1) applies, then

$$\overline{e_n^2(t)} = 2kT_s R_o(\omega)\frac{d\omega}{\pi} \tag{8.9-6}$$

As with a purely resistive source, available power is still independent of the source impedance but depends on the source's temperature

$$dN_{as} = kT_s\frac{d\omega}{2\pi} \tag{8.9-7}$$

We consider two examples that illustrate effective noise temperature.

EXAMPLE 8.9-1. Two different resistors at different physical temperatures are placed in series. The effective noise temperature of the series combination as a noise source is to be found.

Figure 8.9-2 illustrates Thevenin equivalent circuits for the combination. Since the individual resistors as sources may be considered independent, their mean-squared voltages add. Hence,

$$\overline{e_1^2(t)} + \overline{e_2^2(t)} = \overline{e_n^2(t)}$$

By applying (8.9-1) to both sides of the preceding expression, we obtain

$$2k[T_1 R_1 + T_2 R_2]\frac{d\omega}{\pi} = 2k[T_s(R_1 + R_2)]\frac{d\omega}{\pi}$$

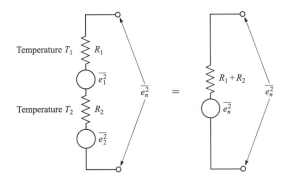

FIGURE 8.9-2
Equivalent circuits for two resistors at different temperatures in series.

or

$$T_s = \frac{T_1 R_1 + T_2 R_2}{R_1 + R_2}$$

Example 8.9-1 clearly shows that effective noise temperature of a source is not necessarily equal to its physical temperature. In the special case where $T_1 = T_2 = T$, then $T_s = T$. More generally, it is true that any passive, two-terminal source that contains only resistors, capacitors, and inductors, all at the same physical temperature T, will have an effective noise temperature $T_s = T$. (Ziemer and Tranter, 1976, p. 471). The next example can be used to illustrate this last point.

EXAMPLE 8.9-2. We reconsider Example 8.9-1, except we now allow a capacitor to be placed across one resistor as shown in Figure 8.9-3.

By superposition, $e_n^2(t)$ is the sum of contributions from each resistor as a noise source. The mean-squared voltage, denoted $\overline{e_{n1}^2(t)}$, due to the first resistor is readily seen to be

$$\overline{e_{n1}^2(t)} = \overline{e_1^2(t)} \left| \frac{1}{1 + j\omega R_1 C_1} \right|^2 = \frac{\overline{e_1^2(t)}}{1 + \omega^2 R_1^2 C_1^2}$$

That due to the second resistor is

$$\overline{e_{n2}^2(t)} = \overline{e_2^2(t)}$$

Thus, by applying (8.9-1) to the two individual resistor mean-squared voltages, we have

$$\overline{e_n^2(t)} = \overline{e_{n1}^2(t)} + \overline{e_{n2}^2(t)} = 2k \left[\frac{T_1 R_1}{1 + \omega^2 R_1^2 C_1^2} + T_2 R_2 \right] \frac{d\omega}{\pi}$$

Next, we find the output impedance of the network as an overall source by imagining the noise sources set to 0. We get

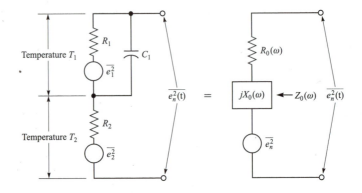

FIGURE 8.9-3

Equivalent circuits for a linear, passive, two-terminal network of two resistors and one capacitor.

$$Z_o(\omega) = R_2 + \frac{R_1(1/j\omega C_1)}{R_1 + (1/j\omega C_1)} = R_2 + \frac{R_1}{1 + j\omega R_1 C_1}$$

$$= R_2 + \frac{R_1(1 - j\omega R_1 C_1)}{1 + \omega^2 R_1^2 C_1^2}$$

which has a resistive part

$$R_o(\omega) = R_2 + \frac{R_1}{1 + \omega^2 R_1^2 C_1^2}$$

By applying (8.9-6) to the equivalent source, we have

$$\overline{e_n^2(t)} = 2kT_s \left[R_2 + \frac{R_1}{1 + \omega^2 R_1^2 C_1^2} \right] \frac{d\omega}{\pi}$$

Finally, we equate $\overline{e_n^2(t)}$ for the actual and equivalent networks to find T_s:

$$T_s = \frac{T_1 R_1 + T_2 R_2(1 + \omega^2 R_1^2 C_1^2)}{R_1 + R_2(1 + \omega^2 R_1^2 C_1^2)}$$

The preceding example shows that effective noise temperature may be a function of frequency. In this case, the available noise power is also frequency dependent.

Again we see that $T_s = T$ in the above example if $T_1 = T_2 = T$, as it must because it is a linear, passive, two-terminal network with only resistors and a capacitor, as noted previously.

An Antenna as a Noise Source

In practice, all antennas produce noise at their output because of reception of electromagnetic radiation from noise sources external to the antenna.† The

---

†There are many sources of external noise; several of these are described by Peebles (1976, pp. 463–464).

amount of available noise power dN_{as} in an incremental band $d\omega$ depends in a rather complicated manner on all the space surrounding the antenna. However, it is possible to model the antenna in a simple way by assigning to it an *antenna temperature* T_a chosen so that dN_{as} and T_a are related by (8.9-4). Thus,

$$dN_{as} = kT_a \frac{d\omega}{2\pi} \qquad (8.9\text{-}8)$$

In general, antenna temperature may vary with frequency. However, in many applications T_a can be considered constant (with respect to ω) because its variation with frequency over a frequency band comparable to that of the desired signal being received is often small.

> **EXAMPLE 8.9-3.** A very sensitive meter that is capable of measuring noise power in a (small) frequency band 1 kHz wide at any frequency $\omega/2\pi$ is attached to a microwave antenna used in a radio relay link. It registers 2.0 (10^{-18}) W when the meter's input impedance is matched to the antenna so that its reading is maximum. We find the antenna temperature T_a.
>
> Since maximum power is extracted from the antenna, the power is its available power and (8.9-8) gives
>
> $$T_a = \frac{2\pi \, dN_{as}}{k \, d\omega} = \frac{2\pi(2)10^{-18}}{1.38(10^{-23})2\pi(10^3)} = \frac{200}{1.38} \approx 144.9 \text{ K}$$

8.10
INCREMENTAL MODELING OF NOISY NETWORKS

In this section we shall show how a noisy network can be modeled as a noise-free network excited by a suitably chosen external noise source. We also develop some measures of the "noisiness" of a network. All our work is applicable to an incremental band $d\omega$.

Available Power Gain

Consider first a linear, noise-free, two-port (4-terminal) network having an input impedance Z_i when the output port is open-circuited. Its output impedance, found by looking back into its output port, is Z_o when being driven by a source with source impedance Z_s. The source open-circuit voltage is $e_s(t)$ and the network's open-circuit output voltage is $e_o(t)$. The applicable network is illustrated in Figure 8.10-1.

The available power, denoted dN_{as}, of the source is

$$dN_{as} = \frac{\overline{e_s^2(t)}}{4R_s} \qquad (8.10\text{-}1)$$

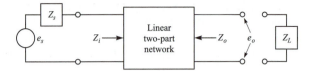

FIGURE 8.10-1
A linear two-port network driven by a source of impedance Z_s.

where R_s is the real part of Z_s. This power is independent of Z_i. The available power, denoted dN_{aos}, in the output due to the source is

$$dN_{aos} = \frac{\overline{e_o^2(t)}}{4R_o} \tag{8.10-2}$$

where R_o is the real part of Z_o. This power does depend on Z_i through its influence on the generation of $e_o(t)$ but does not depend on the load impedance Z_L. We define the *available power gain* denoted G_a of the two-port network as the ratio of the available powers

$$G_a = \frac{dN_{aos}}{dN_{as}} = \frac{R_s \overline{e_o^2(t)}}{R_o \overline{e_s^2(t)}} \tag{8.10-3}$$

When a cascade of M noise-free networks is involved where $M = 1, 2, \ldots,$ it is easy to see that the overall available power gain G_a is the product of available power gains G_m, $m = 1, 2, \ldots, M$, if G_m is the gain of stage m when all preceding stages are connected and treated as its source (see Problem 8.10-1). Thus,

$$G_a = \prod_{m=1}^{M} G_m \tag{8.10-4}$$

Equivalent Networks, Effective Input Noise Temperature

Consider next the case of a linear two-port network *with* internally generated noise. The network is assumed to be driven from a source with effective noise temperature T_s as shown in Figure 8.10-2a. If G_a is the network's available power gain, the available output noise power due to the source alone is

$$dN_{aos} = G_a dN_{as} = G_a k T_s \frac{d\omega}{2\pi} \tag{8.10-5}$$

from (8.10-3) and (8.9-7).

Total available output noise power dN_{ao} is larger than dN_{aos} because of internally generated noise. Let ΔN_{ao} represent the *excess available noise power* at the output. We shall imagine that ΔN_{ao} is generated by the source by defining *effective input noise temperature* T_e as the temprature *increase* that

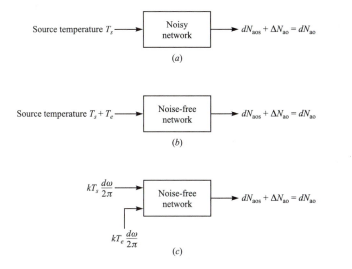

(a)

(b)

(c)

FIGURE 8.10-2
A network with internally generated noise driven from a noise source (a), and
equivalent noise-free networks (b) and (c). [*Reproduced from Peebles (1976) with
permission of publishers Addison-Wesley, Advanced Book Program.*]

the source would require to account for all output available noise power. It
therefore follows that

$$\Delta N_{ao} = G_a k T_e \frac{d\omega}{2\pi} \tag{8.10-6}$$

With this definition, the noisy network is replaced by a noise-free network
driven by a source of temperature $T_s + T_e$ as shown in Figure 8.10-2b.

It is somewhat helpful to model the available source noise power by use of
two inputs, as shown in Figure 8.10-2c. The second input represents the
internally generated noise due to the network. The representation is convenient in visualizing noise effects when networks are cascaded as illustrated in
Figure 8.10-3. By equating expressions for output available noise powers in
the cascade and equivalent network, the effective input noise temperature T_e
of the cascade is determined to be

$$T_e = T_{e1} + \frac{T_{e2}}{G_1} + \frac{T_{e3}}{G_1 G_2} + \cdots + \frac{T_{eM}}{G_1 G_2 \cdots G_{M-1}} \tag{8.10-7}$$

where T_{em} and G_m, $m = 1, 2, \ldots, M$, are the effective input noise temperature
and available power gain, respectively, for the mth stage when all $m - 1$ previous stages are connected and form its source.

An especially useful application of (8.10-7) is to the cascade of stages in an
amplifier. We develop an example.

EXAMPLE 8.10-1. The stages in a three-stage amplifier have effective input
noise temperatures $T_{e1} = 1350\,\text{K}$, $T_{e2} = 1700\,\text{K}$, and $T_{e3} = 2600\,\text{K}$. The
respective available power gains are $G_1 = 16$, $G_2 = 10$, and $G_3 = 6$. We

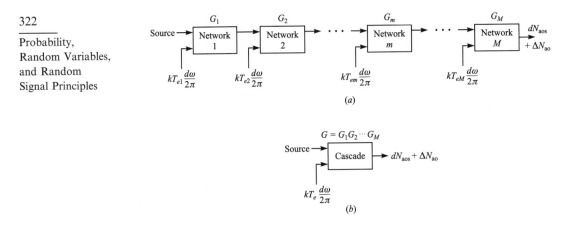

FIGURE 8.10-3
(a) M networks in cascade and (b) the equivalent network. [*Reproduced from Peebles (1976) with permission of publishers Addison-Wesley, Advanced Book Program.*]

find the effective input noise temperature of the overall amplifier by use of (8.10-7):

$$T_e = 1350 + \frac{1700}{16} + \frac{2600}{16(10)} = 1350 + 106.25 + 16.25$$

$$= 1472.5\,\text{K}$$

We see that, even though T_{e2} and T_{e3} are larger than T_{e1}, the contributions to T_e by the second and third stages are much smaller than that of the first stage because of the gain of previous stages. In general, it is clear from (8.10-7) that an amplifier should have its lowest noise, highest gain stage first, followed by its next best stage, etc., for best noise performance.

Spot Noise Figures

Effective input noise temperature T_e of a network is a measure of its noise performance. Better performance corresponds to lower values of T_e. Another measure of performance is *incremental* or *spot noise figure* denoted by F and defined as the total incremental available output noise power dN_{ao} divided by the incremental available output noise power due to the source alone:

$$F = \frac{dN_{ao}}{dN_{aos}} = \frac{dN_{aos} + \Delta N_{ao}}{dN_{aos}} = 1 + \frac{\Delta N_{ao}}{dN_{aos}} \qquad (8.10\text{-}8)$$

An alternative form derives from the substitution of (8.10-5) and (8.10-6):

$$F = 1 + \frac{T_e}{T_s} \qquad (8.10\text{-}9)$$

In an ideal network, $T_e = 0$ so $F = 1$. For any real network, F is larger than unity.

In practice, a given network might be driven by a variety of sources. For example, an amplifier might be driven by an antenna, mixer, attenuator, other amplifier, etc. Its spot noise figure is therefore a function of the effective noise temperature of the source. However, by defining a *standard source* as having a *standard noise temperature* $T_0 = 290\,\text{K}$ and *standard spot noise figure* F_0, given by

$$F_0 = 1 + \frac{T_e}{T_0} \tag{8.10-10}$$

a network can be specified independent of its application.

When a network is used with the source for which it is intended to operate F will be called the *operating spot noise figure* and given the symbol F_{op}. From (8.10-9)

$$F_{op} = 1 + \frac{T_e}{T_s} \tag{8.10-11}$$

Operating and standard spot noise figures can also be developed for a cascade of networks (see Problems 8.10-2 and 8.10-4).

EXAMPLE 8.10-2. An engineer purchases an amplifier that has a narrow bandwidth of 1 kHz and standard spot noise figure of 3.8 at its frequency of operation. The amplifier's available output noise power is 0.1 mW when its input is connected to a radio receiving antenna having an antenna temperature of 80 K. We find the amplifier's input effective noise temperature T_e, its operating spot noise figure F_{op}, and its available power gain G_a.

T_e derives from (8.10-10):

$$T_e = T_0(F_0 - 1) = 290(3.8 - 1) = 812\,\text{K}$$

We can now use (8.10-11) to obtain F_{op}:

$$F_{op} = 1 + \frac{812}{80} = 11.15$$

From (8.10-5) and (8.10-6) we add to get total available output noise power:

$$dN_{ao} = dN_{aos} + \Delta N_{ao} = \frac{k(T_s + T_e)G_a\,d\omega}{2\pi}$$

so

$$G_a = \frac{2\pi\,dN_{ao}}{k(T_s + T_e)\,d\omega} = \frac{2\pi(0.1)10^{-3}}{1.38(10^{-23})(812 + 80)2\pi(10^3)} \approx 8.12(10^{12})$$

8.11
MODELING OF PRACTICAL NOISY NETWORKS

In a realistic network, the frequency band of interest is not incremental. Therefore, such quantities as available power gain, noise temperature, and noise figure are not necessarily constant but become frequency dependent, in general. In this section we extend the earlier concepts based on an incremental frequency band to include practical networks, by defining *average* noise temperatures and *average* noise figures.

Average Noise Figures

We define *average operating noise figure* $\bar{F}_{op}$ as the *total* output available noise power N_{ao} from a network divided by the *total* output available noise power N_{aos} due to the source alone. Thus,

$$\bar{F}_{op} = \frac{N_{ao}}{N_{aos}} \tag{8.11-1}$$

N_{aos} is found by integration of (8.10-5):

$$N_{aos} = \frac{k}{2\pi} \int_0^\infty T_s G_a \, d\omega \tag{8.11-2}$$

We may similarly use (8.10-8) with (8.10-5) to determine N_{ao}:

$$N_{ao} = \int_0^\infty dN_{ao} = \int_0^\infty F_{op} \, dN_{aos} = \frac{k}{2\pi} \int_0^\infty F_{op} T_s G_a \, d\omega \tag{8.11-3}$$

Thus, from (8.11-1)

$$\bar{F}_{op} = \frac{\displaystyle\int_0^\infty F_{op} T_s G_a \, d\omega}{\displaystyle\int_0^\infty T_s G_a \, d\omega} \tag{8.11-4}$$

In many cases the source's temperature is approximately constant. Operating average noise figure then becomes

$$\bar{F}_{op} = \frac{\displaystyle\int_0^\infty F_{op} G_a \, d\omega}{\displaystyle\int_0^\infty G_a \, d\omega} \qquad T_s \text{ constant} \tag{8.11-5}$$

An antenna is an example of a source having an approximately constant noise temperature (so long as the surroundings viewed by the antenna are fixed). Another example is a standard source for which $T_s = T_0 = 290 \, \text{K}$ is constant. We define *average standard noise figure* $\bar{F}_0$ as that for which the source is standard. In this case

$$\bar{F}_0 = \frac{\displaystyle\int_0^\infty F_0 G_a \, d\omega}{\displaystyle\int_0^\infty G_a \, d\omega} \tag{8.11-6}$$

as can be shown by repeating the steps leading to (8.11-4).

Average Noise Temperatures

From the definition of effective input noise temperature T_e, it follows that the incremental available output noise power from a network with available power gain G_a that is driven by a source of temperature T_s is

$$dN_{ao} = G_a k (T_s + T_e) \frac{d\omega}{2\pi} \tag{8.11-7}$$

Total available power is therefore

$$N_{ao} = \int_0^\infty dN_{ao} = \frac{k}{2\pi} \int_0^\infty G_a (T_s + T_e) \, d\omega \tag{8.11-8}$$

Next, we define *average effective source temperature* $\bar{T}_s$ and *average effective input noise temperature* $\bar{T}_e$ as *constant* temperatures that produce the same total available power as given by (8.11-8). Hence

$$N_{ao} = \frac{k}{2\pi} (\bar{T}_s + \bar{T}_e) \int_0^\infty G_a \, d\omega \tag{8.11-9}$$

By equating (8.11-9) and (8.11-8) on a term-by-term basis, we get

$$\bar{T}_s = \frac{\displaystyle\int_0^\infty T_s G_a \, d\omega}{\displaystyle\int_0^\infty G_a \, d\omega} \tag{8.11-10}$$

and

$$\bar{T}_e = \frac{\displaystyle\int_0^\infty T_e G_a \, d\omega}{\displaystyle\int_0^\infty G_a \, d\omega} \tag{8.11-11}$$

If (8.10-10) and (8.10-11) are substituted into (8.11-6) and (8.11-4), respectively, we obtain the interrelationships

$$\bar{F}_0 = 1 + \frac{\bar{T}_e}{T_0} \tag{8.11-12}$$

$$\bar{F}_{op} = 1 + \frac{\bar{T}_e}{\bar{T}_s} \tag{8.11-13}$$

By equating $\bar{T}_e$ from these last two expressions, we obtain alternative interrelationships

$$\bar{F}_0 = 1 + \frac{\bar{T}_s}{T_0}(\bar{F}_{op} - 1) \qquad (8.11\text{-}14)$$

$$\bar{F}_{op} = 1 + \frac{T_0}{\bar{T}_s}(\bar{F}_0 - 1) \qquad (8.11\text{-}15)$$

Average effective noise temperature is a very useful concept for modeling network noise in a simple way. To demonstrate this fact, note that (8.11-9) can be written as

$$N_{ao} = \frac{k}{2\pi}(\bar{T}_s + \bar{T}_e)G_a(\omega_0)\frac{\displaystyle\int_0^\infty G_a(\omega)\,d\omega}{G_a(\omega_0)} \qquad (8.11\text{-}16)$$

where ω_0 is the centerband angular frequency of the function $G_a(\omega)$. Since $G_a(\omega)$ is the available power gain (or power transfer function) of the network, we identify

$$W_N = \frac{\displaystyle\int_0^\infty G_a(\omega)\,d\omega}{G_a(\omega_0)} \qquad (8.11\text{-}17)$$

as the *noise bandwidth* of the network, by analogy with (8.5-6). Equation (8.11-16) becomes

$$N_{ao} = G_a(\omega_0)k(\bar{T}_s + \bar{T}_e)\frac{W_N}{2\pi} \qquad (8.11\text{-}18)$$

which says that actual available output noise power is that due to a source with *constant* temperature $\bar{T}_s + \bar{T}_e$ driving an equivalent noise-free network with an ideal rectangular transfer function of bandwidth W_N (rad/s) and midband available power gain $G_a(\omega_0)$. This result represents a very simple network model.

Modeling of Attenuators

Consider a source of average effective temperature $\bar{T}_s$ driving an impedance-matched lossy attenuator with power loss L (a number not less than one) at all frequencies. The attenuator has a physical temperature T_L. It can be shown (Peebles, 1976, p. 463; Mumford and Scheibe, 1968, p. 23) that the average effective input noise temperature of the attenuator is

$$\bar{T}_e = T_L(L - 1) \qquad (8.11\text{-}19)$$

From (8.11-12) and (8.11-13) the applicable average noise figures are

$$\bar{F}_0 = 1 + \frac{T_L}{T_0}(L - 1) \qquad (8.11\text{-}20)$$

$$\bar{F}_{op} = 1 + \frac{T_L}{\bar{T}_s}(L - 1) \qquad (8.11\text{-}21)$$

Note that if $T_L = T_0$ or if $T_L = \bar{T}_s$, the average noise figure of the attenuator is just equal to its loss.

Model of Example System

One of the most important applications of the theory of this and the preceding two sections is in modeling receiving systems. As illustrated in Figure 8.11-1a, consider a receiving antenna that drives a receiver amplifier through various broad-band components having an overall loss L. These components (which may include microwave transmission lines, isolators, or other devices) are all assumed to have physical temperature T_L. The antenna temperature is T_a and the receiver average effective input noise temperature is $\bar{T}_R$. The receiver's noise bandwidth is W_N and it has a centerband available power gain $G_R(\omega_0)$. We demonstrate that the system is equivalent to that shown in Figure 8.11-1b.

The equivalent system has the same noise bandwidth as the actual system and has a centerband available power gain $G_R(\omega_0)/L$. It is driven by a simple source with *system noise temperature* $\bar{T}_{\text{sys}}$. The available output noise power in the actual system is the sum of the antenna's contribution plus those due to excess noises in the attenuator and receiver. By using earlier models, this noise power is

$$N_{\text{ao}} = k[T_a + T_L(L-1) + \bar{T}_R L]\frac{G_R(\omega_0)W_N}{L(2\pi)} \qquad (8.11\text{-}22)$$

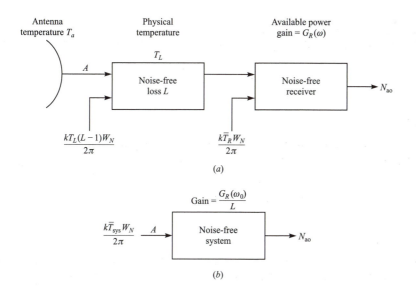

(a)

(b)

FIGURE 8.11-1
A model of a receiving system (a) and its equivalent (b). [*Reproduced from Peebles (1976) with permission of publishers Addison-Wesley, Advanced Book Program.*]

For the equivalent system

$$N_{ao} = k\bar{T}_{sys} \frac{G_R(\omega_0)W_N}{L(2\pi)} \tag{8.11-23}$$

By equating the above two expressions, we obtain

$$\bar{T}_{sys} = T_a + T_L(L-1) + \bar{T}_R L \tag{8.11-24}$$

From (8.11-24), the average effective input noise temperature of the system taken at point A in Figure 8.11-1a is

$$\bar{T}_e = T_L(L-1) + \bar{T}_R L \tag{8.11-25}$$

From (8.11-13) the average system operating noise figure is

$$\bar{F}_{op} = 1 + \frac{T_L}{T_a}(L-1) + \frac{\bar{T}_R}{T_a}L \tag{8.11-26}$$

EXAMPLE 8.11-1. An antenna with temperature $T_a = 150\,\text{K}$ is connected to a receiver by means of a waveguide that is at a physical temperature of 280 K and has a loss of 1.5 (1.76 dB).† The receiver has a noise bandwidth of $W_N/2\pi = 10^6\,\text{Hz}$ and an average effective input noise temperature $\bar{T}_R = 700\,\text{K}$. We determine the system's noise temperature $\bar{T}_{sys}$, its operating average noise figure $\bar{F}_{op}$, and its available output noise power when $G_R(\omega_0) = 10^{12}$ (120 dB).
From (8.11-24)

$$\bar{T}_{sys} = 150 + 280(1.5-1) + 700(1.5) = 1340\,\text{K}$$

From (8.11-26)

$$\bar{F}_{op} = 1 + \frac{280}{150}(1.5-1) + \frac{700}{150}(1.5) \approx 8.93 \qquad \text{or} \qquad 9.51\,\text{dB}$$

Finally, we use (8.11-23) to find N_{ao}:

$$N_{ao} = 1.38(10^{-23})1340.0(10^{12})\frac{10^6}{1.5} \approx 12.3\,\text{mW}$$

8.12
SUMMARY

With temporal (Chapter 6) and spectral (Chapter 7) characteristics of random processes having been defined, this chapter concentrated on using earlier results to examine the behavior of (mainly) linear time-invariant systems to random input signals modeled as random processes. The principal topics and results follow.

---

†A number L expressed in *decibels* (dB), denoted L_{dB}, is related to L as a numeric (power ratio) by $L_{dB} = 10\log_{10}(L)$.

- The basics of linear systems were reviewed as they apply to both deterministic and random signals.
- The response of a linear time-invariant system to a random signal was developed in detail. The temporal characteristics (autocorrelation function, mean-square value of the system's response, and cross-correlation function of input and output) were developed for later use.
- It was shown how random noise can be used to evaluate the impulse response of a linear time-invariant system.
- The all-important power density spectrum was defined. The cross-power spectrum was also developed for the input/output random signals of a linear time-invariant network.
- The measurement of power spectrums was discussed and the noise bandwidth of a random process was defined.
- Special, but important, cases of bandpass, band-limited, and narrowband processes and their properties were discussed in detail.
- Sampling procedures for baseband and bandpass random processes were developed to show how continuous-time processes are related to discrete-time processes and sequences. Discrete-time power spectrums were developed in detail.
- A detailed discussion was given for discrete-time *systems*. Included were developments of A/D conversion. Discussions centered mainly around linear shift-invariant discrete-time systems, with general block diagrams given. Both direct form I and direct form II (canonic form) were shown.
- Detailed developments of discrete-time systems in the sequence domain were given, including impulse response, convolution, stability, and definitions of FIR and IIR systems.
- Discrete-time systems were further developed using transform domain methods. In particular, the discrete-time Fourier transform (DTFT) was introduced and the concept of a transfer function was developed. In addition, the autocorrelation function and the power spectrum of the network's response were discussed.
- The various topics surrounding the discrete-time system were supported by computer examples using MATLAB.
- Practical aspects of modeling real network noises were defined and developed. Of special interest were spot noise temperatures and spot noise figures of noisy networks (for a small incremental frequency band). These topics were extended to the full frequency axis and called average noise temperatures and average noise figures. The concepts were next combined to model the total noise within any typical linear system.

PROBLEMS

8.1-1. A signal $x(t) = u(t)\exp(-\alpha t)$ is applied to a network having an impulse response $h(t) = Wu(t)\exp(-Wt)$. Here α and W are real positive constants and $u(\cdot)$ is the unit-step function. Find the system's response by use of (8.1-10).

8.1-2. Work Problem 8.1-1 by using (8.1-11) to find the spectrum $Y(\omega)$ of the response.

8.1-3. A rectangular pulse of amplitude A and duration T, defined by

$$x(t) = \begin{cases} A & 0 < t < T \\ 0 & \text{elsewhere} \end{cases}$$

is applied to the system of Problem 8.1-1.
(a) Find the time response $y(t)$.
(b) Sketch your response for $W = \pi/T$ and $W = 2\pi/T$.

8.1-4. A filter is called *gaussian* if it has a transfer function

$$H(\omega) = \frac{1}{\sqrt{2\pi}\,W_{\text{rms}}} e^{-\omega^2/2W_{\text{rms}}^2}$$

where W_{rms} is the root-mean-square (rms) bandwidth.
(a) Sketch $H(\omega)$.
(b) How is W_{rms} related to the 3-dB bandwidth?

8.1-5. Two systems have transfer functions $H_1(\omega)$ and $H_2(\omega)$.
(a) Show that the transfer function $H(\omega)$ of the *cascade* of the two, which means that the output of the first feeds the input of the second system, is $H(\omega) = H_1(\omega)H_2(\omega)$.
(b) For a cascade of N systems with transfer functions $H_n(\omega)$, $n = 1, 2, \ldots N$, show that

$$H(\omega) = \prod_{n=1}^{N} H_n(\omega)$$

*8.1-6. Work Problem 8.1-1 if the output of the given network is applied to a second identical network and the response is taken from the second network.

8.1-7. The impulse response of a system is

$$h(t) = \begin{cases} t^3 e^{-t^2} & 0 < t \\ 0 & t < 0 \end{cases}$$

By use of (8.1-8) or (8.1-10), find the response of the network to the pulse

$$x(t) = \begin{cases} A & 0 < t < T \\ 0 & \text{elsewhere} \end{cases}$$

where A and T are real positive constants.

8.1-8. Work Problem 8.1-7 if the network's impulse response is

$$h(t) = \begin{cases} t^3 e^{-t} & 0 < t \\ 0 & t < 0 \end{cases}$$

8.1-9. Given the network shown in Figure P8.1-9.
(a) Find the impulse response $h(t)$.
(b) By Fourier transforming $h(t)$, find $H(\omega)$.
(c) Sketch $h(t)$ and $H(\omega)$.

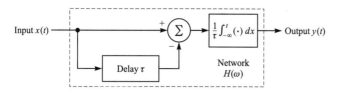

FIGURE P8.1-9
[*Reproduced from Peebles (1976) with permission of publishers Addison-Wesley, Advanced Book Program.*]

8.1-10. Find the transfer function of the network of Figure P8.1-9 by use of (8.1-13).

8.1-11. By using (8.1-13), find the transfer function of the network illustrated in Figure P8.1-11. Assume that no loading is present due to any output circuitry.

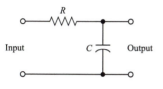

FIGURE P8.1-11

8.1-12. Work Problem 8.1-11 for the network of Figure P8.1-12.

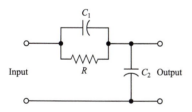

FIGURE P8.1-12

*8.1-13. (*a*) Work Problem 8.1-11 for the network Figure P8.1-13.
  (*b*) Under what conditions will the network behave approximately as a low-pass filter?

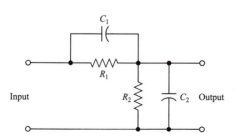

FIGURE P8.1-13
[*Reproduced from Peebles (1976) with permission of publishers Addison-Wesley, Advanced Book Program.*]

(c) Find a relationship between R_1, C_1, R_2, and C_2 such that the network behaves at all frequencies as a pure resistive attenuator.

8.1-14. Given the network shown in Figure P8.1-14.

(a) If the output causes no loading on the network, find the transfer function $H(\omega)$.

(b) Define $\omega_0 = 1/\sqrt{LC}$ and $Q_0 = R/\omega_0 L$. Plot $|H(\omega)|^2$ as a function of $x = (\omega - \omega_0)Q_0/\omega_0$ for Q_0 large and ω near ω_0. (*Hint:* Use the approximation $\omega \approx \omega_0$ for the most significant values of ω when Q_0 is large.)

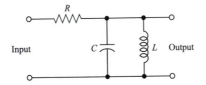

FIGURE P8.1-14

8.1-15. (a) Find the transfer function $H(\omega)$ for the network shown in Figure P8.1-15.

(b) Define $\omega_0 = 1/\sqrt{LC}$ and $Q_0 = 1/\omega_0(R + R_L)C$ and assume $Q_0 \gg 1$, so that the values of ω for which $H(\omega)$ is significant correspond to $\omega \approx \omega_0$. Use these facts to obtain an approximation for $H(\omega)$.

(c) If an impulse is applied to the network, find an expression for the approximate energy absorbed by R_L. (*Hint:* Use Parseval's theorem.)

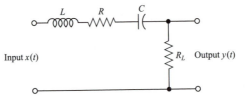

FIGURE P8.1-15

8.1-16. A class of filters called *Butterworth filters* has a power transfer function defined by

$$|H(\omega)|^2 = \frac{1}{1 + (\omega/W)^{2n}}$$

where $n = 1, 2, \ldots$, is a number related to the number of circuit elements and W is the 3-dB bandwidth in radians per second. Sketch $|H(\omega)|^2$ for $n = 1, 2, 4$, and 8 and note the behavior. As $n \to \infty$, what does $|H(\omega)|^2$ become?

8.1-17. Determine which of the following impulse responses do not correspond to a system that is stable, or realizable, or both, and state why.

(a) $h(t) = u(t + 3)$

(b) $h(t) = u(t)e^{-t^2}$

(c) $h(t) = e^t \sin(\omega_0 t)$ ω_0 a real constant

(d) $h(t) = u(t)e^{-3t} \sin(\omega_0 t)$ ω_0 a real constant

8.1-18. Use (8.1-10) and prove (8.1-15).

8.1-19. Show that (8.1-16) must be true if a linear time-invariant system is to be stable.

8.1-20. A system is defined by

$$y(t) = \int_{-\infty}^{t} x(\xi)\, d\xi$$

for all $x(t)$ for which the integral exists. Show that the system is linear, time-invariant, and causal.

8.1-21. A network is driven by a resistive source as shown in Figure P8.1-21. Find: (a) Z_i, (b) Z_o, and (c) G_a. (d) Is the network a matched attenuator?

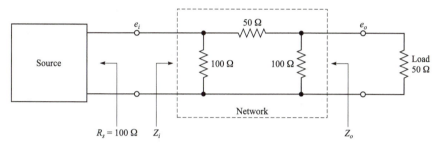

FIGURE P8.1-21

8.1-22. A network has the transfer function

$$H(\omega) = \frac{2e^{j\omega/20}}{(20 + j\omega)^3}$$

(a) Determine and sketch its impulse response. (*Hint*: Use Appendix E.)
(b) Is the network physically realizable?
(c) Determmine if the network is stable by evaluating I in (8.1-16).

*8.1-23. Show that the impulse response of a cascade of N identical networks, each with transfer function

$$H_1(\omega) = 1/(\alpha + j\omega)$$

where $\alpha > 0$ is a constant, is given by

$$h_N(t) = u(t)\left[\frac{t^{N-1}}{(N-1)!}\right]\exp(-\alpha t)$$

8.1-24. If τ in the circuit of Figure P8.1-9 is changed to T, show that the circuit is equivalent to the operation

$$y(t) = \frac{1}{T}\int_{t-T}^{t} x(\xi)\, d\xi$$

8.1-25. Show that the network of Problem 8.1-24 is equivalent to a linear filter with impulse response

$$h(t) = \frac{1}{T}[u(t) - u(t - T)]$$

and find its transfer function.

8.1-26. Work Problem 8.1-11 for the network of Figure P8.1-26.

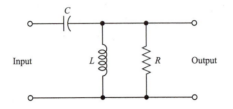

FIGURE P8.1-26

8.1-27. Work Problem 8.1-11 for the network of Figure P8.1-27.

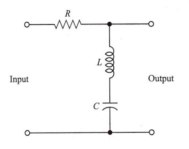

FIGURE P8.1-27

8.2-1. A random process

$$X(t) = A\sin(\omega_0 t + \Theta)$$

where A and ω_0 are real positive constants and Θ is a random variable uniformly distributed on the interval $(-\pi/\pi)$, is applied to the netowrk of Problem 8.1-1. Find an expression for the network's response process using (8.2-3).

8.2-2. Work Problem 8.2-1 for a network with impulse response

$$h(t) = u(t)te^{-t}$$

8.2-3. A random process $X(t)$ is applied to a linear time-invariant system. A response $Y(t) = X(t) - X(t - \tau)$ occurs where τ is a real constant.
(a) Sketch a block diagram of the system.
(b) Find the system's transfer function.

8.2-4. Work Problem 8.2-3 if the response is

$$Y(t) = X(t - \tau) + \int_{t_1}^{t_2} X(t - \xi)\, d\xi$$

where t_1 and t_2 are real constants.

8.2-5. A random process $X(t)$ has an autocorrelation function

$$R_{XX}(\tau) = A^2 + Be^{-|\tau|}$$

where A and B are positive constants. Find the mean value of the response of a system having an impulse response

$$h(t) = \begin{cases} e^{-Wt} & 0 < t \\ 0 & t < 0 \end{cases}$$

where W is a real positive constant, for which $X(t)$ is its input.

8.2-6. Work Problem 8.2-5 for the system for which

$$h(t) = \begin{cases} te^{-Wt} & 0 < t \\ 0 & t < 0 \end{cases}$$

8.2-7. Work Problem 8.2-5 for the system for which

$$h(t) = \begin{cases} e^{-Wt}\sin(\omega_0 t) & 0 < t \\ 0 & t < 0 \end{cases}$$

where W and ω_0 are real positive constants.

8.2-8. White noise with power density $5\,\text{W/Hz}$ is applied to the system of Problem 8.2-5. Find the mean-squared value of the response using (8.2-7).

8.2-9. Work Problem 8.2-8 for the system of Problem 8.2-6.

8.2-10. Work Problem 8.2-8 for the system of Problem 8.2-7.

8.2-11. Let jointly wide-sense stationary processes $X_1(t)$ and $X_2(t)$ cause responses $Y_1(t)$ and $Y_2(t)$, respectively, from a linear time-invariant system with impulse response $h(t)$. If the sum $X(t) = X_1(t) + X_2(t)$ is applied, the response is $Y(t)$. Find expressions, in terms of $h(t)$ and characteristics of $X_1(t)$ and $X_2(t)$, for (*a*) $E[Y(t)]$ (*b*) $R_{YY}(t, t + \tau)$

8.2-12. Show that the cross-correlation function for the output components $Y_1(t)$ and $Y_2(t)$ in Problem 8.2-11 is given by

$$R_{Y_1 Y_2}(t, t + \tau) = \int_{-\infty}^{\infty}\int_{-\infty}^{\infty} R_{X_1 X_2}(\tau + u - v)h(u)h(v)\, du\, dv$$
$$= R_{Y_1 Y_2}(\tau)$$

8.2-13. Two separate systems have impulse responses $h_1(t)$ and $h_2(t)$. A process $X_1(t)$ is applied to the first system and its response is $Y_1(t)$. Similarly, a process $X_2(t)$ invokes a response $Y_2(t)$ from the second system. Find the cross-correlation function of $Y_1(t)$ and $Y_2(t)$ in terms of $h_1(t)$, $h_2(t)$, and the cross-correlation function of $X_1(t)$ and $X_2(t)$. Assume $X_1(t)$ and $X_2(t)$ are jointly wide-sense stationary.

8.2-14. Two systems are cascaded. A random process $X(t)$ is applied to the input of the first system that has impulse response $h_1(t)$; its response $W(t)$ is the input to the second system having impulse response $h_2(t)$. The second system's output is $Y(t)$. Find the cross-correlation function of $W(t)$ and $Y(t)$ in terms of $h_1(t)$ and $h_2(t)$, and the autocorrelation function of $Y(t)$ if $X(t)$ is wide-sense stationary.

8.2-15. Let the two systems of Problem 8.2-14 be identical, each with the impulse response given in Problem 8.2-6. If $E[X(t)] = 2$ and $W = 3\,\text{rad/s}$, find $E[Y(t)]$.

8.2-16. A random process $X(t)$ having autocorrelation function

$$R_{XX}(\tau) = Pe^{-\alpha|\tau|}$$

where P and α are real positive constants, is applied to the input of a system with impulse response

$$h(t) = \begin{cases} We^{-Wt} & 0 < t \\ 0 & t < 0 \end{cases}$$

where W is a real positive constant. Find the autocorrelation function of the network's response $Y(t)$.

8.2-17. Find the cross-correlation function $R_{XY}(\tau)$ for Problem 8.2-16.

8.2-18. A signal

$$x(t) = u(t)\exp(-\alpha t)$$

is applied to a network having an impulse response

$$h(t) = u(t)W^2 t\exp(-Wt)$$

Here $\alpha > 0$ and $W > 0$ are real constants. By use of (8.2-2) find the network's response $y(t)$.

8.2-19. Work Problem 8.2-18 assuming

$$h(t) = u(t)W^3 t^2 \exp(-Wt)$$

8.2-20. A stationary random process $X(t)$ is applied to the input of a system for which

$$h(t) = 3u(t)t^2 \exp(-8t)$$

If $E[X(t)] = 2$ what is the mean value of the system's response $Y(t)$?

8.2-21. Work Problem 8.2-8 for the system of Problem 8.2-20.

8.2-22. White noise with power density $\mathcal{N}_0/2$ is applied to a network with impulse response

$$h(t) = u(t)Wt\exp(-Wt)$$

where $W > 0$ is a constant. Find the cross-correlations of the input and output.

8.2-23. Work Problem 8.2-22 for a network with impulse response

$$h(t) = u(t)Wt \sin(\omega_0 t) \exp(-Wt)$$

where ω_0 is a constant.

8.2-24. A random process $X(t)$ is applied to a network with impulse response

$$h(t) = u(t)t \exp(-bt)$$

where $b > 0$ is a constant. The cross-correlation of $X(t)$ with the output $Y(t)$ is known to have the same form:

$$R_{XY}(\tau) = u(\tau)\tau \exp(-b\tau)$$

(a) Find the autocorrelation of $Y(t)$.
(b) What is the average power in $Y(t)$?

8.2-25. Work Problem 8.2-24 except assume

$$h(t) = u(t)t^2 \exp(-bt)$$

and

$$R_{XY}(\tau) = u(\tau)\tau^2 \exp(-b\tau)$$

8.2-26. Two identical networks are cascaded. Each has impulse response

$$h(t) = u(t)3t \exp(-4t)$$

A wide-sense stationary process $X(t)$ is applied to the cascade's input.
(a) Find an expression for the response $Y(t)$ of the cascade.
(b) If $E[X(t)] = \bar{X} = 6$, find $E[Y(t)]$.

8.2-27. If a "time autocorrelation function" for the impulse response $h(t)$ of a linear filter is defined (for finite-energy impulse responses) by

$$\mathcal{R}_{hh}(\xi) = \int_{-\infty}^{\infty} h(t)h(t + \xi)\, dt$$

show that (8.2-9) can be written as

$$R_{YY}(\tau) = \int_{-\infty}^{\infty} R_{XX}(\alpha)\mathcal{R}_{hh}(\tau - \alpha)\, d\alpha = R_{XX}(\tau) * \mathcal{R}_{hh}(\tau)$$

8.2-28. Use the results of Problem 8.2-27 to find $R_{YY}(\tau)$ applicable to a system defined by the impulse response

$$h(t) = \frac{1}{T}[u(t) - u(t - T)]$$

where $T > 0$ is a constant.

8.3-1. Suppose the system of Figure 8.3-1 defined by $h(t)$ is in operation while the low-level white noise $X(t)$ is applied. That is, suppose the system's input is an operating (random) input signal $S(t)$ added to $X(t)$. $Y(t)$ will then contain a response, the operating output, due to $S(t)$, and a response, the output noise, due to $X(t)$. The cross-correlation measurement system's inputs are still $X(t)$

and $Y(t)$. Show that (8.3-4) still applies provided $S(t)$ is a process that is orthogonal to $X(t)$. Assume $X(t)$ and $S(t)$ are jointly wide-sense stationary.

8.4-1. The random process $X(t)$ of Problem 8.2-1 (the signal) is added to white noise with power density $\mathscr{N}_0/2$, where $\mathscr{N}_0$ is a positive constant, and the sum is applied to the network of Example 8.1-1.
(a) Find the power spectrums of the output signal and output noise.
(b) Find the ratio of output signal average power to output noise average power.
(c) What value of $W = R/L$ will maximize the ratio of part (b)?

8.4-2. For the processes and system of Problem 8.2-11, show that the power spectrum of $Y(t)$ is

$$\mathscr{S}_{YY}(\omega) = |H(\omega)|^2[\mathscr{S}_{X_1X_2}(\omega) + \mathscr{S}_{X_2X_2}(\omega) + \mathscr{S}_{X_1X_1}(\omega) + \mathscr{S}_{X_2X_1}(\omega)]$$

8.4-3. If $X_1(t)$ and $X_2(t)$ are statistically independent random processes in Problem 8.2-11, use the results of Problem 8.4-2 to show that the output power spectrum becomes

$$\mathscr{S}_{YY}(\omega) = |H(\omega)|^2[\mathscr{S}_{X_1X_1}(\omega) + \mathscr{S}_{X_2X_2}(\omega) + 4\pi\bar{X}_1\bar{X}_2\delta(\omega)]$$

8.4-4. Rework Example 8.4-1 when the network is replaced by two identical networks in cascade, that is, when $H(\omega) = [1 + (j\omega L/R)]^{-2}$.

8.4-5. Show that (8.4-7) and (8.4-8) are true.

8.4-6. A network with transfer function $H(\omega) = j\omega$ is a *differentiator*; its input is the wide-sense stationary random process $X(t)$ and its output is $\dot{X}(t) = dX(t)/dt$.
(a) By using (8.4-7), show that

$$R_{X\dot{X}}(\tau) = \frac{dR_{XX}(\tau)}{d\tau}$$

(b) By using (8.4-1), show that

$$R_{\dot{X}\dot{X}}(\tau) = -\frac{d^2R_{XX}(\tau)}{d\tau^2}$$

8.4-7. Given the random process

$$Y(t) = \frac{1}{2T}\int_{t-T}^{t+T} X(\xi)\,d\xi$$

where $X(t)$ is a wide-sense stationary process. Use (8.2-1) to show that the power spectrum of $Y(t)$ is

$$\mathscr{S}_{YY}(\omega) = \mathscr{S}_{XX}(\omega)\left[\frac{\sin(\omega T)}{\omega T}\right]^2$$

8.4-8. A stationary random process $X(t)$, having an autocorrelation function

$$R_{XX}(\tau) = 2\exp(-4|\tau|)$$

is applied to the network of Figure P8.4-8. Find: (a) $\mathscr{S}_{XX}(\omega)$, (b) $|H(\omega)|^2$, and (c) $\mathscr{S}_{YY}(\omega)$.

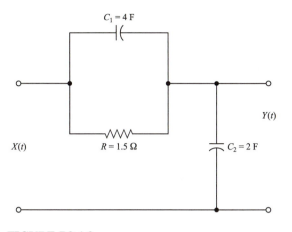

FIGURE P8.4-8

8.4-9. A wide-sense stationary process $X(t)$, with mean value 5 and power spectrum

$$\mathscr{S}_{XX}(\omega) = 50\pi\delta(\omega) + 3/[1 + (\omega/2)^2]$$

is applied to a network with impulse response

$$h(t) = 4\exp(-4|t|)$$

(a) Find $H(\omega)$ for the network.
Determine: (b) the mean $\bar{Y}$, and (c) the power spectrum of the response $Y(t)$.

8.4-10. White noise, for which $R_{XX}(\tau) = 10^{-2}\delta(\tau)$, is applied to a network with impulse response

$$h(t) = u(t)3t\exp(-4t)$$

(a) Use (8.2-9) to obtain the network's output noise power (in a 1-ohm resistor).
(b) Obtain an expression for the output power spectrum.

8.4-11. White noise with power density $\mathscr{N}_0/2 = 6(10^{-6})\,\text{W/Hz}$ is applied to an ideal filter (gain = 1) with bandwidth W (rad/s). Find W so that the output's average noise power is 15 watts.

8.4-12. An ideal filter with a midband power gain of 8 and bandwidth of 4 rad/s has noise $X(t)$ at its input with power spectrum

$$\mathscr{S}_{XX}(\omega) = \left(50/\sqrt{8\pi}\right)\exp(-\omega^2/8)$$

What is the noise power at the network's output?

8.4-13. A stationary random signal $X(t)$ has an autocorrelation function $R_{XX}(\tau) = 10\exp(-|\tau|)$. It is added to white noise [independent of $X(t)$] for which $\mathscr{N}_0/2 = 10^{-3}$ and the sum is applied to a filter having a transfer function

$$H(\omega) = \frac{2}{(1+j\omega)^3}$$

(a) Find the signal component of the output power spectrum and the average power in the output signal.
(b) Find the power spectrum of, and average power in, the output noise.
(c) What is the ratio of the output signal's power to the output average noise power?

8.4-14. A random noise $X(t)$, having a power spectrum

$$\mathscr{S}_{XX}(\omega) = \frac{3}{49 + \omega^2}$$

is applied to a differentiator that has a transfer function $H_1(\omega) = j\omega$. The differentiator's output is applied to a network for which

$$h_2(t) = u(t)t^2 \exp(-7t).$$

The network's response is a noise denoted by $Y(t)$.
(a) What is the average power in $X(t)$?
(b) Find the power spectrum of $Y(t)$.
(c) Find the average power in $Y(t)$.

8.5-1. Prove (8.5-6).

8.5-2. A random process $X(t)$ has a power spectrum $\mathscr{S}_{XX}(\omega)$ that is nonzero only for $-W_X < \omega < W_X$, where W_X is a real positive constant. $X(t)$ is applied to a system with transfer function

$$H(\omega) = 1 + j(\omega/W_H) \qquad - W_X < \omega < W_X$$

Find the average power P_{YY} in the network's response $Y(t)$ in terms of the rms bandwidth of $\mathscr{S}_{XX}(\omega)$, the constant W_H, and the average power P_{XX} in $X(t)$. Discuss the effect of letting $W_X \to \infty$.

8.5-3. Find the noise bandwidth of the system having the power transfer function

$$|H(\omega)|^2 = \frac{1}{1 + (\omega/W)^4}$$

where W is a real positive constant.

8.5-4. Work Problem 8.5-3 for the function

$$|H(\omega)|^2 = \frac{1}{[1 + (\omega/W)^2]^2}$$

8.5-5. Work Problem 8.5-3 for the function

$$|H(\omega)|^2 = \frac{1}{[1 + (\omega/W)^2]^3}$$

8.5-6. White noise with power density $\mathscr{N}_0/2$ is applied to a lowpass network for which $|H(0)| = 2$; it has a noise bandwidth of $2\,\text{MHz}$. If the average output noise power is $0.1\,\text{W}$ in a $1\text{-}\Omega$ resistor, what is $\mathscr{N}_0$?

8.5-7. White noise with power density $\mathcal{N}_0/2$, $\mathcal{N}_0 > 0$ a constant, is applied to a lowpass network for which $H(0) = 8$ and its noise bandwidth is 12 MHz. If average output noise power is 0.5 W in a 1-ohm resistor, what is $\mathcal{N}_0$?

8.5-8. A system's power transfer function is

$$|H(\omega)|^2 = 16/[256 + \omega^4]$$

(a) What is its noise bandwidth?
(b) If white noise with power density $6(10^{-3})$ W/Hz is applied to the input, find the noise power in the system's output.

8.5-9. Suppose

$$|H(\omega)|^2 = \frac{\omega^2}{[1 + (\omega/W)^2]^4}$$

for a network where $W > 0$ is a constant. Treat $|H(\omega)|^2$ as a bandpass function.
(a) Find ω_0, the value of ω where $|H(\omega)|^2$ is maximum.
(b) Find the network's noise bandwidth.

8.5-10. Work Problem 8.5-9 except for a network defined by

$$|H(\omega)|^2 = \frac{\omega^4}{[1 + (\omega/W)^2]^4}$$

8.6-1. White noise with power density $\mathcal{N}_0/2$ is applied to an ideal lowpass filter with bandwidth W.
(a) Find and sketch the autocorrelation function of the response.
(b) If samples of the output noise taken at times $t_n = n\pi/W$, $n = 0, \pm1, \pm2$, ..., are considered as values of random variables, what can you say about these random variables?

8.6-2. Work Problem 8.6-1 for an ideal bandpass filter centered on a frequency ω_0 $/2\pi$ that has a bandwidth W. Assume sample times are now $t_n = n2\pi/W$, $n = 0, \pm1, \pm2, \ldots$.

8.6-3. A band-limited random process $N(t)$ has the power density spectrum

$$\mathcal{S}_{NN}(\omega) = \begin{cases} P\cos[\pi(\omega - \omega_0)/W] & -W/2 \le \omega - \omega_0 \le W/2 \\ P\cos[\pi(\omega + \omega_0)/W] & -W/2 \le \omega + \omega_0 \le W/2 \\ 0 & \text{elsewhere} \end{cases}$$

where P, W, and $\omega_0 > W$ are real positive constants.
(a) Find the power in $N(t)$.
(b) Find the power spectrum $\mathcal{S}_{XX}(\omega)$ of $X(t)$ when $N(t)$ is represented by (8.6-2).
(c) Find the cross-correlation function $R_{XY}(\tau)$.
(d) Are $X(t)$ and $Y(t)$ orthogonal processes?

8.6-4. A band-limited random process is given by (8.6-2) and has the power density spectrum shown in Figure P8.6-4.

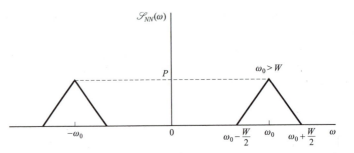

FIGURE P8.6-4

(a) Sketch $\mathscr{S}_{XX}(\omega)$.
(b) Sketch $\mathscr{S}_{XY}(\omega)$, if a sketch is possible.

8.6-5. Work Problem 8.6-4 for the power spectrum of Figure P8.6-5.

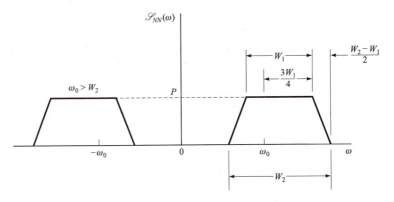

FIGURE P8.6-5

*8.6-6. Use (8.6-2) and derive (8.6-23).

*8.6-7. Assume a band-limited random process $N(t)$ has a power spectrum

$$\mathscr{S}_{NN}(\omega) = B[u(\omega - \omega_0 + W_1) - u(\omega - \omega_0 - W_2)]\exp[-a(\omega - \omega_0 + W_1)]$$
$$+ B[u(-\omega - \omega_0 + W_1) - u(-\omega - \omega_0 - W_2)]\exp[-a(-\omega - \omega_0 + W_1)]$$

where B, ω_0, W_1, and W_2 are positive constants, and a is a constant.
Assume $2\omega_0 > W_1 + W_2$ and find analytical expressions for (a) the power spectrum $\mathscr{S}_{XX}(\omega)$ and (b) the cross-power spectrum $\mathscr{S}_{XY}(\omega)$ for the processes $X(t)$ and $Y(t)$ involved in the representation of (8.6-2) for $N(t)$.
(c) Sketch $\mathscr{S}_{XX}(\omega)$ and $\mathscr{S}_{XY}(\omega)$ for $W_1 = W_2/2$ and $a = 1/W_1$.
(d) Repeat part (c) except with $a = -1/W_1$.

8.6-8. Find the functions $R_{XX}(\tau)$ and $R_{XY}(\tau)$ applicable in Problem 8.6-7.

8.6-9. Work Problem 8.6-3 except assume $N(t)$ has the power density spectrum

$$\mathcal{S}_{NN}(\omega) = \begin{cases} K_1 + K_2\left(\dfrac{\omega - \omega_0}{W}\right)^2 & -W/2 < \omega - \omega_0 < W/2 \\[2mm] K_1 + K_2\left(\dfrac{\omega + \omega_0}{W}\right)^2 & -W/2 < \omega + \omega_0 < W/2 \\[2mm] 0 & \text{elsewhere in } \omega \end{cases}$$

where $K_1 > 0$, $K_2 > 0$, and $W > 0$ are real constants.

8.6-10. A bandpass band-limited noise $N(t)$ has the power density spectrum of Problem 8.6-9. Find its autocorrelation function by using (8.6-11).

8.7-1. Show that (8.7-3) is the Fourier transform of (8.7-2).

8.7-2. Solve the convolution of (8.7-4) and prove its last form.

*8.7-3. Inverse Fourier transform (8.7-5) and show that (8.7-6) is valid.

8.7-4. Show that the response $Y(t)$ of the network in Figure 8.7-6 is $X(t)$, the band-limited bandpass random process having baseband components $X_I(t)$ and $X_Q(t)$ which are recovered from their I and Q samples.

8.7-5. By use of the lowpass sampling theorem show that the band-limited signal

$$x(t) = \frac{\pi \cos(W_X t)}{(\pi/2)^2 - (W_X t)^2}$$

is the sum of only two sampling functions separated in time by $T_s = \pi/W_X$. (*Hint*: Choose $T_s/2$ and $-T_s/2$ as the sample times.)

8.7-6. A signal $x(t) = B\mathrm{Sa}(W_X t)$, where B is a constant, is band-limited to a band $|\omega| < W_X$. Use sampling theory to justify that only *one* sample is adequate to reconstruct $x(t)$ for all time.

8.7-7. A nonrandom signal $x(t)$ is band-limited to $|\omega| < 2\pi(17.5)10^3$ rad/s. It is to be reconstructed exactly from its samples using an ideal lowpass filter. (*a*) What is the filter's minimum allowed bandwidth? (*b*) At what rate must the signal be sampled?

8.7-8. A band-limited nonrandom signal $x(t)$ is passed through a square-law non-linearity to produce a signal $y(t) = x^2(t)$. Give arguments to justify that $y(t)$ can be represented by the sampling theorem.

8.7-9. A music signal (considered random) is often classified as "high fidelity" if its maximum spectral extent is 20 kHz. If a compact disk player is to reconstruct the music using samples at four times the minimum sample rate, what is the sample rate?

8.7-10. Begin with (8.7-24) and let $T_p \to 0$ so that the rectangular functions become impulses. Next, use (A-29) to obtain a modified form. Finally, use (7.5-3) and Fourier transform the result to show that (8.7-25) can be expressed in the equivalent form

$$\mathscr{S}_{X_s X_s}(\omega) = \frac{1}{T_s} \sum_{n=-\infty}^{\infty} \mathscr{S}_{XX}(\omega - n\omega_s)$$

8.7-11. Assume a random process $X(t)$ has a power spectrum $\mathscr{S}_{XX}(\omega) = A_0/[1 + (\omega/W)^2]^2$, where A_0 and W are real positive constants. If the replicas in the power spectrum of the sampled version of $X(t)$ are to fall to 5% of their peak value when frequency is halfway between a pair of replicas, what must the sampling rate ω_s be in terms of W?

8.7-12. Work Problem 8.7-11 except for the power spectrum $\mathscr{S}_{XX}(\omega) = A_0/[1 + (\omega/W)^2]^4$.

8.7-13. Use the results of Problem 8.7-10 to sketch the sampled process' power spectrum, assuming the autocorrelation of the unsampled process is defined in Problem 8.7-11. Note the effect of aliasing. Assume $\omega_s = 2W$ for the sketch.

8.7-14. A nonrandom signal is given by

$$g(t) = \frac{KW_X}{8\pi} \left\{ 3\mathrm{Sa}\left[W_X\left(t - \frac{\pi}{2W_X} \right) \right] + 3\mathrm{Sa}\left[W_X\left(t + \frac{\pi}{2W_X} \right) \right] \right.$$

$$\left. + \mathrm{Sa}\left[W_X\left(t - \frac{3\pi}{2W_X} \right) \right] + \mathrm{Sa}\left[W_X\left(t + \frac{3\pi}{2W_X} \right) \right] \right\} \qquad (1)$$

where $K > 0$ and $W_X > 0$ are real constants. (a) Show that the Fourier transform $G(\omega)$ of $g(t)$ is the band-limited function

$$G(\omega) = K \cos^3\left(\frac{\pi\omega}{2W_X} \right) \mathrm{rect}\left(\frac{\omega}{2W_X} \right) \qquad (2)$$

(b) Assume $g(t)$ is to be sampled with sampling interval $T_s = \pi/W_X$. Write an expression for $g[t - (T_s/2)]$ in terms of T_s and $\omega_s = 2\pi/T_s$. (c) Discuss sampling of $g[t - (T_s/2)]$ and define where in time only four samples are needed to completely define the function. (d) Assume $T_s = \pi/W_X$ as in part (b) and define $T_0 = T_s/2$. Use (8.7-8) to write an expression for $g(t)$ in terms of T_0 and T_s. (e) Repeat part (c) except use the expression for $g(t)$ of part (d).

8.7-15. The system of Problem 8.7-14 is band-limited and nonnegative, so (2) could represent the power spectrum of some process. That is, a process $X(t)$ may be assumed to have the power spectrum $\mathscr{S}_{XX}(\omega)$ and autocorrelation function equal, respectively, to the right sides of (2) and (1) (with t replaced by τ) of Problem 8.7-14. Assume a sampling interval of $T_s = \pi/W_X$ and take $N = 11$ samples of $R_{XX}(\tau)$ at times $\tau_n = nT_s$, $-5 \leq n \leq 5$. Use these samples in a truncated version of (8.7-26) to calculate $\mathscr{S}_{X_s X_s}(\omega)$ for several values of $-\omega_s/2 < \omega < \omega_s/2$. Plot $T_s\mathscr{S}_{X_s X_s}(\omega)/K$ for these values and compare to a similarly normalized plot of $\mathscr{S}_{XX}(\omega)$.

8.7-16. In (8.6-2) suppose $X(t)$ and $Y(t)$ are baseband, band-limited, statistically independent, zero-mean, processes representing two sources of information. The spectral extent of $X(t)$ is 1.6 MHz and that of $Y(t)$ is 3.2 MHz. Discuss how I and Q sampling can be applied to the right side of (8.6-2). What minimum total sampling rate can be used, in principle?

8.7-17. Rework Example 8.7-2 except use $N = 20$. Is there a significant change in results from doubling N? Discuss your results.

8.8-1. Show that the DTFT of $\delta[n]$ is unity.

8.8-2. Show that the DTFT of $\delta[n - r]$, where r is any integer shift in $\delta[n]$, is $\exp(-jr\Omega)$.

8.8-3. If the DTFT of $h[n]$ is $\boldsymbol{H}(e^{j\Omega})$, show that the DTFT of $h[n - r]$, for r an integer, is $\boldsymbol{H}\{\exp(j\Omega)\}\exp(-jr\Omega)$.

8.8-4. A DT system is defined by

$$h[n] = \begin{cases} e^{-\alpha n} & n \geq 0 \\ 0 & n < 0 \end{cases}$$

for $\alpha > 0$ a constant. Determine if this sytem is BIBO stable. [*Hint*: Use (C-62).]

8.8-5. Find the transfer function for the DT system of Problem 8.8-4.

8.8-6. A DT random sequence is defined by

$$R_{XX}[n] = e^{-W_X|n|}$$

for $W_X > 0$ a constant. Find the power spectrum of this sequence. Is it band-limited? If not, what value of W_X is required for a given sample interval T_s so that the power spectrum is 5% of its maximum value when $\omega = \omega_s/2$?

*8.8-7. A DT system is to satisfy (8.8-21) with only b_0 and b_1 nonzero constants. (*a*) Find an expression for the magnitude squared $|\boldsymbol{H}[\exp(j\Omega)]|^2$ of the system's transfer function (in terms of b_0 and b_1). Determine its value for $\Omega = 0, \pi/2$, and π. (*b*) If the DT system is to approximate the lowpass filter function

$$\boldsymbol{H}_f(e^{j\Omega}) = \frac{1}{1 + j(\Omega/\Omega_0)}$$

where Ω_0 is a real positive constant, find b_0 and b_1 in terms of Ω_0 such that $|\boldsymbol{H}[\exp(j\Omega)]|^2 = |\boldsymbol{H}_f[\exp(j\Omega)]|^2$ at $\Omega = 0$ and $\Omega = \pi/2$. To what values must Ω_0 be restricted for b_0 and b_1 to be real? (*c*) For the smallest value of Ω_0 found in part (*b*), find $|\boldsymbol{H}[\exp(j\pi)]|^2$ and $|\boldsymbol{H}_f[\exp(j\pi)]|^2$ and observe the accuracy of the approximation.

8.8-8. A DT system has an impulse response $h[n] = \delta[n] + 2\delta[n - 1] + (1/2)\delta[n - 2]$. (*a*) Find all coefficients a_r and b_r in (8.8-21). Is this an FIR or IIR system? What are M and N? (*b*) If an input random process to the system corresponds to the autocorrelation sequence $R_{XX}[n] = a^{|n|}$, $0 < a < 1$, find the response sequence $Y[n]$ and its autocorrelation sequence $R_{YY}[k]$.

8.8-9. A white noise sequence $X[n]$, having an autocorrelation sequence $R_{XX}[n] = \sigma_X^2 \delta[n]$, is applied to an FIR system defined by (8.8-11) with

345

$M = 3$. (*a*) Find the autocorrelation sequence of the response sequence $Y[n]$. (*b*) Find the power spectrum of the response.

8.8-10. If a stationary white noise sequence $X[n]$ for which $R_{XX}[n] = \sigma_X^2 \delta[n]$ is applied to a DT system, use (8.8-22) to show that the response sequence $Y[n]$ has the autocorrelation sequence

$$R_{YY}[n] = \sigma_X^2 \sum_{m=-\infty}^{\infty} h[m]h[m+n] \tag{1}$$

The sum of (1) can be taken as the autocorrelation function of the system's impulse response.

8.8-11. Use the results of Problem 8.8-10 and (8.8-12) to show that an FIR DT system's response to white input noise has the autocorrelation sequence

$$R_{YY}[n] = \sigma_X^2 \sum_{m=0}^{M-|n|} b_m b_{m+|n|} \qquad 0 \le |n| \le M$$

$$= 0 \qquad\qquad\qquad \text{all other } n$$

8.8-12. Assume that a DT FIR system is to approximate a lowpass filter where

$$h(t) = \alpha \, u(t) \, e^{-\alpha t} \leftrightarrow H(\omega) = \frac{\alpha}{\alpha + j\omega}$$

From (8.8-12) the sequence of samples of $h(t)$ define the system's coefficients to be

$$h[n] = h(nT_s) = \alpha \, e^{-n\alpha T_s} = b_n \qquad 0 \le n \le M$$

(*a*) Use MATLAB and the results of Problem 8.8-11 to find and plot the normalized autocorrelation sequence $R_{YY}[n]/(\alpha^2 \sigma_X^2)$ when $\alpha T_s = \pi/4$ and $M = 8$. (*b*) Rework part (*a*) except assume $M = 2$ and note the effect of reducing M.

8.8-13. Use (8.8-15) and MATLAB to find the normalized transfer function $H[\exp(j\Omega)]/\alpha$ and plot its magnitude for the DT system defined by the impulse response sequence of Problem 8.8-12. Plot the magnitude versus $-\pi < \Omega < \pi$. Observe the effect of changing M from 8 to 2.

8.8-14. Work Problem 8.8-12 except assume the DT system is to approximate the following impulse response

$$h(t) = \alpha^2 u(t) t e^{-\alpha t} \leftrightarrow H(\omega) = \frac{\alpha^2}{(\alpha + j\omega)^2}$$

and use $\alpha T_s = \pi/[4(2^{0.5} - 1)^{0.5}]$.

8.8-15. Work Problem 8.8-13 except assume the DT system defined in Problem 8.8-14.

8.8-16. Work Problem 8.8-12 except assume the DT system is to approximate the following impulse response

$$h(t) = \frac{\alpha^3}{2} u(t) t^2 e^{-\alpha t} \leftrightarrow H(\omega) = \frac{\alpha^3}{(\alpha + j\omega)^3}$$

and use $\alpha T_s = \pi/[4(2^{1/3} - 1)^{0.5}]$.

8.8-17. Work Problem 8.8-13 except assume the DT system defined in Problem 8.8-16.

8.8-18. The impulse response of a DT system is $h[k] = u[k]\exp[-\alpha k]$, where $\alpha > 0$ is a real constant and $u[k]$ is the unit-step function given by

$$u[k] = \begin{cases} 1 & k \geq 0 \\ 0 & k < 0 \end{cases}$$

The system's input is defined by the sequence

$$X[m] = \begin{cases} a^m & m \geq 0 \text{ and } 0 < a < 1 \\ 0 & \text{otherwise} \end{cases}$$

Find the response sequence $Y[n]$.

8.8-19. Determine if the DT system of Problem 8.8-18 is BIBO stable.

8.8-20. Use (8.8-10) to prove whether or not the DT system of Problem 8.8-14 is BIBO stable.

8.8-21. Assume the only nonzero coefficients in (8.8-21) are $b_0 = 1$, $a_1 = 0.5$, and $a_2 = 0.2$. Use MATLAB with $N = 32$ to calculate the autocorrelation sequence $h[n]$ from (7.5-12). Plot both $h[n]$ and $|H(e^{j\Omega})|$.

8.8-22. Work Problem 8.8-21 except assume the only nonzero coefficients of (8.8-21) are $b_0 = 1$, $b_1 = 0.5$, and $b_2 = 0.2$.

8.9-1. A sonar echo system on a submarine transmits a random noise $n(t)$ to determine the distance to another "target" submarine. Distance R is given by v $\tau_R/2$ where v is the speed of the sound waves in water and τ_R is the time it takes the reflected version of $n(t)$ to return. Its block diagram is shown in Figure P8.9-1. Assume that $n(t)$ is a sample function of an ergodic random process $N(t)$ and T is very large.
(a) Find V in terms of a correlation function of $N(t)$.
(b) What value of the delay τ_T will cause V to be maximum?
(c) State in words how the submarine can determine the distance to the target.

8.9-2. Two resistors with resistances R_1 and R_2 are connected in parallel and have physical temperatures T_1 and T_2, respectively.
(a) Find the effective noise temperature of T_s of an equivalent resistor with resistance equal to the parallel combination of R_1 and R_2.
(b) If $T_1 = T_2 = T$, what is T_s?

8.9-3. Work Problem 8.9-2 for three resistances R_1, R_2, and R_3 in parallel when they have physical temperatures T_1, T_2, and T_3, respectively.

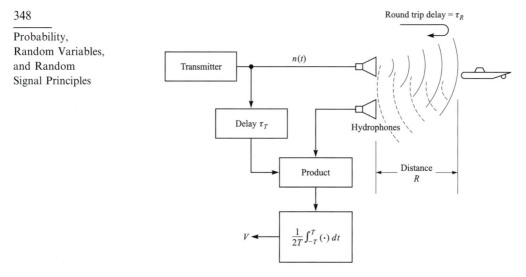

FIGURE P8.9-1

8.9-4. Work Example 8.9-2 if a second capacitor is placed across the resistance R_2. Is it possible to choose C_2 so that T_s is independent of frequency?

*8.9-5. Find the effective noise temperature of the network of Figure P8.9-5 if R_1 and R_2 are at physical temperatures T_1 and T_2, respectively.

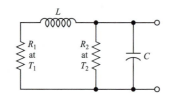

FIGURE P8.9-5

8.9-6. A two-port network is illustrated in Figure P8.9-6. Find its available power gain.

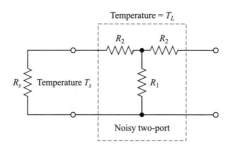

FIGURE P8.9-6

8.9-7. If the two-port network of Problem 8.9-6 has a physical temperature T_L and is driven by a source of resistance R_s and effective noise temperature T_s, what is the effective input noise temperature of the network?

8.9-8. If the output of the network of Problem 8.9-6 is connected to the input of a second identical network, what is the available power gain of the cascade if $R_1 = 5\Omega$, $R_2 = 3\Omega$, and $R_s = 7\Omega$?

8.9-9. Determine the effective noise temperature of the network of Figure P8.9-9 if resistors R_1 and R_2 are at different physical temperatures T_1 and T_2, respectively.

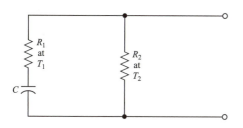

FIGURE P8.9-9

8.9-10. Two resistors in series have different physical temperatures as in Example 8.9-1. Let R_1 and R_2 be independent random variables uniformly distributed on $(1000, 1500)$ and $(2200, 2700)$, respectively. Their average resistances are then $\bar{R}_1 = 1250\,\Omega$ and $\bar{R}_2 = 2450\,\Omega$.
  (a) What is the effective noise temperature of the two resistors as a source if $T_1 = 250\,\mathrm{K}$ and $T_2 = 330\,\mathrm{K}$ and average resistors are used?
  (b) What is the mean effective noise temperature of the source for the same values of T_1 and T_2?

8.9-11. Find the effective noise temperature T_s of the network of Figure P8.9-11. What values does T_s assume for $\omega = 1/\sqrt{LC}$, $\omega = -\infty$, and $\omega = \infty$?

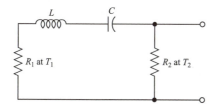

FIGURE P8.9-11

8.9-12. Work Problem 8.9-11, except replace the series $L–C$ circuit by a parallel $L–C$ circuit.

8.10-1. Show that (8.10-4) is valid.

8.10-2. In a cascade of M network stages for which the mth stage has available power gain G_m and operating spot noise figure F_{opm} when driven by all previous stages as its source, show that the overall cascade's operating spot noise figure is

$$F_{op} = F_{op1} + \frac{T_{s1}(F_{op2} - 1)}{T_s G_1} + \cdots + \frac{T_{s(M-1)}(F_{opM} - 1)}{T_s G_1 G_2 \cdots G_{M-1}}$$

where $T_{s(m-1)}$ is the temperature of all stages prior to stage m treated as a source.

8.10-3. An amplifier has a standard spot noise figure $F_0 = 6.31$ (8.0 dB). An engineer uses the amplifier to amplify the output of an antenna that is known to have antenna temperature of $T_a = 180$ K.
(a) What is the effective input noise temperature of the amplifier?
(b) What is the operating spot noise figure?

8.10-4. In a cascade of M stages for which F_{0m}, $m = 1, 2, \ldots, M$, is the standard spot noise figure of stage m which has available power gain G_m, show that the standard spot noise figure of the cascade of networks is

$$F_0 = F_{01} + \frac{F_{02} - 1}{G_1} + \frac{F_{03} - 1}{G_1 G_2} + \cdots + \frac{F_{0M} - 1}{G_1 G_2 \cdots G_{M-1}}$$

8.10-5. An amplifier has three stages for which $T_{e1} = 200$ K (first stage), $T_{e2} = 450$ K, and $T_{e3} = 1000$ K (last stage). If the available power gain of the second stage is 5, what gain must the first stage have to guarantee an effective input noise temperature of 250 K?

8.10-6. An amplifier has an operating spot noise figure of 10 dB when driven by a source of effective noise temperature 225 K?
(a) What is the standard spot noise figure of the amplifier?
(b) If a matched attenuator with a loss of 3.2 dB is placed between the source and the amplifier's input, what is the operating spot noise figure of the attenuator-amplifier cascade if the attenuator's physical temperature is 290 K?
(c) What is the standard spot noise figure of the cascade in (b)?

8.10-7. One manufacturer sells a microwave receiver having an operating spot noise figure of 10 dB when driven by a source with effective noise temperature 130 K. Another sells a receiver with a standard spot noise figure of 6 dB.
(a) Find the effective input noise temperatures of the two receivers.
(b) All other parameters, such as gain, cost, etc., being the same, which receiver would be the best to purchase?

8.10-8. An amplifier has three stages for which $T_{e1} = 150$ K (first stage), $T_{e2} = 350$ K, and $T_{e3} = 600$ K (output stage). Available power gain of the first stage is 10 and overall input effective noise temperature is 190 K.
(a) What is the available power gain of the second stage?
(b) What is the cascade's standard spot noise figure?
(c) What is the cascade's operating spot noise figure when used with a source of noise temperature $T_s = 50$ K?

8.10-9. Three networks are cascaded. Available power gains are $G_1 = 8$ (input stage), $G_2 = 6$, and $G_3 = 20$ (output stage). Respective input effective spot noise temperatures are $T_{e1} = 40\,K$, $T_{e2} = 100\,K$, and $T_{e3} = 280\,K$.

(a) What is the input effective spot noise temperature of the cascade?

(b) If the cascade is used with a source of noise temperature $T_s = 30\,K$, find the percentage of total available output noise power (in a band $d\omega$) due to each of the following: (1) source, and the excess noises of (2) network 1, (3) network 2, and (4) network 3.

8.10-10. An antenna with effective noise temperature $T_a = 90\,K$ is connected to an attenuator that is at a physical temperature of $270\,K$ and has a loss of 1.9. What is the effective spot noise temperature of the antenna-attenuator cascade if its output is considered as a noise source?

8.10-11. Three identical amplifiers, each having a spot effective input noise temperature of $125\,K$ and available power gain G, are cascaded. The overall spot effective input noise temperature of the cascade is $155\,K$. What is G?

8.10-12. Three amplifiers that may be connected in any order in a cascade are defined as follows:

TABLE P8.10-12

Amplifier	Effective input noise temperature	Available power gain
A	110 K	4
B	120 K	6
C	150 K	12

What sequence of connections will give the lowest overall effective input noise temperature for the cascade?

8.10-13. In an amplifier the first stage in a cascade of 5 stages has $T_{e1} = 75\,K$ and $G_1 = 0.5$. Each succeeding stage has an effective input noise temperature and an available power gain that are each 1.75 times that of the stage preceding it. What is the cascade's effective input noise temperature?

*8.10-14. Generalize Problem 8.10-13 by letting T_{e1} and G_1 be arbitrary and letting each succeeding stage have an effective input noise temperature and available power gain of K times that of the stage before it, where $K > 0$. Find a value of K that minimizes T_e for the cascade. Use the value of K found to determine the minimum value of T_e (for any G_1).

8.10-15. A designer requires an amplifier to give an operating spot noise figure of not more than 1.8 when operating with a 160-K source. What is the largest value of standard spot noise figure that will be acceptable in a purchased amplifier?

8.10-16. Two amplifiers have standard spot noise figures of $F_{01} = 1.6$ (unit 1) and $F_{02} = 1.4$ (unit 2). They have respective available power gains of $G_{a1} = 12$ and $G_{a2} = 8$. The two amplifiers are to be used in a cascade driven from an antenna to obtain an overall available power gain of $(8)12 = 96$.

(a) For best performance, which unit should be driven by the antenna?

(b) What is the standard spot noise figure of the best cascade?

8.10-17. An antenna with an effective noise temperature of 80 K drives a cascade of two amplifiers. The first (fed by the antenna) has an available power gain of 15 while the second has an input effective noise temperature of 600 K. The input effective noise temperature of the cascade is 140 K. The available power at the cascade's output in a small 1000-Hz band is $4.14(10^{-16})$ W.

(a) What is the input effective noise temperature of the first amplifier?

(b) What is the available power gain of the second stage?

(c) What is the cascade's operating spot noise figure?

(d) What is the cascade's standard spot noise figure?

8.11-1. What is the maximum average effective input noise temperature that an amplifier can have if its average standard noise figure is to not exceed 1.7?

8.11-2. An amplifier has an average standard noise figure of 2.0 dB and an average operating noise figure of 6.5 dB when used with a source of average effective source temperature $\bar{T}_s$. What is $\bar{T}_s$?

8.11-3. An antenna with average noise temperature 60 K connects to a receiver through various microwave elements that can be modeled as an impedance-matched attenuator with an overall loss of 2.4 dB and a physical temperature of 275 K. The overall system noise temperature is $\bar{T}_{sys} = 820$ K.

(a) What is the average effective input noise temperature of the receiver?

(b) What is the average operating noise figure of the attenuator-receiver cascade?

(c) What is the available output noise power of the receiver if it has an available power gain of 110 dB and a noise bandwidth of 10 MHz?

8.11-4. If the antenna-attenuator cascade of Problem 8.11-3 is considered as a noise source, what is its average effective noise temperature?

8.11-5. The loss L in Figure 8.11-1a is replaced by two cascaded matched attenuators, one with loss L_1 at temperature T_1 attached to the antenna output, and one with loss L_2 at temperature T_2 that connects to the receiver. Derive a new expression for $\bar{T}_{sys}$ analogous to (8.11-24).

8.11-6. An amplifier, when used with a source of average noise temperature 60 K, has an average operating noise figure of 5.

(a) What is $\bar{T}_e$?

(b) If the amplifier is sold to the engineering public, what noise figure would be quoted in a catalog (give a numerical answer)?

(c) What average operating noise figure results when the amplifier is used with an antenna of temperature 30 K?

8.11-7. An engineer purchases an amplifier with average operating noise figure of 1.8 when used with a 50-Ω broadband source having average source temperature of 80 K. When used with a different 50-Ω source the average operating noise figure is 1.25. What is the average noise temperature of the source?

8.11-8. An amplifier with a noise bandwidth of at least 1.8 MHz is needed by an
engineer. Two units from which he can choose are: unit 1—average standard
noise figure = 3.98, noise bandwidth = 2.0 MHz, and available power gain
= 10^6; unit 2—average standard noise figure = 2.82, noise bandwidth = 2.9
MHz, and available power gain = 10^6.

Find: (a) $\bar{T}_e$ for unit 1, (b) $\bar{T}_e$ for unit 2, (c) excess noise power of unit 1,
and (d) excess noise power of unit 2.

(e) If the source's noise temperature $\bar{T}_s$ is very small, which unit is the best to
purchase and why7?

(f) if $\bar{T}_s \gg \bar{T}_e$, which is best and why?

*8.11-9. A resistor is cooled to 75 K and serves as a noise source for a network with
available power gain

$$G_a(\omega) = 10^{36}/(10^6 + \omega^2)^4$$

(a) Write an expression for the power spectrum of the network's output
noise that is due to the source.

(b) Compute the available output noise power that is due to the source
alone.

8.11-10. A broadband antenna, for which $T_a = 120$ K, connects through an attenuator with loss 2.5 to a receiver with average input effective noise temperature
80 K, available power gain 10^{12}, and noise bandwidth 20 MHz. The antenna
and attenuator both have a physical temperature of 200 K.

(a) What is the attenuator's input effective noise temperature?

(b) What is the system's noise temperature?

(c) Find the average standard noise figure of the receiver by itself.

(d) What is the available noise power at the receiver's output (in system
operation)?

(e) Determine the input effective noise temperature of the attenuator-receiver taken as a unit.

(f) What is the average operating noise figure of this system when the
antenna is the source?

8.11-11. An antenna with average noise temperature 120 K connects to a receiver
through an impedance-matched attenuator having a loss of 1.5 and physical
temperature 75 K. For the overall system $\bar{T}_{sys} = 500$ K.

(a) What is the average effective input noise temperature of the receiver?

(b) What is the average operating noise figure of the attenuator-receiver
cascade?

(c) What is the available output noise power of the receiver if its available
power gain is 120 dB and its noise bandwidth is 20 MHz (system is connected)?

8.11-12. A receiving system consists of an antenna with noise temperature 80 K that
feeds a matched attenuator with physical temperature 220 K and loss 2.6.
The attenuator drives an amplifier with average effective noise temperature
170 K, noise bandwidth 4 MHz, and available power gain 10^8.

Find: (a) the overall system's average noise temperature $\bar{T}_{sys}$, (b) the available noise power N_{ao} at the system's output, (c) the total noise power available at the attenuator's output (within the noise bandwidth) and how much

of the total (as a percentage) is due to the antenna alone, and (d) the average operating noise figure $\bar{F}_{op}$ of the system.

8.11-13. Assume a source has an effective noise temperature of

$$T_s(\omega) = \frac{8000}{100 + \omega^2}$$

and feeds an amplifier that has an available power gain of

$$G_a(\omega) = \left| \frac{8}{10 + j\omega} \right|^2$$

(a) Find $\bar{T}_s$ for this source.
(b) Find the amplifier's noise bandwidth.
(c) What is the noise power available at the amplifier's output due to the source?

8.11-14. The available power gain of a network is

$$G_a(\omega) = \frac{K\omega^2}{(W^2 + \omega^2)^3}$$

where K and W are positive constants.
(a) At what value of ω, denoted by ω_0, does $G_a(\omega)$ reach a maximum?
(b) If $G_a(\omega)$ is considered to be a bandpass function with nominal (center) frequency ω_0, what is its noise bandwidth?

8.11-15. Work Problem 8.11-14, except assume an available power gain

$$G_a(\omega) = \frac{K\omega^4}{(W^2 + \omega^2)^4}$$

8.11-16. For the network of Problem 8.11-14, assume its input spot effective noise temperature varies as $T_e = 50 + (4\omega/W)^2$ and find its average input effective noise temperature.

8.11-17. A receiving system can be modeled as in Figure 8.11-1 if $T_a = 130\,\text{K}$, $L = 1.6$, $T_L = 200\,\text{K}$, $W_N/2\pi = 8\,\text{MHz}$, $G_r(\omega_0) = 5(10^9)$, and $\bar{T}_{sys} = 558\,\text{K}$. A sinusoidal signal with an angular frequency ω_0 is also being received that produces an available power of $55(10^{-12})\,\text{W}$ at the antenna's output. Find: (a) $\bar{T}_R$, (b) N_{ao}, (c) available output signal power S_{ao}, (d) the signal-to-noise ratio S_{ao}/N_{ao}, (e) $\bar{F}_0$ for the receiver, and (f) the effective input noise temperature of the loss, $\bar{T}_{e(loss)}$.

8.11-18. A receiving system has an antenna, for which $T_a = 120\,\text{K}$, driving two broadband matched-impedance attenuators in cascade, which then drive a receiver for which $\bar{T}_R = 100\,\text{K}$, $W_N/2\pi = 5\,\text{MHz}$, and $G_R(\omega_0) = 10^{12}$. The attenuator to which the antenna is connected has a physical temperature of $70\,\text{K}$ and a loss of 1.6. The other attenuator's physical temperature is $250\,\text{K}$ and its loss is 1.9.
(a) What is the receiver's available output noise power?
(b) What available signal power at the antenna's output will produce a signal-to-noise power ratio of 1000 (or 30 dB) at the receiver's output?
(c) What is $\bar{T}_{sys}$?

8.11-19. An antenna, for which $T_a = 60$ K, feeds a cascade of two impedance-matched attenuators. The first, connected to the antenna, has a physical temperature of 75 K and a loss of 1.9. The second attenuator, at a physical temperature of 290 K and with a loss of 1.4, drives a broadband mixer that has an available power gain of 0.5 and $\bar{T}_e = 500$ K. Finally, the mixer drives an amplifier for which $G_a(\omega_0) = 10^7$, $\bar{F}_0 = 5$, and $W_N = 2\pi(10^6)$ rad/s.

(a) What is the input effective noise temperature of the attenuator that is at physical temperature 75 K?

(b) Repeat (a) for the second attenuator.

(c) What is the average input effective noise temperature of the whole cascade?

(d) What is $\bar{T}_{sys}$?

(e) What is the average operating noise figure of the cascade having the antenna as its source?

(f) What average noise power is available at the amplifier's output?

Optimum Linear Systems

9.0
INTRODUCTION

The developments of the preceding chapter related entirely to the *analysis* of a linear system. In this chapter we do an about-face and concentrate only on the *synthesis* of a linear system. In particular, we choose the system in such a way that is satisfies the certain rules that make it *optimum*.

In designing any optimum system we must consider three things: *input specficiation*, *system constraints*, and *criterion of optimality*.

Input specification means that at least some knowledge must be available about the input to the system. For example, we might specify the input to consist of the sum of a random signal and a noise. Alternatively, the input could be the sum of a deterministic signal and a noise. In addition, we may be able to specify signal and noise correlation functions, power spectrums, or probability densities. Thus, we may know a great deal about the inputs in some cases or little in others. Regardless, however, there is some minimum knowledge required of the characteristics of the input for any given problem.

System constraints define the form of the resulting system. For example, we might allow the system to be linear, nonlinear, time-invariant, realizable, etc. In our work we shall be exclusively concerned with linear time-invariant systems but will not necessarily require that they be realizable. By relaxing the realizability constraint, we shall be able to introduce the most important topics of interest without undue mathematical complexity.

In principle, there is great latitude available in choosing the criterion of optimality. In a practical sense, however, it should be a meaningful measure of "goodness" for the problem at hand and should correspond to equations that are mathematically tractable. We shall be concerned with only two criteria. One will involve the minimization of a suitably defined error quantity. The other will relate to maximization of the ratio of a signal power to a noise power. This last criterion leads us to an optimum system often called a *matched filter*.

357

CHAPTER 9:
Optimum Linear
Systems

9.1
SYSTEMS THAT MAXIMIZE SIGNAL-TO-NOISE RATIO

An important class of systems involves the transmission of a deterministic signal of known form in noise. A digital communication system is one example† where, during a time interval T, a known signal may arrive at the receiver in the presence of additive noise. The presence of the signal corresponds to transmission of a digital "1," while absence of the signal occurs when a digital "0" is transmitted (noise is always present). It would seem reasonable that some system (or filter‡) could be found that would enhance its output signal power at some instant in time while reducing its output average noise power. Indeed, such a filter that maximizes this output *signal-to-noise ratio* can be found and it is called a *matched filter*. It can be shown that decisions made as to whether the signal was present or not during time interval T have the smallest probability of being in error if they are based on samples taken at the times of maximum signal-to-noise ratio. Although our comments here are directed toward a digital communication system, we shall find as we progress that the matched filter concept is a broad one, applying to many situations.

In this section we shall consider the optimization of a linear time-invariant system when the input consists of the sum of a Fourier-transformable deterministic signal $x(t)$ of known form and continuous noise $n(t)$. If we denote by $x_o(t)$ and $n_o(t)$ the output signal and noise, the criterion of optimality we choose is the maximization of the ratio of the output signal power at some time t_o to the output average noise power. Thus, with $n_o(t)$ assumed to be a sample function of a wide-sense stationary random process§ $N_o(t)$, we maximize

$$\left(\frac{\hat{S}_o}{N_o}\right) = \frac{|x_o(t_o)|^2}{E[N_o^2(t)]} \tag{9.1-1}$$

where

$$\hat{S}_o = |x_o(t_o)|^2 \tag{9.1-2}$$

is the output signal power at time t_o and

$$N_o = E[N_o^2(t)] \tag{9.1-3}$$

is the output average noise power.

---

†Although we discuss only this example, many systems such as radars, sonars, radio altimeters, ionospheric sounders, and automobile crash avoidance systems are other examples.
‡We often use the words system, filter, or network in this chapter to convey the same meaning.
§This assumption is equivalent to assuming the input noise is from a wide-sense stationary random process since the system is assumed to be linear and time-invariant (see Section 8.2).

Matched Filter for Colored Noise

Define $X(\omega)$ as the Fourier transform of $x(t)$, and $H(\omega)$ as the transfer function of the system. The output signal at any time t is

$$x_o(t) = \frac{1}{2\pi} \int_{-\infty}^{\infty} X(\omega)H(\omega)e^{j\omega t}\, d\omega \tag{9.1-4}$$

From (8.4-6), the output average noise power can be written in the form

$$N_o = E[N_o^2(t)] = \frac{1}{2\pi} \int_{-\infty}^{\infty} \mathscr{S}_{NN}(\omega)|H(\omega)|^2\, d\omega \tag{9.1-5}$$

where $\mathscr{S}_{NN}(\omega)$ is the power density spectrum of the random process, denoted $N(t)$, that represents the input noise $n(t)$. By use of (9.1-4) at time t_o and (9.1-5), we can write (9.1-1) as

$$\left(\frac{\hat{S}_o}{N_o}\right) = \frac{\left|\dfrac{1}{2\pi} \displaystyle\int_{-\infty}^{\infty} X(\omega)H(\omega)e^{j\omega t_o}\, d\omega\right|^2}{\dfrac{1}{2\pi} \displaystyle\int_{-\infty}^{\infty} \mathscr{S}_{NN}(\omega)|H(\omega)|^2\, d\omega} \tag{9.1-6}$$

To find $H(\omega)$ that maximizes (9.1-6), we shall apply the *Schwarz†ineequality*. If $A(\omega)$ and $B(\omega)$ are two possibly complex functions of the real variable ω, the inequality states that

$$\left|\int_{-\infty}^{\infty} A(\omega)B(\omega)\, d\omega\right|^2 \le \int_{-\infty}^{\infty} |A(\omega)|^2\, d\omega \int_{-\infty}^{\infty} |B(\omega)|^2\, d\omega \tag{9.1-7}$$

The equality holds only when $B(\omega)$ is proportional to the compelx conjugate of $A(\omega)$; that is, when

$$B(\omega) = CA^*(\omega) \tag{9.1-8}$$

where C is any *arbitrary* real constant.

By making the substitutions

$$A(\omega) = \sqrt{\mathscr{S}_{NN}(\omega)}H(\omega) \tag{9.1-9}$$

$$B(\omega) = \frac{X(\omega)e^{j\omega t_o}}{2\pi\sqrt{\mathscr{S}_{NN}(\omega)}} \tag{9.1-10}$$

in (9.1-7) we obtain

$$\left|\frac{1}{2\pi}\int_{-\infty}^{\infty} X(\omega)H(\omega)e^{j\omega t_o}\, d\omega\right|^2 \le \int_{-\infty}^{\infty} \mathscr{S}_{NN}(\omega)|H(\omega)|^2\, d\omega \frac{1}{(2\pi)^2}\int_{-\infty}^{\infty} \frac{|X(\omega)|^2}{\mathscr{S}_{NN}(\omega)}\, d\omega \tag{9.1-11}$$

---

†Named for the German mathematician Hermann Amandus Schwarz (1843–1921).

With this last result, we write (9.1-6) as

$$\left(\frac{\hat{S}_o}{N_o}\right) \leq \frac{1}{2\pi} \int_{-\infty}^{\infty} \frac{|X(\omega)|^2}{\mathscr{S}_{NN}(\omega)} d\omega \qquad (9.1\text{-}12)$$

The maximum value of $(\hat{S}_o/N_o)$ occurs when the equality holds in (9.1-12), which implies that (9.1-8) is true. Denote the optimum filter transfer function by $H_{\text{opt}}(\omega)$. We find this function by solving (9.1-8) using (9.1-9) and (9.1-10); the result is

$$H_{\text{opt}}(\omega) = \frac{1}{2\pi C} \frac{X^*(\omega)}{\mathscr{S}_{NN}(\omega)} e^{-j\omega t_o} \qquad (9.1\text{-}13)$$

From (9.1-13), we find that the optimum filter is proportional to the complex conjugate of the input signal's spectrum; we might say that the system is therefore *matched* to the specified signal since it depends so intimately on it. $H_{\text{opt}}(\omega)$ is also inversely proportional to the power spectrum of the input noise. In general, this noise has been assumed nonwhite; that is, colored. Because of these facts, an optimum filter given by (9.1-13) is called a *matched filter for colored noise*.

$H_{\text{opt}}(\omega)$ is also proportional to the inverse of the arbitrary constant C. In other words, $H_{\text{opt}}(\omega)$ has an arbitrary absolute magnitude. This fact allows the optimal system to have arbitrary gain. Intuitively, we feel that this should be true because gain affects both input signal and input noise in the same way, and, in the ratio of (9.1-1), gain cancels.

The time t_o at which the output ratio $(\hat{S}_o/N_o)$ is maximum enters into the optimum system transfer function only through the factor $\exp(-j\omega t_o)$. Such a factor only represents an ideal delay. Since t_o is a parameter that a designer may have some latitude in choosing, its value may be selected in some cases to make the optimum filter causal.

In general, the system defined by (9.1-13) may not be realizable. For certain forms of colored noise realizable filters may be found (Thomas, 1969, Chapter 5). In practice, one can always approximate (9.1-13) by a suitably chosen real filter.

Matched Filter for White Noise

If the input noise is white with power density $\mathscr{N}_0/2$, the optimum filter of (9.1-13) becomes

$$H_{\text{opt}}(\omega) = KX^*(\omega)e^{-j\omega t_o} \qquad (9.1\text{-}14)$$

where $K = 1/\pi C \mathscr{N}_0$ is an arbitrary constant. Here the optimum filter is related only to the input signal's spectrum and the time that $(\hat{S}_o/N_o)$ is maximum. Thus, the name *matched filter* is very appropriate. Indeed, the name was originally attached to the filter in white noise; we have liberalized the name to include the preceding colored noise case.

The impulse response denoted $h_{\text{opt}}(t)$ of the optimum filter is the inverse Fourier transform of $H_{\text{opt}}(\omega)$. From (9.1-14), it is easily found that

$$h_{opt}(t) = Kx^*(t_o - t) \tag{9.1-15}$$

For real signals $x(t)$, (9.1-15) reduces to

$$h_{opt}(t) = Kx(t_o - t) \tag{9.1-16}$$

Equation (9.1-16) indicates that the impulse response is equal to the input signal displaced to a new origin at $t = t_o$ and folded about this point so as to "run backward."

> **EXAMPLE 9.1-1.** We shall find the matched filter for the signal of Figure 9.1-1a when received in white noise. From (9.1-16), the matched filter's impulse response is as shown in (b). By Fourier transformation of the waveform in (b), we readily obtain
>
> $$H_{opt}(\omega) = KA\tau \frac{\sin(\omega\tau/2)}{(\omega\tau/2)} e^{-j\omega[t_o + \tau_o - (\tau/2)]}$$
>
> An alternative development consists of Fourier-transforming the input signal to get $X(\omega)$ and then using (9.1-14).
>   Whether or not any chance exists for the matched filter to be realizable may be determined from the impulse response of Figure 9.1-1b.

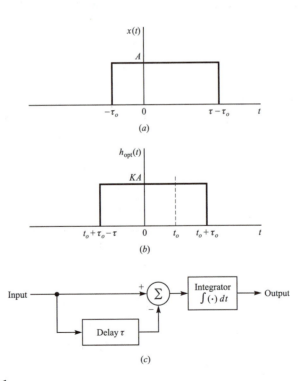

FIGURE 9.1-1
A matched filter and its related signals. (a) Input signal, (b) the filter's impulse response, and (c) the filter's block diagram. [*Reproduced from Peebles (1976), with permission of publishers Addison-Wesley, Advanced Book Program.*]

Clearly, to be causal, and therefore realizable, the delay must be at least $\tau - \tau_o$; that is

$$t_o \geq \tau - \tau_o$$

If we assume this last condition is satisfied, the optimum filter is illustrated in (c) where the arbitrary constant K is set equal to $1/A$. This filter still requires that perfect integrators be possible. Of course, they are not. However, very good approximations are possible using modern operational amplifiers with feedback, so for all practical purposes matched filters for rectangular pulses in white noise may be constructed.[†]

9.2
SYSTEMS THAT MINIMIZE MEAN-SQUARED ERROR

A second class of optimum systems is concerned with causing the output to be a good estimate of some function of the input signal which arrives along with additive noise. One example corresponds to the output being a good estimate of the *derivative* of the input signal. In another case, the system could be designed so that its output is a good estimate of either the past, present, or future value of the input signal. We shall concern ourselves with only this last case. The optimum system or filter that results is called a *Wiener filter*.[‡]

Wiener Filters

The basic problem to be studied is depicted by Figure 9.2-1. The input signal $x(t)$ is now assumed to be *random*; it is therefore modeled as a sample function

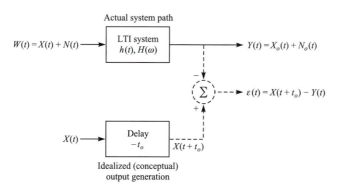

FIGURE 9.2-1
Operations that define the Wiener filter problem.

[†]Other techniques using *integrate-and-dump* methods exist. See Peebles (1976), pp. 361–362.
[‡]After Norbert Wiener (1894–1964), a great American mathematician whose work has tremendously affected many areas of science and engineering.

of a random process $X(t)$. It is applied to the input of the system along with additive noise $n(t)$ that is a sample function of a noise process $N(t)$. We assume $X(t)$ and $N(t)$ are jointly wide-sense stationary processes and that $N(t)$ has zero mean. The sum of signal and noise is denoted $W(t)$:

$$W(t) = X(t) + N(t) \qquad (9.2\text{-}1)$$

The system is assumed to be linear and time-invariant with a real impulse response $h(t)$ and a transfer function $H(\omega)$. The output of the system is denoted $Y(t)$.

In general, we shall select $H(\omega)$ so that $Y(t)$ is the best possible estimate of the input signal $X(t)$ at a time $t + t_o$; that is, the best estimate of $X(t + t_o)$. If $t_o > 0$, $Y(t)$ is an estimate of a *future* value of $X(t)$ corresponding to a *prediction filter*. If $t_o < 0$, $Y(t)$ is an estimate of a *past* value of $X(t)$ and we have a *smoothing filter*. If $t_o = 0$, $Y(t)$ is an estimate of the current value of $X(t)$.

Now if $Y(t)$ differs from the desired true value of $X(t + t_o)$, we make an error of

$$\varepsilon(t) = X(t + t_o) - Y(t) \qquad (9.2\text{-}2)$$

This error is illustrated conceptually in Figure 9.2-1 by dashed lines. The optimum filter will be chosen so as to minimize the mean-squared value of $\varepsilon(t)$.† We shall not be concerned with obtaining a system that is realizable. Some information is given by Thomas (1969) on the more difficult problem where $H(\omega)$ must be realizable. Thus, we seek to find $H(\omega)$ that minimizes

$$
\begin{aligned}
E[\varepsilon^2(t)] &= E[\{X(t + t_o) - Y(t)\}^2] \\
&= E[X^2(t + t_o) - 2Y(t)X(t + t_o) + Y^2(t)] \\
&= R_{XX}(0) - 2R_{YX}(t_o) + R_{YY}(0) \qquad (9.2\text{-}3)
\end{aligned}
$$

From the Fourier transform relationship between an autocorrelation function and a power spectrum, we have

$$R_{XX}(0) = \frac{1}{2\pi} \int_{-\infty}^{\infty} \mathscr{S}_{XX}(\omega)\, d\omega \qquad (9.2\text{-}4)$$

where $\mathscr{S}_{XX}(\omega)$ is the power density spectrum of $X(t)$. From a similar relationship and (8.4-1) we have

$$R_{YY}(0) = \frac{1}{2\pi} \int_{-\infty}^{\infty} \mathscr{S}_{WW}(\omega)|H(\omega)|^2\, d\omega \qquad (9.2\text{-}5)$$

where $\mathscr{S}_{WW}(\omega)$ is the power spectrum of $W(t)$. By substitution of (9.2-4) and (9.2-5) into (9.2-3), we have

$$E[\varepsilon^2(t)] = -2R_{YX}(t_o) + \frac{1}{2\pi} \int_{-\infty}^{\infty} [\mathscr{S}_{XX}(\omega) + \mathscr{S}_{WW}(\omega)|H(\omega)|^2]\, d\omega \qquad (9.2\text{-}6)$$

---

†We could elect to minimize the average error, or even force such an error to be zero. This approach does not prevent large positive errors from being offset by large negative errors, however. By minimizing the squared error, we eliminate such possibilities.

To reduce (9.2-6) further, we develop the cross-correlation function:

$$R_{YX}(t_o) = E[Y(t)X(t+t_o)] = E\left[X(t+t_o)\int_{-\infty}^{\infty} h(\xi)W(t-\xi)\,d\xi\right]$$

$$= \int_{-\infty}^{\infty} R_{WX}(t_o + \xi)h(\xi)\,d\xi \tag{9.2-7}$$

where $R_{WX}(\cdot)$ is the cross-correlation function of $W(t)$ and $X(t)$. After replacing $R_{WX}(t_o + \xi)$ by its equivalent, the inverse Fourier transform of the cross-power spectrum $\mathscr{S}_{WX}(\omega)$, we obtain

$$R_{YX}(t_o) = \int_{-\infty}^{\infty}\frac{1}{2\pi}\int_{-\infty}^{\infty}\mathscr{S}_{WX}(\omega)e^{j\omega(t_o+\xi)}\,d\omega\, h(\xi)\,d\xi$$

$$= \frac{1}{2\pi}\int_{-\infty}^{\infty}\mathscr{S}_{WX}(\omega)e^{j\omega t_o}\left\{\int_{-\infty}^{\infty} h(\xi)e^{j\omega\xi}\,d\xi\right\}d\omega$$

$$= \frac{1}{2\pi}\int_{-\infty}^{\infty}\mathscr{S}_{WX}(\omega)H(-\omega)e^{j\omega t_o}\,d\omega \tag{9.2-8}$$

Substitution of this expression into (9.2-6) allows it to be written as

$$E[\varepsilon^2(t)] = \frac{1}{2\pi}\int_{-\infty}^{\infty}\{\mathscr{S}_{XX}(\omega) - 2\mathscr{S}_{WX}(\omega)H(-\omega)e^{j\omega t_o} + \mathscr{S}_{WW}(\omega)|H(\omega)|^2\}\,d\omega \tag{9.2-9}$$

The transfer function that minimizes $E[\varepsilon^2(t)]$ is now found. We may write $H(\omega)$ in the form

$$H(\omega) = A(\omega)e^{jB(\omega)} \tag{9.2-10}$$

where $A(\omega)$ is the magnitude of $H(\omega)$, and $B(\omega)$ is its phase. Next we observe that $\mathscr{S}_{XX}(\omega)$ and $\mathscr{S}_{WW}(\omega)$ are real nonnegative functions, since they are power spectrums, while the cross-power spectrum $\mathscr{S}_{WX}(\omega)$ is complex in general and can be written as

$$\mathscr{S}_{WX}(\omega) = C(\omega)e^{jD(\omega)} \tag{9.2-11}$$

After using (9.2-10) and (9.2-11) in (9.2-9) and invoking the fact that

$$H(-\omega) = H^*(\omega) \tag{9.2-12}$$

for filters having a real impulse response $h(t)$, we obtain

$$E[\varepsilon^2(t)] = \frac{1}{2\pi}\int_{-\infty}^{\infty}\{\mathscr{S}_{XX}(\omega) + \mathscr{S}_{WW}(\omega)A^2(\omega)\}\,d\omega$$

$$- \frac{1}{2\pi}\int_{-\infty}^{\infty} 2C(\omega)A(\omega)e^{j[\omega t_o + D(\omega) - B(\omega)]}\,d\omega \tag{9.2-13}$$

We minimize $E[\varepsilon^2(t)]$ by first selecting the phase of $H(\omega)$ to maximize the second integral in (9.2-13) and then, with the optimum phase substituted, minimize the resulting expression by choice of $A(\omega)$. Clearly, choosing

$$B(\omega) = \omega t_o + D(\omega) \tag{9.2-14}$$

will maximize the second integral and give the expression

$$E[\varepsilon^2(t)] = \frac{1}{2\pi} \int_{-\infty}^{\infty} \{ \mathscr{S}_{XX}(\omega) - 2C(\omega)A(\omega) + \mathscr{S}_{WW}(\omega)A^2(\omega) \} \, d\omega$$

$$= \frac{1}{2\pi} \int_{-\infty}^{\infty} \left\{ \mathscr{S}_{XX}(\omega) - \frac{C^2(\omega)}{\mathscr{S}_{WW}(\omega)} + \mathscr{S}_{WW}(\omega) \left[A(\omega) - \frac{C(\omega)}{\mathscr{S}_{WW}(\omega)} \right]^2 \right\} \, d\omega$$

$$(9.2\text{-}15)$$

In writing the last form of (9.2-15), we have completed the square in $A(\omega)$. Finally, it is clear that choosing

$$A(\omega) = \frac{C(\omega)}{\mathscr{S}_{WW}(\omega)} \tag{9.2-16}$$

will minimize the right side of (9.2-15). By combining (9.2-16), (9.2-14), and (9.2-11) with (9.2-10) we have the optimum filter transfer function which we denote $H_{\text{opt}}(\omega)$:

$$H_{\text{opt}}(\omega) = \frac{\mathscr{S}_{WX}(\omega)}{\mathscr{S}_{WW}(\omega)} e^{j\omega t_o} \tag{9.2-17}$$

For the special case where input signal and noise are uncorrelated, it is easy to show that

$$\mathscr{S}_{WW}(\omega) = \mathscr{S}_{XX}(\omega) + \mathscr{S}_{NN}(\omega) \tag{9.2-18}$$

$$\mathscr{S}_{WX}(\omega) = \mathscr{S}_{XX}(\omega) \tag{9.2-19}$$

where $\mathscr{S}_{NN}(\omega)$ is the power spectrum of $N(t)$. Hence, for this special case

$$H_{\text{opt}}(\omega) = \frac{\mathscr{S}_{XX}(\omega)}{\mathscr{S}_{XX}(\omega) + \mathscr{S}_{NN}(\omega)} e^{j\omega t_o} \tag{9.2-20}$$

EXAMPLE 9.2-1. We find the optimum filter for estimating $X(t + t_o)$ when there is no input noise. We let $\mathscr{S}_{NN}(\omega) = 0$ in (9.2-20):

$$H_{\text{opt}}(\omega) = e^{j\omega t_o}$$

This expression corresponds to an ideal delay line with delay $-t_o$. If $t_o > 0$, corresponding to prediction, we require an unrealizable negative delay line. If $t_o > 0$, corresponding to a smoothing filter, the required delay is positive and realizable. Of course, $t_o = 0$ results in $H_{\text{opt}}(\omega) = 1$. In other words, the optimum filter for estimating $X(t)$ when no noise is present is just a direct connection from input to output, a result that is intuitively agreeable.

Minimum Mean-Squared Error

On substitution of (9.2-17) into (9.2-15), we readily find the mean-squared error of the optimum filter

$$E[\varepsilon^2(t)]_{min} = \frac{1}{2\pi} \int_{-\infty}^{\infty} \frac{\mathscr{S}_{XX}(\omega)\mathscr{S}_{WW}(\omega) - |\mathscr{S}_{WX}(\omega)|^2}{\mathscr{S}_{WW}(\omega)} \, d\omega \qquad (9.2\text{-}21)$$

For the special case where input signal and noise are uncorrelated, this equation reduces to

$$E[\varepsilon^2(t)]_{min} = \frac{1}{2\pi} \int_{-\infty}^{\infty} \frac{\mathscr{S}_{XX}(\omega)\mathscr{S}_{NN}(\omega)}{\mathscr{S}_{XX}(\omega) + \mathscr{S}_{NN}(\omega)} \, d\omega \qquad (9.2\text{-}22)$$

9.3
OPTIMIZATION BY PARAMETER SELECTION

We conclude our discussions of optimum linear systems by briefly considering a second approach that minimizes mean-squared error. The problem we undertake is identical to that of the last section up to (9.2-9), which defines the mean-squared error. Now, however, rather than seeking the filter that minimizes this error, we *specify* the *form* of the filter in terms of a number of unknown parameters and then determine the parameter values that minimize the mean-squared error. This procedure necessarily leads to a real filter so long as the form we choose corresponds to such a filter.

If we assume the special case where the input signal $X(t)$ and noise $N(t)$ are uncorrelated, (9.2-9) can be written as

$$E[\varepsilon^2(t)] = \frac{1}{2\pi} \int_{-\infty}^{\infty} \mathscr{S}_{\varepsilon\varepsilon}(\omega) \, d\omega \qquad (9.3\text{-}1)$$

where

$$\mathscr{S}_{\varepsilon\varepsilon}(\omega) = \mathscr{S}_{XX}(\omega) - 2\mathscr{S}_{XX}(\omega)H(-\omega)e^{j\omega t_o} + [\mathscr{S}_{XX}(\omega) + \mathscr{S}_{NN}(\omega)]|H(\omega)|^2 \qquad (9.3\text{-}2)$$

Since the imaginary part of $H(-\omega)\exp(j\omega t_o)$ is an odd function of ω when $h(t)$ is real (as assumed), the only contribution to the integral of (9.3-1) due to the middle term in (9.3-2) results from the real part of $H(-\omega)\exp(j\omega t_o)$. Thus, the error-contributing part of (9.3-2) can be written as†

$$\mathscr{S}_{\varepsilon\varepsilon} = \mathscr{S}_{XX}(\omega)[1 - H(\omega)e^{-j\omega t_o} - H(-\omega)e^{j\omega t_o} + |H(\omega)|^2] + \mathscr{S}_{NN}(\omega)|H(\omega)|^2$$
$$= \mathscr{S}_{XX}(\omega)|1 - H(-\omega)e^{j\omega t_o}|^2 + \mathscr{S}_{NN}(\omega)|H(\omega)|^2 \qquad (9.3\text{-}3)$$

because $H(-\omega) = H^*(\omega)$.

We summarize the synthesis procedure. First, a filter form is chosen for a real filter. The applicable transfer function $H(\omega)$ will depend on a number of unknown parameters. $H(\omega)$ is next substituted into (9.3-3), to obtain $\mathscr{S}_{\varepsilon\varepsilon}(\omega)$, the power density spectrum of the error $\varepsilon(t)$. Finally, the error $E[\varepsilon^2(t)]$ is calculated from (9.3-1) and the parameters are then found by formally minimizing this error. Although this procedure is direct and conceptually simple to

---

†In writing (9.3-3), we also use the fact that $2\,\text{Re}\,(z) = z + z^*$ for any complex number z.

apply, the solution of the integral of (9.3-1) may be tedious. For the case where $\mathscr{S}_{XX}(\omega)$ and $\mathscr{S}_{NN}(\omega)$ are rational functions of ω and $H(\omega)$ corresponds to a real filter form, the resulting integral has been tabulated for a number of functions $\mathscr{S}_{\varepsilon\varepsilon}(\omega)$ involving orders of ω up to 14 (Thomas, 1969, pp. 249 and 636, and James, et al., 1947, p. 369).

All the preceding discussion has related to the special case where the input signal and input zero-mean noise are jointly wide-sense stationary and uncorrelated. For the more general case of correlated signal and noise, the choice of form for $H(\omega)$ must be substituted into (9.2-9) and the integral solved. The unknown filter coefficients are then determined that minimize $E[\varepsilon^2(t)]$.

9.4
SUMMARY

All earlier chapters concentrated on definitions and *analysis* of random processes and the system responses to such processes. This chapter has given a very *brief* introduction to the *synthesis* of *optimum* systems. In essence, we sought to find the best system for a given problem. Two main categories of problem were introduced and the specific topics covered were:

- The system (filter) was derived that causes the maximum ratio of peak signal power (at a point in time) to the average noise power at the system's output. The optimum filter is called a matched filter; two results were developed, one for arbitrary noise and one for white noise. Results were general and the filters were not constrained to be realizable.
- The second system category involved the optimum filter that minimized a suitable mean-squared-error criterion. One such filter, called the Wiener filter, was chosen for development. It minimized the square of the difference between the filter's response and a desired response defined as the input signal at any chosen past, present, or future time (a prediction filter).
- The Wiener and matched filters that were found were not necessarily realizable. The chapter closed with a brief discussion of a method of selecting a realizable filter, defined by a finite number of parameters, to approximate the optimum filter. The procedure was called parameter selection.

PROBLEMS

9.1-1. A matched filter is to be found for a signal defined by

$$x(t) = \begin{cases} A(\tau + t)/\tau & -\tau < t < 0 \\ A(\tau - t)/\tau & 0 < t < \tau \\ 0 & \text{elsewhere} \end{cases} \leftrightarrow X(\omega) = A\tau \left[\frac{\sin(\omega\tau/2)}{\omega\tau/2} \right]^2$$

when added to noise having a power density spectrum

$$\mathscr{S}_{NN}(\omega) = \frac{W_2}{W_2^2 + \omega^2}$$

where A, τ, and W_2 are real positive constants.

(a) Find the matched filter's transfer function $H_{opt}(\omega)$.
(b) Find the filter's impulse response $h_{opt}(t)$. Plot $h_{opt}(t)$.
(c) Is there a value of t_o for which the filter is causal? If so, find it.
(d) Sketch the block diagram of a network that has $H_{opt}(\omega)$ as its transfer function.

9.1-2. Work Problem 9.1-1 (a), (b), and (c) for the signal

$$x(t) = u(t)[e^{-W_2 t} - e^{-\alpha W_2 t}]$$

if $\alpha > 1$ is a real constant.

9.1-3. Work Problem 9.1-1 (a), (b), and (c) for the signal

$$x(t) = u(-t)[e^{W_2 t} - e^{\alpha W_2 t}]$$

if $\alpha > 1$ is a real constant.

*9.1-4. By proper inverse Fourier transformation of (9.1-13), show that the impulse response $h_{opt}(t)$ of the matched filter for signals in colored noise satisfies

$$\int_{-\infty}^{\infty} h_{opt}(\xi) R_{NN}(t - \xi)\, d\xi = x^*(t_o - t)$$

9.1-5. A signal $x(t)$ and colored noise $N(t)$ are applied to the network of Figure P9.1-5. We select $|H_1(\omega)|^2 = 1/\mathscr{S}_{NN}(\omega)$ so that the noise $N_1(t)$ is white. We also make $H_2(\omega)$ a matched filter for the signal $x_1(t)$ in the white noise $N_1(t)$. Show that the cascade is a matched filter for $x(t)$ in the noise $N(t)$.

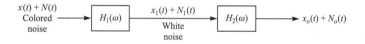

FIGURE P9.1-5

9.1-6. For the matched filter of Example 9.1-1, find and sketch the output signal. [*Hint*: Fourier-transform $x(t)$ and use a transform pair from Appendix E to obtain $x_o(t)$.]

9.1-7. Assume the power density of the white noise at the input to the matched filter of Example 9.1-1 is $\mathscr{N}_0/2$ with $\mathscr{N}_0 > 0$ a real constant. Find the output signal-to-noise ratio of the filter at time t_o.

9.1-8. Show that the maximum output signal-to-noise ratio obtainable from a filter matched to a signal $x(t)$ in white noise with power density $\mathscr{N}_0/2$ is

$$\left(\frac{\hat{S}_o}{N_o}\right)_{max} = \frac{2}{\mathscr{N}_0} \int_{-\infty}^{\infty} |x(t)|^2\, dt = \frac{2E}{\mathscr{N}_0}$$

where E is the energy in $x(t)$ and $\mathscr{N}_0 > 0$ is a real constant.

9.1-9. Let τ be a positive real constant. A pulse

$$x(t) = \begin{cases} A\cos(\pi t/\tau) & |t| < \tau/2 \\ 0 & |t| > \tau/2 \end{cases}$$

is added to white noise with a power density of $\mathcal{N}_0/2$. Find $(\hat{S}_o/N_o)_{max}$ for a filter matched to $x(t)$ by using the result of Problem 9.1-8.

9.1-10. Find the matched filter's transfer function applicable to Problem 9.1-9.

9.1-11. Show that the output signal $x_o(t)$ from a filter matched to a signal $x(t)$ in white noise is

$$x_o(t) = K\int_{-\infty}^{\infty} x^*(\xi)x(\xi + t - t_o)\,d\xi$$

That is, $x_o(t)$ is proportional to the *correlation integral* of $x(t)$.

9.1-12. Show that the output signal $x_o(t)$ from a filter matched to a signal in white noise reaches its maximum magnitude at $t = t_o$ if the filter impulse response is given by (9.1-15). (*Hint*: Use the result of Problem 9.1-11.)

9.1-13. Fourier-transform the signal of Figure 9.1-1a, and use (9.1-14) to verify the optimum system transfer function given in Example 9.1-1.

9.1-14. The signal

$$x(t) = u(t)e^{-Wt}$$

where $W > 0$ is a real constant, is applied to a filter along with white noise with power density $\mathcal{N}_0/2$, $\mathcal{N}_0 > 0$ being a real constant.
(a) Find the transfer function of the filter matched to $x(t)$ at time t_o.
(b) Find and sketch the filter's impulse response.
(c) Is there any value of t_o that will make the filter causal?
(d) Find the output maximum signal-to-noise ratio.

9.1-15. Work Problem 9.1-14 for the signal

$$x(t) = u(-t)e^{Wt}$$

9.1-16. Work Problem 9.1-14 for the signal

$$x(t) = u(t)te^{-Wt}$$

9.1-17. Work Problem 9.1-14 for the signal

$$x(t) = -u(-t)te^{Wt}$$

9.1-18. If a real signal $x(t)$ exists only in the interval $0 < t < T$, show that the *correlation receiver* of Figure P9.1-18 is a matched filter at time $t = T$; that is, show that the ratio of peak signal power to average noise power, both at time T, is the same as the ordinary matched filter. Assume white input noise.

9.1-19. Find the matched filter for the signal

$$x(t) = Ae^{-\alpha t^2}$$

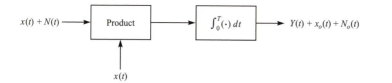

FIGURE P9.1-18

in white noise with power density $\mathcal{N}_0/2$ where $\mathcal{N}_0 > 0$, $\alpha > 0$, and A are real constants.

9.1-20. A signal $x(t) = u(t)5t^2 \exp(-2t)$ is added to white noise for which $\mathcal{N}_0/2 = 10^{-2}$ W/Hz. The sum is applied to a matched filter.
(a) What is the filter's transfer function?
(b) What is $(\hat{S}_o/N_o)$?
(c) Sketch the impulse response of the filter.
(d) Is the filter realizable?

9.1-21. A signal

$$x(t) = u(t)t^2 \exp(-Wt)$$

is added to noise with power spectrum

$$\mathcal{S}_{NN}(\omega) = P/(W_N^2 + \omega^2)$$

where W, P, and W_N are positive constants. The sum is applied to a matched filter.
(a) Find the filter's transfer function.
(b) Find the filter's impulse response.
(c) What is the maximum signal-to-noise ratio at the output?

9.1-22. A pulse of amplitude $A > 0$ and duration $\tau > 0$ is $x(t) = A \operatorname{rect}(t/\tau)$. The pulse is added to white noise of power density $\mathcal{N}_0/2$ when it arrives at a receiver. For some practical reasons the receiver (filter) is not a matched filter but is a simple lowpass filter with transfer function

$$H(\omega) = W/(W + j\omega)$$

$W > 0$ a constant.
(a) Find the ratio of instantaneous output signal power $x_o^2(t)$ at any time t to average noise power $E[N_o^2(t)]$ at the filter's output. At what time, denoted by t_o, is the ratio maximum?
(b) At time t_o what bandwidth W will maximize signal-to-noise ratio?
(c) Plot the loss in output signal-to-noise ratio that results, compared to a matched filter, for various values of $0 < W \leq 5/\tau$. What is the minimum loss?

*9.1-23. Reconsider the system of Problem 9.1-22 except assume

$$H(\omega) = W^2/(W + j\omega)^2$$

(a) Find the time t_o at which output signal-to-noise ratio is largest.
(b) For the t_o found in (a) determine the output signal-to-noise ratio. Plot this result versus $W\tau$ for $0 < W\tau \le 6$ and determine what value of W gives the best performance.
(c) What minimum loss in signal-to-noise ratio occurs compared to a matched filter?

9.1-24. A pulse

$$x(t) = A \, \text{rect}(t/2\tau)[1 - (t/\tau)^2]$$

where A and $\tau > 0$ are constants, is added to white noise.
(a) Find the output signal $x_o(t)$ of a filter matched to the pulse.
(b) Sketch $x(t)$ and $x_o(t)$.
(c) What is the matched filter's output signal-to-noise ratio?
(d) What is the transfer function if K in (9.1-16) is chosen so that $|H_{\text{opt}}(0)| = 1$? Is there a value of t_o that makes the filter causal?

*9.1-25. A deterministic waveform $\psi(t)$ is defined by

$$\psi(t) = a(t)e^{j\phi(t)+j\omega_0 t} = v(t)e^{j\omega_0 t}$$

where $a(t)$ and $\phi(t)$ are "slowly" varying amplitude and phase "modulation" functions and $\omega_0 > 0$ is a large constant. The white-noise matched filter for $\psi(t)$ is defined by

$$h_{\text{opt}}(t) = \psi^*(t_o - t)$$

if $K = 1$ in (9.1-15). Now let $\psi(t)$ be offset in frequency by an amount ω_d before being applied to the "matched filter" so that

$$\psi_R(t) = \psi(t)\exp(-j\omega_d t)$$

is applied with noise to the filter.
(a) Show that the filter's response to $\psi_R(t)$ is

$$\chi(t_o - t, \omega_d) = \int_{-\infty}^{\infty} \psi(\xi)\psi^*(t_o - t + \xi)e^{-j\omega_d \xi} \, d\xi$$

The function $|\chi(\alpha, \omega_d)|^2$ is called the *ambiguity function* of the waveform $\psi(t)$.
(b) Show that the volume under the ambiguity function does not depend on the form of $\psi(t)$ but only on $|\chi(0, 0)|^2$.
(c) Show that

$$\chi(t_o - t, \omega_d) = e^{j\omega_0(t-t_o)}\int_{-\infty}^{\infty} v(\xi)v^*(t_o - t + \xi)e^{-j\omega_d \xi} \, d\xi$$

*9.1-26. Reconsider the ambiguity function of Problem 9.1-25.
(a) Show that $|(\tau, \omega_d)|^2 \le |\chi(0, 0)|^2$.
(b) Show that another form for $\chi(\tau, \omega_d)$ is

$$\chi(\tau, \omega_d) = \frac{1}{2\pi}\int_{-\infty}^{\infty} \Psi^*(\omega)\Psi(\omega + \omega_d)e^{-j\omega\tau} \, d\omega$$

where $\Psi(\omega)$ is the Fourier transform of $\psi(t)$.

(c) Show that

$$\chi(\tau, 0) = \int_{-\infty}^{\infty} \psi(\xi)^*(\xi + \tau)\, d\xi$$

$$= \frac{1}{2\pi} \int_{-\infty}^{\infty} |\Psi(\omega)|^2 e^{-j\omega\tau}\, d\omega$$

$$\chi(0, \omega_d) = \int_{-\infty}^{\infty} |\psi(\xi)|^2 e^{-j\omega_d\xi}\, d\xi$$

$$= \frac{1}{2\pi} \int_{-\infty}^{\infty} \Psi^*(\omega)\Psi(\omega + \omega_d)\, d\omega$$

(d) Show that the symmetry of $\chi(\tau, \omega_d)$ is given by

$$\chi(\tau, \omega_d) = e^{j\omega_d\tau} \chi^*(-\tau, -\omega_d)$$

9.1-27. The deterministic signal

$$x(t) = \text{rect}(t/T) \exp(j\omega_0 t + j\mu t^2/2)$$

is a pulse having a linearly varying frequency with time during the pulse's duration T. The nominal frequency is ω_0 (rad/s). The matched filter for white noise has the impulse response of (9.1-15) which, for $t_o = 0$, is

$$h_{\text{opt}}(t) = K \, \text{rect}(t/T) \exp(j\omega_o t - j\mu t^2/2)$$

(a) If instantaneous frequency is to increase by a total amount $\Delta\omega$ (rad/s) during the pulse's duration T, how is the constant μ related to $\Delta\omega$ and T?
(b) Find the value of K such that $|H_{\text{opt}}(\omega_0)| = 1$ when μ is large. [*Hint:* Note that

$$C(x) = \int_0^x \cos(\pi\xi^2/2)\, d\xi$$

and

$$S(x) = \int_0^x \sin(\pi\xi^2/2)\, d\xi$$

called *Fresnel integrals*, approach $\frac{1}{2}$ as $x \to \infty$.]
(c) For the K found in (b), determine the output $x_o(t)$ of the filter. Sketch the envelopes of the signals $x(t)$ and $x_o(t)$ for $\Delta\omega T = 80\pi$ using the same time-voltage axes. What observations can you make about what has happened to $x(t)$ as it passes through the filter?

*9.1-28. (a) Find the transfer function $H_{\text{opt}}(\omega)$ of the matched filter of Problem 9.1-27. (*Hint:* Put the epxression in terms of Fresnel integrals having arguments

$$x_1 = \sqrt{\Delta\omega T/2\pi}\{1 - [2(\omega - \omega_0)/\Delta\omega]\}/\sqrt{2}$$

and

$$x_2 = \sqrt{\Delta\omega T/2\pi}\{1 + [2(\omega - \omega_0)/\Delta\omega]\}/\sqrt{2}$$

where $\mu = \Delta\omega/T$.)
(b) Sketch the approximate form of $|H_{\text{opt}}(\omega)|$ that results when $\Delta\omega T$ is large.

9.1-29. A signal

$$x(t) = \begin{cases} \dfrac{1}{2}e^{-t/6} & 0 < t < 3/2 \\ 0 & \text{elsewhere in } t \end{cases}$$

is added to white noise of power density $\mathcal{N}_0/2$ and the sum is applied to the input of a matched filter. The output peak signal-to-noise power ratio is 14. What is $\mathcal{N}_0/2$? (*Hint*: Use the results of Problem 9.1-8.)

9.1-30. White noise, for which $\mathcal{N}_0/2 = 10^{-8}/(24\pi)$, and a signal

$$x(t) = \begin{cases} Wte^{-Wt} & 0 < t < 2/W \\ 0 & \text{elsewhere in } t \end{cases}$$

are applied to a matched filter. What ratio of output peak signal power to average noise power can be achieved if $W = 5(10^6)$ rad/s? (*Hint*: Use results of Problem 9.1-8.)

9.1-31. In trying to build the matched filter required in Problem 9.1-30 an engineer encounters difficulties and builds, instead, a filter matched to the signal

$$x_a(t) = u(t)Wte^{-Wt}$$

which is the unlimited-time version of the signal $x(t)$. What ratio of output peak signal power to average noise power can be achieved for the same values of W and $\mathcal{N}_0/2$ as assumed in Problem 9.1-30?

*9.1-32. Assume the signal and noise of Problem 9.1-30 are applied to the filter used by the engineer in Problem 9.1-31. Since the filter is not matched, optimum performance is not achieved.
(*a*) Use convolution to find the output signal $x_o(t)$ of the filter at any time t.
(*b*) Find the value of t_o for which the output signal's amplitude is maximum at $t = 0$.
(*c*) Find the maximum peak power in the output signal.
(*d*) Find the average noise power and the maximum output signal-to-noise power ratio.

9.1-33. The sum of a signal

$$x(t) = \begin{cases} 0 & t < -3 \\ 6 + 2t & -3 < t < 5 \\ 0 & 5 < t \end{cases}$$

and white noise, for which $\mathcal{N}_0/2 = 0.1$ W/Hz, is applied to a matched filter.
(*a*) What is the smallest value of t_o required for the filter to be causal?
(*b*) For the value of t_0 found in (*a*), sketch the impulse response of the matched filter.
(*c*) Find the maximum output signal-to-noise ratio it provides. (*Hint*: Use the results of Problem 9.1-8.)

9.1-34. Work Problem 9.1-33 except assume the signal shown in Figure P9.1-34.

9.1-35. Find the transfer function of the white-noise matched filter corresponding to the signal

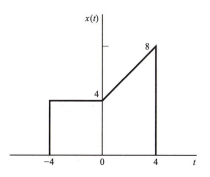

FIGURE P9.1-34

$$x(t) = (A + Bt)\left[u\left(t + \frac{A}{B}\right) - u\left(t - \frac{A}{B}\right)\right]$$

where $A > 0$ and $B > 0$ are constants.

9.1-36. Work Problem 9.1-22, except assume the input pulse is

$$x(t) = \frac{A}{\tau}t[u(t) - u(t - \tau)]$$

9.2-1. A random signal $X(t)$ and uncorrelated white noise $N(t)$ have autocorrelation functions

$$R_{XX}(\tau) = \frac{WP}{2}e^{-W|\tau|}$$
$$R_{NN}(\tau) = (\mathcal{N}_0/2)\delta(\tau)$$

where $W > 0$, $P > 0$, and $\mathcal{N}_0 > 0$ are real constants.
(a) Find the transfer function of the optimum Wiener filter.
(b) Find and sketch the impulse response of the filter when $t_o < 0$, $t_o > 0$, and $t_o = 0$.

9.2-2. Find the minimum mean-squared error of the filter in Problem 9.2-1.

9.2-3. Work Problem 9.2-1 for colored noise defined by

$$R_{NN}(\tau) = W_N e^{-W_N|\tau|}$$

where $W_N > 0$ is a real constant.

9.2-4. Work Problem 9.2-2 for the noise defined in Problem 9.2-3.

9.2-5. A random signal $X(t)$ and additive uncorrelated noise $N(t)$ have respective power spectrums

$$\mathcal{S}_{XX}(\omega) = \frac{9}{9 + \omega^4} \qquad \text{and} \qquad \mathcal{S}_{NN}(\omega) = \frac{3}{6 + \omega^4}$$

(a) Find the transfer function of the Wiener filter for the given signal and noise.
(b) Find the minimum value of the error in predicting $X(t + t_o)$.

9.2-6. Work Problem 9.2-5 for signal and uncorrelated white noise defined by

$$\mathscr{S}_{XX}(\omega) = \frac{A}{W^2 + \omega^4}$$

$$\mathscr{S}_{NN}(\omega) = \mathscr{N}_0/2$$

where $A > 0$, $W > 0$, and $\mathscr{N}_0 > 0$ are real constants.

9.2-7. A random signal $X(t)$ and uncorrelated white noise have respective power spectrums

$$\mathscr{S}_{XX}(\omega) = 2\sqrt{2}P_{XX}(W_X\omega^2/(W_X^4 + \omega^4))$$

and

$$\mathscr{S}_{NN}(\omega) = \mathscr{N}_0/2$$

Here P_{XX} is the average power in $X(t)$, while W_X and $\mathscr{N}_0$ are positive constants.
(a) Find the transfer function of the Wiener filter for this signal and noise.
(b) What is the minimum mean-squared filter error?
(c) Evaluate the result of (b) for $P_{XX} = 2\,\text{W}$, $W_X = 15\,\text{rad/s}$, and $\mathscr{N}_0/2 = 0.1\,\text{W/Hz}$. [*Hint*: Use the known integral (Thomas, 1969, p. 249)

$$I_2 = \frac{1}{2\pi} \int_{-\infty}^{\infty} \frac{(b_1 - b_0\omega^2)d\omega}{a_2^2 + (a_1^2 - 2a_0a_2)\omega^2 + a_0^2\omega^4} = \frac{a_0b_1 - a_2b_0}{2a_0a_1a_2}$$

where b_0, b_1, a_0, a_1, and a_2 are constants and $a_0\lambda^2 + a_1\lambda + a_2$ has no roots in the lower half-plane when $\lambda = \omega + j\sigma$.]

9.2-8. Work Problem 9.2-7 for the signal with the power spectrum

$$\mathscr{S}_{XX}(\omega) = A/(W_X^2 + \omega^2)^2$$

Put results in terms of the average power P_{XX} in $X(t)$.

9.2-9. The respective power spectrums of a random signal $X(t)$ and uncorrelated noise $N(t)$ are

$$\mathscr{S}_{XX}(\omega) = (1/20)/(10^2 + \omega^2)$$

and

$$\mathscr{S}_{NN}(\omega) = \omega^2/(16^2 + \omega^2)^2$$

(a) What is the transfer function of the Wiener filter?
(b) What is the minimum mean-squared prediction error? (*Hint*: Use results from Problem 9.2-7.)

*9.2-10. Generalize the random signal of Problem 9.2-9 by assuming its power spectrum is

$$\mathscr{S}_{XX}(\omega) = (W_X^2/2000)/(W_X^2 + \omega^2)$$

where W_X is the signal's 3-dB bandwidth. Find the minimum mean-squared prediction error and plot the result for $W_X > 9.5$. What does an increase in W_X mean in a physical sense?

9.2-11. A random signal $X(t)$ plus uncorrelated noise $N(t)$, having respective power
spectrums

$$\mathscr{S}_{XX}(\omega) = 2P_{XX}W_X/(W_X^2 + \omega^2)$$

and

$$\mathscr{S}_{NN}(\omega) = 4P_{NN}W_N^3/(W_N^2 + \omega^2)^2$$

is applied to a Wiener filter. Here P_{XX} and P_{NN} are the average signal and
noise powers, respectively, while W_X and W_N are positive constants.
(a) Use (9.2-22) and find the filter's minimum mean-squared prediction error.
(b) Show that as $P_{XX} \to \infty$, $E[\varepsilon^2(t)]_{min} \to P_{NN}$, and that $E[\varepsilon^2(t)]_{min} \to P_{XX}$ if
$P_{NN} \to \infty$.
(c) From a graphical plot of $E[\varepsilon^2(t)]_{min}/P_{NN}$ versus W_X/W_N, determine if
there is a preferred bandwidth ratio when $P_{NN}/P_{XX} = 8$. Is there a ratio
that should be avoided? Discuss. (*Hint:* Use the integral given in Problem
9.2-7.)

9.2-12. Work Problem 9.2-7 except assume the signal $X(t)$ has a power spectrum

$$\mathscr{S}_{XX}(\omega) = \frac{P_{XX}4W_X\omega^2}{(W_X^2 + \omega^2)^2}$$

9.2-13. Use (9.2-20) and give arguments to justify that the Wiener filter emphasizes
those frequencies where the ratio of signal power to noise power is largest.

9.2-14. A random signal $X(t)$ has a power density spectrum

$$\mathscr{S}_{XX}(\omega) = \frac{\mathscr{N}_X}{2} \operatorname{rect}\left(\frac{\omega}{W_X}\right)$$

where $\mathscr{N}_X > 0$ and $W_X > 0$ are constants.
(a) Find the average power P_{XX} in $X(t)$.
(b) Find the optimum (Wiener) filter's transfer function when input noise is
independent of $X(t)$ and white with power density $\mathscr{N}_0/2$.
(c) Find the ratio of the minimum mean-squared error to the power P_{XX}.
Evaluate the result for $\mathscr{N}_X/\mathscr{N}_0 = 16$.

9.3-1. A deterministic signal $x(t) = A\cos(\omega_0 t)$ and white noise with power density
$\mathscr{N}_0/2$ are applied to a one-section lowpass filter with transfer function
$H(\omega) = W/(W + j\omega)$. Here $W > 0$, $\mathscr{N}_0 > 0$, ω_0, and A are all real constants.
What value of W will cause the ratio of output *average* signal power to
average noise power to be maximum?

9.3-2. Work Problem 9.3-1 if the network consists of two identical one-section filters
in cascade.

9.3-3. Work Problem 9.3-1 if $x(t) = A\cos(\omega_0 t + \Theta)$, where Θ is a random variable
uniformly distributed on the interval $(0, 2\pi)$.

9.3-4. A random signal $X(t)$ having the autocorrelation function

$$R_{XX}(\tau) = W_X e^{-W_X|\tau|}$$

and uncorrelated noise with power density $\mathscr{N}_0/2$ are applied to a lowpass filter with transfer function

$$H(\omega) = \frac{W}{W + j\omega}$$

Here $W > 0$ and $W_X > 0$ are real constants.

(a) What value of W will minimize the mean-squared error if the output is to be an estimate of $X(t)$?

(b) Calculate the minimum mean-squared error.

*9.3-5. Work Problem 9.3-4 by finding the real constants $G > 0$ and $W > 0$ for the filter defined by

$$H(\omega) = \frac{GW}{W + j\omega}$$

Some Practical Applications of the Theory

10.0
INTRODUCTION

The main purpose of this book has been to introduce the reader to the basic principles necessary to model random signals and noise. The principles were broad enough to include the descriptions of waveforms modified by passage through linear networks. In this chapter we shall apply the basic principles to a few practical problems that involve random signals, noise, and networks. Obviously, the list of practical applications is almost limitless and it is necessary to select only a finite few. Although the applications discussed here may not necessarily serve the main interests of all readers, they do represent important applications and do serve to illustrate the use of the book's theory.

In the following sections we shall describe two practical communication systems, two control systems (one with application to one of the communication systems), an application involving a computer-type signal, and two applications that relate to radar. In every case we are primarily interested in how these applications are affected by the presence of random noise. We begin by considering the common broadcast AM (amplitude modulation) communication system.

10.1
NOISE IN AN AMPLITUDE MODULATION COMMUNICATION SYSTEM

The communication system most familiar to the general public is probably the AM (amplitude modulation) system. In this system the amplitude of a high-frequency "carrier" is made to vary (be modulated) as a linear function of the message waveform, usually derived from music, speech, or other audio source. The carrier frequency assigned to a broadcast station in the United States is one of the values from 540 to 1600 kHz in 10-kHz steps. Each station must contain its radiated power to a 10-kHz band centered on its assigned frequency.

In this section we shall give a very brief introduction to the AM broadcast system and illustrate how the noise principles of the preceding chapters can be used to analyze the system's performance.

AM System and Waveforms

Figure 10.1-1 illustrates the basic *functions* that must be present in an AM system. In this figure we include only those functions necessary to the study of noise performance. A practical system would include many other devices such as amplifiers, mixers, oscillators, and antennas that do not directly affect our performance calculations.

The transmitted AM signal has the form

$$s_{AM}(t) = [A_0 + x(t)] \cos[\omega_0 t + \theta_0] \qquad (10.1\text{-}1)$$

where $A_0 > 0$, ω_0, and θ_0 are constants, while $x(t)$ represents a message that we model as a sample function of a random process $X(t)$. Note that the amplitude $[A_0 + x(t)]$ of the carrier $\cos(\omega_0 t + \theta_0)$ is a linear function of $x(t)$. Now, in general, one has no control over θ_0 because the turn-on time of a transmitter is random and the channel itself may introduce a phase angle that is random (which we presume is absorbed in the value of θ_0). Thus, we may properly model θ_0 as a value of a random variable Θ_0 independent of $X(t)$ and uniformly distributed on $(0, 2\pi)$. These considerations allow $s_{AM}(t)$ to be modeled as a sample function of a transmitted random process $S_{AM}(t)$ given by

$$S_{AM}(t) = [A_0 + X(t)] \cos(\omega_0 t + \Theta_0) \qquad (10.1\text{-}2)$$

The transmitted signal arrives at the receiver after passing through a channel with voltage gain G_{ch}. The channel is assumed to add no signal distortion but does add zero-mean white gaussian noise of power density $\mathcal{N}_0/2$. A practical channel typically adds delay but this effect does not modify the noise performance. A receiver bandpass filter passes the received signal $s_R(t) = G_{ch}s_{AM}(t)$ with negligible distortion but has no wider bandwidth than necessary so as to not pass excessive noise.† The noise $n(t)$ at the filter's output is a bandpass noise so the theory of Section 8.6 applies.

We model waveforms $s_R(t)$ and $n(t)$ as sample functions of processes $S_R(t)$ and $N(t)$, respectively. Thus, we may write

$$S_R(t) = G_{ch}S_{AM}(t)$$
$$= G_{ch}[A_0 + X(t)] \cos(\omega_0 t + \Theta_0) \qquad (10.1\text{-}3)$$

$$N(t) = N_c(t) \cos(\omega_0 t + \Theta_0) - N_s(t) \sin(\omega_0 t + \Theta_0) \qquad (10.1\text{-}4)$$

---

†The required bandwidth W_{rec} must be at least twice the spectral extent W_X of $X(t)$.

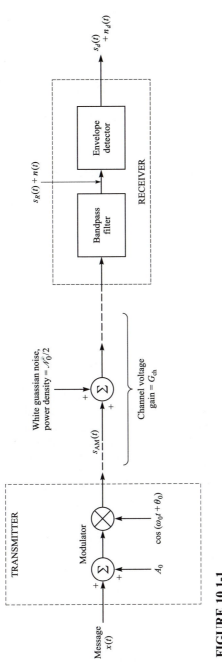

FIGURE 10.1-1
Functional block diagram of a broadcast AM system.

where $N_c(t)$ and $N_s(t)$ are lowpass noises with average powers $\overline{N_c^2(t)} = \overline{N_s^2(t)} = \overline{N^2(t)}$ from Section 8.6.

Noise Performance

A good measure of noise performance is the ratio of the average power in the output signal $s_d(t)$ of the system to the average power in the output noise $n_d(t)$. In the AM receiver an envelope detector is used to recover the transmitted message.

The total waveform applied to the envelope detector becomes

$$S_R(t) + N(t) = \{G_{ch}[A_0 + X(t)] + N_c(t)\}\cos(\omega_0 t + \Theta_0) - N_s(t)\sin(\omega_0 + \Theta_0)$$
$$= A(t)\cos[\omega_0 t + \Theta_0 + \psi(t)] \tag{10.1-5}$$

where†

$$\psi(t) = \tan^{-1}\left\{\frac{N_s(t)}{G_{ch}[A_0 + X(t)] + N_c(t)}\right\} \tag{10.1-6}$$

$$A(t) = \left\langle\{G_{ch}[A_0 + X(t)] + N_c(t)\}^2 + N_x^2(t)\right\rangle^{1/2}$$

$$= G_{ch}[A_0 + X(t)]\left\langle 1 + \frac{2N_c(t)}{G_{ch}[A_0 + X(t)]} + \frac{N_c^2(t) + N_s^2(t)}{G_{ch}^2[A_0 + X(t)]^2}\right\rangle^{1/2} \tag{10.1-7}$$

Now only (10.1-7) is of interest because $A(t)$ is the envelope of $S_R(t) + N(t)$. The detector output is this envelope.

Since $N_c^2(t) + N_s^2(t)$ is the instantaneous envelope of the square of $N(t)$ (related to received noise power), while $G_{ch}^2[A_0 + X(t)]^2$ is the instantaneous envelope of the detector's input signal (related to received signal power), we make the assumption that input (received) signal-to-noise power ratio is large so that $[N_c^2(t) + N_s^2(t)]/G_{ch}^2[A_0 + X(t)]^2$ is small *most of the time*. The assumption allows

$$A(t) \approx G_{ch}[A_0 + X(t)] + N_c(t) \tag{10.1-8}$$

from (10.1-7). Only when this condition is true do we obtain quality performance anyway, so other situations are not usually of interest.

If we model $s_d(t)$ and $n_d(t)$ in Figure 10.1-1 as sample functions of processes $S_d(t)$ and $N_d(t)$, respectively, then (10.1-8) clearly gives

$$S_d(t) = G_{ch}[A_0 + X(t)] \tag{10.1-9}$$
$$N_d(t) = N_c(t) \tag{10.1-10}$$

The *useful* output signal average power, denoted by S_o, is that due to $X(t)$ in (10.1-9). If output average noise power is denoted by N_o then

---

†Typically, *overmodulation* where $|X(t)|_{max}$, the maximum magnitude of $X(t)$, exceeds A_0 is undesirable in AM, so $[A_0 + X(t)] > 0$ is assumed in (10.1-7).

$$S_o = G_{ch}^2 \overline{X^2(t)} \tag{10.1-11}$$

$$N_o = \overline{N_c^2(t)} = \overline{N^2(t)} \tag{10.1-12}$$

and performance is measured by

$$\left(\frac{S_o}{N_o}\right)_{AM} = \frac{G_{ch}^2 \overline{X^2(t)}}{\overline{N^2(t)}} \tag{10.1-13}$$

Next, we model the bandpass filter in Figure 10.1-1 as an ideal filter with bandwidth $W_{rec}(rad/s)$. Noise power readily follows

$$\overline{N^2(t)} = \frac{1}{2\pi} 2 \int_{\omega_0-(W_{rec}/2)}^{\omega_0+(W_{rec}/2)} (\mathcal{N}_0/2)\, d\omega = \frac{\mathcal{N}_0 W_{rec}}{2\pi} \tag{10.1-14}$$

From (10.1-13) we have

$$\left(\frac{S_o}{N_o}\right)_{AM} = \frac{2\pi G_{ch}^2 \overline{X^2(t)}}{\mathcal{N}_0 W_{rec}} \tag{10.1-15}$$

Equation (10.1-15) is the principal result of this section. It describes the performance of the AM system. It is helpful to demonstrate the use of (10.1-15) by means of an example.

EXAMPLE 10.1-1. Assume an AM system uses an unmodulated carrier of peak amplitude $A_0 = 10\sqrt{95}\,V$ and a message of power $\overline{X^2(t)} = 500\,W$. Its channel has a gain $G_{ch} = \sqrt{32}/100$ with a noise density $\mathcal{N}_0/2 = (10^{-8})$ W/Hz. The receiver uses a filter with bandwidth $W_{rec} = 2\pi(10^4)\,rad/s$. We compute various signal powers and system performance.

From Problem 10.1-1 the average power in the transmitted carrier is $A_0^2 = 4750\,W$; the transmitted power due to message modulation is $R_{XX}(0)/2 = \overline{X^2(t)}/2 = 250\,W$. Total average transmitted power is, therefore, 5000 W.

From (10.1-15) we compute

$$\left(\frac{S_o}{N_o}\right)_{AM} = \frac{2\pi(32)10^{-14}(500)}{2(10^{-8})2\pi(10^4)} = 8000 \qquad (\text{or } 39.03 \text{ dB})$$

This signal-to-noise ratio represents fairly good performance.

At the input to the envelope detector the received average signal power is 5000 W decreased by the loss incurred in passing over the channel: $5000(\sqrt{32}/100)^2 = 16\,W$. From (10.1-14) and (10.1-12) the input average noise power is $10^{-8}2\pi(10^4)/\pi = 2(10^{-4})\,W$. Input signal-to-noise ratio becomes $16/2(10^{-4}) = 80,000$ (or 49.03 dB). This value is well above the minimum for performance as required for (10.1-15) to be valid; in fact, if the performance of an AM system is satisfactory then (10.1-15) will always be valid (the reader should justify this fact by examining the *efficiency* of an AM system—see Problems 10.1-4 and 10.1-2.

10.2
NOISE IN A FREQUENCY MODULATION COMMUNICATION SYSTEM

Another communication system with which the reader is familiar is the broadcast FM (frequency modulation) system. Here the instantaneous frequency of a sinusoidal "carrier" waveform is made to vary as a linear function of the message waveform. If $X(t)$ is a process representing the message, the FM transmitted waveform can be represented by the process

$$S_{FM}(t) = A \cos\left[\omega_0 t + \Theta_0 + k_{FM} \int X(t)\, dt\right] \qquad (10.2\text{-}1)$$

where A, ω_0, and $k_{FM} > 0$ are constants† and Θ_0 is a random variable independent of $X(t)$ and uniformly distributed on $(0, 2\pi)$. In a practical station $\omega_0/2\pi$ is the station's assigned frequency and is one of 100 possible frequencies from 88.1 to 107.9 MHz. Each station transmits power in a 200-kHz "channel" centered on its assigned frequency.

The constant k_{FM} in (10.2-1) is the transmitter's modulation constant. Its unit is rad/second per volt when $X(t)$ is a voltage. Transmitted signal bandwidth is difficult to compute in FM because FM is a nonlinear modulation. If k_{FM} is large enough, this bandwidth can readily be much larger than the bandwidth of the message process $X(t)$. If $X(t)$ is presumed to be bounded at $|X(t)|_{max}$ and have a crest factor defined by (Problem 10.1-3)

$$K_{cr}^2 = \frac{|X(t)|_{max}^2}{E[X^2(t)]} = \frac{|X(t)|_{max}^2}{\overline{X^2(t)}} \qquad (10.2\text{-}2)$$

the bandwidth of $S_{FM}(t)$ for the broadband case is approximated by (Peebles, 1976)

$$W_{FM} \approx 2\Delta\omega = 2k_{FM}|X(t)|_{max}$$
$$= 2k_{FM}K_{cr}\sqrt{\overline{X^2(t)}} \qquad (10.2\text{-}3)$$

Here

$$\Delta\omega = k_{FM}|X(t)|_{max} \qquad (10.2\text{-}4)$$

is the peak frequency deviation that instantaneous frequency can make from ω_0 (on either side).

Although difficult to prove, the average transmitted waveform power is

$$P_{FM} = E[S_{FM}^2(t)] = \frac{A^2}{2} \qquad (10.2\text{-}5)$$

which is independent of the modulation.

---

†If k_{FM} is negative, its sign can be absorbed into the definition of $X(t)$.

FM System and Waveforms

383

CHAPTER 10:
Some Practical
Applications of the
Theory

Figure 10.2-1 illustrates the basic functions present in a typical FM system. The transmitted waveform passes over the channel modeled as a power gain G_{ch}^2 without distortion or delay (as also assumed in Section 10.1 above). The receiver's bandpass filter (BPF) is wide enough to pass $G_{ch}S_{FM}(t)$ with little distortion but not so wide as to pass excess noise. Its bandwidth is, therefore, $W_{FM} = 2\Delta\omega$.

The purpose of the limiter is to remove amplitude fluctuations in the received waveform. The limiter is necessary so that the receiver responds only to frequency variations (that contain the message) and not to amplitude variations that are mainly due to noise. The discriminator is the actual demodulation device; it produces a voltage proportional (constant of proportionality K_D) to instantaeous deviations of the frequency of its input waveform from a nominal value ω_0. Ideally, with no noise, the discriminator's output signal is $K_D k_{FM} X(t)$. The lowpass filter must pass this waveform with low distortion so that its output is proportional to $X(t)$

$$S_d(t) = K_D k_{FM} X(t) \tag{10.2-6}$$

It should have a bandwidth no wider than the spectral extent of $X(t)$, denoted by W_X, so as to not allow excessive output noise.

If the receiver's "input" is defined as the input to the limiter, the input signal's average power S_i is

$$S_i = G_{ch}^2 \frac{A^2}{2} \tag{10.2-7}$$

while the output signal power is

$$S_o = E[S_d^2(t)] = K_D^2 k_{FM}^2 \overline{X^2(t)} \tag{10.2-8}$$

By modeling the BPF in Figure 10.2-1 as an ideal filter the input noise power is readily found to be

$$N_i = \frac{1}{2\pi} 2 \int_{\omega_0 - \Delta\omega}^{\omega_0 + \Delta\omega} \frac{\mathcal{N}_0}{2} \, d\omega = \frac{\mathcal{N}_0 \Delta\omega}{\pi} \tag{10.2-9}$$

Input signal-to-noise power ratio is

$$\left(\frac{S_i}{N_i}\right)_{FM} = \frac{\pi G_{ch}^2 A^2}{2\mathcal{N}_0 \Delta\omega} \tag{10.2-10}$$

from (10.2-7) and (10.2-9).

Computation of output noise power is less straightforward than the preceding computations. However, its development forms the most interesting problem in computing system performance.

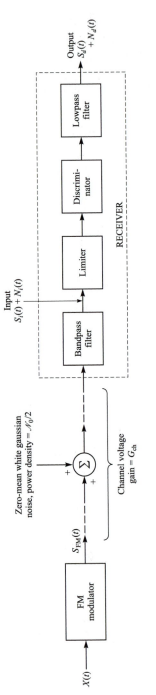

FIGURE 10.2-1
Functional block diagram of an FM communication system.

FM System Performance

385

CHAPTER 10:
Some Practical
Applications of the
Theory

Care must be exercised in finding output noise power because FM is a non-linear operation. For relatively large $(S_i/N_i)_{FM}$ and wideband operation (developed above), signal and noise powers may be independently found. Signal power is found assuming noise zero (above). Noise power is found assuming the message is zero but carrier is still transmitted. In this latter case the waveform at the limiter is

$$G_{ch}A\cos[\omega_0 t + \Theta_0] + N_c(t)\cos(\omega_0 t + \Theta_0) - N_s(t)\sin(\omega_0 t + \Theta_0)$$
$$= A(t)\cos[\omega_0 t + \Theta_0 + \psi(t)] \tag{10.2-11}$$

where the bandpass noise $N_i(t)$ is modeled as in (10.1-4) (see also Section 8.6) and

$$A(t) = \{[G_{ch}A + N_c(t)]^2 + N_s^2(t)\}^{1/2} \tag{10.2-12}$$

$$\psi(t) = \tan^{-1}\left\{\frac{N_s(t)}{G_{ch}A + N_c(t)}\right\} \tag{10.2-13}$$

For large input signal-to-noise ratio we have $|G_{ch}A| \gg |N_c(t)|$ and $|G_{ch}A| \gg |N_s(t)|$ *most of the time*, so (10.2-13) becomes

$$\psi(t) \approx \tan^{-1}\left[\frac{N_s(t)}{G_{ch}A}\right] \approx \frac{N_s(t)}{G_{ch}A} \tag{10.2-14}$$

Equation (10.2-11) is now approximated by

$$A(t)\cos[\omega_0 t + \Theta_0 + \psi(t)] \approx A(t)\cos\left[\omega_0 t + \Theta_0 + \frac{N_s(t)}{G_{ch}A}\right] \tag{10.2-15}$$

Because the limiter removes $A(t)$ and the discriminator responds only to instantaneous frequency deviations from ω_0, the input to the lowpass filter is

$$\left(\frac{K_D}{G_{ch}A}\right)\frac{dN_s(t)}{dt} \tag{10.2-16}$$

If $\mathscr{S}_{N_sN_s}(\omega)$ is the power spectrum of $N_s(t)$ the power spectrum of (10.2-16) is

$$\left(\frac{K_D}{G_{ch}A}\right)^2 \omega^2 \mathscr{S}_{N_sN_s}(\omega) \tag{10.2-17}$$

However, we may use (8.6-17) and (8.6-16) to write this power spectrum as

$$\left(\frac{K_D}{G_{ch}A}\right)^2 \omega^2 [\mathscr{S}_{N_iN_i}(\omega - \omega_0) + \mathscr{S}_{N_iN_i}(\omega + \omega_0)] \qquad |\omega| < \Delta\omega \tag{10.2-18}$$

where $\mathscr{S}_{N_iN_i}(\omega)$ is the power spectrum of $N_i(t)$; it is constant at $\mathscr{N}_0/2$ over bands of width $2\Delta\omega$ centered at ω_0 and $-\omega_0$.

Final output noise power results from the action of the lowpass filter on (10.2-18). We have

$$N_o = E[N_d^2(t)] = \frac{1}{2\pi} \int_{-W_X}^{W_X} \left(\frac{K_D}{G_{ch}A}\right)^2 \omega^2 [\mathscr{S}_{N_i N_i}(\omega - \omega_0) + \mathscr{S}_{N_i N_i}(\omega + \omega_0)]\,d\omega$$

$$= \frac{K_D^2}{2\pi G_{ch}^2 A^2} \int_{-W_X}^{W_X} \omega^2 \left[\frac{\mathscr{N}_0}{2} + \frac{\mathscr{N}_0}{2}\right] d\omega = \frac{K_D^2 \mathscr{N}_0 W_X^3}{3\pi G_{ch}^2 A^2} \qquad (10.2\text{-}19)$$

Output performance is determined by

$$\left(\frac{S_o}{N_o}\right)_{FM} = \frac{3\pi G_{ch}^2 A^2 k_{FM}^2 \overline{X^2(t)}}{\mathscr{N}_0 W_X^3} \qquad (10.2\text{-}20)$$

from (10.2-8) and (10.2-19). An alternative form of (10.2-20) is

$$\left(\frac{S_o}{N_o}\right)_{FM} = \frac{6}{K_{cr}^2} \left(\frac{\Delta\omega}{W_X}\right)^3 \left(\frac{S_i}{N_i}\right)_{FM} \qquad (10.2\text{-}21)$$

An important observation derives from (10.2-21). Since FM bandwidth is $2\Delta\omega$, we see that performance increases as the *cube* of bandwidth relative to $(S_i/N_i)_{FM}$. However, $(S_i/N_i)_{FM}$ decreases as the reciprocal of bandwidth from (10.2-10), so the *net* performance increases as the *square* of bandwidth. By simply increasing bandwidth at the transmitter, system performance rapidly increases. There is a limit to this procedure, unfortunately, that occurs when conditions under which the performance equations were derived are no longer valid. The break point, or *threshold*, occurs approximately where $(S_i/N_i)_{FM}$ drops below about 10 (or 10 dB). For a more detailed discussion of FM threshold the reader is refered to Peebles (1976). We shall emphasize FM system performance through an example.

> **EXAMPLE 10.2-1.** An FM system uses a message with crest factor 3 and bandwidth $W_X/2\pi = 3\,\text{kHz}$. The FM modulator's bandwidth is $2\Delta\omega/2\pi = 20\,\text{kHz}$ and the receiver's input signal-to-noise ratio is 81. From (10.2-21) $(S_o/N_o)_{FM} = 2000$ (or 33.01 dB). We determine how much performance can be increased by raising $\Delta\omega$.
>
> From (10.2-10) $(S_i/N_i)_{FM}$ decreases to 10 from 81 if $\Delta\omega$ increases by a factor of 8.1. Next, we again use (10.2-21) but now with $\Delta\omega/2\pi = 8.1(10)\,\text{kHz}$ and $(S_i/N_i)_{FM} = 10$:
>
> $$\left(\frac{S_o}{N_o}\right)_{FM} = \frac{6}{9}\left(\frac{81}{3}\right)^3 (10) = 131,220$$
>
> (or 51.18 dB). The bandwidth increase of 8.1 times has improved $(S_o/N_o)_{FM}$ by 65.61 times.

10.3
NOISE IN A SIMPLE CONTROL SYSTEM

In this section we shall briefly consider the noise response of a simple control system modeled by the block diagram shown in Figure 10.3-1. The following section will then illustrate how a very practical network can be analyzed by applying the results developed here.

Transfer Function

Typical loop behavior in Figure 10.3-1 is to force the feedback signal F to approximate the command C so that the error $C–F$ is small. The control loop's response R may be conveniently chosen. For example, if R in the time domain is to be the derivative of the command then $H_2(\omega) = 1/j\omega$, the transfer function of an integrator. If R is to approximate C then $H_2(\omega) = 1$.

From Figure 10.3-1 it is clear that

$$R(\omega) = H_1(\omega)[C(\omega) - H_2(\omega)R(\omega)] \qquad (10.3\text{-}1)$$

so

$$R(\omega) = C(\omega)\left[\frac{H_1(\omega)}{1 + H_1(\omega)H_2(\omega)}\right] \qquad (10.3\text{-}2)$$

We define the *transfer function* of the control loop as

$$H(\omega) = \frac{R(\omega)}{C(\omega)} = \frac{H_1(\omega)}{1 + H_1(\omega)H_2(\omega)} \qquad (10.3\text{-}3)$$

The transfer function (10.3-3) is not always stable. There are combinations of $H_1(\omega)$ and $H_2(\omega)$ that can cause instability. In general, if $H_1(\omega)$ and $H_2(\omega)$ are stable and $|H_1(\omega)H_2(\omega)|$ falls below unity, as a function of ω, before the phase of $H_1(\omega)H_2(\omega)$ becomes $-\pi$, and if the phase of $H_1(\omega)H_2(\omega)$ equals $-\pi$ at only one frequency, the transfer function $H(\omega)$ is stable. The product $H_1(\omega)H_2(\omega)$ is called the *open-loop transfer function* of the control system. Stability is a deep subject in control systems, and we shall not develop it further because it detracts from the simple points to be made here.

Now suppose the command waveform in Figure 10.3-1 is the sum of a signal $S_c(t)$ and noise $N_c(t)$. Because the system is linear its responses to signal and noise may be computed separately. If $\mathscr{S}_{N_cN_c}(\omega)$ is the power spectrum of $N_c(t)$ then the power spectrum of the response noise $N_R(t)$ is

$$\mathscr{S}_{N_RN_R}(\omega) = \mathscr{S}_{N_cN_c}(\omega)\left|\frac{H_1(\omega)}{1 + H_1(\omega)H_2(\omega)}\right|^2 \qquad (10.3\text{-}4)$$

whenever the network is stable.

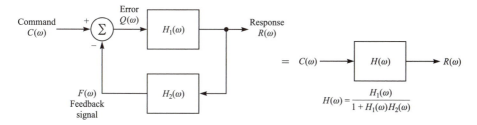

FIGURE 10.3-1
Block diagram of a simple control system.

An example serves to illustrate the use of (10.3-4).

EXAMPLE 10.3-1. Let a signal

$$S_c(t) = Au(t)e^{-Wt}$$

plus white noise of power density $\mathcal{N}_0/2$ be applied to the control network where

$$H_1(\omega) = \frac{K_1 W_1}{W_1 + j\omega} \qquad K_1 \gg 1$$

$$H_2(\omega) = 1$$

This choice means that we desire the response to equal the command. We find the output signal and the output noise power.

From (10.3-3)

$$H(\omega) = \frac{K_1 W_1/(W_1 + j\omega)}{1 + [K_1 W_1/(W_1 + j\omega)]} = \frac{K_1 W_1}{(1 + K_1)W_1 + j\omega}$$

From pair 15 of Appendix E the inverse transform of $H(\omega)$ is

$$h(t) = K_1 W_1 u(t)e^{-(1+K_1)W_1 t}$$

The response signal becomes

$$S_R(t) = \int_{-\infty}^{\infty} h(\xi)S_c(t - \xi)\,d\xi$$

$$= K_1 W_1 A \int_{-\infty}^{\infty} u(\xi)u(t - \xi)e^{-[(1+K_1)W_1 - W]\xi}\,d\xi e^{-Wt}$$

$$= K_1 W_1 A u(t)e^{-Wt} \int_0^t e^{-[(1+K_1)W_1 - W]\xi}\,d\xi$$

$$= \frac{K_1 W_1}{(1 + K_1)W_1 - W}\langle 1 - \exp\{-[(1 + K_1)W_1 - W]t\}\rangle S_c(t)$$

For $K_1 \gg 1$ so that $(1 + K_1)W_1 \gg W$ this result becomes

$$S_R(t) \approx S_c(t)$$

The approximation is more accurate as t becomes large.

From (10.3-4) the output noise power spectrum is

$$\mathcal{S}_{N_R N_R}(\omega) = \frac{\mathcal{N}_0 (K_1 W_1)^2/2}{[(1 + K_1)W_1]^2 + \omega^2}$$

Output noise power is found using (C-25):

$$P_{N_R N_R} = \frac{1}{2\pi}\int_{-\infty}^{\infty} \mathcal{S}_{N_R N_R}(\omega)\,d\omega$$

$$= \frac{\mathcal{N}_0 K_1^2 W_1}{4(1 + K_1)} \approx \frac{\mathcal{N}_0 K_1 W_1}{4}$$

We observe in passing that this control loop is stable and its transfer function is equivalent to a simple lowpass filter of gain $K_1/(1 + K_1) \approx 1$ and 3-dB bandwidth $(1 + K_1)W_1 \approx K_1 W_1$. This unity-gain large-bandwidth filter resulted from a narrowband (bandwidth W_1) high gain filter (gain K_1) inside the loop.

Error Function

The error $Q = C - F$ in Figure 10.3-1 is readily found. From

$$Q(\omega) = C(\omega) - F(\omega) = C(\omega) - H_2(\omega)H_1(\omega)Q(\omega) \tag{10.3-5}$$

we have

$$Q(\omega) = \frac{C(\omega)}{1 + H_1(\omega)H_2(\omega)} \tag{10.3-6}$$

Wiener Filter Application

By comparing (10.3-3) with the transfer function of a Wiener filter for uncorrelated signal and noise as given by (9.2-20) we see that the Wiener filter can be implemented as a loop. From (9.2-20)

$$H_{\text{opt}}(\omega) = \frac{e^{j\omega t_o}}{1 + [\mathscr{S}_{NN}(\omega)/\mathscr{S}_{XX}(\omega)]} \tag{10.3-7}$$

Thus

$$H(\omega) = H_{\text{opt}}(\omega) \tag{10.3-8}$$

if

$$H_1(\omega) = e^{j\omega t_o} \tag{10.3-9}$$

$$H_2(\omega) = [\mathscr{S}_{NN}(\omega)/\mathscr{S}_{XX}(\omega)]e^{-j\omega t_o} \tag{10.3-10}$$

Of course these functions $H_1(\omega)$ and $H_2(\omega)$ may not be realizable even for realizable signal and noise power spectrums. Other choices for $H_1(\omega)$ and $H_2(\omega)$ are also possible (Problem 10.3-2).

10.4
NOISE IN A PHASE-LOCKED LOOP

The phase-locked loop (PLL) is a practical system to which the noise theory of this book can be applied as a good example. The PLL is also an example of the control system of the preceding section.

Figure 10.4-1 depicts the block diagram of a PLL. Broadly, the action of the loop is to force the phase of the output of the voltage-controlled oscillator (VCO) to closely follow the phase of the input signal. This action leads to one of the most important uses of the PLL, that of demodulating a frequency-modulated signal. If there is no input noise $N_i(t)$ and the VCO's phase follows that of the input FM signal, then the VCO's signal has the same FM as that transmitted. Since the VCO is just a frequency modulator, its input waveform (loop's output waveform) has to be proportional to the original message used at the transmitter. When input noise is present there is noise on the output signal. In this section we shall develop this output noise power and find the available output signal-to-noise power ratio.

Phase Detector

Consider first the phase detector. Although there are many forms of phase detector [Blanchard (1976) and Klapper et al. (1972)] they all provide an output response proportional to the difference between the phases of the two input waveforms for small difference phases. Thus

$$e_L(t) \approx K_p[\theta_1(t) - \theta_2(t)] \tag{10.4-1}$$

if the two input waveform's phases are defined as $\theta_1(t)$ and $\theta_2(t)$. The constant K_p is the phase detector's sensitivity constant; its unit is volts per radian for $e_L(t)$ a voltage. In some phase detectors the response is also proportional to the amplitudes of the two input waveforms. Others depend only on one input amplitude because the other is large enough to saturate the device giving a type of limiting. Another type allows both inputs to limit in the detector and the output is not a function of either waveform's level. We shall assume either this last form of detector or that an actual limiter is in the path of the signal's input when a detector is used with limiting in the feedback path's input. Thus our phase detector is described by (10.4-1).

Loop Transfer Function

Since the VCO in Figure 10.4-1 acts like a frequency modulator for the "message" $s_R(t)$, its output can be written as

$$\text{VCO output} = A_V \cos[\omega_0 t + \theta_0 + \theta_V(t)]$$

$$= A_V \cos\left[\omega_0 t + \theta_0 + k_V \int s_R(t)\, dt\right]$$

$$= A_V \cos[\theta_2(t)] \tag{10.4-2}$$

where k_V is the VCO's modulation constant,

$$\theta_2(t) = \omega_0 t + \theta_0 + k_V \int s_R(t)\, dt \tag{10.4-3}$$

and

$$\theta_V(t) = k_V \int s_R(t)\, dt \tag{10.4-4}$$

The other phase detector input signal, from Figure 10.4-1, is the input waveform. If we define its phase as

$$\theta_1(t) = \omega_0 t + \theta_0 + \theta_i(t) \tag{10.4-5}$$

then the phase detector's response (10.4-1) becomes

$$
\begin{aligned}
e_L(t) &= K_P\left[\omega_0 t + \theta_0 + \theta_i(t) - \omega_0 t - \theta_0 - k_V \int s_R(t)\,dt\right] \\
&= K_P\left[\theta_i(t) - k_V \int s_R(t)\,dt\right]
\end{aligned}
\tag{10.4-6}
$$

Next, if we define Fourier transforms as follows

$$e_L(t) \leftrightarrow E_L(\omega) \tag{10.4-7}$$

$$\theta_i(t) \leftrightarrow \Theta_i(\omega) \tag{10.4-8}$$

$$s_R(t) \leftrightarrow S_R(\omega) \tag{10.4-9}$$

we may write (10.4-6) as

$$E_L(\omega) = K_P\left[\Theta_i(\omega) - \frac{k_V S_R(\omega)}{j\omega}\right] \tag{10.4-10}$$

From Figure 10.4-1

$$E_L(\omega) = \frac{S_R(\omega)}{H_L(\omega)} \tag{10.4-11}$$

On equating (10.4-10) and (10.4-11) we find the PLL's transfer function, denoted by $H_T(\omega)$, to be

$$H_T(\omega) = \frac{S_R(\omega)}{\Theta_i(\omega)} = \frac{K_P j\omega H_L(\omega)}{j\omega + K_P k_V H_L(\omega)} = \frac{j\omega}{k_V} H(\omega) \tag{10.4-12}$$

where we also define†

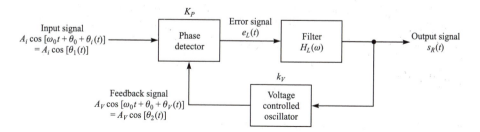

FIGURE 10.4-1
Block diagram of a phase-locked loop (PLL).

---

†In many texts $H(\omega)$ is called the PLL transfer function but the loop's output is defined at a different point. (Where would it be?)

$$H(\omega) = \frac{K_P k_V H_L(\omega)}{j\omega + K_P k_V H_L(\omega)} \tag{10.4-13}$$

It should be noted that the above definition of transfer function relates the output *signal* to the input signal's *phase modulation* $\theta_i(t)$ according to

$$S_R(\omega) = H_T(\omega)\Theta_i(\omega) \tag{10.4-14}$$

or

$$s_R(t) = \int_{-\infty}^{\infty} h_T(t - \xi)\theta_i(\xi)\,d\xi \tag{10.4-15}$$

where $h_T(t)$ denotes the inverse transform of $H_T(\omega)$

$$h_T(t) \leftrightarrow H_T(\omega) \tag{10.4-16}$$

The above developments show, in effect, that Figure 10.4-2 is an equivalent form for the loop of Figure 10.4-1.

Loop Noise Performance

We shall apply the preceding results to the case where the input to the PLL is the sum of an FM signal plus bandpass noise $N_i(t)$ modeled as

$$N_i(t) = N_c(t)\cos(\omega_0 t) - N_s(t)\sin(\omega_0 t) \tag{10.4-17}$$

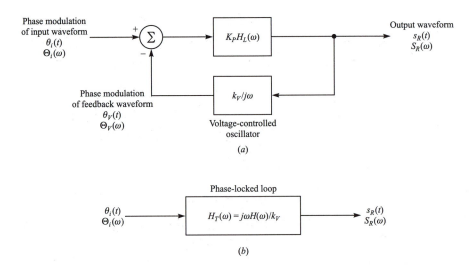

FIGURE 10.4-2
(a) Equivalent block diagram of the linear PLL of Figure 10.4-1, and (b) the transfer function equivalent of the loop in (a).

The representation (10.4-17) follows developments of Section 8.6 where $N_c(t)$
and $N_s(t)$ are lowpass random processes having the properties defined in
(8.6-7) through (8.6-19). The actual input to the PLL is, therefore,

$$A_i \cos\left[\omega_0 t + \theta_0 + k_{FM} \int X(t)\, dt\right] + N_c(t)\cos(\omega_0 t) - N_s(t)\sin(\omega_0 t) \quad (10.4\text{-}18)$$

where k_{FM} is the FM modulator's constant, $X(t)$ is the message process, and A_i, ω_0, and θ_0 are the input FM signal's peak amplitude, frequency, and phase, respectively.

The exact analysis of the PLL's response to the waveform of (10.4-18) is very involved. However, it can be shown that the waveform of (10.4-18) can be put in the form (Problem 10.4-1)

$$R(t)\cos[\omega_0 t + \theta_0 + \theta_{FM}(t) + \theta_N(t)] \quad (10.4\text{-}19)$$

where

$$\theta_{FM}(t) = k_{FM} \int X(t)\, dt \quad (10.4\text{-}20)$$

and $\theta_N(t)$ is a phase angle caused by noise. For large-input signal-to-noise ratio $(A_i^2/2)/E[N_i^2(t)]$ and input noise $N_i(t)$ broadband relative to the FM signal, the autocorrelation function of $\theta_N(t)$ is approximately $1/A_i^2$ times the autocorrelation function of $N_c(t)$ (Problem 10.4-2). This fact means that within a reasonable approximation, $\theta_N(t)$ can be replaced by the *equivalent* angle $N_c(t)/A_i$.

With the above noise equivalence used, the input phase modulation to the PLL from (10.4-19) is

$$\theta_i(t) = \theta_{FM}(t) + \theta_N(t)$$
$$= \theta_{FM}(t) + \frac{N_c(t)}{A_i} \quad (10.4\text{-}21)$$

The component $\theta_{FM}(t)$ is due to the signal. If $X(t)$ is a random process with power spectrum $\mathcal{S}_{XX}(\omega)$, we use (10.4-20) in (10.4-21) and find that the power spectrum of $\theta_i(t)$ is

$$\mathcal{S}_{\theta_i \theta_i}(\omega) = \frac{k_{FM}^2 \mathcal{S}_{XX}(\omega)}{\omega^2} + \frac{\mathcal{S}_{N_c N_c}(\omega)}{A_i^2} \quad (10.4\text{-}22)$$

After using the PLL's transfer function (10.4-12), the output waveform's power spectrum becomes

$$\mathcal{S}_{s_R s_R}(\omega) = \mathcal{S}_{\theta_i \theta_i}(\omega)|H_T(\omega)|^2$$
$$= \mathcal{S}_{XX}(\omega)\left(\frac{k_{FM}}{k_V}\right)^2 |H(\omega)|^2 + \mathcal{S}_{N_c N_c}(\omega)\frac{\omega^2}{A_i^2 k_V^2}|H(\omega)|^2 \quad (10.4\text{-}23)$$

The first right-side term in (10.4-23) is due to the desired message while the second is due to noise. Loop design in typically chosen so that $|H(\omega)|^2 \approx 1$ for all frequencies of interest in $\mathcal{S}_{XX}(\omega)$. In fact, if the message is to be preserved with very small distortion, the bandwidth of the transfer function $H(\omega)$ may be significantly *larger* than the frequencies of interest in $\mathscr{S}_{XX}(\omega)$. Thus, if W_X is the spectral extent of the message $X(t)$ then the power in the output signal component is

$$S_o = \frac{1}{2\pi} \int_{-\infty}^{\infty} \mathscr{S}_{XX}(\omega) \left(\frac{k_{FM}}{k_V}\right)^2 |H(\omega)|^2 \, d\omega \approx \left(\frac{k_{FM}}{k_V}\right)^2 \frac{1}{2\pi} \int_{-\infty}^{\infty} \mathscr{S}_{XX}(\omega) \, d\omega$$

$$= \left(\frac{k_{FM}}{k_V}\right)^2 \overline{X^2(t)} \tag{10.4-24}$$

In some loops (see example to follow) $|H(\omega)|^2$ does not decrease rapidly enough to remove high-frequency noise due to the factor ω^2 in $\omega^2|H(\omega)|^2$ in (10.4-23). In these cases it may be necessary to follow the loop with a separate filter to better remove noise spectral components at frequencies $|\omega| > W_X$. As long as either the loop or a separate filter removes these components, the overall output noise power is approximately

$$N_o \approx \frac{1}{2\pi} \int_{-W_X}^{W_X} \mathscr{S}_{N_c N_c}(\omega) \frac{\omega^2}{A_i^2 k_V^2} |H(\omega)|^2 \, d\omega$$

$$\approx \frac{\mathscr{N}_0}{2\pi A_i^2 k_V^2} \int_{-W_X}^{W_X} \omega^2 \, d\omega = \frac{\mathscr{N}_0 W_X^3}{3\pi A_i^2 k_V^2} \tag{10.4-25}$$

Finally, we determine output signal-to-noise power ratio from (10.4-25) and (10.4-24). As in Section 10.2, we let A be the peak amplitude of the transmitted FM signal and let G_{ch} be the gain of the channel, so that

$$A_i = A G_{ch} \tag{10.4-26}$$

Thus,

$$\left(\frac{S_o}{N_o}\right)_{FM} = \frac{3\pi G_{ch}^2 A^2 k_{FM}^2 \overline{X^2(t)}}{\mathscr{N}_0 W_X^3} \tag{10.4-27}$$

On comparing (10.4-27) with (10.2-20) we find that both the discriminator and PLL forms of FM receiver have the same performance when the received (input) signal-to-noise ratio is large.

> **EXAMPLE 10.4-1.** As an example of a practical PLL's transfer function let the loop filter be a simple lowpass function with 3-dB bandwidth W_L where
>
> $$H_L(\omega) = \frac{W_L}{W_L + j\omega}$$
>
> The function $H(\omega)$, from (10.4-13), becomes
>
> $$H(\omega) = \frac{1}{1 - \left(\dfrac{\omega}{\omega_n}\right)^2 + j2\zeta\left(\dfrac{\omega}{\omega_n}\right)}$$

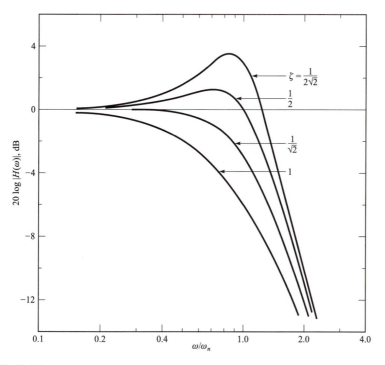

FIGURE 10.4-3
$|H(\omega)|$ for the loop of Example 10.4-1.

where the quantities defined by

$$\omega_n = (K_P k_V W_L)^{1/2}$$

$$\zeta = \frac{1}{2}\sqrt{\frac{W_L}{K_P k_V}}$$

are called the *natural frequency* and *damping factor*, respectively, of the loop. Figure 10.4-3 illustrates how $|H(\omega)|$ behaves with ω/ω_n for ζ as a parameter. The curve for $\zeta = 1/\sqrt{2}$ is most flat in the sense that the largest number of derivatives of $|H(\omega)|$ are zero at $\omega = 0$.

For $\zeta = 1/\sqrt{2}$ and $\omega_n = W_X$, the signal's spectral extent, we have

$$|H(\omega)|^2 = \frac{W_X^4}{W_X^4 + \omega^4}$$

The more exact power in the noise term of (10.4-23) becomes

$$N_o = \frac{\mathscr{N}_0 W_X^4}{2\pi A_i^2 k_V^2} \int_{-\infty}^{\infty} \frac{\omega^2 \, d\omega}{W_X^4 + \omega^4} = \frac{\mathscr{N}_0 W_X^3}{2\sqrt{2} A_i^2 k_V^2}$$

after using (C-38). On comparing this result with (10.4-25) we see the noise in the loop output is $3\pi/2\sqrt{2} \approx 3.33$ times that of a broadband loop followed by an abrupt-cutoff filter of bandwidth W_X.

10.5
CHARACTERISTICS OF RANDOM COMPUTER-TYPE WAVEFORM

As another example of the practical application of the theory of this book we examine a waveform not unlike those encountered in binary computers. The waveform is shown in Figure 10.5-1; it consists of a sequence of rectangular pulses of durations T_b having amplitudes that randomly may equal A or $-A$. Amplitudes A and $-A$ are assumed to occur with equal probability and the amplitude of any pulse interval is assumed to be statistically independent of the amplitudes of all other intervals. The random process from which this type of waveform is modeled as a sample function is called a *semirandom binary process* (see also Problem 6.1-4); in the remainder of this section we shall examine the description, power spectrum, and autocorrelation function of this process.

Process Description

The semirandom binary process $X(t)$ can be described by

$$X(t) = \sum_{k=-\infty}^{\infty} A_k \, \text{rect}\left[\frac{t - kT_b}{T_b}\right] \tag{10.5-1}$$

where $\{A_k\}$ is a set of statistically independent random variables and rect $(\cdot)$ is defined by (E-2). The A_k satisfy

$$E[A_k] = 0 \qquad k = 0, \pm 1, \pm 2, \ldots \tag{10.5-2}$$

$$E[A_k A_m] = \begin{cases} A^2 & k = m \\ 0 & k \neq m \end{cases} \tag{10.5-3}$$

The truncated version of $X(t)$ is needed in calculating power spectrum. We truncate to a time interval $2T$ centered on $t = 0$ that is a discrete multiple of T_b according to

$$2T = (2K + 1)T_b \tag{10.5-4}$$

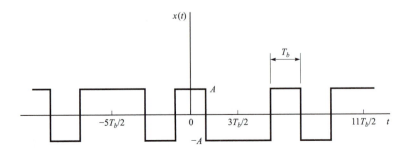

FIGURE 10.5-1
Typical waveform of a semirandom binary random process.

Thus, the truncated process $X_T(t)$ is

$$X_T(t) = \sum_{k=-K}^{K} A_k \text{ rect}\left[\frac{t - kT_b}{T_b}\right] \qquad (10.5\text{-}5)$$

Power Spectrum

We compute the power spectrum $\mathscr{S}_{XX}(\omega)$ of $X(t)$ by use of (7.1-11). The Fourier transform of $X_T(t)$, denoted by $X_T(\omega)$, is

$$X_T(\omega) = T_b \sum_{k=-K}^{K} A_k \text{Sa}(\omega T_b/2) e^{-jk\omega T_b}$$

$$= T_b \text{Sa}(\omega T_b/2) \sum_{k=-K}^{K} A_k e^{-jk\omega T_b} \qquad (10.5\text{-}6)$$

from (10.5-5) and pair 5 of Table E-1. Next,

$$\frac{E[|X_T(\omega)|^2]}{2T} = \frac{T_b \text{Sa}^2(\omega T_b/2)}{(2K + 1)} \sum_{k=-K}^{K} \sum_{m=-K}^{K} E[A_k A_m] e^{-j(k-m)\omega T_b}$$

$$= A^2 T_b \text{Sa}^2(\omega T_b/2) \qquad (10.5\text{-}7)$$

Now because (10.5-7) does not depend on K, and therefore not on T through (10.5-4), we have

$$\mathscr{S}_{XX}(\omega) = \lim_{T\to\infty} \frac{E[|X_T(\omega)|^2]}{2T} = A^2 T_b \text{Sa}^2(\omega T_b/2) \qquad (10.5\text{-}8)$$

The bandwidth of this power spectrum at its -3-dB point is $0.4429(2\pi/T_b) = 0.4429\omega_b$.

Autocorrelation Function

It follows from (10.5-1) through (10.5-3) that $E[X(t)X(t + \tau)]$ is zero unless both t and $t + \tau$ fall in the same pulse interval. The autocorrelation function is, therefore,

$$R_{XX}(t, t + \tau) = E[X(t)X(t + \tau)]$$

$$= \begin{cases} A^2 & (k - \tfrac{1}{2})T_b < (t \text{ and } t + \tau) < (k + \tfrac{1}{2})T_b \\ 0 & \text{elsewhere} \end{cases} \qquad (10.5\text{-}9)$$

Thus, the process $X(t)$ is not even wide-sense stationary since (10.5-9) depends on absolute time t.

The time-averaged autocorrelation function is readily obtained by inverse Fourier transforming (10.5-8) according to (7.2-9). After using pair 7 of Table E-1 we obtain

$$R_{XX}(\tau) = \lim_{T \to \infty} \frac{1}{2T} \int_{-T}^{T} R_{XX}(t, t + \tau)\, dt = A^2 \operatorname{tri}\left(\frac{\tau}{T_b}\right) \qquad (10.5\text{-}10)$$

The direct computation of $R_{XX}(\tau)$ by time-averaging $R_{XX}(t, t + \tau)$ is possible, but a bit more complicated than the inverse transform procedure used here (see Thomas, 1969, p. 107).

10.6
ENVELOPE AND PHASE OF A SINUSOIDAL SIGNAL PLUS NOISE

Many practical problems involve the probability density function of the envelope of the sum of a sinuosidal signal and noise. A radar, for example, may be interested in determining if a short segment (pulse) of a sinusoidal waveform is being received at some time or if only noise is being received. This problem is one of detection based on observing the received waveform's envelope; if the envelope is large enough (because of the signal's presence) the radar decides both the signal and the noise are present. We examine radar detection further in Section 10.7.

In this section we discuss probability densities involved in describing the envelope and phase of the sum of the sinusoidal signal and noise.

Waveforms

Let the signal be

$$s(t) = A_0 \cos(\omega_0 t + \theta_0) = A_0 \cos(\theta_0) \cos(\omega_0 t) - A_0 \sin(\theta_0) \sin(\omega_0 t) \qquad (10.6\text{-}1)$$

where A_0, ω_0, and θ_0 are constants. We assume the noise $n(t)$ to be added to $s(t)$ is a sample function of a zero-mean, wide-sense stationary gaussian bandpass process $N(t)$ with power $E[N^2(t)] = \sigma^2$. From (8.6-2), the sum can be written as

$$s(t) + N(t) = [A_0 \cos(\theta_0) + X(t)] \cos(\omega_0 t) - [A_0 \sin(\theta_0) + Y(t)] \sin(\omega_0 t)$$
$$= R(t) \cos[\omega_0 t + \Theta(t)] \qquad (10.6\text{-}2)$$

where $X(t)$ and $Y(t)$ are zero-mean, gaussian, lowpass processes having the same powers $E[X^2(t)] = E[Y^2(t) = E[N^2(t)] = \sigma^2$. Other properties of $X(t)$ and $Y(t)$ are given in (8.6-7) through (8.6-19). The envelope and phase of the sum are $R(t)$ and $\Theta(t)$, respectively. We may think of $R(t)$ and $\Theta(t)$ as transformations of $X(t)$ and $Y(t)$ as follows:

$$R = T_1(X, Y) = \{[A_0 \cos(\theta_0) + X]^2 + [A_0 \sin(\theta_0) + Y]^2\}^{1/2} \qquad (10.6\text{-}3a)$$

$$\Theta = T_2(X, Y) = \tan^{-1}\left[\frac{A_0 \sin(\theta_0) + Y}{A_0 \cos(\theta_0) + X}\right] \qquad (10.6\text{-}3b)$$

Inverse transformations are:

$$X = T_1^{-1}(R, \Theta) = R\cos(\Theta) - A_0\cos(\theta_0) \qquad (10.6\text{-}4a)$$

$$Y = T_2^{-1}(R, \Theta) = R\sin(\Theta) - A_0\sin(\theta_0) \qquad (10.6\text{-}4b)$$

The functional dependence on t has been suppressed in writing (10.6-3) and (10.6-4) with the implied understanding that the quantities X, Y, R, and Θ are random variables defined from the respective processes at time t.

Probability Density of the Envelope

From (8.6-15), processes $X(t)$ and $Y(t)$ are statistically independent (at the same time t) because they are gaussian and uncorrelated. The joint density of random variables X and Y is, therefore,

$$f_{X,Y}(x, y) = \frac{e^{-(x^2+y^2)/2\sigma^2}}{2\pi\sigma^2} \qquad (10.6\text{-}5)$$

From (5.4-6) the jacobian of the transformations (10.6-4) is readily found to be R. We next apply (5.4-8) to obtain the joint density of random variables R and Θ:

$$f_{R,\Theta}(r, \theta) = \frac{u(r)r}{2\pi\sigma^2} \exp\left\{-\frac{1}{2\sigma^2}[r^2 - 2rA_0\cos(\theta - \theta_0) + A_0^2]\right\} \qquad (10.6\text{-}6)$$

The density of R alone is obtained by integrating over all values of Θ:

$$f_R(r) = \int_0^{2\pi} f_{R,\Theta}(r, \theta)\, d\theta$$

$$= \frac{u(r)r}{\sigma^2} e^{-(r^2+A_0^2)/2\sigma^2} \frac{1}{2\pi} \int_0^{2\pi} e^{rA_0\cos(\theta-\theta_0)/\sigma^2}\, d\theta \qquad (10.6\text{-}7)$$

The integral is known to equal the modified Bessel function of order zero

$$I_0(\beta) = \frac{1}{2\pi} \int_0^{2\pi} e^{\beta\cos(\theta)}\, d\theta \qquad (10.6\text{-}8)$$

Thus,

$$f_R(r) = \frac{u(r)}{\sigma^2} r I_0\left(\frac{rA_0}{\sigma^2}\right) e^{-(r^2+A_0^2)/2\sigma^2} \qquad (10.6\text{-}9)$$

which is known as the *Rice* probability density.

Equation (10.6-9) is our principal result; it is the density of the envelope $R(t)$ at any time t. Figure 10.6-1 illustrates the behavior of (10.6-9). For $A_0/\sigma = 0$, the case of no signal, the density is Rayleigh. For A_0/σ large the density becomes gaussian. To show this last fact we note that

$$I_0(\beta) \approx \frac{e^\beta}{\sqrt{2\pi\beta}} \qquad \beta \gg 1 \qquad (10.6\text{-}10)$$

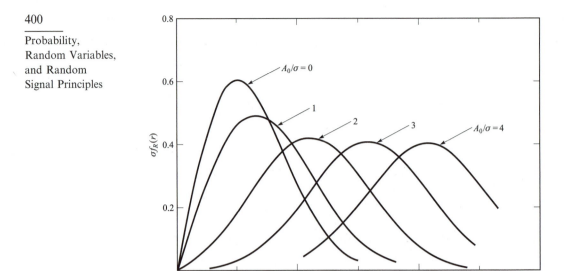

FIGURE 10.6-1
Probability densities of the envelope of a sinusoidal signal (amplitude A_0) plus noise
(power σ^2) for various ratios A_0/σ.

so for rA_0/σ^2 large

$$f_R(r) \approx u(r)\sqrt{\frac{r}{2\pi A_o \sigma^2}}\ \exp\left[\frac{-(r-A_0)^2}{2\sigma^2}\right] \qquad (10.6\text{-}11)$$

This function peaks for r near A_0, and since $A_0 \gg \sigma$, the most significant
values of r exist only near A_0. Therefore, with $r \approx A_0$ (10.6-11) becomes

$$f_R(r) \approx \frac{e^{-(r-A_0)^2/2\sigma^2}}{\sqrt{2\pi\sigma^2}} \qquad (10.6\text{-}12)$$

which is a gaussian function with mean A_0 and variance σ^2.

Although difficult to derive, the mean and variance of R as found from
(10.6-9) are known (Appendix F).

Probability Density of Phase

The density of the phase Θ of (10.6-2) derives by integrating (10.6-6) over all
values of R. We shall leave the detailed steps for the reader as an exercise
(Problem 10.6-1). The procedure is to first complete the square in r in the
exponent, and, after a suitable variable change, integrate the sum of two
terms. The result becomes (Middleton, 1960, p. 417)

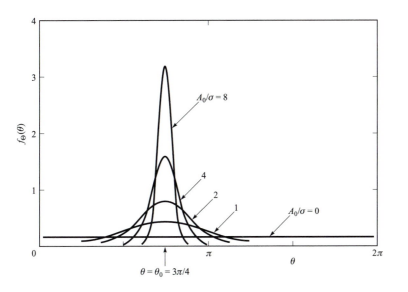

FIGURE 10.6-2
Probability density function of the phase of the sum of a sinusoidal signal and gaussian noise. Curves are plotted for a signal phase of $\theta_0 = 3\pi/4$.

$$f_\Theta(\theta) = (1/2\pi) \exp(-A_0^2/2\sigma^2)$$
$$+ \frac{A_0 \cos(\theta - \theta_0)}{\sqrt{2\pi}\sigma} \exp\left[\frac{-A_0^2 \sin^2(\theta - \theta_0)}{2\sigma^2}\right]$$
$$\cdot F\left[\frac{A_0 \cos(\theta - \theta_0)}{\sigma}\right] \tag{10.6-13}$$

where the function $F(\cdot)$ is given by (B-3). Figure 10.6-2 illustrates the behavior of $f_\Theta(\theta)$ for various values of A_0/σ when $\theta_0 = 3\pi/4$.

For noise only, which is the case of $A_0/\sigma = 0$, Figure 10.6-2 shows that the density of Θ is uniform on $(0, 2\pi)$. As A_0/σ becomes large the density approaches an impulse function located at the signal's phase (at $\theta = \theta_0$). Thus,

$$\lim_{A_0/\sigma \to \infty} [f_\Theta(\theta)] = \delta(\theta - \theta_0) \tag{10.6-14}$$

(Problem 10.6-2).

10.7
RADAR DETECTION USING A SINGLE OBSERVATION

Radar† can be used to detect the presence (and distance) of a nearby object (called the radar *target*). A representative problem might be to detect the presence of an aircraft approaching an airport. Here the airport's radar radi-

---

†For a general introduction to radar see Peebles (1998).

ates a pulse of radio frequency (RF) energy. The pulse propagates outward until it strikes the target (aircraft), whereupon some of the energy is reflected back toward the radar. The target's presence can be detected at the radar simply by detecting the presence of the reflected RF pulse. Once the received pulse is detected the delay between the time of the radiated pulse and the received pulse is proportional to the target's distance from the radar. After a sufficient time interval (called the *pulse repetition frequency*, or PRF, *interval*, chosen for the most distant detection of interest) the radar transmits another RF pulse and the entire "detection" process is repeated.

A straightforward implementation within the radar receiver to achieve detection is depicted in Figure 10.7-1. During any PRF interval, noise is always being received (mainly due to the radar's own self-generated noise). A reflected pulse is received with this noise only when a target is present. The envelope detector produces an output $W(t)$ that is some monotonic function $g(\cdot)$ of the envelope $R(t)$ of the received signal-plus-noise waveform. The first-order probability density function of $R(t)$ was developed in the preceding Section 10.6. On the average $R(t)$, and therefore $W(t)$, with a target present will be larger than $R(t)$ when only noise is being received. A suitable detection logic compares $W(t)$ to a *threshold* W_T; if $W(t) > W_T$ the receiver decides that a target is present; if $W(t) \leq W_T$ it assumes only noise is being received. These tests amount to determining when $D > 0$ in Figure 10.7-1; when $D > 0$ a target is declared to be present.

On the *average* the detection logic is valid. On any one PRF interval, however, it is possible for the receiver to make mistakes. For example, if no target is truly present it may occur that noise could become large enough at some time to make $W(t)$ exceed W_T and cause a false detection; this type of detection is called a *false alarm*. The probability of a false alarm, denoted by P_{fa}, is

$$P_{\mathrm{fa}} = \int_{W_T}^{\infty} f_0(w)\, dw \qquad (10.7\text{-}1)$$

where $f_0(w)$ is the probability density of $W(t)$ given that there is no target present. Generally, a radar wants P_{fa} to be small.

Another type of error occurs when a target is actually present but noise is such as to cancel its effect during the signal's duration and force $W(t) < W_T$. The radar usually is designed such that the probability of this event, called the

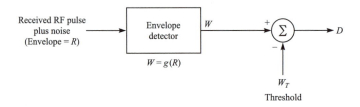

FIGURE 10.7-1
Simple radar detection network.

probability of a miss, is small; it equals one minus the *detection* probability, denoted by P_d, given by

$$P_d = \int_{W_T}^{\infty} f_1(w)\, dw \qquad (10.7\text{-}2)$$

Here $f_1(w)$ is the probability density of $W(t)$ when a target is present.

In most radars P_d and P_{fa} are parameters of greatest importance. W_T is usually chosen to give a prescribed value of P_{fa}. P_d then depends on the amplitude of the target's returned signal. In this section we shall develop expressions for P_{fa} and P_d when the radar makes detection decisions based on a signal observation (uses only one PRF interval). Our results can be extended to multiple observations, but the details are complicated and we only refer the reader to the literature (Difranco and Rubin, 1968).

False Alarm Probability and Threshold

When there is no target only noise is present at the input to the envelope detector. From (10.6-9) the density of the envelope of the noise is

$$f_R(r) = \frac{u(r)}{\sigma^2}\, r e^{-r^2/2\sigma^2} \qquad (10.7\text{-}3)$$

where σ^2 is the power in the input noise. Because the detector characteristic $g(R)$ is assumed monotonic, there is an *equivalent threshold* R_T on R that is related to W_T by

$$W_T = g(R_T) \qquad (10.7\text{-}4)$$

$$R_T = g^{-1}(W_T) \qquad (10.7\text{-}5)$$

where $g^{-1}(\cdot)$ is the inverse function of $g(\cdot)$. We may then compute P_{fa} from the envelope as follows:

$$P_{\text{fa}} = \int_{W_T}^{\infty} f_0(w)\, dw = \int_{R_T}^{\infty} f_R(r)\, dr$$

$$= \int_{R_T}^{\infty} \frac{r}{\sigma^2} e^{-r^2/2\sigma^2}\, dr = e^{-R_T^2/2\sigma^2} \qquad (10.7\text{-}6)$$

Thus,

$$R_T = \left\{ 2\sigma^2 \ln\!\left(\frac{1}{P_{\text{fa}}}\right) \right\}^{1/2} \qquad (10.7\text{-}7)$$

and

$$W_T = g\!\left[\left\{ 2\sigma^2 \ln\!\left(\frac{1}{P_{\text{fa}}}\right) \right\}^{1/2} \right] \qquad (10.7\text{-}8)$$

where $\ln(\cdot)$ represents the natural logarithm.

Equation (10.7-8) gives the threshold W_T that is to be used to realize a specified value of P_{fa} when the noise power level is σ^2 at the detector's input.

EXAMPLE 10.7-1. A radar receiver uses a square-law envelope detector defined by $W = 3R^2$. We find what threshold is required when noise power at the detector's input is $\sigma^2 = 0.025$ W and $P_{fa} = 10^{-6}$ is required. From (10.7-8)

$$W_T = 3\left[2(0.025)\ln\left(\frac{1}{10^{-6}}\right)\right] \approx 2.07 \text{ V}$$

Detection Probability

When a target signal is present the density of the received waveform's envelope is given by (10.6-9). Again using the idea of an equivalent threshold R_T on the envelope R we expand (10.7-2) to get

$$P_d = \int_{W_T}^{\infty} f_1(w)\,dw = \int_{R_T}^{\infty} f_R(r)\,dr$$

$$= \int_{\sqrt{2\sigma^2 \ln(1/P_{fa})}}^{\infty} \frac{r}{\sigma^2} I_0\left(\frac{rA_0}{\sigma^2}\right) e^{-(r^2+A_0^2)/2\sigma^2}\,dr$$

$$= Q\left[\sqrt{\frac{A_0^2}{\sigma^2}}, \sqrt{2\ln\left(\frac{1}{P_{fa}}\right)}\right] \qquad (10.7\text{-}9)$$

where

$$Q(\alpha, \beta) = \int_{\beta}^{\infty} \xi I_0(\alpha\xi) e^{-(\xi^2+\alpha^2)/2}\,d\xi \qquad (10.7\text{-}10)$$

is called Marcum's Q-function (Marcum, 1950, 1960). Figure 10.7-2 illustrates P_d for various values of $A_0^2/2\sigma^2$ with P_{fa} as a parameter. Generally, the smaller P_{fa} is required to be, the larger is the necessary signal strength to achieve a given value of P_d.

When P_{fa} is small while P_d is relatively large so that the threshold W_T is large and signal strength is relatively large, the approximation of (10.6-12) can be used in (10.7-9) to obtain

$$P_d \approx F\left[\frac{A_0}{\sigma} - \sqrt{2\ln\left(\frac{1}{P_{fa}}\right)}\right] \qquad (10.7\text{-}11)$$

where $F(\cdot)$ is given by (B-3).

EXAMPLE 10.7-2. We find the value of P_d in a receiver having $P_{fa} = 10^{-10}$ when the received signal-to-noise power ratio at the detector's input is 16.0 dB. Here $A_0^2/2\sigma^2 = 39.811$ (16 dB). Thus, $(A_0/\sigma) - \sqrt{2\ln(1/P_{fa})} \approx 2.137$. From Table B-1 and (10.7-11), $P_d \approx F(2.137) \approx 0.9837$ or 98.37%, which is in agreement with Figure 10.7-2.

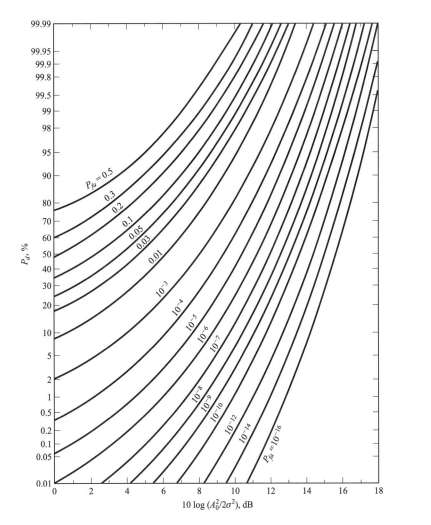

FIGURE 10.7-2
Radar detection probabilities for various false alarm probabilities when detection is based on a single observation. [*Adapted from Barton (1964) with permission.*]

10.8
SUMMARY

This chapter introduced seven practical applications of the theory presented in the earlier parts of the book. Although the main thrust of the work was to show how noise (a random waveform) enters into practical systems, one example of a computer waveform was to demonstrate direct modeling of the signal as a random process. The specific examples developed were:

• Amplitude modulation (AM) broadcast radio system.
• Frequency modulation (FM) broadcast radio system.

- A simple control system.
- A phase-locked loop.
- A computer-type waveform.
- The envelope and phase behavior of the sum of a sinusoidal (desired) signal and (undesired) noise.
- A simple example of detection in a radar system.

PROBLEMS

10.1-1. Show that (*a*) the time-averaged autocorrelation function of $S_{AM}(t)$, as given by (10.1-2) is

$$R_{AM}(\tau) = \tfrac{1}{2}[A_0^2 + R_{XX}(\tau)]\cos(\omega_0\tau)$$

if $X(t)$ is a zero-mean process, and (*b*) the power spectrum is

$$\mathscr{S}_{AM}(\omega) = \frac{A_0^2\pi}{2}[\delta(\omega - \omega_0) + \delta(\omega + \omega_0)]$$
$$+ \tfrac{1}{4}[\mathscr{S}_{XX}(\omega - \omega_0) + \mathscr{S}_{XX}(\omega + \omega_0)]$$

where $\mathscr{S}_{XX}(\omega)$ is the power spectrum of $X(t)$.

10.1-2. Define transmitter efficiency η_{AM} in an amplitude modulation communication system as the ratio of transmitted power due to the message to the total power. For a zero-mean stationary random message show that

$$\eta_{AM} = \frac{R_{XX}(0)}{A_0^2 + R_{XX}(0)} = \frac{\displaystyle\int_{-\infty}^{\infty}\mathscr{S}_{XX}(\omega)\,d\omega}{2\pi A_0^2 + \displaystyle\int_{-\infty}^{\infty}\mathscr{S}_{XX}(\omega)\,d\omega} = \frac{\overline{X^2(t)}}{A_0^2 + \overline{X^2(t)}}$$

where $R_{XX}(\tau)$ and $\mathscr{S}_{XX}(\omega)$ are the autocorrelation function and power spectrum, respectively, of the message $X(t)$.

10.1-3. Define *crest factor* K_{cr} for a zero-mean, bounded, random signal by

$$K_{cr}^2 = |X(t)|_{max}^2 / \overline{X^2(t)}$$

If no *overmodulation* is to occur, such that $|X(t)|_{max} \leq A_0$ in the transmitted signal of an amplitude modulation system, show that the transmitter efficiency (Problem 10.1-2) is

$$\eta_{AM} \leq \frac{1}{1 + K_{cr}^2}$$

What is the maximum efficiency for a message $X(t) = A_m\cos(\omega_m t + \Theta_m)$, where A_m and ω_m are constants while Θ_m is a random variable uniform on $(0, 2\pi)$?

10.1-4. Use (10.1-3), (10.1-4), and (10.1-14) to show that the input signal-to-noise power ratio at the envelope detector of Figure 10.1-1 is

$$\left(\frac{S_i}{N_i}\right)_{AM} = \frac{\overline{S_R^2(t)}}{\overline{N^2(t)}} = \frac{\pi G_{ch}^2 [A_0^2 + \overline{X^2(t)}]}{\mathcal{N}_0 W_{rec}}$$

Use this result to show that (10.1-15) can be written in the form

$$\left(\frac{S_o}{N_o}\right)_{AM} = 2\eta_{AM} \left(\frac{S_i}{N_i}\right)_{AM}$$

where η_{AM} is defined in Problem 10.1-2.

10.1-5. In an AM broadcast system the total average transmitted power is 1 kW. The channel gain is $G_{ch} = 3\sqrt{2}(10^{-3})$. Average noise power at the envelope detector's input is 10^{-5} W and the output signal-to-noise power ratio of the receiver is 180 (or 22.55 dB).
(a) What is the average signal power at the input to the envelope detector?
(b) find $(S_i/N_i)_{AM}$.
(c) What is the transmitter's efficiency?
(*Hint*: Use results of Problem 10.1-4.)

10.1-6. In an AM system the message's power is $\overline{X^2(t)} = 0.1$ W. The modulator's efficiency is $\eta_{AM} = 0.159$. Noise power at the receiver's envelope detector is $N_i = 10^{-4}$ W, and the signal-to-noise power ratio at the output is known to be $(S_o/N_o)_{AM} = 5000$. If $|X(t)|_{max} = A_0$, find: (a) K_{cr} (defined in Problem 10.1-3), (b) A_0, (c) $(S_i/N_i)_{AM}$, and (d) G_{ch}.

10.1-7. An AM communication system transmits an average power of 2 kW when using a message having a crest factor (defined in Problem 10.1-3) of $\sqrt{24}$ and a peak amplitude equal to A_0. The channel's voltage gain is $3(10^{-6})$ and in the receiver $(S_i/N_i)_{AM} = 5(10^5)$.
(a) What is the system's efficiency as defined in Problem 10.1-2?
(b) What is S_i?
(c) Find $(S_o/N_o)_{AM}$.

10.2-1. When the message in an FM system is a sinusoid, such as $x(t) = A_m \cos(\omega_m t)$ where $A_m > 0$ and ω_m are constants, *modulation index* β_{FM} is defined by $\beta_{FM} = \Delta\omega/\omega_m$.
(a) Write an expression for the instantaneous frequency (rad/s) of the FM waveform in terms of β_{FM}.
(b) What is the approximate bandwidth of the FM signal in terms of β_{FM} if $\Delta\omega$ is large relative to ω_m?
(c) For the specific waveform $x(t) = 0.1\cos(10^3 t)$, what are β_{FM} and the transmitter's constant k_{FM} if the approximate bandwidth is to be 200 kHz?

10.2-2. Find an expression for the autocorrelation function of $S_{FM}(t)$, as given by (10.2-1), when $X(t)$ is a gaussian, zero-mean process. Formulate the expression in terms of the correlation coefficient and variance of the process

$$\Gamma(t) = k_{FM} \int X(t)\, dt$$

[*Hint*: Note that the expectation involving $X(t)$ leads to a characteristic function.]

10.2-3. In an FM system the transmitted signal has $10\,\mathrm{kW}$ of average power and a bandwidth of approximately $150\,\mathrm{kHz}$ when a random message with a crest factor of 4 is used (Problem 10.1-3). The signal passes over a channel for which $G_{\mathrm{ch}} = 10^{-6}$ and $\mathcal{N}_0/2 = 5(10^{-15})/3$.
(a) Find the signal and noise average powers and the signal-to-noise ratio at the receiver's input.
(b) What is the message's spectral extent if the output signal-to-noise power ratio of the receiver is found to be 25,000?

10.2-4. A random message has spectral extent $W_X = \pi(10^4)\,\mathrm{rad/s}$, a crest factor of 4.2, and average power of $\overline{X^2(t)} = 0.02\,\mathrm{W}$. The message is transmitted over an FM system for which $k_{\mathrm{FM}} = 1.06(10^6)\,\mathrm{rad/s}$ per volt. At the receiver $(S_i/N_i)_{\mathrm{FM}} = 10$. Find: (a) $\Delta\omega$, and (b) $(S_o/N_o)_{\mathrm{FM}}$.

10.2-5. In an FM system $\Delta\omega = 8\pi(10^4)\,\mathrm{rad/s}$, $(S_i/N_i)_{\mathrm{FM}} = 14$ and $(S_o/N_o)_{\mathrm{FM}} = 4200$ when the message has a crest factor of 3.2 and $\mathcal{N}_0 = 10^{-11}\,\mathrm{W/Hz}$.
(a) Find W_X.
(b) What is the peak voltage of the FM signal at the input to the receiver?
(c) Is the system operating above threshold?

10.2-6. An FM signal of peak amplitude $0.1\,\mathrm{V}$ at the input to an FM demodulator results in an input signal-to-noise power ratio of 20. If $\Delta\omega = \pi(10^5)\,\mathrm{rad/s}$ in the FM signal and the effective transmitted signal voltage is $A = 2\,\mathrm{kV}$, find: (a) G_{ch} and (b) $\mathcal{N}_0$. If the transmitted message is an audio signal for which $|X(t)|_{\max} = 1.4\,\mathrm{V}$, $W_X = \pi(10^4)\,\mathrm{rad/s}$, and $K_{\mathrm{cr}} = 3$, find: (c) $(S_o/N_o)_{\mathrm{FM}}$ and (d) k_{FM}.

10.3-1. Let $H_1(\omega) = K_1 W_1/(W_1 + j\omega)$ and $H_2(\omega) = 1/j\omega$ in (10.3-3) where $K_1 > 0$ and $W_1 > 0$ are constants.
(a) Are there any values of K_1 and/or W_1 that will make the loop of Figure 10.3-1 unstable?
(b) If $W_1 = 200$ and $K_1 = 40$ find the loop's output noise power if white noise of power density $\mathcal{N}_0/2 = 10^{-4}\,\mathrm{W/Hz}$ is applied at the input. (*Hint*: Use the integral given in Problem 9.2-7.)

10.3-2. Show that the transfer function of the control system of Figure 10.3-1 is the same as the Wiener filter of (9.2-20) if

$$H_1(\omega) = \left[\frac{\mathcal{S}_{XX}(\omega)}{\mathcal{S}_{NN}(\omega)}\right] e^{j\omega t_o}$$

and

$$H_2(\omega) = e^{-j\omega t_o}$$

10.3-3. A signal and white noise of power density $\mathcal{N}_0/2 = 10^{-3}$ are applied to the input of a simple linear control system for which

$$H_1(\omega) = \frac{Kj\omega}{(W_1 + j\omega)(W_2 + j\omega)}$$

$$H_2(\omega) = 1$$

Here K, W_1, and W_2 are positive constants.

(a) Find an expression for the noise power at the loop's output. (*Hint*: Use results of Problem 9.2-7.)

(b) Evaluate the result found in (a) for $K = 10$, $W_1 = 5$, and $W_2 = 25$.

*10.4-1. Show that the sum of an FM waveform plus noise as given by (10.4-18) can be written in the form

$$R(t)\cos[\omega_0 t + \theta_0 + \theta_{FM}(t) + \theta_N(t)]$$

where

$$\theta_{FM}(t) = k_{FM}\int X(t)\,dt$$

and

$$R(t) = \left(\{N_c(t) + A_i\cos[\theta_0 + \theta_{FM}(t)]\}^2 + \{N_s(t) + A_i\sin[\theta_0 + \theta_{FM}(t)]\}^2\right)^{1/2}$$

$$\theta_N(t) = \tan^{-1}\left\{\frac{N_s(t)\cos[\theta_0 + \theta_{FM}(t)] - N_c(t)\sin[\theta_0 + \theta_{FM}(t)]}{A_i + N_c(t)\cos[\theta_0 + \theta_{FM}(t)] + N_s(t)\sin[\theta_0 + \theta_{FM}(t)]}\right\}$$

*10.4-2. Assume the bandpass noise $N_i(t)$ in Problem 10.4-1 is wide-sense stationary and gaussian and note that if $|A_i| \gg |N_c(t)|$ and $|A_i| \gg N_s(t)|$ *most of the time*, then

$$\theta_N(t) \approx \frac{N_s(t)}{A_i}\cos[\theta_0 + \theta_{FM}(t)] - \frac{N_c(t)}{A_i}\sin[\theta_0 + \theta_{FM}(t)]$$

(a) Show that the autocorrelation function of the process $\Theta_N(t)$, for which $\theta_N(t)$ is a sample function, is

$$R_{\Theta_N\Theta_N}(t, t+\tau) = \frac{1}{A_i^2}R_{N_cN_c}(\tau)E\left[\cos\left\{k_{FM}\int_t^{t+\tau}X(\xi)\,d\xi\right\}\right]$$

$$+ \frac{1}{A_i^2}R_{N_cN_s}(\tau)E\left[\sin\left\{k_{FM}\int_t^{t+\tau}X(\xi)\,d\xi\right\}\right]$$

where $R_{N_cN_c}(\tau)$ and $R_{N_cN_s}(\tau)$ are the correlation functions of $N_c(t)$ and $N_s(t)$, and the expectations are with respect to the message process $X(t)$ assumed statistically independent of the noises. (*Hint*: Use the results of Section 8.6.)

(b) If noises $N_c(t)$ and $N_s(t)$ are broadband relative to the FM signal, justify that

$$R_{\Theta_N\Theta_N}(t, t+\tau) \approx \frac{1}{A_i^2}R_{N_cN_c}(\tau) = R_{\Theta_N\Theta_N}(\tau)$$

(c) If the message process varies slowly enough for values of τ that are important to $R_{N_cN_c}(\tau)$ such that

$$k_{FM}\int_t^{t+\tau}X(\xi)\,d\xi \approx k_{FM}X(t)\tau$$

is valid, show that the expression of part (a) reduces to

$$R_{\Theta_N\Theta_N}(t, t+\tau) \approx \frac{1}{A_i^2}\exp\left[\frac{-\sigma_X^2 k_{FM}^2\tau^2}{2}\right]R_{N_cN_c}(\tau)$$

if $X(t)$ is a zero-mean, wide-sense stationary gaussian message of power σ_X^2. (*Hint*: Make use of characteristic functions.)

10.4-3. In Example 10.4-1 let $\zeta = \frac{1}{2}$ instead of $1/\sqrt{2}$ and recompute the loop's output noise power N_o. Compare the result with that of (10.4-25). Is there any improvement over the case where $\zeta = 1/\sqrt{2}$? (*Hint*: Make use of the integral given in Problem 9.2-7.)

10.4-4. Assume white noise is added to an FM signal and the sum is applied to a phase-locked loop for message recovery. Thus, $\mathscr{S}_{N_c n_c}(\omega) = \mathscr{N}_0$ in (10.4-23).
(*a*) If

$$H_L(\omega) = \frac{W_1 W_3(W_2 + j\omega)}{W_2(W_1 + j\omega)(W_3 + j\omega)}$$

where W_1, W_2, and W_3 are positive constants, find an expression for the power contained in the noise part of (10.4-23).
(*b*) Assume the loop is designed so that $W_3 = 2\omega_0$, $W_2 = \omega_0/5$, and $W_1 = \omega_0^2/5K$, where $K = K_P k_V$ and ω_0 is called the loop's *crossover frequency* (rad/s); it is the frequency where $|KH_L(\omega)/j\omega| = 1$. Evaluate the result found in part (*a*) when ω_0 equals the message's spectral extent W_X.
(*c*) If K is very large, to what does the evaluation of part (*b*) approach? [*Hint*: Use the known integral

$$I_3 = \frac{1}{2\pi} \int_{-\infty}^{\infty} \frac{(b_0\omega^4 - b_1\omega^2 + b_2)\, d\omega}{a_0^2\omega^6 + (a_1^2 - 2a_0a_2)\omega^4 + (a_2^2 - 2a_1a_3)\omega^2 + a_3^2}$$
$$= \frac{a_0b_1 - a_2b_0 - (a_0a_1b_2/a_3)}{2a_0(a_0a_3 - a_1a_2)}$$

where $a_0, a_1, a_2, a_3, b_0, b_1$, and b_2 are constants and $a_0\lambda^3 + a_1\lambda^2 + a_2\lambda + a_3$ has no roots in the lower half-plane when $\lambda = \omega + j\sigma$ (Thomas, 1969, p. 249).]

10.5-1. A sample function of a semirandom binary process is to be passed through a lowpass filter with transfer function $H(\omega) = W_L/(W_L + j\omega)$ where W_L is its 3-dB bandwidth. If the rise and fall times of the pulses in the output waveform are not to exceed 5% of the pulse duration T_b, what minimum value of W_L is required? (*Hint*: Assume the input waveform has been at level $-A$ for many pulse intervals and suddenly makes a transition to level A; determine rise time as that required for the output to rise from $-A$ to $0.9A$.)

*10.6-1. Carry out the steps suggested in the text and show that (10.6-13) derives from (10.6-6).

*10.6-2. If $A_0/\sigma \to \infty$ in (10.6-13) show that (10.6-14) is true.

10.6-3. A sinusoidal signal plus gaussian bandpass noise are applied to an envelope detector having an output R defined by (10.6-9). If the input signal-to-noise power ratio is $A_0^2/(2\sigma^2) = 2$, find the ratio of the mean of R to σ. Also find the variance of R relative to σ^2. (*Hint*: Use results from Appendix F.)

10.7-1. A radar receiver uses a linear envelope detector where $W = R$. Find an expression for false alarm probability P_{fa} in terms of W_T, the threshold voltage level.

10.7-2. Work Problem 10.7-1 for a square-law detector defined by $W = KR^2$, where $K > 0$ is a constant.

10.7-3. A radar uses a linear envelope detector defined by $W = R/4$. The threshold voltage is $W_T = 0.7$ volt. Measurements show that $P_{fa} = 4(10^{-7})$. What is the noise power at the envelope detector's input?

10.7-4. Work Problem 10.7-3 for a square-law detector with characteristic $W = R^2/4$.

10.7-5. False alarm probability is 10^{-8} in a radar that must have a detection probability of 0.9901. When target is present what signal-to-noise power ratio is necessary at the envelope detector's input? [*Hint*: Assume (10.7-11) applies.]

*10.7-6. A radar receiver as shown in Figure 10.7-1 uses a square-law detector defined by $W = KR^2$ where $K > 0$ is a constant. Find an expression for the probability density of W.

*10.7-7. A radar receiver uses a binary detection logic based on observing N PRF intervals (multiple observations). If the observations in the N intervals are statistically independent and the detection and false alarm probabilities on any one observation are P_{d1} and P_{fa1}, respectively, find P_d and P_{fa} that correspond to an overall detection logic based on obtaining at least n detections in N intervals.

10.7-8. In a simple radar that uses a single observation for detection, the threshold voltage is $W_T = 0.92$ V. The rms noise voltage at the detector's input is $\sigma = 0.33$ V and the detector's characteristic is defined by $W = R^2/3$. When a target is present $P_d = 0.9015$.
(*a*) Find P_{fa}.
(*b*) What is the target's signal-to-noise ratio at the detector's input when expressed in dB?

10.7-9. A radar's envelope detector has the characteristic $g(R) = 0.4R^3$. False alarm probability is 10^{-8} for a single pulse when the noise power at the detector's input is $\sigma^2 = 0.01$ W.
(*a*) What is the threshold voltage W_T?
(*b*) What peak signal amplitude A_0 is needed at the detector's input if the detection probability is to be 0.998 on one pulse?

10.7-10. A radar uses a square-law envelope detector for which $g(r) = 0.4R^2$. It is known that $P_{fa} = 10^{-4}$ when the input noise power is $\sigma^2 = 0.03$ W. When signal is present, $P_d = 0.92$.
(*a*) Find the threshold voltage W_T.
(*b*) What is A_0, the received signal's peak amplitude?

APPENDIX A

Review of the Impulse Function

BASIC REVIEW

There are several ways of defining what is known as the impulse function (Papoulis, 1962) denoted $\delta(x)$. The most mathematically sound approach is to define $\delta(x)$ on the basis of its integral property. If $\phi(x)$ is any arbitrary function of x,† $x_1 < x_2$ are two values of x, and x_0 is the point of "occurrence" of the impulse, then $\delta(x)$ satisfies (Korn and Korn, 1961, p. 742)

$$\int_{x_1}^{x_2} \phi(x)\delta(x - x_0)\, dx = \begin{cases} 0 & x_2 < x_0 \quad \text{or} \quad x_0 < x_1 \\ \dfrac{1}{2}[\phi(x_0^+) + \phi(x_0^-)] & x_1 < x_0 < x_2 \\ \dfrac{1}{2}\phi(x_0^+) & x_0 = x_1 \\ \dfrac{1}{2}\phi(x_0^-) & x_0 = x_2 \end{cases}$$

$$(\text{A-1})$$

It can be shown, using (A-1), that $\delta(x)$ behaves as a function having even symmetry, an area of unity, a vanishingly small "duration," and an infinite "amplitude" (Peebles, 1976, pp. 34–35).

A simpler form of (A-1) is often applicable to many practical situations. If $x_1 = -\infty$, $x_2 = \infty$, and $\phi(x)$ is arbitrary except that it is continuous at $x = x_0$, then

$$\int_{-\infty}^{\infty} \phi(x)\delta(x - x_0)\, dx = \phi(x_0) \qquad (\text{A-2})$$

---

†The function is also assumed to have *bounded variation* in the neighborhood of $x = x_0$ (see footnote, page 425).

A useful fact that is easily obtained from (A-1) is

$$\int_{-\infty}^{x} \delta(\xi - x_0) \, d\xi = u(x - x_0) \tag{A-3}$$

or, equivalently

$$\frac{du(x)}{dx} = \delta(x) \tag{A-4}$$

where $u(x)$ is the unit-step function defined by

$$u(x) = \begin{cases} 1 & 0 < x \\ 0 & x < 0 \end{cases} \tag{A-5}$$

The impulse function can be generalized to N-dimensional space (Korn and Korn, 1961, p. 745). If we assume a cartesiean coordinate system with axes $\xi_1, \xi_2, \ldots, \xi_N$, and a function $\phi(\xi_1, \xi_2, \ldots, \xi_N)$ that is continuous at the point $(\xi_1 = x_1, \xi_2 = x_2, \ldots, \xi_N = x_N)$, then an N-*dimensional impulse function* $\delta(\xi_1, \xi_2, \ldots, \xi_N)$ is defined by

$$\int_{-\infty}^{\infty} \cdots \int_{-\infty}^{\infty} \phi(\xi_1, \xi_2, \ldots, \xi_N) \delta(\xi_1 - x_1, \xi_2 - x_2, \ldots, \xi_N - x_N) \, d\xi_1 \cdots d\xi_N$$

$$= \phi(x_1, x_2, \ldots, x_N) \tag{A-6}$$

Of special interest is the two-dimensional case; it is known that $\delta(\xi_1, \xi_2)$ can be written as (Bracewell, 1965, p. 85)

$$\delta(\xi_1, \xi_2) = \delta(\xi_1)\delta(\xi_2) \tag{A-7}$$

so (A-6) becomes

$$\int_{-\infty}^{\infty} \int_{-\infty}^{\infty} \phi(\xi_1, \xi_2)\delta(\xi_1 - x_1)\delta(\xi_2 - x_2) \, d\xi_1 \, d\xi_2 = \phi(x_1, x_2) \tag{A-8}$$

By using (A-7) with an appropriate choice of $\phi(\xi_1, \xi_2)$ we readily find that for $N = 2$, (A-6) can be written as

$$\int_{-\infty}^{y} \int_{-\infty}^{x} \delta(\xi_1 - x_0, \xi_2 - y_0) \, d\xi_1 \, d\xi_2$$

$$= \int_{-\infty}^{y} \delta(\xi_2 - y_0) \, d\xi_2 \int_{-\infty}^{x} \delta(\xi_1 - x_0) \, d\xi_1$$

$$= u(x - x_0)u(y - y_0) \tag{A-9}$$

If $u(x - x_0)u(y - y_0)$ is interpreted as a *two-dimensional unit-step function*

$$u(x - x_0, y - y_0)$$

we have

$$\frac{\partial^2 u(x - x_0, y - y_0)}{\partial x \, \partial y} = \delta(x - x_0, y - y_0) \tag{A-10}$$

where

$$\delta(x - x_0, y - y_0) = \delta(x - x_0)\delta(y - y_0) \tag{A-11}$$

$$u(x - x_0, y - y_0) = u(x - x_0)u(y - y_0) \tag{A-12}$$

The preceding discussion concerned the "symmetrical" impulse for which a more precise statement of its integral of (A-5) is

$$u(x) = \begin{cases} 1 & x > 0 \\ 1/2 & x = 0 \\ 0 & x < 0 \end{cases} \tag{A-13}$$

However, there are two other forms of impulses, called *asymmetrical impulses*, that we consider. In one, the impulse is denoted by $\delta_+(x)$ and its integral is

$$u_+(x) = \int_{-\infty}^{x} \delta_+(\xi)\,d\xi = \begin{cases} 1 & x > 0 \\ 0 & x \leq 0 \end{cases} \tag{A-14}$$

The other, denoted by $\delta_-(x)$, has the integral

$$u_-(x) = \int_{-\infty}^{x} \delta_-(\xi)\,d\xi = \begin{cases} 1 & x \geq 0 \\ 0 & x < 0 \end{cases} \tag{A-15}$$

These impulses differ in the way the value at $x = 0$ is assigned. Strictly, the impulse $\delta_-(x)$ and unit step $u_-(x)$ are the ones to be used in probability problems because of the way probabilities are assigned for discrete random variables. However, since the distribution for discrete variables has been justified through proper definitions in the text, there is no need there to be concerned about any other differences between symmetrical and asymmetrical impulses.

PROPERTIES OF IMPULSES

Many properties exist that simplify problems involving impulses. Most of these are proved rigorously only through the *theory of distributions* (sometimes referred to as the theory of *generalized functions*). Peebles (1998) has reviewed some of this theory and given historical references to its development. Here we shall only present some of the most appropriate properties without proofs. In most cases it is sufficient to consider the functions $\phi(x)$ and $g(x)$ to follow as continuous at the value of x where the impulse "occurs." The properties all relate to the symmetrical impulse.

Definition

$$\int_{-\infty}^{\infty} \delta(x)\phi(x)\,dx = \phi(0) \tag{A-16}$$

Derivatives

The first derivative $d\delta(x)/dx$ is usually called a *unit-doublet*:

$$\int_{-\infty}^{\infty} \frac{d\delta(x)}{dx} \phi(x)\, dx = -\frac{d\phi(x)}{dx}\bigg|_{x=0} \tag{A-17}$$

$$\int_{-\infty}^{\infty} \frac{d^n\delta(x)}{dx^n} \phi(x)\, dx = (-1)^n \frac{d^n\phi(x)}{dx^n}\bigg|_{x=0} \tag{A-18}$$

Linearity

For a set of functions $\phi_n(x)$ and constants α_n, $n = 1, 2, \ldots, N$, with N possibly infinite, let

$$\phi(x) = \sum_{n=1}^{N} \alpha_n \phi_n(x) \tag{A-19}$$

then

$$\int_{-\infty}^{\infty} \delta(x)\phi(x)\, dx = \sum_{n=1}^{N} \alpha_n \int_{-\infty}^{\infty} \delta(x)\phi_n(x)\, dx = \sum_{n=1}^{N} \alpha_n \phi_n(0) \tag{A-20}$$

Fourier Transforms of Impulses

Let $(\cdot) \leftrightarrow (\cdot)$ denote a Fourier transform pair. Then

$$\int_{-\infty}^{\infty} \delta(x)e^{-j\omega x}\, dx = 1 \qquad \text{so} \qquad \delta(x) \leftrightarrow 1 \tag{A-21}$$

Also

$$\frac{1}{2\pi}\int_{-\infty}^{\infty} \delta(\omega)e^{j\omega x}\, d\omega = \frac{1}{2\pi} \qquad \text{so} \qquad \frac{1}{2\pi} \leftrightarrow \delta(\omega) \tag{A-22}$$

From these results we have useful mathematical representations for impulses:

$$\delta(x) = \frac{1}{2\pi}\int_{-\infty}^{\infty} e^{j\omega x}\, d\omega \tag{A-23}$$

$$\delta(\omega) = \frac{1}{2\pi}\int_{-\infty}^{\infty} e^{-j\omega x}\, dx \tag{A-24}$$

Fourier Transforms of Derivatives of Impulses

$$\int_{-\infty}^{\infty} \frac{d^n \delta(x)}{dx^n} e^{-j\omega x} \, dx = (j\omega)^n \qquad \text{so} \qquad \frac{d^n \delta(x)}{dx^n} \leftrightarrow (j\omega)^n \tag{A-25}$$

$$\int_{-\infty}^{\infty} \frac{d^n \delta(\omega)}{d\omega^n} e^{j\omega x} \, d\omega = (-jx)^n \qquad \text{so} \qquad (-jx)^n \leftrightarrow 2\pi \frac{d^n \delta(\omega)}{d\omega^n} \tag{A-26}$$

Shifting

For a real constant x_0

$$\int_{-\infty}^{\infty} \delta(x - x_0)\phi(x) \, dx = \phi(x_0) \tag{A-27}$$

$$\int_{-\infty}^{\infty} \delta(x)\phi(x - x_0) \, dx = \phi(-x_0) \tag{A-28}$$

Product of a Function and Impulse

For a function $g(x)$

$$g(x)\delta(x) = g(0)\delta(x) \tag{A-29}$$

Product of a Function and Derivative of Impulse

$$g(x)\frac{d\delta(x)}{dx} = g(0)\frac{d\delta(x)}{dx} - \frac{dg(x)}{dx}\bigg|_{x=0} \delta(x) \tag{A-30}$$

For $g(x) = x$ (A-30) becomes

$$x\frac{d\delta(x)}{dx} = -\delta(x) \tag{A-31}$$

Convolution of Two Impulses

$$\int_{-\infty}^{\infty} \delta(x)\delta(x_0 - x) \, dx = \delta(x_0) \tag{A-32}$$

APPENDIX B

Gaussian Distribution Function

The general gaussian or normal probability density and distribution functions are:

$$f_X(x) = \frac{1}{\sqrt{2\pi\sigma_X^2}} e^{-(x-a_X)^2/2\sigma_X^2} \tag{B-1}$$

$$F_X(x) = \int_{-\infty}^{x} f_X(\xi)\, d\xi = F\left(\frac{x - a_X}{\sigma_X}\right) \tag{B-2}$$

where $-\infty < a_X < \infty$, $0 < \sigma_X$ are constants and $F(\cdot)$ is the "normalized" distribution function for $a_X = 0$ and $\sigma_X = 1$; that is

$$F(x) = \int_{-\infty}^{x} \frac{1}{\sqrt{2\pi}} e^{-\xi^2/2}\, d\xi \tag{B-3}$$

$F(x)$ is listed in Table B-1. When $a_X \neq 0$ and $\sigma_X \neq 1$, $F_X(x)$ can be found from $F(x)$ by use of (B-2). For negative values of x, use

$$F(-x) = 1 - F(x) \tag{B-4}$$

A function closely related to $F(x)$ is the Q-function defined by

$$Q(x) = \frac{1}{\sqrt{2\pi}} \int_{x}^{\infty} e^{-\xi^2/2}\, d\xi \tag{B-5}$$

For negative values of x, use

$$Q(-x) = 1 - Q(x) \tag{B-6}$$

$Q(x)$ is related to $F(x)$ by

$$F(x) = 1 - Q(x) \tag{B-7}$$

TABLE B-1
Values of $F(x)$ for $0 \le x \le 3.89$ in steps of 0.01

x	.00	.01	.02	.03	.04	.05	.06	.07	.08	.09
0.0	.5000	.5040	.5080	.5120	.5160	.5199	.5239	.5279	.5319	.5359
0.1	.5398	.5438	.5478	.5517	.5557	.5596	.5636	.5675	.5714	.5753
0.2	.5793	.5832	.5871	.5910	.5948	.5987	.6026	.6064	.6103	.6141
0.3	.6179	.6217	.6255	.6293	.6331	.6368	.6406	.6443	.6480	.6517
0.4	.6554	.6591	.6628	.6664	.6700	.6736	.6772	.6808	.6844	.6879
0.5	.6915	.6950	.6985	.7019	.7054	.7088	.7123	.7157	.7190	.7224
0.6	.7257	.7291	.7324	.7357	.7389	.7422	.7454	.7486	.7517	.7549
0.7	.7580	.7611	.7642	.7673	.7704	.7734	.7764	.7794	.7823	.7852
0.8	.7881	.7910	.7939	.7967	.7995	.8023	.8051	.8078	.8106	.8133
0.9	.8159	.8186	.8212	.8238	.8264	.8289	.8315	.8340	.8365	.8389
1.0	.8413	.8438	.8461	.8485	.8508	.8531	.8554	.8577	.8599	.8621
1.1	.8643	.8665	.8686	.8708	.8729	.8749	.8770	.8790	.8810	.8830
1.2	.8849	.8869	.8888	.8907	.8925	.8944	.8962	.8980	.8997	.9015
1.3	.9032	.9049	.9066	.9082	.9099	.9115	.9131	.9147	.9162	.9177
1.4	.9192	.9207	.9222	.9236	.9251	.9265	.9279	.9292	.9306	.9319
1.5	.9332	.9345	.9357	.9370	.9382	.9394	.9406	.9418	.9429	.9441
1.6	.9452	.9463	.9474	.9484	.9495	.9505	.9515	.9525	.9535	.9545
1.7	.9554	.9564	.9573	.9582	.9591	.9599	.9608	.9616	.9625	.9633
1.8	.9641	.9649	.9656	.9664	.9671	.9678	.9686	.9693	.9699	.9706
1.9	.9713	.9719	.9726	.9732	.9738	.9744	.9750	.9756	.9761	.9767
2.0	.9773	.9778	.9783	.9788	.9793	.9798	.9803	.9808	.9812	.9817
2.1	.9821	.9826	.9830	.9834	.9838	.9842	.9846	.9850	.9854	.9857
2.2	.9861	.9864	.9868	.9871	.9875	.9878	.9881	.9884	.9887	.9890
2.3	.9893	.9896	.9898	.9901	.9904	.9906	.9909	.9911	.9913	.9916
2.4	.9918	.9920	.9922	.9925	.9927	.9929	.9931	.9932	.9934	.9936
2.5	.9938	.9940	.9941	.9943	.9945	.9946	.9948	.9949	.9951	.9952
2.6	.9953	.9955	.9956	.9957	.9959	.9960	.9961	.9962	.9963	.9964
2.7	.9965	.9966	.9967	.9968	.9969	.9970	.9971	.9972	.9973	.9974
2.8	.9974	.9975	.9976	.9977	.9977	.9978	.9979	.9979	.9980	.9981
2.9	.9981	.9982	.9982	.9983	.9984	.9984	.9985	.9985	.9986	.9986
3.0	.9987	.9987	.9987	.9988	.9988	.9989	.9989	.9989	.9990	.9990
3.1	.9990	.9991	.9991	.9991	.9992	.9992	.9992	.9992	.9993	.9993
3.2	.9993	.9993	.9994	.9994	.9994	.9994	.9994	.9995	.9995	.9995
3.3	.9995	.9995	.9996	.9996	.9996	.9996	.9996	.9996	.9996	.9997
3.4	.9997	.9997	.9997	.9997	.9997	.9997	.9997	.9997	.9998	.9998
3.5	.9998	.9998	.9998	.9998	.9998	.9998	.9998	.9998	.9998	.9998
3.6	.9998	.9999	.9999	.9999	.9999	.9999	.9999	.9999	.9999	.9999
3.7	.9999	.9999	.9999	.9999	.9999	.9999	.9999	.9999	.9999	.9999
3.8	.9999	.9999	.9999	.9999	.9999	.9999	.9999	1.0000	1.0000	1.0000

Although a closed-form solution for $Q(x)$ is not known, an excellent approximation is

$$Q(x) \approx \left[\frac{1}{0.661x + 0.339\sqrt{x^2 + 5.51}} \right] \frac{e^{-x^2/2}}{\sqrt{2\pi}} \qquad x \ge 0 \qquad \text{(B-8)}$$

which is due to Börjesson and Sundberg, 1979. The maximum absolute relative error in the approximation for $Q(x)$ is given as 0.27 percent for any $x \ge 0$. By using the approximation (B-8) for $Q(x)$ in (B-7), an excellent approximation for $F(x)$ is realized.

APPENDIX C

Useful Mathematical Quantities

TRIGONOMETRIC IDENTITIES

$$\cos(x \pm y) = \cos(x)\cos(y) \mp \sin(x)\sin(y) \tag{C-1}$$

$$\sin(x \pm y) = \sin(x)\cos(y) \pm \cos(x)\sin(y) \tag{C-2}$$

$$\cos\left(x \pm \frac{\pi}{2}\right) = \mp \sin(x) \tag{C-3}$$

$$\sin\left(x \pm \frac{\pi}{2}\right) = \pm \cos(x) \tag{C-4}$$

$$\cos(2x) = \cos^2(x) - \sin^2(x) \tag{C-5}$$

$$\sin(2x) = 2\sin(x)\cos(x) \tag{C-6}$$

$$2\cos(x) = e^{jx} + e^{-jx} \tag{C-7}$$

$$2j\sin(x) = e^{jx} - e^{-jx} \tag{C-8}$$

$$2\cos(x)\cos(y) = \cos(x-y) + \cos(x+y) \tag{C-9}$$

$$2\sin(x)\sin(y) = \cos(x-y) - \cos(x+y) \tag{C-10}$$

$$2\sin(x)\cos(y) = \sin(x-y) + \sin(x+y) \tag{C-11}$$

$$2\cos^2(x) = 1 + \cos(2x) \tag{C-12}$$

$$2\sin^2(x) = 1 - \cos(2x) \tag{C-13}$$

$$4\cos^3(x) = 3\cos(x) + \cos(3x) \tag{C-14}$$

$$4\sin^3(x) = 3\sin(x) - \sin(3x) \tag{C-15}$$

$$8\cos^4(x) = 3 + 4\cos(2x) + \cos(4x) \tag{C-16}$$

$$8\sin^4(x) = 3 - 4\cos(2x) + \cos(4x) \tag{C-17}$$

$$A\cos(x) - B\sin(x) = R\cos(x + \theta) \tag{C-18}$$

where

$$R = \sqrt{A^2 + B^2} \tag{C-19a}$$

$$\theta = \tan^{-1}(B/A) \tag{C-19b}$$

$$A = R\cos(\theta) \tag{C-19c}$$

$$B = R\sin(\theta) \tag{C-19d}$$

INDEFINITE INTEGRALS

Rational Algebraic Functions

$$\int (a + bx)^n \, dx = \frac{(a + bx)^{n+1}}{b(n+1)} \qquad 0 < n \tag{C-20}$$

$$\int \frac{dx}{a + bx} = \frac{1}{b}\ln|a + bx| \tag{C-21}$$

$$\int \frac{dx}{(a + bx)^n} = \frac{-1}{(n-1)b(a+bx)^{n-1}} \qquad 1 < n \tag{C-22}$$

$$\int \frac{dx}{c + bx + ax^2} = \frac{2}{\sqrt{4ac - b^2}} \tan^{-1}\left(\frac{2ax + b}{\sqrt{4ac - b^2}}\right) \qquad b^2 < 4ac$$

$$= \frac{1}{\sqrt{b^2 - 4ac}} \ln\left|\frac{2ax + b - \sqrt{b^2 - 4ac}}{2ax + b + \sqrt{b^2 - 4ac}}\right| \qquad b^2 > 4ac$$

$$= \frac{-2}{2ax + b} \qquad b^2 = 4ac \tag{C-23}$$

$$\int \frac{x \, dx}{c + bx + ax^2} = \frac{1}{2a}\ln|ax^2 + bx + c| - \frac{b}{2a}\int \frac{dx}{c + bx + ax^2} \tag{C-24}$$

$$\int \frac{dx}{a^2 + b^2 x^2} = \frac{1}{ab}\tan^{-1}\left(\frac{bx}{a}\right) \tag{C-25}$$

$$\int \frac{x \, dx}{a^2 + x^2} = \frac{1}{2}\ln(a^2 + x^2) \tag{C-26}$$

$$\int \frac{x^2 \, dx}{a^2 + x^2} = x - a\tan^{-1}\left(\frac{x}{a}\right) \tag{C-27}$$

$$\int \frac{dx}{(a^2 + x^2)^2} = \frac{x}{2a^2(a^2 + x^2)} + \frac{1}{2a^3}\tan^{-1}\left(\frac{x}{a}\right) \tag{C-28}$$

$$\int \frac{x \, dx}{(a^2 + x^2)^2} = \frac{-1}{2(a^2 + x^2)} \tag{C-29}$$

$$\int \frac{x^2 \, dx}{(a^2 + x^2)^2} = \frac{-x}{2(a^2 + x^2)} + \frac{1}{2a} \tan^{-1}\left(\frac{x}{a}\right) \tag{C-30}$$

$$\int \frac{dx}{(a^2 + x^2)^3} = \frac{x}{4a^2(a^2 + x^2)^2} + \frac{3x}{8a^4(a^2 + x^2)} + \frac{3}{8a^5} \tan^{-1}\left(\frac{x}{a}\right) \tag{C-31}$$

$$\int \frac{x^2 \, dx}{(a^2 + x^2)^3} = \frac{-x}{4(a^2 + x^2)^2} + \frac{x}{8a^2(a^2 + x^2)} + \frac{1}{8a^3} \tan^{-1}\left(\frac{x}{a}\right) \tag{C-32}$$

$$\int \frac{x^4 \, dx}{(a^2 + x^2)^3} = \frac{a^2 x}{4(a^2 + x^2)^2} - \frac{5x}{8(a^2 + x^2)} + \frac{3}{8a} \tan^{-1}\left(\frac{x}{a}\right) \tag{C-33}$$

$$\int \frac{dx}{(a^2 + x^2)^4} = \frac{x}{6a^2(a^2 + x^2)^3} + \frac{5x}{24a^4(a^2 + x^2)^2} + \frac{5x}{16a^6(a^2 + x^2)}$$
$$+ \frac{5}{16a^7} \tan^{-1}\left(\frac{x}{a}\right) \tag{C-34}$$

$$\int \frac{x^2 \, dx}{(a^2 + x^2)^4} = \frac{-x}{6(a^2 + x^2)^3} + \frac{x}{24a^2(a^2 + x^2)^2} + \frac{x}{16a^4(a^2 + x^2)}$$
$$+ \frac{1}{16a^5} \tan^{-1}\left(\frac{x}{a}\right) \tag{C-35}$$

$$\int \frac{x^4 \, dx}{(a^2 + x^2)^4} = \frac{a^2 x}{6(a^2 + x^2)^3} - \frac{7x}{24(a^2 + x^2)^2} + \frac{x}{16a^2(a^2 + x^2)}$$
$$+ \frac{1}{16a^3} \tan^{-1}\left(\frac{x}{a}\right) \tag{C-36}$$

$$\int \frac{dx}{a^4 + x^4} = \frac{1}{4a^3\sqrt{2}} \ln\left(\frac{x^2 + ax\sqrt{2} + a^2}{x^2 - ax\sqrt{2} + a^2}\right) + \frac{1}{2a^3\sqrt{2}} \tan^{-1}\left(\frac{ax\sqrt{2}}{a^2 - x^2}\right) \tag{C-37}$$

$$\int \frac{x^2 \, dx}{a^4 + x^4} = -\frac{1}{4a\sqrt{2}} \ln\left(\frac{x^2 + ax\sqrt{2} + a^2}{x^2 - ax\sqrt{2} + a^2}\right) + \frac{1}{2a\sqrt{2}} \tan^{-1}\left(\frac{ax\sqrt{2}}{a^2 - x^2}\right) \tag{C-38}$$

Trigonometric Functions

$$\int \cos(x) \, dx = \sin(x) \tag{C-39}$$

$$\int x \cos(x) \, dx = \cos(x) + x \sin(x) \tag{C-40}$$

$$\int x^2 \cos(x) \, dx = 2x \cos(x) + (x^2 - 2) \sin(x) \tag{C-41}$$

$$\int \sin(x)\,dx = -\cos(x) \tag{C-42}$$

$$\int x\sin(x)\,dx = \sin(x) - x\cos(x) \tag{C-43}$$

$$\int x^2 \sin(x)\,dx = 2x\sin(x) - (x^2 - 2)\cos(x) \tag{C-44}$$

Exponential Functions

$$\int e^{ax}\,dx = \frac{e^{ax}}{a} \qquad a \text{ real or complex} \tag{C-45}$$

$$\int xe^{ax}\,dx = e^{ax}\left[\frac{x}{a} - \frac{1}{a^2}\right] \qquad a \text{ real or complex} \tag{C-46}$$

$$\int x^2 e^{ax}\,dx = e^{ax}\left[\frac{x^2}{a} - \frac{2x}{a^2} + \frac{2}{a^3}\right] \qquad a \text{ real or complex} \tag{C-47}$$

$$\int x^3 e^{ax}\,dx = e^{ax}\left[\frac{x^3}{a} - \frac{3x^2}{a^2} + \frac{6x}{a^3} - \frac{6}{a^4}\right] \qquad a \text{ real or complex} \tag{C-48}$$

$$\int e^{ax}\sin(x)\,dx = \frac{e^{ax}}{a^2 + 1}[a\sin(x) - \cos(x)] \tag{C-49}$$

$$\int e^{ax}\cos(x)\,dx = \frac{e^{ax}}{a^2 + 1}[a\cos(x) + \sin(x)] \tag{C-50}$$

DEFINITE INTEGRALS

$$\int_{-\infty}^{\infty} e^{-a^2 x^2 + bx}\,dx = \frac{\sqrt{\pi}}{a}e^{b^2/(4a^2)} \qquad a > 0 \tag{C-51}$$

$$\int_0^{\infty} x^2 e^{-x^2}\,dx = \sqrt{\pi}/4 \tag{C-52}$$

$$\int_0^{\infty} \text{Sa}(x)\,dx = \int_0^{\infty} \frac{\sin(x)}{x}\,dx = \frac{\pi}{2} \tag{C-53}$$

$$\int_0^{\infty} \text{Sa}^2(x)\,dx = \pi/2 \tag{C-54}$$

FINITE SERIES

$$\sum_{n=1}^{N} n = \frac{N(N+1)}{2} \tag{C-55}$$

$$\sum_{n=1}^{N} n^2 = \frac{N(N+1)(2N+1)}{6} \tag{C-56}$$

$$\sum_{n=1}^{N} n^3 = \frac{N^2(N+1)^2}{4} \tag{C-57}$$

$$\sum_{n=0}^{N} x^n = \frac{x^{N+1} - 1}{x - 1} \tag{C-58}$$

$$\sum_{n=0}^{N} \frac{N!}{n!(N-n)!} x^n y^{N-n} = (x+y)^N \tag{C-59}$$

$$\sum_{n=0}^{N} e^{j(\theta+n\phi)} = \frac{\sin[(N+1)\phi/2]}{\sin(\phi/2)} e^{j[\theta+(N\phi/2)]} \tag{C-60}$$

$$\sum_{n=0}^{N} \binom{N}{n} = \sum_{n=0}^{N} \frac{N!}{n!(N-n)!} = 2^N \tag{C-61}$$

$$\sum_{n=N_1}^{N_2} w^n = \frac{w^{N_1} + w^{N_2+1}}{1 - w} \qquad \left\{ \begin{array}{l} N_2 > N_1 \text{ and } w \\ \text{real or complex} \end{array} \right. \tag{C-62}$$

INFINITE SERIES

$$e^x = 1 + x + \frac{x^2}{2!} + \frac{x^3}{3!} + \cdots = \sum_{n=0}^{\infty} \frac{x^n}{n!} \tag{C-63}$$

APPENDIX D

Review of Fourier Transforms

The *Fourier transform*† or spectrum $X(\omega)$ of a signal $x(t)$ is given by

$$X(\omega) = \int_{-\infty}^{\infty} x(t)e^{-j\omega t}\, dt \qquad \text{(D-1)}$$

The *inverse Fourier transform* allows the recovery of $x(t)$ from its spectrum $X(\omega)$. It is given by

$$x(t) = \frac{1}{2\pi}\int_{-\infty}^{\infty} X(\omega)e^{j\omega t}\, d\omega \qquad \text{(D-2)}$$

Together, (D-1) and (D-2) form a *Fourier transform pair*. Extensive tables of transform pairs exist (Campbell and Foster, 1948). A transform pair is often symbolized by use of a double-ended arrow:

$$x(t) \leftrightarrow X(\omega) \qquad \text{(D-3)}$$

The Fourier transform $X(\omega)$ is valid for real or complex signals and, in general, is a complex function of ω even for real signals $x(t)$. $X(\omega)$ describes the relative complex voltages (amplitudes and phases) as a function of ω that are present in a waveform $x(t)$. From (D-1), we see that the unit of $X(\omega)$ is volts per hertz if $x(t)$ is a voltage-time waveform. Thus, $X(\omega)$ can be considered as the *density of voltage* in $x(t)$ as a function of angular frequency ω.

---

†Named for the great French mathematician and physicist Baron Jean Baptiste Joseph Fourier (1768–1830).

Conditions that guarantee the existence of the Fourier transform of a waveform $x(t)$ are:

1. that $x(t)$ be bounded with at most a finite number of maxima and minima and a finite number of discontinuities in any finite open time interval,† and
2.

$$\int_{-\infty}^{\infty} |x(t)|\, dt < \infty \qquad (D-4)$$

These conditions are only *sufficient* for $X(\omega)$ to exist; they are *not necessary*. Many signals of practical interest do not satisfy these conditions but do have transforms. Examples are: the unit-impulse function $\delta(t)$ that has the transform $X(\omega) = 1$; and the unit-step function $u(t)$, defined by $u(t) = 1$ for $0 < t$ and $u(t) = 0$ for $t < 0$, that has the transform $X(\omega) = \pi\delta(\omega) + (1/j\omega)$.

PROPERTIES

A number of extremely useful properties of Fourier transforms may be stated. We give these without proofs since the proofs may readily be found in the literature (Peebles, 1976, p. 29; Papoulis, 1962, p. 14). In these properties, we assume the Fourier transform of some signal $x(t)$ is $X(\omega)$, while the notation $X_n(\omega)$ implies the transform of a signal $x_n(t)$ with $n = 1, 2, \ldots, N$.

Linearity

For constants α_n (that may be complex):

$$x(t) = \sum_{n=1}^{N} \alpha_n x_n(t) \leftrightarrow \sum_{n=1}^{N} \alpha_n X_n(\omega) = X(\omega) \qquad (D-5)$$

Time and Frequency Shifting

With t_0 and ω_0 real constants:

$$x(t - t_0) \leftrightarrow X(\omega)e^{-j\omega t_0} \qquad (D-6)$$

$$x(t)e^{j\omega_0 t} \leftrightarrow X(\omega - \omega_0) \qquad (D-7)$$

---

†These are known as the *Dirichlet conditions*, after the German mathematician Peter Gustov Lejeune Dirichlet (1805–1859). A signal satisfying them is said to have *bounded variation* (Thomas, 1969, p. 579).

Scaling

With α a real constant:

$$x(\alpha t) \leftrightarrow \frac{1}{|\alpha|} X\left(\frac{\omega}{\alpha}\right) \tag{D-8}$$

Duality

$$X(t) \leftrightarrow 2\pi x(-\omega) \tag{D-9}$$

Differentiation

$$\frac{d^n x(t)}{dt^n} \leftrightarrow (j\omega)^n X(\omega) \tag{D-10}$$

$$(-jt)^n x(t) \leftrightarrow \frac{d^n X(\omega)}{d\omega^n} \tag{D-11}$$

Integration

$$\int_{-\infty}^{t} x(\tau) \, d\tau \leftrightarrow \pi X(0)\delta(\omega) + \frac{X(\omega)}{j\omega} \tag{D-12}$$

$$\pi x(0)\delta(t) - \frac{x(t)}{jt} \leftrightarrow \int_{-\infty}^{\omega} X(\xi) \, d\xi \tag{D-13}$$

Conjugation

$$x^*(t) \leftrightarrow X^*(-\omega) \tag{D-14}$$

$$x^*(-t) \leftrightarrow X^*(\omega) \tag{D-15}$$

Convolution

$$x(t) = \int_{-\infty}^{\infty} x_1(\tau) x_2(t - \tau) \, d\tau \leftrightarrow X_1(\omega) X_2(\omega) = X(\omega) \tag{D-16}$$

$$x(t) = x_1(t) x_2(t) \leftrightarrow \frac{1}{2\pi} \int_{-\infty}^{\infty} X_1(\xi) X_2(\omega - \xi) \, d\xi = X(\omega) \tag{D-17}$$

$$x(t) = \int_{-\infty}^{\infty} x_1^*(\tau)x_2(\tau+t)\,d\tau \leftrightarrow X_1^*(\omega)X_2(\omega) = X(\omega) \qquad \text{(D-18)}$$

$$x(t) = x_1^*(t)x_2(t) \leftrightarrow \frac{1}{2\pi}\int_{-\infty}^{\infty} X_1^*(\xi)X_2(\xi+\omega)\,d\xi = X(\omega) \qquad \text{(D-19)}$$

Parseval's† Theorem

$$\int_{-\infty}^{\infty} x_1^*(\tau)x_2(\tau)\,d\tau = \frac{1}{2\pi}\int_{-\infty}^{\infty} X_1^*(\omega)X_2(\omega)\,d\omega \qquad \text{(D-20)}$$

An alternative form occurs when $x_1(t) = x_2(t) = x(t)$:

$$\int_{-\infty}^{\infty} |x(t)|^2\,dt = \frac{1}{2\pi}\int_{-\infty}^{\infty} |X(\omega)|^2\,d\omega \qquad \text{(D-21)}$$

MULTIDIMENSIONAL FOURIER TRANSFORMS

The Fourier transform $X(\omega_1, \omega_2)$ of a function $x(t_1, t_2)$ of two "time" variables t_1 and t_2 is defined as the iterated double transform. Upon Fourier transforming $x(t_1, t_2)$ first with respect to t_1 we have

$$X(\omega_1, t_2) = \int_{-\infty}^{\infty} x(t_1, t_2)e^{-j\omega_1 t_1}\,dt_1 \qquad \text{(D-22)}$$

$X(\omega_1, \omega_2)$ results from Fourier transformation of $X(\omega_1, t_2)$ with respect to t_2:

$$X(\omega_1, \omega_2) = \int_{-\infty}^{\infty} X(\omega_1, t_2)e^{-j\omega_2 t_2}\,dt_2 \qquad \text{(D-23)}$$

or

$$X(\omega_1, \omega_2) = \int_{-\infty}^{\infty}\int_{-\infty}^{\infty} x(t_1, t_2)e^{-j\omega_1 t_1 - j\omega_2 t_2}\,dt_1\,dt_2 \qquad \text{(D-24)}$$

By use of similar logic, the two-dimensional inverse Fourier transform is

$$x(t_1, t_2) = \frac{1}{(2\pi)^2}\int_{-\infty}^{\infty}\int_{-\infty}^{\infty} X(\omega_1, \omega_2)e^{j\omega_1 t_1 + j\omega_2 t_2}\,d\omega_1\,d\omega_2 \qquad \text{(D-25)}$$

The extension of the above procedures to an N-dimensional function is direct; we obtain the Fourier transform pair

---

†Named for M. A. Parseval (1755–1836), a French mathematician.

$$X(\omega_1, \ldots, \omega_N) = \int_{-\infty}^{\infty} \cdots \int_{-\infty}^{\infty} x(t_1, \ldots, t_N) e^{-j\omega_1 t_1 - \cdots - j\omega_N t_N} \, dt_1 \cdots dt_N \quad \text{(D-26)}$$

$$x(t_1, \ldots, t_N) = \frac{1}{(2\pi)^N} \int_{-\infty}^{\infty} \cdots \int_{-\infty}^{\infty} X(\omega_1, \ldots, \omega_N) e^{j\omega_1 t_1 + \cdots + j\omega_N t_N} \, d\omega_1 \cdots d\omega_N$$

$$\text{(D-27)}$$

PROBLEMS

D-1. Find the Fourier transform of a pulse $x(t)$ defined by

$$x(t) = \begin{cases} A & -\tau/2 < t < \tau/2 \\ 0 & \text{elsewhere} \end{cases}$$

where $\tau > 0$ and A are real constants.

D-2. If a signal $y(t)$ is the product of $x(t)$ of Problem D-1 with a cosine wave, that is, if

$$y(t) = x(t) \cos(\omega_0 t + \theta_0)$$

where ω_0 and θ_0 are real constants, what is the Fourier transform of $y(t)$?

D-3. Find the Fourier transform of the waveform

$$x(t) = \begin{cases} A\left(1 - \dfrac{|t|}{\tau}\right) & |t| \leq \tau \\ 0 & |t| > \tau \end{cases}$$

where $\tau > 0$ and A are real constants.

D-4. By direct use of (D-1), find the Fourier transform of the waveform

$$x(t) = \begin{cases} A \cos(\pi t/2\tau) & |t| \leq \tau \\ 0 & |t| > \tau \end{cases}$$

where $\tau > 0$ and A are real constants.

D-5. The waveform of Problem D-4 can be written in the form

$$x(t) = A \, \text{rect}(t/2\tau) \cos(\pi t/2\tau)$$

where $\text{rect}(t/2\tau)$ is defined by (E-2). By using (D-19), find the Fourier transform of $x(t)$.

D-6. The complex form of the *Fourier series* of an arbitrary periodic signal $y(t)$ of period T is

$$y(t) = \sum_{n=-\infty}^{\infty} C_n e^{jn2\pi t/T}$$

where the *Fourier series coefficients* are given by

$$C_n = \frac{1}{T} \int_{-T/2}^{T/2} y(t) e^{-jn2\pi t/T} \, dt$$

for $n = 0, \pm 1, \pm 2, \ldots$. Show that the Fourier transform of this arbitrary periodic signal is

$$Y(\omega) = 2\pi \sum_{n=-\infty}^{\infty} C_n \delta\left(\omega - \frac{n2\pi}{T}\right)$$

where $\delta(\cdot)$ is the unit-impulse function of Appendix A.

*D-7. Prove the Fourier transform pair

$$\sum_{n=-\infty}^{\infty} \delta(t - nT) \leftrightarrow \frac{2\pi}{T} \sum_{n=-\infty}^{\infty} \delta\left(\omega - \frac{n2\pi}{T}\right)$$

where $T > 0$ is a real constant and $\delta(\cdot)$ is the impulse function of Appendix A. (*Hint*: Represent the time function by a complex Fourier series as in Problem D-6, find the Fourier coefficients of the series, and then Fourier-transform the series.)

*D-8. From the expression in Problem D-7, it is readily shown that

$$\sum_{n=-\infty}^{\infty} e^{-jn\omega T} = \frac{2\pi}{T} \sum_{n=-\infty}^{\infty} \delta\left(\omega - \frac{n2\pi}{T}\right)$$

Use this result to prove that the periodic signal

$$y(t) = \sum_{n=-\infty}^{\infty} x(t - nT)$$

comprised of repetitions in each period T of a basic waveform $x(t)$, has the Fourier transform $Y(\omega)$ given by

$$Y(\omega) = \frac{2\pi}{T} \sum_{n=-\infty}^{\infty} X\left(\frac{n2\pi}{T}\right) \delta\left(\omega - \frac{n2\pi}{T}\right)$$

where $X(\omega)$ is the Fourier transform of $x(t)$. By using the result of Problem D-6, we see that the coefficient C_n of the Fourier series of $y(t)$ is related to the Fourier transform of its component waveform $x(t)$ by

$$C_n = \frac{1}{T} X\left(\frac{n2\pi}{T}\right)$$

D-9. Find the Fourier transform of the waveform

$$x(t) = u(t)e^{j\omega_0 t}$$

where $u(\cdot)$ is the unit-step function of (A-5) and ω_0 is a real constant.

D-10. Find the Fourier transform of a sequence of $2N + 1$ pulses of the form given in Problem D-1, where $N = 0, 1, 2, \ldots$. That is, find the transform of

$$y(t) = \sum_{n=-N}^{N} x(t - nT)$$

with $T > 0$ a real constant and $\tau < T$.

D-11. Determine the Fourier transform of the signal

$$x(t) = \begin{cases} At^2 & 0 < t < \tau \\ 0 & \text{elsewhere} \end{cases}$$

where $\tau > 0$ and A are real constants.

D-12. Show that the inverse Fourier transform of the function

$$X(\omega) = \begin{cases} K & -W < \omega < W \\ 0 & \text{elsewhere} \end{cases}$$

is

$$x(t) = (KW/\pi)\text{Sa}(Wt)$$

where $W > 0$ and K are real constants and $\text{Sa}(\cdot)$ is the *sampling function* defined by (E-3).

D-13. The transfer function $H(\omega)$ of a lowpass filter can be approximated by

$$H(\omega) = \begin{cases} K_0 + 2 \sum_{n=1}^{N} K_n \cos(n\pi\omega/W) & -W < \omega < W \\ 0 & \text{elsewhere} \end{cases}$$

Here $W > 0$, $K_0, K_1, \ldots, K_N$ are real constants and $N \geq 0$ is an integer. Find the inverse Fourier transform $h(t)$ of $H(\omega)$ which is the *impulse response* of the network, in terms of sampling functions (see Problem D-12).

D-14. Let $x(t)$ have the Fourier transform $X(\omega)$. Find the transforms of the following functions in terms of $X(\omega)$:

    (a) $x(t - 2)\exp(j\omega_0 t)$ (b) $\dfrac{dx(t)}{dt}\exp[j\omega_0(t - 3)]$ (c) $x(t - 3) - 3x(2t)$

Here ω_0 is a real constant.

D-15. If $x(t) \leftrightarrow X(\omega)$, find the inverse transforms of the following functions in terms of $x(t)$:

    (a) $X(\omega)X^*(\omega + \omega_0)$ (b) $X(\omega - \omega_0)\dfrac{dX(\omega)}{d\omega}$ (c) $X^*(-\omega) + X(\omega)$

Here * represents complex conjugation and ω_0 is a real constant.

D-16. A voltage $x(t)$ exists across a resistor of resistance R. Show that the real energy E expended in the resistance is

$$E = \frac{1}{2\pi R} \int_{-\infty}^{\infty} |X(\omega)|^2 \, d\omega$$

where $X(\omega)$ is the Fourier transform of $x(t)$.

D-17. It is known that

$$x(t) = e^{-\alpha|t|} \leftrightarrow \frac{2\alpha}{\alpha^2 + \omega^2} = X(\omega)$$

where $\alpha > 0$ is a real constant. Use (D-9) and find the Fourier transform $Y(\omega)$

$$y(t) = \frac{6}{4+t^2}$$

D-18. Use the definition (A-2) of an impulse function to prove that the impulse has the Fourier transform 1. That is, show that

$$\delta(t) \leftrightarrow 1$$

D-19. By use of various Fourier transform properties, show that the following are true:
(a) $A \leftrightarrow A(2\pi)\delta(\omega)$ where a is a constant
(b) $\cos(\omega_0 t) \leftrightarrow \pi[\delta(\omega - \omega_0) + \delta(\omega + \omega_0)]$ where ω_0 is a real constant.

D-20. Use the facts that

$$u(t)e^{-\alpha t} \leftrightarrow \frac{1}{\alpha + j\omega}$$

and

$$\cos(\omega_0 t) \leftrightarrow \pi[\delta(\omega - \omega_0) + \delta(\omega + \omega_0)]$$

where $\alpha > 0$ and ω_0 are real constants, to prove that

$$u(t)e^{-\alpha t}\cos(\omega_0 t) \leftrightarrow \frac{\alpha + j\omega}{(\alpha^2 + \omega_0^2 - \omega^2) + j(2\alpha\omega)}$$

D-21. Prove (D-6) and (D-10).

*D-22. Prove (D-12). [*Hint*: Use (D-16).]

D-23. Prove (D-18).

D-24. Find the Fourier transform of the signal

$$x(t_1, t_2) = \begin{cases} A & -\tau_1 < t_1 < \tau_1 \\ 0 & \text{elsewhere} \end{cases} \quad \text{and} \quad -\tau_2 < t_2 < \tau_2$$

where $\tau_1 > 0$, $\tau_2 > 0$, and A are real constants.

D-25. By direct transformation, find the Fourier transform of

$$x(t) = \begin{cases} At & 0 < t < \tau \\ 0 & t < 0 \text{ and } t > \tau \end{cases}$$

D-26. Show that a frequency-domain impulse can be represented by

$$\delta(\omega) = \frac{1}{2\pi} \int_{-\infty}^{\infty} e^{-j\omega t}\, dt$$

D-27. Show that a time-domain impulse can be represented by

$$\delta(t) = \frac{1}{2\pi} \int_{-\infty}^{\infty} e^{j\omega t}\, d\omega$$

D-28. First find the spectrum of the signal $x(t) = A \exp[-W|t|]$, where $A > 0$ and $W > 0$ are constants, and then use the result with the frequency shifting property of Fourier transforms to determine the spectrum of $y(t) = A \exp[-W|t|] \cos(\omega_0 t)$, where $\omega_0 > 0$ is a constant.

D-29. A waveform $x(t)$ and its Fourier transform are defined by the pair

$$u(t)t^3 e^{-\alpha t} \leftrightarrow \frac{6}{(\alpha + j\omega)^4}$$

where $\alpha > 0$ is a constant. Use the duality property to develop a dual transform pair.

D-30. A waveform $x(t)$ has a derivative $y(t) = dx(t)/dt$ and spectrum $Y(\omega)$ defined by the Fourier transform pair

$$y(t) = \left\{ \begin{array}{c} \delta(t + 3\tau) + \delta(t + \tau) \\ -\delta(t - 3\tau) - \delta(t - \tau) \end{array} \right\} \leftrightarrow Y(\omega) = \left\{ \begin{array}{c} e^{j3\omega\tau} + e^{j\omega\tau} \\ -e^{-j3\omega\tau} - e^{-j\omega\tau} \end{array} \right\}$$

where τ is a positive constant. (*a*) Sketch $x(t)$. (*b*) Find $X(\omega)$, the spectrum of $x(t)$, by use of the integral property of Fourier transforms.

D-31. A waveform $x(t)$ has a Fourier transform

$$X(\omega) = A \operatorname{rect}\left[\frac{\omega - \omega_0}{W}\right] \cos^2\left[\frac{\pi(\omega - \omega_0)}{W}\right]$$

$$+ A \operatorname{rect}\left[\frac{\omega + \omega_0}{W}\right] \cos^2\left[\frac{\pi(\omega + \omega_0)}{W}\right]$$

where A, W, and $\omega_0 > W/2$ are all positive constants. (*a*) Find $x(t)$. (*b*) Find the energy in $x(t)$ by use of Parseval's theorem.

D-32. Prove that (D-20) is valid.

APPENDIX E

Table of Useful Fourier Transforms

In Table E-1 of Fourier transform pairs, we define

$$u(\xi) = \begin{cases} 1 & \xi > 0 \\ 0 & \xi < 0 \end{cases} \tag{E-1}$$

$$\text{rect}\,(\xi) = \begin{cases} 1 & |\xi| < \tfrac{1}{2} \\ 0 & |\xi| > \tfrac{1}{2} \end{cases} \tag{E-2}$$

$$\text{Sa}(\xi) = \frac{\sin(\xi)}{\xi} \tag{E-3}$$

$$\text{tri}\,(\xi) = \begin{cases} 1 - |\xi| & |\xi| < 1 \\ 0 & |\xi| > 1 \end{cases} \tag{E-4}$$

$$x(t) \leftrightarrow X(\omega) \tag{E-5}$$

and let α, τ, σ, ω_0, and W be real constants.

TABLE E-1
Fourier Transform Pairs

Pair	$x(t)$	$X(\omega)$	Notes
1	$\alpha\delta(t)$	α	
2	$\alpha/2\pi$	$\alpha\delta(\omega)$	
3	$u(t)$	$\pi\delta(\omega) + (1/j\omega)$	
4	$\dfrac{1}{2}\delta(t) - \dfrac{1}{j2\pi t}$	$u(\omega)$	
5	$\text{rect}(t/\tau)$	$\tau\,\text{Sa}(\omega\tau/2)$	$\tau > 0$
6	$(W/\pi)\text{Sa}(Wt)$	$\text{rect}\,(\omega/2W)$	$W > 0$
7	$\text{tri}\,(t/\tau)$	$\tau\,\text{Sa}^2(\omega\tau/2)$	$\tau > 0$
8	$(W/\pi)\text{Sa}^2(Wt)$	$\text{tri}\,(\omega/2W)$	$W > 0$
9	$e^{j\omega_0 t}$	$2\pi\delta(\omega - \omega_0)$	
10	$\delta(t - \tau)$	$e^{-j\omega\tau}$	
11	$\cos(\omega_0 t)$	$\pi[\delta(\omega - \omega_0) + \delta(\omega + \omega_0)]$	
12	$\sin(\omega_0 t)$	$-j\pi[\delta(\omega - \omega_0) - \delta(\omega + \omega_0)]$	
13	$u(t)\cos(\omega_0 t)$	$\dfrac{\pi}{2}[\delta(\omega - \omega_0) + \delta(\omega + \omega_0)] + \dfrac{j\omega}{\omega_0^2 - \omega^2}$	
14	$u(t)\sin(\omega_0 t)$	$-j\dfrac{\pi}{2}[\delta(\omega - \omega_0) - \delta(\omega + \omega_0)] + \dfrac{\omega_0}{\omega_0^2 - \omega^2}$	
15	$u(t)e^{-\alpha t}$	$\dfrac{1}{\alpha + j\omega}$	$\alpha > 0$
16	$u(t)te^{-\alpha t}$	$\dfrac{1}{(\alpha + j\omega)^2}$	$\alpha > 0$
17	$u(t)t^2 e^{-\alpha t}$	$\dfrac{2}{(\alpha + j\omega)^3}$	$\alpha > 0$
18	$u(t)t^3 e^{-\alpha t}$	$\dfrac{6}{(\alpha + j\omega)^4}$	$\alpha > 0$
19	$e^{-\alpha|t|}$	$\dfrac{2\alpha}{\alpha^2 + \omega^2}$	$\alpha > 0$
20	$e^{-t^2/(2\sigma^2)}$	$\sigma\sqrt{2\pi}e^{-\sigma^2\omega^2/2}$	$\sigma > 0$

APPENDIX F

Some Probability Densities and Distributions

For convenience of reference we list below the probability density $f_X(x)$ and distribution function $F_X(x)$ for some well-known distributions. Where appropriate, we also give the mean $\bar{X}$, variance σ_X^2, and characteristic function $\Phi_X(\omega)$.

A number of constants and functions are used as defined below:†

$$a, a_1, a_2, b, b_1, b_2, \sigma, \text{ and } p \text{ are real constants} \tag{F-1a}$$

$$N \text{ is a positive integer} \tag{F-1b}$$

$$\delta(\xi) = \text{impulse function of (2.3-2)} \tag{F-1c}$$

$$u(\xi) = \text{unit-step function of (2.2-4)} \tag{F-1d}$$

$$\text{rect } (\xi) = \text{rectangular function of (E-2)} \tag{F-1e}$$

$$\Gamma(x) = \int_0^\infty \xi^{x-1} e^{-\xi} \, d\xi \qquad \text{Re}\,(x) > 0$$

$$= \text{gamma function} \tag{F-1f}$$

$$P(\alpha, \beta) = \frac{1}{\Gamma(\alpha)} \int_0^\beta \xi^{\alpha-1} e^{-\xi} \, d\xi \qquad \text{Re}\,(\alpha) > 0$$

$$= \text{incomplete gamma function} \tag{F-1g}$$

---

†Re (z) denotes the real part of z.

$$P(x|N) = \frac{1}{2^{N/2}\Gamma(N/2)} \int_0^x \xi^{(N/2)-1} e^{-\xi/2} \, d\xi$$

$$= \text{chi-square probability function}$$

$$= P\left(\frac{N}{2}, \frac{x}{2}\right) \tag{F-1h}$$

$$I(u, p) = \frac{1}{\Gamma(p+1)} \int_0^{u\sqrt{p+1}} \xi^p e^{-\xi} \, d\xi$$

$$= \text{Pearson's form of incomplete gamma function (Pearson, 1934)}$$

$$= P(p+1, u\sqrt{p+1}) \tag{F-1i}$$

$$I_x(a, b) = \frac{\Gamma(a+b)}{\Gamma(a)\Gamma(b)} \int_0^x \xi^{a-1}(1-\xi)^{b-1} \, d\xi$$

$$= \text{incomplete beta function} \tag{F-1j}$$

$$F(x) = \text{gaussian distribution of (B-3)} \tag{F-1k}$$

$$I_n(x) = (x/2)^n \sum_{k=0}^\infty \frac{(x/2)^{2k}}{k!(n+k)!}$$

$$= \frac{1}{\pi} \int_0^\pi e^{x\cos(\theta)} \cos(n\theta) \, d\theta$$

$$= \text{modified Bessel function of first kind of order } n = 0, 1, 2, \ldots \tag{F-1l}$$

$$Q(\alpha, \beta) = \int_\beta^\infty \xi I_0(\alpha\xi) \exp\left[\frac{-(\xi^2 + \alpha^2)}{2}\right] d\xi \tag{F-1m}$$

The functions of (F-1f) through (F-1j) and that of (F-1l) are discussed in detail in Abramowitz and Stegun, editors (1964). $Q(\alpha, \beta)$ is Marcum's Q-function; it is tabulated in Marcum (1950).

DISCRETE FUNCTIONS

Bernoulli

For $0 < p < 1$

$$f_X(x) = (1-p)\delta(x) + p\delta(x-1) \tag{F-2}$$

$$F_X(x) = (1-p)u(x) + pu(x-1) \tag{F-3}$$

$$\bar{X} = p \tag{F-4}$$

$$\sigma_X^2 = p(1-p) \tag{F-5}$$

$$\Phi_X(\omega) = 1 - p + pe^{j\omega} \tag{F-6}$$

Binomial

For $0 < p < 1$ and $N = 1, 2, \ldots$

$$f_X(x) = \sum_{k=0}^{N} \binom{N}{k} p^k (1-p)^{N-k} \delta(x-k) \qquad \text{(F-7)}$$

$$F_X(x) = \sum_{k=0}^{N} \binom{N}{k} p^k (1-p)^{N-k} u(x-k) \qquad \text{(F-8)}$$

$$\bar{X} = Np \qquad \text{(F-9)}$$

$$\sigma_X^2 = Np(1-p) \qquad \text{(F-10)}$$

$$\Phi_X(\omega) = [1 - p + pe^{j\omega}]^N \qquad \text{(F-11)}$$

Pascal†

For $0 < p < 1$ and $N = 1, 2, \ldots$

$$f_X(x) = \sum_{k=N}^{\infty} \binom{k-1}{N-1} p^N (1-p)^{k-N} \delta(x-k) \qquad \text{(F-12)}$$

$$F_X(x) = \sum_{k=N}^{\infty} \binom{k-1}{N-1} p^N (1-p)^{k-N} u(x-k) \qquad \text{(F-13)}$$

$$\bar{X} = \frac{N}{p} \qquad \text{(F-14)}$$

$$\sigma_X^2 = \frac{N(1-p)}{p^2} \qquad \text{(F-15)}$$

$$\Phi_X(\omega) = p^N e^{jN\omega} [1 - (1-p)e^{j\omega}]^{-N} \qquad \text{(F-16)}$$

Poisson

For $b > 0$

$$f_X(x) = e^{-b} \sum_{k=0}^{\infty} \frac{b^k}{k!} \delta(x-k) \qquad \text{(F-17)}$$

$$F_X(x) = e^{-b} \sum_{k=0}^{\infty} \frac{b^k}{k!} u(x-k) \qquad \text{(F-18)}$$

$$\bar{X} = b \qquad \text{(F-19)}$$

$$\sigma_X^2 = b \qquad \text{(F-20)}$$

$$\Phi_X(\omega) = \exp[b(e^{j\omega} - 1)] \qquad \text{(F-21)}$$

---

†Blaise Pascal (1623–1662) was a French mathematician.

CONTINUOUS FUNCTIONS

Arcsine

For $a > 0$

$$f_X(x) = \frac{\text{rect }(x/2a)}{\pi\sqrt{a^2 - x^2}} \tag{F-22}$$

$$F_X(x) = \begin{cases} 0 & -\infty < x < -a \\ \dfrac{1}{2} + \dfrac{1}{\pi}\sin^{-1}\left(\dfrac{x}{a}\right) & -a \le x < a \\ 1 & a \le x < \infty \end{cases} \tag{F-23}$$

$$\bar{X} = 0 \tag{F-24}$$

$$\sigma_X^2 = \frac{a^2}{2} \tag{F-25}$$

Beta

For $a > 0$ and $b > 0$

$$f_X(x) = \frac{\Gamma(a+b)}{\Gamma(a)\Gamma(b)}[u(x) - u(x-1)]x^{a-1}(1-x)^{b-1} \tag{F-26}$$

$$F_X(x) = \begin{cases} I_x(a, b)u(x) & x < 1 \\ 1 & x \ge 1 \end{cases} \tag{F-27}$$

$$\bar{X} = \frac{a}{a+b} \tag{F-28}$$

$$\sigma_X^2 = \frac{ab}{(a+b)^2(a+b+1)} \tag{F-29}$$

Cauchy

For $b > 0$ and $-\infty < a < \infty$

$$f_X(x) = \frac{(b/\pi)}{b^2 + (x-a)^2} \tag{F-30}$$

$$F_X(x) = \frac{1}{2} + \frac{1}{\pi}\tan^{-1}\left(\frac{x-a}{b}\right) \tag{F-31}$$

$$\bar{X} = \text{is undefined} \tag{F-32}$$

$$\sigma_X^2 = \text{is undefined} \tag{F-33}$$

$$\Phi_X(\omega) = e^{ja\omega - b|\omega|} \tag{F-34}$$

Chi-Square with N Degrees of Freedom

For $N = 1, 2, \ldots$

$$f_X(x) = \frac{x^{(N/2)-1}}{2^{N/2}\Gamma(N/2)} e^{-x/2} u(x) \qquad \text{(F-35)}$$

$$F_X(x) = P(x|N) = P\left(\frac{N}{2}, \frac{x}{2}\right) \qquad \text{(F-36)}$$

$$\bar{X} = N \qquad \text{(F-37)}$$

$$\sigma_X^2 = 2N \qquad \text{(F-38)}$$

$$\Phi_X(\omega) = (1 - j2\omega)^{-N/2} \qquad \text{(F-39)}$$

Erlang

For $N = 1, 2, \ldots$ and $a > 0$

$$f_X(x) = \frac{a^N x^{N-1} e^{-ax}}{(N-1)!} u(x) \qquad \text{(F-40)}$$

$$F_X(x) = \left[1 - e^{-ax} \sum_{n=0}^{N-1} \frac{(ax)^n}{n!}\right] u(x) \qquad \text{(F-41)}$$

$$\bar{X} = \frac{N}{a} \qquad \text{(F-42)}$$

$$\sigma_x^2 = \frac{N}{a^2} \qquad \text{(F-43)}$$

$$\Phi_X(\omega) = \left(\frac{a}{a - j\omega}\right)^N \qquad \text{(F-44)}$$

Exponential

For $a > 0$

$$f_X(x) = ae^{-ax} u(x) \qquad \text{(F-45)}$$

$$F_X(x) = [1 - e^{-ax}] u(x) \qquad \text{(F-46)}$$

$$\bar{X} = \frac{1}{a} \qquad \text{(F-47)}$$

$$\sigma_X^2 = \frac{1}{a^2} \qquad \text{(F-48)}$$

$$\Phi_X(\omega) = \frac{a}{a - j\omega} \qquad \text{(F-49)}$$

Gamma

For $a > 0$ and $b > 0$

$$f_X(x) = \frac{a^b x^{b-1} e^{-ax}}{\Gamma(b)} u(x) \tag{F-50}$$

$$F_X(x) = I\left(\frac{ax}{\sqrt{b}}, b - 1\right) u(x) \tag{F-51}$$

$$\bar{X} = \frac{b}{a} \tag{F-52}$$

$$\sigma_X^2 = \frac{b}{a^2} \tag{F-53}$$

$$\Phi_X(\omega) = \left(\frac{a}{a - j\omega}\right)^b \tag{F-54}$$

Note that if b is a positive integer the gamma density becomes the Erlang density. Also if $b = N/2$, for $N = 1, 2, \ldots$, and $a = \frac{1}{2}$ the gamma density becomes the chi-square density.

Gaussian-Univariate

For $b > 0$ and $-\infty < a < \infty$

$$f_X(x) = (\pi b)^{-1/2} e^{-(x-a)^2/b} \tag{F-55}$$

$$F_X(x) = F\left(\frac{x - a}{\sqrt{b/2}}\right) \tag{F-56}$$

$$\bar{X} = a \tag{F-57}$$

$$\sigma_X^2 = \frac{b}{2} \tag{F-58}$$

$$\Phi_X(\omega) = e^{j\omega a - (\omega^2 b/4)} \tag{F-59}$$

Gaussian-Bivariate

For $-\infty < a_1 < \infty$, $-\infty < a_2 < \infty$, $b_1 > 0$, $b_2 > 0$ and $-1 \le \rho \le 1$

$$f_{X_1, X_2}(x_1, x_2) = [\pi^2 b_1 b_2 (1 - \rho^2)]^{-1/2}$$

$$\cdot \exp\left\{\frac{-1}{(1 - \rho^2)}\left[\frac{(x_1 - a_1)^2}{b_1}\right.\right.$$

$$\left.\left. - \frac{2\rho(x_1 - a_1)(x_2 - a_2)}{\sqrt{b_1 b_2}} + \frac{(x_2 - a_2)^2}{b_2}\right]\right\} \tag{F-60}$$

$$F_{X_1,X_2}(x_1, x_2) = L\left(-\left[\frac{x_1 - a_1}{\sqrt{b_1/2}}\right], -\left[\frac{x_2 - a_2}{\sqrt{b_2/2}}\right], \rho\right) \qquad \text{(F-61)}$$

where $L(x_1, x_2, \rho)$ is a probability function discussed extensively and graphed in Abramowitz and Stegun, editors (1964), p. 936. Also

$$\bar{X}_1 = a_1 \qquad \text{(F-62)}$$

$$\bar{X}_2 = a_2 \qquad \text{(F-63)}$$

$$\sigma_{X_1}^2 = b_1/2 \qquad \text{(F-64)}$$

$$\sigma_{X_2}^2 = b_2/2 \qquad \text{(F-65)}$$

$$\Phi_{X_1,X_2}(\omega_1, \omega_2) = \exp\{j\omega_1 a_1 + j\omega_2 a_2 - \tfrac{1}{4}[\omega_1^2 b_1 + 2\rho\omega_1\omega_2\sqrt{b_1 b_2} + \omega_2^2 b_2]\}$$
$$\text{(F-66)}$$

Laplace

For $b > 0$ and $-\infty < a < \infty$

$$f_X(x) = \frac{b}{2}e^{-b|x-a|} \qquad \text{(F-67)}$$

$$F_X(x) = \begin{cases} \frac{1}{2}e^{b(x-a)} & -\infty < x < a \\ 1 - \frac{1}{2}e^{-b(x-a)} & a \leq x < \infty \end{cases} \qquad \text{(F-68)}$$

$$\bar{X} = a \qquad \text{(F-69)}$$

$$\sigma_X^2 = \frac{2}{b^2} \qquad \text{(F-70)}$$

$$\Phi_X(\omega) = b^2\frac{e^{j a \omega}}{b^2 + \omega^2} \qquad \text{(F-71)}$$

Log-Normal

For $-\infty < a < \infty$, $-\infty < b < \infty$, and $\sigma > 0$

$$f_X(x) = \frac{u(x - b)e^{-[\ln(x-b)-a]^2/2\sigma^2}}{\sqrt{2\pi}(x - b)\sigma} \qquad \text{(F-72)}$$

$$F_X(x) = u(x - b)F\{\sigma^{-1}[\ln(x - b) - a]\} \qquad \text{(F-73)}$$

$$\bar{X} = b + \exp\left(a + \frac{\sigma^2}{2}\right) \qquad \text{(F-74)}$$

$$\sigma_X^2 = [\exp(\sigma^2) - 1]\exp(2a + \sigma^2) \qquad \text{(F-75)}$$

Rayleigh

For $-\infty < a < \infty$ and $b > 0$

$$f_X(x) = \frac{2}{b}(x - a)e^{-(x-a)^2/b}u(x - a)$$ (F-76)

$$F_X(x) = [1 - e^{-(x-a)^2/b}]u(x - a)$$ (F-77)

$$\bar{X} = a + \sqrt{\frac{\pi b}{4}}$$ (F-78)

$$\sigma_X^2 = \frac{b(4 - \pi)}{4}$$ (F-79)

Rice [Thomas (1969), Middleton (1960)]

For $a > 0$ and $b > 0$

$$f_X(x) = \frac{x}{b^2}e^{-(a^2+x^2)/2b^2}I_0\left(\frac{ax}{b^2}\right)u(x)$$ (F-80)

$$F_X(x) = \left[1 - Q\left(\frac{a}{b}, \frac{x}{b}\right)\right]u(x)$$ (F-81)

$$\bar{X} = b\sqrt{\frac{\pi}{2}}e^{-k^2/4}\left[\left(1 + \frac{k^2}{2}\right)I_0\left(\frac{k^2}{4}\right) + \frac{k^2}{2}I_1\left(\frac{k^2}{4}\right)\right]$$ (F-82)

$$\sigma_X^2 = b^2(2 + k^2) - (\bar{X})^2$$ (F-83)

$$k^2 = \frac{a^2}{b^2}$$ (F-84)

Uniform

For $-\infty < a < b < \infty$

$$f_X(x) = \frac{u(x - a) - u(x - b)}{b - a}$$ (F-85)

$$F_X(x) = \begin{cases} \dfrac{(x - a)u(x - a)}{b - a} & x < b \\ 1 & x \geq b \end{cases}$$ (F-86)

$$\bar{X} = \frac{a + b}{2}$$ (F-87)

$$\sigma_X^2 = \frac{(b - a)^2}{12}$$ (F-88)

$$\Phi_X(\omega) = \frac{e^{j\omega b} - e^{j\omega a}}{j\omega(b - a)}$$ (F-89)

Weibull

For $a > 0$ and $b > 0$

$$f_X(x) = abx^{b-1}e^{-ax^b}u(x) \tag{F-90}$$

$$F_X(x) = [1 - e^{-ax^b}]u(x) \tag{F-91}$$

$$\bar{X} = \frac{\Gamma(1 + b^{-1})}{a^{1/b}} \tag{F-92}$$

$$\sigma_X^2 = \frac{\Gamma(1 + 2b^{-1}) - [\Gamma(1 + b^{-1})]^2}{a^{2/b}} \tag{F-93}$$

Note that if $b = 2$ the Weibull density becomes a Rayleigh density.

APPENDIX G

Some Mathematical Topics of Interest

In this appendix we list several useful mathematical theorems. No formal proofs are presented. Rather, the reader is referred to the literature for proofs.

Leibniz's Rule†

Let $G(u)$ represent the integral

$$G(u) = \int_{\alpha(u)}^{\beta(u)} H(x, u)\, dx \tag{G-1}$$

where it is assumed that $\alpha(u)$, and $\beta(u)$ are real differentiable functions of a real parameter u, and $H(x, u)$ and its derivative $dH(x, u)/du$ are both continuous functions in x and u. Then the derivative of the integral with respect to parameter u is given by

$$\frac{dG(u)}{du} = H[\beta(u), u]\frac{d\beta(u)}{du} - H[\alpha(u), u]\frac{d\alpha(u)}{du}$$
$$+ \int_{\alpha(u)}^{\beta(u)} \frac{\partial H(x, u)}{\partial u}\, dx \tag{G-2}$$

Equation (G-2) is called *Leibniz's rule*. A proof is given by Wylie (1951), p. 591.

---

†After the great German mathematician Gottfried Wilhelm von Leibniz (1646–1716).

Interchange of Derivative and Integral

A special case of Leibniz's rule occurs when the integral's limits are constants relative to the parameter u. Equations (G-1) and (G-2) become

$$G(u) = \int_a^b H(x, u) \, dx \tag{G-3}$$

$$\frac{dG(u)}{du} = \frac{d}{du} \int_a^b H(x, u) \, dx = \int_a^b \frac{\partial H(x, u)}{\partial u} \, dx \tag{G-4}$$

for constant limits a and b. This last result indicates the derivative and integral operations may be interchanged.

Interchange of Integrals

Fubini's theorem states that if any one of the conditions

$$\int_{-\infty}^{\infty} \int_{-\infty}^{\infty} |x(t, u)| \, du \, dt < \infty \tag{G-5a}$$

$$\int_{-\infty}^{\infty} \left[\int_{-\infty}^{\infty} |x(t, u)| \, du \right] dt < \infty \tag{G-5b}$$

$$\int_{-\infty}^{\infty} \left[\int_{-\infty}^{\infty} |x(t, u)| \, dt \right] du < \infty \tag{G-5c}$$

is true, then [Sakrison (1968), p. 45]

$$\int_{-\infty}^{\infty} \left[\int_{-\infty}^{\infty} x(t, u) \, dt \right] du = \int_{-\infty}^{\infty} \left[\int_{-\infty}^{\infty} x(t, u) \, du \right] dt$$

$$= \int_{-\infty}^{\infty} \int_{-\infty}^{\infty} x(t, u) \, dt \, du \tag{G-6}$$

The proof is involved but is in Burkhill (1963), Chapter 5. More general versions of Fubini's theorem exist, and it has versions for integral multiples above 2 also [Korn and Korn (1961), p. 105].

Continuity of Random Processes

A real random process $X(t)$ defined on $-\infty < t < \infty$ is said to be mean-square continuous at any point t if

$$\lim_{\epsilon \to 0} E\{[X(t + \epsilon) - X(t)]^2\} = 0 \tag{G-7}$$

$X(t)$ will be mean-square continuous at any value of t if its autocorrelation function $R_{XX}(t_1, t_2)$ is continuous at any point (t, t). For a proof see Viniotis (1998), p. 410. Prabhu (1965), p. 25, gives a result similar to the above but requires $X(t)$ to be a *regular process*, that is, one for which $E[|X(t)|^2] < \infty$, all t.

Differentiation of Random Processes

A real, regular, random process $X(t)$ is said to be differentiable in a mean-square sense if there exists a process $X'(t)$ such that

$$\lim_{\epsilon \to 0} E\{[\hat{X}(t) - X'(t)]^2\} = 0 \qquad \text{(G-8)}$$

where

$$\hat{X}(t) = \frac{X(t + \epsilon) - X(t)}{\epsilon} \qquad \text{(G-9)}$$

The process $X(t)$ will be mean-square differentiable at any time t if the autocorrelation function has the second-order derivative

$$\left. \frac{\partial^2 R_{XX}(t_1, t_2)}{\partial t_1 \partial t_2} \right|_{t_1 = t_2 = t} < \infty \qquad \text{(G-10)}$$

For proofs see Viniotis (1998), p. 411, or Helstrom (1991), pp. 410–411. Parzen (1962), p. 83, gives (G-10) without the condition $t_1 = t_2 = t$ and states that if (G-10) is true, then also true are:

$$E\left[\frac{dX(t)}{dt}\right] = \frac{d}{dt}\{E[X(t)]\} \qquad \text{(G-11)}$$

$$C_{X'X'}(t_1, t_2) = \frac{\partial^2}{\partial t_1 \partial t_2} C_{XX}(t_1, t_2) \qquad \text{(G-12)}$$

$$C_{X'X}(t_1, t_2) = \frac{\partial}{\partial t_1} C_{XX}(t_1, t_2) \qquad \text{(G-13)}$$

Here a prime indicates time differentiation. We note that these results allow operations of expectation and differentiation to be interchanged with processes for which (G-10) is true.

Integration of Random Processes

Let $X(t)$ be a real, regular, random process and define I as the integral of $X(t)$ on interval $[a, b]$ according to

$$I = \int_a^b X(t)\, dt \qquad \text{(G-14)}$$

If the interval $[a, b]$ is subdivided into N contiguous subintervals with boundary times t_k such that $a = t_0 < t_1 < \cdots < t_N = b$ and widths $\Delta t_k = t_k - t_{k-1}$, $k = 1, 2, \ldots, N$, then, for $t_{k-1} \le \xi_k \le t_k$, if the Riemann sum

$$I_N = \sum_{k=1}^{N} X(\xi_k)\, \Delta t_k \qquad \text{(G-15)}$$

converges in the mean-square sense to a limit as $N \to \infty$ in a way that the largest subinterval $\Delta t_k \to 0$, we say that this limit is equal to I. In other words, if

$$\lim_{N \to \infty} \{E[(I - I_N)^2]\} = 0 \qquad \text{(G-16)}$$

then $X(t)$ is mean-square Riemann† integrable on $[a, b]$.

If, and only if,

$$\int_a^b \int_a^b R_{XX}(t_1, t_2) \, dt_1 \, dt_2 < \infty \qquad \text{(G-17)}$$

will $X(t)$ be Riemann integrable. If (G-17) is true then

$$E\left[\int_a^b X(t) \, dt\right] = \int_a^b E[X(t)] \, dt \qquad \text{(G-18)}$$

$$E\left[\int_a^b \int_a^b X(t_1)X(t_2) \, dt_1 \, dt_2\right] = \int_a^b \int_a^b E[X(t_1)X(t_2)] \, dt_1 \, dt_2 \qquad \text{(G-19)}$$

which say that the expectation and either single or double integral operations may be interchanged.

For proofs of the above see Viniotis (1998), pp. 412–413, or Prabhu (1965), pp. 27–28. It is also known that if $X(t)$ is a mean-square continuous process [see (G-7)], then its integral exists (Viniotis, 1998, p. 413).

Interchange of Expectation and Integration

Let $X(t)$ represent a random process bounded on (a, b) where a and b may be finite or infinite. Also let $h(t)$ be a real or complex function of t. The integral of $X(t)h(t)$ forms a random variable I

$$I = \int_a^b X(t)h(t) \, dt \qquad \text{(G-20)}$$

Under an appropriate measurability condition,‡ this integral exists with probability 1 [for all sample functions of $X(t)$ except possibly some with probability zero] if

$$\int_a^b E[|X(t)h(t)|] \, dt = \int_a^b E[|X(t)|]|h(t)| \, dt < \infty \qquad \text{(G-21)}$$

---

†Named for Georg Bernhard Riemann (1826–1866), a German mathematician. One of the greatest of all time.

‡See Davenport and Root (1958), pp. 65–66, who refer to Theorem 2.7 of Doob (1953) for details on measurability constraints. See also Integration of Random Processes above. Cooper and McGillem (1986), p. 288, also give (G-22) and state that, more generally, $X(t)$ may be replaced by some function, such as the square, of a process $X(t)$ which is not restricted to be a stationary process.

As a consequence of (G-21)

$$E\left[\int_a^b X(t)h(t)\,dt\right] = \int_a^b E[X(t)]h(t)\,dt \tag{G-22}$$

Hölder's Inequality

Let $x_1(t)$ and $x_2(t)$ be continuous time functions of the real variable t, and be nonvanishing such that $x_1(t)x_2(t) \neq 0$ and integrable on (a, b) according to

$$\int_a^b |x_1(t)|^p\,dt < \infty \tag{G-23}$$

$$\int_a^b |x_2(t)|^q\,dt < \infty \tag{G-24}$$

where $p > 1$ and $(1/p) + (1/q) = 1$. Then a tight form of Hölder's inequality states that

$$\int_a^b |x_1(t)x_2(t)|\,dt \leq \left[\int_a^b |x_1(t)|^p\,dt\right]^{1/p}\left[\int_a^b |x_2(t)|^q\,dt\right]^{1/q} < \infty \tag{G-25a}$$

where

$$(1/p) + (1/q) = 1 \tag{G-25b}$$

For a proof of (G-25) see Thomas (1969), p. 615. The proof also verifies the fact that

$$\int_a^b |x_1(t)x_2(t)|\,dt < \infty \tag{G-26}$$

when (G-23) and (G-24) are true.

A weaker form of Hölder's inequality may be more convenient for some purposes. It is

$$\left|\int_a^b x_1(t)x_2(t)\,dt\right| \leq \left[\int_a^b |x_1(t)|^p\,dt\right]^{1/p}\left[\int_a^b |x_2(t)|^q\,dt\right]^{1/q} \tag{G-27a}$$

where

$$(1/p) + (1/q) = 1 \tag{G-27b}$$

Hölder's inequality can be applied to discrete-time sequences $x_1[n]$ and $x_2[n]$. The tight form is [Thomas (1969), p. 617]

$$\sum_n |x_1[n]x_2[n]| \leq \left(\sum_n |x_1[n]|^p\right)^{1/p}\left(\sum_n |x_2[n]|^q\right)^{1/q} \tag{G-28a}$$

where the sums are over all values of n and

$$(1/p) + (1/q) = 1 \tag{G-28b}$$

The weak form is

$$\left| \sum_{n} x_1[n] x_2[n] \right| \leq \left(\sum_{n} |x_1[n]|^p \right)^{1/p} \left(\sum_{n} |x_2[n]|^q \right)^{1/q} \qquad \text{(G-29a)}$$

where

$$(1/p) + (1/q) = 1 \qquad \text{(G-29b)}$$

In both (G-28a) and (G-29a) we assume

$$\sum_{n} |x_1[n]|^p < \infty \qquad \text{(G-30)}$$

$$\sum_{n} |x_2[n]|^q < \infty \qquad \text{(G-31)}$$

Schwarz's Inequality

This inequality derives from Hölder's inequality with $p = q = 2$. For continuous waveforms

$$\left| \int_a^b x_1(t) x_2(t)\, dt \right| \leq \left[\int_a^b |x_1(t)|^2\, dt \right]^{1/2} \left[\int_a^b |x_2(t)|^2\, dt \right]^{1/2} \qquad \text{(G-32)}$$

For discrete-time sequences

$$\left| \sum_{n} x_1[n] x_2[n] \right| \leq \left(\sum_{n} |x_1[n]|^2 \right)^{1/2} \left(\sum_{n} |x_2[n]|^2 \right)^{1/2} \qquad \text{(G-33)}$$

By squaring on both sides of (G-32) and (G-33), the more frequently seen forms of Schwartz's inequality are obtained.

Minkowski's Inequality

For $x_1(t)$ and $x_2(t)$ each satisfying (G-23) with $p \geq 1$, this inequality states that [Thomas (1969), p. 618]

$$\left[\int_a^b |x_1(t) \pm x_2(t)|^p\, dt \right]^{1/p} \leq \left[\int_a^b |x_1(t)|^p\, dt \right]^{1/p} + \left[\int_a^b |x_2(t)|^p\, dt \right]^{1/p} \qquad \text{(G-34)}$$

for continuous signals, or

$$\left(\sum_{n} |x_1[n] \pm x_2[n]|^p \right)^{1/p} \leq \left(\sum_{n} |x_1[n]|^p \right)^{1/p} + \left(\sum_{n} |x_2[n]|^p \right)^{1/p} \qquad \text{(G-35)}$$

for discrete-time sequences.

Bibliography

Abramowitz, M., and I. Stegun, editors (1964): *Handbook of Mathematical Functions with Formulas, Graphs, and Mathematical Tables*, National Bureau of Standards Applied Mathematics Series 55, U.S. Government Printing Office, Washington, D.C.

Barton, D. K. (1964): *Radar System Analysis*, Prentice-Hall, Englewood Cliffs, New Jersey.

Bendat, J. S., and A. G. Piersol (1986): *Random Data: Analysis and Measurement Procedures*, 2d ed., Wiley Interscience, New York.

Blackman, R. B., and J. W. Tukey (1958): *The Measurement of Power Spectra*, Dover, New York.

Blanchard, A. (1976): *Phase-Locked Loops, Application to Coherent Receiver Design*, Wiley, New York.

Börjesson, P. O., and C.-E. W. Sundberg (1979): "Simple Approximations of the Error Function $Q(x)$ for Communications Applications," *IEEE Transactions on Communications*, vol. COM-27, no. 3, March, 1979, pp. 639-643.

Bracewell, R. (1965): *The Fourier Transform and Its Applications*, McGraw-Hill, New York.

Burkhill, J. C. (1963): *The Lebesque Integral*, Cambridge University Press, New York.

Campbell, G. A., and R. M. Foster (1948): *Fourier Integrals for Practical Applications*, Van Nostrand, Princeton, New Jersey.

Carlson, A. B. (1975): *Communication Systems, An Introduction to Signals and Noise in Electrical Communication*, 2d ed., McGraw-Hill, New York. (See also 3d ed., 1986.)

Childers, D. G. (1997): *Probability and Random Processes*, Irwin, Chicago.

Clarke, A. B., and R. L. Disney (1970): *Probability and Random Processes for Engineers and Scientists*, Wiley, New York.

Cooper, G. R., and C. D. McGillem (1971): *Probabilistic Methods of Signal and System Analysis*, Holt, Rinehart and Winston, New York. See also 2d ed., 1986.

Cramér, H. (1946): *Mathematical Methods of Statistics*, Princeton University Press, Princeton, New Jersey.

Davenport, W. B., Jr. (1970): *Probability and Random Processes, An Introduction for Applied Scientists and Engineers*, McGraw-Hill, New York.

Davenport, W. B., Jr., and W. L. Root (1958): *An Introduction to the Theory of Random Signals and Noise*, McGraw-Hill, New York.

DiFranco, J. V., and W. L. Rubin (1968): *Radar Detection*, Prentice-Hall, Englewood Cliffs, New Jersey.

Dillard, G. M. (1967): "Generating Random Numbers Having Probability Distributions Occurring in Signal Detection Problems," *Transactions of IEEE*, vol. IT-13, no. 4, October 1967, pp. 616–617.

Doob, J. L. (1953): *Stochastic Processes*, Wiley, New York.

Dubes, R. C. (1968): *The Theory of Applied Probability*, Prentice-Hall, Englewood Cliffs, New Jersey.

Dwight, H. B. (1961): *Tables of Integrals and Other Mathematical Data*, 4th ed., Macmillan, New York.

Gardner, W. A. (1990): *Introduction to Random Processes with Applications to Signals and Systems*, 2d ed., McGraw-Hill, New York.

Gray, R. M., and L. D. Davisson (1986): *Random Processes: A Mathematical Approach for Engineers*, Prentice-Hall, Englewood Cliffs, New Jersey.

Helstrom, C. W. (1984): *Probability and Stochastic Processes for Engineers*, Macmillan, New York. See also 2d ed., 1991.

James, H. M., N. B. Nichols, and R. S. Phillips (1947): *Theory of Servomechanisms*, M.I.T. Radiation Laboratory Series, vol. 25, McGraw-Hill, New York.

Kamen, E. W., and B. S. Heck (1997): *Fundamentals of Signals and Systems Using MATLAB*, Prentice-Hall, Upper Saddle River, New Jersey.

Kay, S. M. (1986): *Modern Spectral Estimation—Theory and Applications*, Prentice-Hall, Englewood Cliffs, New Jersey.

Kennedy, J. B., and A. M. Neville (1986): *Basic Statistical Methods for Engineers and Scientists*, 3d ed., Harper & Row, New York.

Klapper, J., and J. T. Frankle (1972): *Phase-Locked and Frequency-Feedback Systems*, Academic Press, New York.

Kohlenberg, A. (1953): "Exact Interpolation of Band-Limited Functions," *Journal of Applied Physics*, December 1953, pp. 1432–1436.

Korn, G. A., and T. M. Korn (1961): *Mathematical Handbook for Scientists and Engineers*, McGraw-Hill, New York.

Lathi, B. P. (1968): *An Introduction to Random Signals and Communication Theory*, International Textbook, Scranton, Pennsylvania.

Leon-Garcia, A. (1989): *Probability and Random Processes for Electrical Engineering*, Addison-Wesley, Reading, Massachusetts.

Marcum, J. I. (1960): "Studies of Target Detection by Pulsed Radar," Special Monograph Issue *IRE Transactions on Information Theory*, vol. IT-6, no. 2, April 1960.

Marcum, J. I. (1950): "Table of Q Functions," U.S. Air Force Project RAND Research Memorandom RM-339, January 1, 1950 (ASTIA Document AD116551).

McFadden, M. (1963): *Sets, Relations, and Functions*, McGraw-Hill, New York.

Melsa, J. L., and A. P. Sage (1973): *An Introduction to Probability and Stochastic Processes*, Prentice-Hall, Englewood Cliffs, New Jersey.

Middleton, D. (1960): *An Introduction to Statistical Communication Theory*, McGraw-Hill, New York.

Miller, K. S. (1974): *Complex Stochastic Processes, An Introduction to Theory and Application*, Addison-Wesley, Reading, Massachusetts.

Milton, J. S., and C. P. Tsokos (1976): *Probability Theory with the Essential Analysis*, Addison-Wesley, Reading, Massachusetts.

Mumford, W. W., and E. H. Scheibe (1968): *Noise Performance Factors in Communication Systems*, Horizon House-Microwave, Dedham, Massachusetts.

Oppenheim, A. V., and R. W. Schafer (1989): *Discrete-Time Signal Processing*, Prentice-Hall, Englewood Cliffs, New Jersey.

Papoulis, A. (1962): *The Fourier Integral and Its Applications*, McGraw-Hill, New York.

Papoulis, A. (1965): *Probability, Random Variables, and Stochastic Processes*, McGraw-Hill, New York. See also 2d ed., 1984, and 3d ed., 1991.

Parzen, E. (1962): *Stochastic Processes*, Holden-Day, San Francisco, California.

Pearson, K., editor (1934): *Tables of the Incomplete Gamma Functions*, Cambridge University Press, Cambridge, England.

Peebles, P. Z., Jr. (1976): *Communication System Principles*, Addison-Wesley, Reading, Massachusetts.

Peebles, P. Z., Jr. (1987): *Digital Communication Systems*, Prentice-Hall, Englewood-Cliffs, New Jersey.

Peebles, P. Z., Jr. (1998): *Radar Principles*, Wiley, New York.

Peebles, P. Z., Jr., and T. A. Giuma (1991): *Principles of Electrical Engineering*, McGraw-Hill, New York.

Prabhu, N. U. (1965): *Stochastic Processes, Basic Theory and Its Applications*, Macmillan, New York.

Ralston, A., and H. S. Wilf (1967): *Mathematical Methods for Digital Computers*, vol. II, Wiley, New York.

Rosenblatt, M. (1974): *Random Processes*, 2d ed., Springer-Verlag, New York.

Ross, S. M. (1972): *Introduction to Probability Models*, Academic Press, New York.

Sakrison, D. J. (1968): *Communication Theory: Transmission of Waveforms and Digital Information*, Wiley, New York.

Shanmugan, K. S., and A. M. Breipohl (1988): *Random Signals: Detection, Estimation and Data Analysis*, Wiley, New York.

Spiegel, M. R. (1963): *Theory and Problems of Advanced Calculus*, Schaum, New York.

Taylor, F. J. (1994): *Principles of Signals and Systems*, McGraw-Hill, New York.

Thomas, J. B. (1969): *An Introduction to Statistical Communication Theory*, Wiley, New York.

van der Ziel, A. (1970): *Noise: Sources, Characterization, Measurement*, Prentice-Hall, Englewood Cliffs, New Jersey.

Viniotis, Y. (1998): *Probability and Random Processes for Electrical Engineers*, WCB/McGraw-Hill, Boston, Massachusetts.

Wilks, S. S. (1962): *Mathematical Statistics*, Wiley, New York.

Wozencraft, J. M., and I. M. Jacobs (1965): *Principles of Communication Engineering*, Wiley, New York.

Wylie, C. R., Jr. (1951): *Advanced Engineering Mathematics*, McGraw-Hill, New York.

Ziemer, R. E., and W. H. Tranter (1976): *Principles of Communications Systems, Modulation, and Noise*, Houghton Mifflin, Boston, Massachusetts.

Index